Microforming Technology

Microforming Technology

Theory, Simulation, and Practice

Zhengyi Jiang, Jingwei Zhao, and Haibo Xie
University of Wollongong, NSW, Australia

ELSEVIER

ACADEMIC PRESS
An imprint of Elsevier
elsevier.com

Academic Press is an imprint of Elsevier
125 London Wall, London EC2Y 5AS, United Kingdom
525 B Street, Suite 1800, San Diego, CA 92101-4495, United States
50 Hampshire Street, 5th Floor, Cambridge, MA 02139, United States
The Boulevard, Langford Lane, Kidlington, Oxford OX5 1GB, United Kingdom

Library of Congress Cataloging-in-Publication Data
A catalog record for this book is available from the Library of Congress

British Library Cataloguing-in-Publication Data
A catalogue record for this book is available from the British Library

ISBN: 978-0-12-811212-0

Publisher: Joe Hayton
Acquisition Editor: Brian Guerin
Editorial Project Manager: Edward Payne
Production Project Manager: Anusha Sambamoorthy
Cover Designer: Alan Studholme

Typeset by MPS Limited, Chennai, India

Contents

16. Practice of Micro Ultrathin Strip Rolling

17. Practice of Micro Deep Drawing

18. Practice of Micro Hydromechanical Deep Drawing

19. Practice of Micro Bending

Foreword

Product miniaturization is a trend for facilitating product usage, enabling product functions to be implemented in microscale geometries, and aimed at reducing product weight, volume, cost, and pollution. Increased demands for miniaturized products have led to rapid technological development of microforming, one of the most popular micromanufacturing processes for manufacturing very small metallic parts in the submillimeter range. As a promising micromanufacturing technology, microforming possesses lots of advantages including high productivity, ease in manufacturing complex geometries, applicable to a wide range of materials, excellent product properties, short cycle, and low cost. From the viewpoint of production engineering, microforming is considered as an effective process to economically and precisely manufacture micro products that are essential when bulk production is required.

However, due to the unique characteristic of size effects existed in the forming of metals in microscale, research on microforming becomes comprehensive because forming technology established in the macro world cannot be simply miniaturized according to the theory of similarity. Size effects have significant impact on the forming results, and the quality of formed parts should be carefully taken into account in the forming of metals in microscale. As a result, development of technology that is very applicable to microforming process would be essential and significant for the forming of high-quality micro products with excellent dimensional tolerance, required mechanical properties and improved surface quality.

It is my great pleasure to witness this major work of *Microforming Technology: Theory, Simulation and Practice* by Prof. Zhengyi Jiang, Dr. Jingwei Zhao and Dr. Haibo Xie. The global microforming domain faces major challenges which may be summarized by the common problem of size effects in microscale, increasing complexity of micro products and processes, as well as increasing requirement towards quality. This book is a timely capture of the state-of-the-art development of microforming technology, covering theory, simulation and practice of the latest research findings and achievements in micro cross wedge rolling, micro flexible rolling, micro ultrathin strip rolling, micro deep drawing, micro hydromechanical deep drawing, micro bending and micro compression, and has addressed as many issues as possible involved in microforming to help getting a thorough grasp of microforming technology by following a step-by-step process. I have beyond doubt that this contribution will be invaluable to researchers as well as graduate students in the field of microforming worldwide.

I sincerely congratulate the authors on having produced this splendid new scholarly book, and I am very confident it will become a worthy and widely recognized work. I look forward to this book's success in proving that "microforming" represents promising manufacturing technology which has a bright future and will take a long time for research with the tremendously increasing demand for micro products.

My best wishes to the authors.

Peter Robinson
University of Wollongong, NSW, Australia
October 2016

Preface

The booming development of the micro products market has promoted the rapid development of microtechnology production techniques. The microforming of metals is a cost effective approach to the batch production of complex, high performance micro components for a diverse range of applications including medical devices, precision equipment, communication devices, micro-electromechanical systems and micro fluidics systems.

Microforming differs from the conventional forming technology in terms of material, processes, tools, and machines and equipment, due to the miniaturization nature of the whole microforming system. Forming technology established in the macro world cannot be simply scaled down to the micro world to make high-quality miniaturized products. This is because it is impossible to scale down all parameters in the microforming process according to the theory of similarity, due to the existence of size effects in microforming processes. A number of unexpected problems in key aspects of mechanical behavior, tribology, and scatter of material behavior are encountered. As a result, low quality micro products, with high reject rates may be manufactured if there is no appropriate strategy available for the design and control of microforming processes. At present, there is still lack of a comprehensive knowledge based on microforming technology. Challenges remain in the high efficiency manufacturing of high-quality micro products due to the common problem of microscale size effects, complexity of processes for making micro products and the ever increasing requirement to improve product quality and performance. A book that systematically addresses the complex range of issues involved in microforming is thus very much in demand to provide readers with a technical documentation that can be used in the research, development and application of microforming technology.

Microforming Technology: Theory, Simulation and Practice is designed to provide engineers, scientists, academics and graduate students with the necessary comprehensive knowledge and information on microforming fundamentals, microforming theory, simulation models, numerical modeling of microforming process, equipment and tools design, practical microforming procedure, and microforming-related supporting system. Emphasis has been placed on the simulation and practice of seven different types of microforming methods, including micro cross wedge rolling (MCWR), micro flexible rolling, micro ultrathin strip rolling, micro deep drawing (MDD), micro hydromechanical deep drawing (MHDD), micro bending and micro compression to help readers obtain a thorough grasp of the lasted research developments in microforming by following a coherent step-to-step process. This requires an interdisciplinary approach drawing on relevant knowledge from materials science, mechanical engineering, materials processing engineering, and materials manufacturing engineering in order to clearly understand the technical issues involved in this book. We have included as many references as possible at the end of each chapter to provide readers a way to access relevant information in relation to the contents presented.

The book consists of four parts with a total of 20 chapters.

In Part I Introductory Overview. Chapter 1, Fundamentals of Microforming states that the fundamentals of microforming including microforming concept and microforming system followed by an introduction on some typical microforming methods. Chapter 2, Size Effects in Microforming describes the definition and categories of size effects, and discusses the problems caused by size effects in microforming. Strategies for control of size effects in microforming are also presented.

In Part II Theory of Microforming. Chapter 3, Scaling Laws presents the scaling laws that are applicable to microforming process, including scaling in geometry, dynamics, mechanics, hydrodynamics, heat transfer, electromagnetic force, electrostatic force and electricity. In Chapter 4, Strain Gradient Plasticity Theory, higher-order strain gradient plasticity theories, including couple stress theory, phenomenological strain gradient plasticity theory, and mechanism-based strain gradient plasticity theory that are widely studied are introduced. A conventional theory of mechanism-based strain gradient plasticity which does not involve the higher-order stress is also outlined. In Chapter 5, Crystal Plasticity Theory, two crystal plasticity (CP) theories including rate dependent and rate independent theories which have extensive applications in microforming are presented. A method for simplifying the rate dependent CP theory is introduced in order to reduce calculation process. The use of CP theory in combination with finite element method (FEM) in commercial software ABAQUS is also introduced to illustrate the application of CPFEM in simulation.

In Part III Simulation of Microforming Process. In Chapter 6, Simulation Models in Microforming, simulation models, including surface layer model, mesoscopic model and Voronoi model, that are popular in the investigation of microforming process are introduced, and the implementation methods of Voronoi model in FE software ABAQUS/CAE is also described. Chapter 7, Simulation of Micro Cross Wedge Rolling introduces the simulation process of MCWR, using a developed novel material model based on Hall–Petch relationship, surface layer and composited model with grained heterogeneity in 3D Voronoi tessellation. In Chapter 8, Simulation of Micro Flexible Rolling, a FE model is established to numerically study the springback in thickness direction during micro flexible rolling process, where 3D Voronoi tessellation has been applied to describe the grain boundary and generation process of grains in the workpiece. The influences of processing parameters on surface asperity flattening of workpieces are also numerically studied and analyzed. Chapter 9, Simulation of Micro Ultrathin Strip Rolling presents the simulation of micro ultrathin strip rolling. The rolling deformation characteristics of foil have been theoretically analyzed and a 2D elastic–plastic FE model is established to investigate the roll bite behavior in cold foil rolling process where the contact pressure distribution and roll contour in the roll bite are also discussed. Numerical investigation on the micro asymmetrical rolling process for ultrathin strip is conducted, and the surface roughness evolution and the effect of speed ratio in rolling are identified. In Chapter 10, Simulation of Micro Deep Drawing, a Voronoi blank model is developed for simulating the MDD process. Real microstructures and Voronoi structures are applied in microstructural models through image-based modeling method. Chapter 11, Simulation of Micro Hydromechanical Deep Drawing presents the simulation of MHDD where the closed and open lubricant pockets theory has been developed, and different coefficients of friction are applied in different lubricant pockets. A critical hydraulic pressure value that influences the wrinkling and earing occurrence and tendency is identified. In Chapter 12, Simulation of Micro Bending, the scatter of grain mechanical properties is calculated based on the results from micro tensile tests and the mechanical properties of these grains are assigned to each Voronoi tessellation to achieve the grain heterogeneity in ABAQUS/CAE. A new FE model is built with Voronoi tessellation and grain heterogeneity to simulate the micro V-bending process. Chapter 13, Simulation of Micro Compression develops a material constitutive model based on Hall–Petch relationship, composite model, surface layer model and grain heterogeneity. FE simulation associated with Voronoi tessellation is carried out to simulate the micro compression process with a consideration of temperature.

In Part IV Practice of Microforming. Chapter 14, Practice of Micro Cross Wedge Rolling introduces the design of MCWR machine and the forming process. The surface wear of forming tool and the surface toughness evolution are investigated. The influences of springback and friction behavior on the quality of rolled products are discussed, and the size effects on flow stress are investigated in order to improve the mechanical properties of micro parts. Chapter 15, Practice of Micro Flexible Rolling deals with the flexible rolling of aluminum alloy thin strip into different thickness ratios and various transition zone lengths, using an innovative micro-rolling mill. The microstructure and mechanical properties of the flexibly-rolled strip are systematically investigated and interpreted. Chapter 16, Practice of Micro Ultrathin Strip Rolling presents the development of a micro ultrathin strip rolling apparatus for manufacturing ultrathin foils with high accuracy. The rolling limit is analyzed, and the lubrication behavior of foil rolling is analyzed based on the feature size effects of rolled material and lubricating status. In Chapter 17, Practice of Micro Deep Drawing, MDD practice is described, and the influences of grain size, foil thickness, punch size, and initial gap on the punch force and displacement, friction and product's quality are discussed. Chapter 18, Practice of Micro Hydromechanical Deep Drawing develops a MHDD apparatus for manufacturing a micro-complex-shape component. The developed MHDD system can prevent wrinkling by applying an appropriate constant gap and the generated counter pressure. An appropriate counter pressure applied in MHDD can reduce the frictional drawing force and improve the forming limit and dimensions. Chapter 19, Practice of Micro Bending introduces the practice of micro bending and investigates the grain size effects on texture evolution of annealed and micro V-bended phosphor bronze foil with various thickness/average grain size ratios. In Chapter 20, Practice of Micro Compression, the practice of micro compression of pure copper at both room and elevated temperatures using the desk-top servo-press machine is introduced. Micro compression tests with assistance of laser heating are conducted to investigate the influences of temperature in micro compression. The effects of feature size on the material response in micro compression are also discussed.

This book is designed to provide a comprehensive overview of the theoretical, numerical and practical issues involved in developing microforming technology for those who are developing and producing metallic micro products and their forming technologies worldwide. We would like to thank all the people including academic staff, higher degree research students and technicians who have made contributions to this book.

Zhengyi Jiang, Jingwei Zhao, and **Haibo Xie**
University of Wollongong, NSW, Australia
October 2016

Part I

Introductory Overview

Chapter 1

Fundamentals of Microforming

1.1 MICROFORMING CONCEPT

Product miniaturization is a global trend for facilitating product usage, enabling product functions to be implemented in microscale geometries, and aimed at reducing product weight, volume, cost, and pollution. Driven by ongoing miniaturization in diverse areas including medical devices, precision equipment, communication devices, micro-electromechanical systems (MEMS), and micro fluidics systems (MFS), the demands for micro products, as shown in Fig. 1.1, have been tremendously increased. Such a trend requires development of efficient theories and technologies for manufacturing high-quality micro products with excellent dimensional tolerances, required mechanical properties and improved surface quality.

For manufacturing micro products, the conventional metal forming technology has been scaled down to microscale due to its process simplicity, high production capability, minimized or zero material waste, and near-net-shape production. Microforming, in the context of metal forming, is defined as the forming of parts or geometrical features with at least two dimensions in the submillimeter range by using mechanically based processing technologies. Products of this size are commonly used in electronics production and other fields including the medical sector, MEMS, and MFS. As a promising micromanufacturing technology, microforming has a lot of advantages including high productivity, ease in manufacturing complex geometries, applicable to a wide range of materials and excellent product properties, as compared to other processing methods such as turning, milling, and lithographic technologies. From the viewpoint of production engineering, microforming is considered as an effective process to economically and precisely manufacture micro products that are essential when bulk production is required. With the booming development of micro products market, more and more energy and resource will flow to the field of microforming, and the manufacturing of micro products will absorb more attention.

Microforming can be regarded as a forming technology processed in the so-called microfactory. A microfactory is a downsized manufacturing system which represents an approach to manufacture miniaturized products with less space and operational costs, and as well as reduced consumption

Microforming Technology. DOI: http://dx.doi.org/10.1016/B978-0-12-811212-0.00001-7

FIGURE 1.1 Cold headed micro products [1].

of resources and energy [2]. It has been demonstrated that in the case of 1/10 size reduction of production machines, the total energy consumption in the factory could be decreased to approximately 1/100 of that of the conventional factory [3]. In microforming, the size reduction of production machines can be realized by geometrically scaling down of the whole system, or by the combination and substitution of several components by just one component. However, a straightforward approach of a simple geometrical scaling of the forming system from macroscale to microscale may not result in the desired processing outcome due to the restrict transfer of existing know-how from macroscale to microscale [4]. The machines for microforming have some special requirements, such as high accuracy in positioning in the range of several micrometers or even less, which cannot be achieved through simply scaling down the conventional equipment for the microforming process.

In metal forming, the theories and technologies on forming processes in macroscale have been well established and widely used to manufacture high-quality complex-shaped parts with highly sophisticated multi-stage tools, making the forming of several geometrical features at an individual part possible. The conventional forming process has been thought to be an appropriate choice for the economical and reliable manufacturing of a variety of metal components in the large-scale production. Unfortunately, metal forming theories and technologies established in the macro world cannot be simply scaled down to the micro world, because it is impossible to scale down all parameters in the microforming process according to the theory of similarity due to the size effects, which are well known as a unique characteristic with miniaturization and will be detailed in the following Chapter 2, Size Effects in Microforming. The formability and forming behavior of a miniaturized part in microforming no longer follow the conventional rules established in macroscale metal forming process, and new knowledge on microforming should be established in order to produce high-quality micro products. At present, a challenge in microforming still exists because of geometrical sources, physical sources, change in the ratio between the surface and volume, as well as the increasing influence of Van der Waals and gravity forces [5].

1.2 MICROFORMING SYSTEM

A microforming system can be divided into four primary groups including material, processes, tools, and machines/equipment [6], as shown in Fig. 1.2.

Material. When a forming process is scaled down to microscale, the material to be deformed cannot be considered as a continuum due to the large share of volume occupied by an individual grain. Different from the conventional forming process, only limited grains are located in the forming area during microforming. Fig. 1.3 shows an example of the difference of a material between macroscale and microscale compressions. Grain and grain boundaries, which may have less impact on the material's mechanical behavior in macroscale compression, will have more significant impact on the mechanical behavior with the scaling down of dimensions. With miniaturization, the characteristics of an individual grain, including its size, shape, orientation, and position, and as well as other microstructural characteristics including grain boundaries, precipitations, and other phase constituents will have a close relevance to the material's mechanical conversion processes. The type of the material usable for microforming process is still

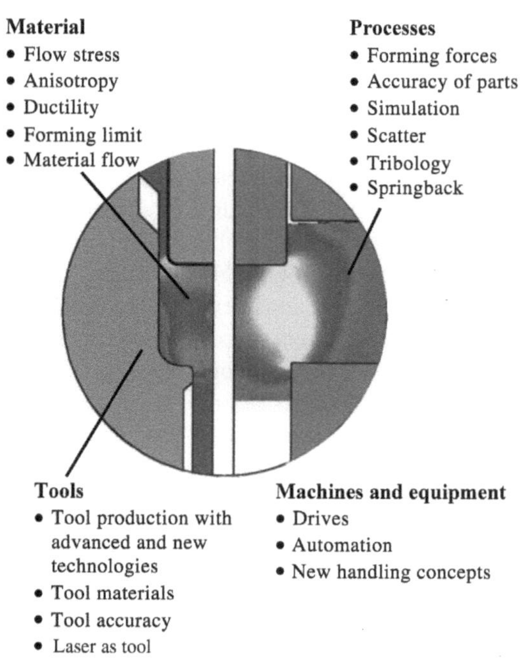

FIGURE 1.2 Components of a microforming system [6].

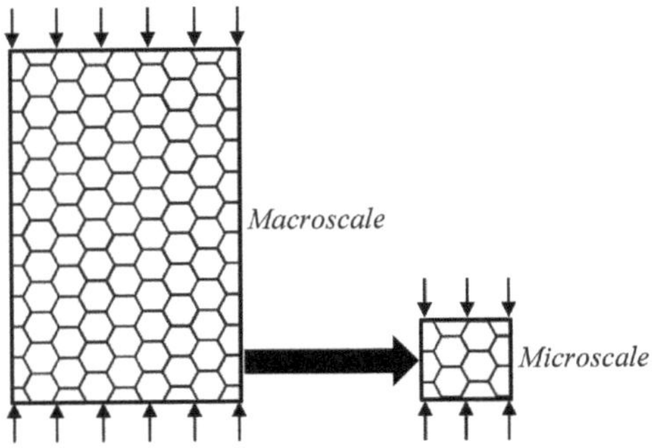

FIGURE 1.3 Schematic illustration of the difference between macroscale and microscale compressions.

limited due to the difficulties in the control of material flowability, anisotropy, ductility, and forming limit, which are important factors need to be taken into account in the evaluation of the mechanical response of the material during microforming process in order to manufacture high-quality micro products. Seeking for new technological solutions to improve the performance of current materials or developing new materials that are suitable for microforming are two ways to break the constraint by the limited materials that are processable in microscale with required quality and efficiency.

Processes. In the conventional metal forming, appropriate design of forming processes is essential for economical and reliable manufacturing of high-quality products. At present, it is possible to form complex-shaped parts with different geometrical features by combining optimized forming processes with appropriate tools and machines/equipment in large-scale production in industry. However, the conventional metal forming methods and strategies are not applicable to the reliable forming processes in microscale. Microforming largely uses nonconventional processes or scaling down or modifying the conventional processes, as appropriate, to fully address the issues related to the forming in the micro world. The microforming processes are, of course, strongly coupled with the material to be formed. Further, other factors, including forming forces, tribology, springback, scatter of the results, and accuracy of formed parts, should be considered in the design of microforming processes. Also, due to the differences between the conventional forming and microforming, simulation models should be improved by addressing as many issues as possible involved in the modeling of microforming processes. An optimized design of microforming processes is the premise of successful manufacturing of high-quality micro products.

Tools. Forming tools are core parts of the metal forming processes. As the size of a component is reduced to microscale, the demands for precise tools with high surface quality and close tolerance are increased. Unlike the manufacturing of tools for the conventional metal forming, the manufacturing processes of microforming tools are quite complex due to the difficulties in the creation of very small contour with high accuracy, which is, however, necessary for microforming operations. Especially, the tools with complex inner shapes, e.g., extrusion dies, are difficult to manufacture because close tolerance and adequate surface quality are simultaneously required [7]. In order to overcome these difficulties, development of advanced and new manufacturing technologies of microforming tools becomes essential. In the manufacturing of microforming tools, two main factors including tool materials and tool accuracy must be taken into account. The materials for tools should have superior performance so that the service life of tools can be prolonged. A guarantee of the tool accuracy is essential for the manufacturing of high precision micro products in the following microformig process. In addition, flexibility of tools is also a very important factor that should be considered in the development of tooling system in order to achieve efficient microforming processes. With the ongoing trend toward miniaturization, forming of micro products with more complex contours and shapes will become more common, and therefore advanced technologies involving innovative manufacturing methods of high-quality microforming tools are always expected.

Machines and equipment. Many problems occurred in microforming process are associated with machines and equipment with miniaturization. In the conventional forming processes, the clearance or backlash between machine parts can be negligible, which, however, may have a detrimental influence on the accuracy of the produced products in the range of a few 100 μm [6,7]. Therefore, it is necessary to develop machines and equipment that are particularly suitable for the microforming processes. Currently, the main trend in the development of machines and equipment for microforming is to reduce their scales which could, in turn, result in reduced energy consumption, pollution, and equipment cost due to the use of fewer materials for the miniaturized forming system. Also, as the scales of machines and equipment are reduced, the weight of the mechanical parts will be markedly reduced. As a result, the speed of the forming tools and the rate of production could be increased. However, as the size of micro part is very small and the part weight is quite low as compared with adhesion forces, appropriate strategies should be developed to separate the part from a gripper when handling. Driving, automation control, and measurement systems should also be considered in the development of machines and equipment in order to achieve a high productivity, ensure the product quality and enable an easy control of the forming processes. In addition, the extremely small-sized products with high precision might make it necessary to produce in a clean environment, which is cost-intensive and must lead to new machine concepts [6]. With advances in microforming technology, it is possible to develop high-performance microforming machines and equipment that address all of the above-mentioned technical issues in the future, which will then greatly contribute to the manufacturing of complex-shaped micro products with high productivity, high precision, and low cost.

1.3 MICROFORMING METHODS AND PROCESSES

Microforming methods and processes can be classified into a number of categories based on the way in which components/products are to be formed. Typical microforming methods include micro rolling (e.g., micro cross wedge rolling (MCWR), micro flexible rolling, and micro ultrathin strip rolling), micro deep drawing (MDD), micro hydromechanical deep drawing (MHDD), micro bending, and micro compression.

1.3.1 Micro Rolling

In metalworking, rolling is a metal forming process in which workpiece is passed through one or more pairs of rolls or featured tools to reduce the thickness or to form specifically shaped products. A schematic illustration of flat rolling process is shown in Fig. 1.4. Micro rolling differs from the conventional rolling that the dimensions of workpiece along the deformation direction in micro rolling are in the submillimeter range, and rolling deformation is processed in microscale. Typical micro rolling technologies against the way of producing products include MCWR, micro flexible rolling, and micro ultrathin strip rolling.

1.3.1.1 Micro Cross Wedge Rolling

Before the introduction of MCWR technology, the concept of cross wedge rolling (CWR) should be first understood. CWR is a metal forming technology in which a cylindrical billet is plastically deformed into an axisymmetrical part by the action of wedge-shaped dies moving tangentially relative to the workpiece. CWR is an efficient metal forming technology applied mainly for the manufacturing of stepped axles and shafts, and has been proven to be an economical and reproducible means of producing preforms for forging operations. There are five different variants of CWRs in which two or three wedges mounted on rolls or flat or concave plates of rolling mills are used, as shown in Fig. 1.5 [8]. Among these techniques, the most common methods are forming with two rolls, three rolls, and flat wedges, as indicated in Fig. 1.5A, C, and D, respectively.

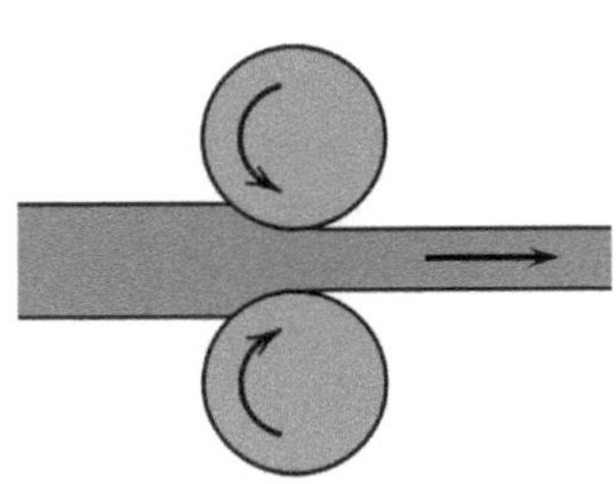

FIGURE 1.4 Schematic illustration of flat rolling process.

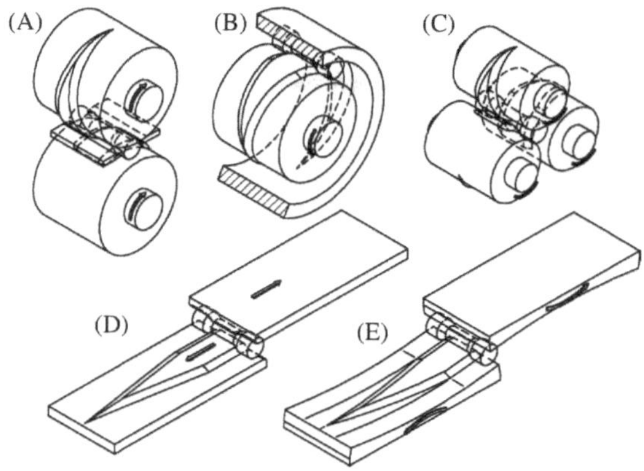

FIGURE 1.5 Methods of cross-wedge rolling in configurations: (A) two wedges, (B) wedge-concave segment, (C) three wedges, (D) two flat wedges, and (E) two concave wedges [8].

CWR has a significant potential in technology and applications in the field of metal forming. Compared to the conventional machining, forging, and casting processes, CWR offers several innovative and unique features. These features include:

Higher productivity: One or more parts can be produced in each pass of the forming dies. If the product's geometry is not extremely complex, pairs can be made in one operation by a symmetrical tool layout. Normally, 6–20 parts could be produced per minute making productivity 3–10 times higher than that of machining and forging.

Higher material utilization: In CWR, less than 10% of the raw material is wasted in end-cutting and grinding so that CWR greatly reduces the cost of raw materials. In contrast, 40% of the material is generally lost in machining processes in the form of chips.

Better working environment: The average noise received for hammer operators under the working condition is about 100 dB. For long-term exposures (10 years or more), it is indicated that hearing losses resulting from impact noise in the drop-forging industry are as great as or greater than those resulting from continuous noise. Sounds of less than 80 dB, even after long exposure, are unlikely to cause hearing loss. With the CWR process, the working noise level is about 65–75 dB. Compared to the conventional forging, CWR provides significantly less noise and impact. Therefore, it is a more environment-friendly process. In addition, by simplifying the overall manufacturing processes, much less energy will be consumed for completing transmission shafts.

At present, fundamental research on CWR is still in its developing stages in the conventional scale. When CWR technology is scaled down into micro domain, the equipment cannot be simply minimized according to similarity theory due to the restrict requirement of processing accuracy and derived

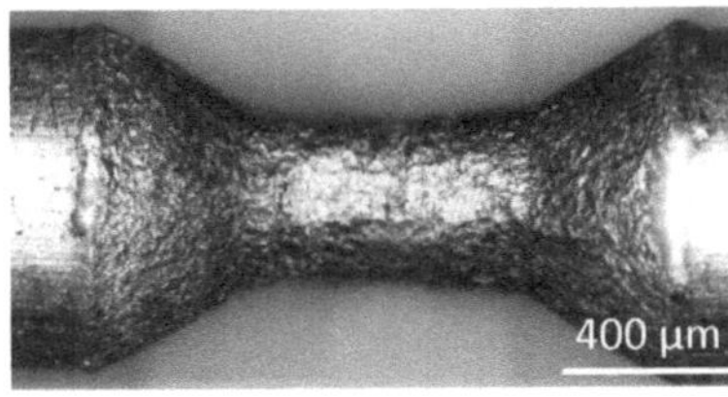

FIGURE 1.6 A product manufactured by MCWR.

phenomena from miniaturization, such as micro gravity and increased adhesion. In the manufacture of micro products, the development of sustainable manufacturing requires the disengagement of the link between increased micro products manufacture and the rise in energy consumption. It requires the replacement of the old technology and manufacturing processes with new and advanced technology that is less energy intensive per unit of manufactured products. Therefore, it is promising to apply CWR in the area of microforming because of its high production rate, minimized material consumption, and excellent mechanical properties when producing stepped shafts [5]. Fig. 1.6 shows an example of product manufactured by MCWR technology. However, as MCWR will encounter the common problems involved in microforming processes including size effects, it is inevitable to confront with the various technical challenges in the development of MCWR technology. Studies have indicated that size effects have significant influence on the processing factors in MCWR, for instance, springback and surface topology of workpiece [9]. The springback in micro domain cannot be predicted using the conventional mechanical theories and the springback behavior in MCWR is relevant to the material microstructure and deformation amount [10]. The distribution and evolution of strain and stress will become irregular in the processing of MCWR due to grain heterogeneity, which will induce inhomogeneous deformation. Lu [11] found that a homogeneous phenomenon of deformability of workpiece could be achieved at elevated temperatures, which could in turn lead to improvements in the edge quality of the products manufactured by MCWR due to the activation of more slip systems with thermal activation. In order to take full advantage of MCWR technology, comprehensive investigations on the influence of miniaturization, especially size effects, on process, accuracy control, and product quality are still needed even though staged achievements of MCWR have been currently achieved. Additionally, research on the interaction of tools and workpiece that may influence workpiece defects and tool wear is also essential.

1.3.1.2 Micro Flexible Rolling

Thin strip with varying thickness is an innovative strip product with thickness transition along the rolling or lateral direction or along both rolling and lateral directions. Thin strip with varying thickness such as tailor rolled blank (TRB) is manufactured by flexible rolling or the combination of flexible and profile rolling. It has many advantages compared with the tailor welded blank which is widely used in the automotive and aerospace industries. There is a wide range of sheet thickness gradient and position which are freely selectable, whereas for tailor welded blank the position of welding seams always needs careful consideration. Due to the absence of welding seams, the formability of thin strip with varying thickness (i.e., TRB) in subsequent sheet metal forming processes can be significantly improved with no stress peaks in the thickness transition, reduced abrasive tool wear, and improved surface quality. The application of strips with varying thickness not only reduces the weight of components, but also increases formability, such as the drawing depth. Tailor rolled strip with varying thickness exhibits an improvement of structural behavior which can result in 15.6% less intrusions at a weight reduction of 13.5%, and furthermore, the crash boxes may display 15.9% less average deformation, while absorbing 48% more buckling forces [12]. Therefore, thin strip with varying thickness has been regarded as a product of the future for manufacturing lightweight structures in industries such as the automotive and aerospace industries due to its advanced characteristics.

Flexible rolling is a novel forming method for the production of strip with longitudinal thickness variation. In flexible rolling, a gap control system is applied to adjust the roll gap online deliberately during rolling operations, and to manufacture the thickness transition along the rolling direction, as illustrated in Fig. 1.7. At present, study on flexible rolling mainly targets at the macro world and plenty of technologies and theories have been set up and adopted for mass production of flexibly rolled strips in automotive, construction, and energy industries. Due to the ongoing trend toward miniaturization of systems and devices, particularly in the field of electronics production, micro flexible rolling technology has been developed and applied to produce ultrathin strip with varying longitudinal thickness in microscale [13]. However, when the thickness of workpiece to be flexibly rolled is scaled down to the submillimeter range, the so-called size effects that

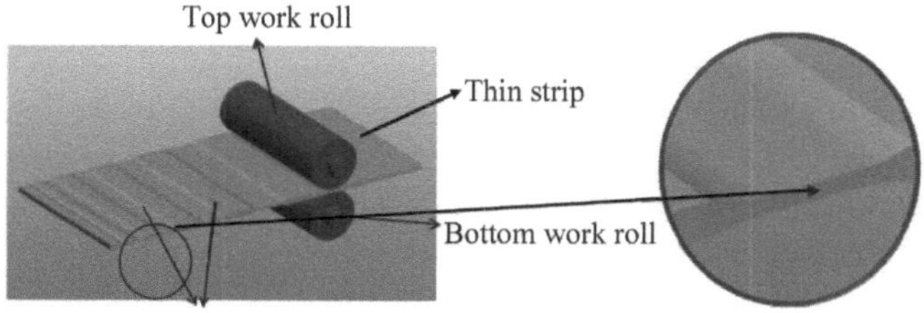

FIGURE 1.7 Illustration of rolling of thin strip with varying longitudinal thickness.

characterize the micro world will have a significant influence on the forming process. A typical work conducted by Qu et al. [14] indicates that significant springback problem is involved in micro flexible rolling due to size effects, which has a great impact on the rolled products.

In order to produce ultrathin strip with varying thickness and perfect shape, flatness and optimal thickness transition length, new and advanced micro flexible rolling technologies need to be continuously developed because the technologies for the conventional flexible rolling cannot be simply transferred to the microscale. Such technological development is necessary not only to meet the high demands of the metal microforming/micromanufacturing industry, but also to enhance the advanced rolling theory for the international metal manufacturing community. As the thickness of workpiece in micro flexible rolling is in the submillimeter range, reduction will become difficult due to the existence of a limit on the thickness of workpiece which can be reduced on a given mill. This minimum gage phenomenon is the result of a complex interaction of effects between the rolls and the strip rather than a function of the strip behavior or of the mill behavior along [15]. As a result, control of the reduction of the transition length of the varying strip thickness zone, and as well as the strip shape, profile, and flatness for thin strip, will be challenges for researchers and manufacturers in micro flexible rolling. This requires development of advanced micro flexible rolling equipment with high stiffness rolls and high-precision control system that have the capacity to realize rapid dynamic response to change the roll gap and roll speed when the rolling mill is running.

1.3.1.3 Micro Ultrathin Strip Rolling

Rolled ultrathin strip has a wide application in electronic and instrument industries, and its production has always been of major interest to the manufacturers and researchers in the area of metal rolling. Micro ultrathin strip rolling is a metal forming method in microscale by using rolling technology. In this method, the thickness of strip to be rolled is usually in the range of hundreds or tens of micrometer, which is a challenge in rolling due to the involvement of the control of strip dimensional accuracy, strip shape and profile, flatness, and surface finish.

The deformation of ultrathin strip in micro rolling is affected by size effects which are embodied in the flow stress and the coefficient of friction (COF). In micro rolling of ultrathin strip, flow stress is decreased with the scaling down of specimen size, and COF will increase with miniaturization and with sliding distance as well [16]. The rolling force, intermediate force, and edge contact force increase significantly with an increase of COF. However, when the COF increases, the strip profile will be improved [17]. If the strip edge friction increases, the rolling force and intermediate force will increase. However, the strip profile will be improved when the strip edge friction increases.

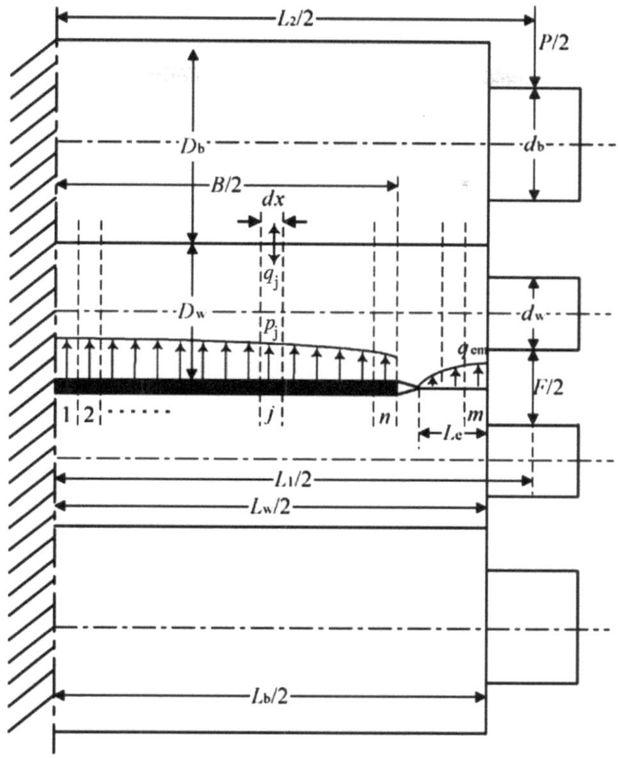

FIGURE 1.8 Edge kiss of work rolls for a 4-high rolling mill [18].

In practical rolling of thin strip, there is a phenomenon that the upper and lower work rolls may contact each other beyond the edges of strip if the strip is very thin, as shown in Fig. 1.8. In the figure, m and n are the numbers of the slab elements along the half-roll barrel and half-strip width, respectively, dx is the width of each element, D_w the diameter of work roll, D_b the diameter of backup roll, d_w the diameter of work roll neck, d_b the diameter of backup roll neck, L_w the half-width of work roll barrel, L_b the half-width of backup roll barrel, F the bending force, P the rolling force, q_j the intermediate force between the work roll and backup roll at element j, p_j the rolling force at element j, q_{em} the edge kiss force between the upper and lower work rolls acting on slab element m, B the strip width, L_1 the central distance between the work roll bending cylinders, L_2 the central distance between the housing screws, and L_e the roll edge contact length. The work roll edge contact can increase work roll edge wear, which will shorten the work roll service life. The edge kiss of work rolls can improve the strip shape when no work roll bending force is applied. The edge kiss force between the two work rolls will increase when the COF increases, which can improve the shape of ultrathin strip in rolling [18]. When suitable friction conditions along the longitudinal and transverse directions are applied to the micro rolling process, a good shape and profile of ultrathin strip can be obtained [19].

Lubrication in micro ultrathin strip rolling process plays an important role in the cooling of rolled material, reducing of rolling force, decreasing of wear of both rolls and rolled material, and lowering of friction. However, as the conventional models on lubrication are not reliable in the scaled-down process, study on lubrication characterization in micro ultrathin strip rolling is getting increasing attractive to researchers [20]. During micro ultrathin strip rolling, the oil film thickness decreases with increasing the pass reduction and decreasing the rolling speed. When the oil film thickness is increased, rolling force will be reduced, the surface of the product, however, will become dull. Lubricating condition is remarkably influenced by roll-coating formation, especially by using smooth surface of rolls. For micro ultrathin strip rolling, the most preferable lubricant is emulsion.

Tribology plays a significant role for process feasibility and quality, especially in microforming where the influence of tribological condition is very essential due to an extensive increase in friction when the process is scaled from macro to micro case. Asperity flattening in surface roughness affects the friction behavior at the interface and the transfer of surface roughness during rolling process. Surface roughness affects the friction behavior, stress distribution, and surface quality, which will become more significant with the scaling down of strip size. Therefore, research on surface roughness is extremely important in micro ultrathin strip rolling. For studying the surface roughness evolution in micro ultrathin strip rolling, Xie et al. [21] conducted a research on the evolution of surface roughness in asymmetrical rolling based on real surface features. Asymmetrical rolling is an effective process for the manufacturing of metal strip with small size, and can be achieved by using rolls with unequal diameters, by prescribing different speeds of rotation to the rolls, by imposing different COFs between the two rolls and the rolled strip, or by a combination of the above various factors. Asymmetrical rolling can be an alternative way for micro ultrathin strip rolling, however attention should be drawn as the rolling force across strip width and the surface quality of rolled ultrathin strip will be significantly affected by a number of rolling parameters, especially the asymmetrical factor of speed.

1.3.2 Micro Deep Drawing

Deep drawing is one of the most widely used processes in sheet metal forming, in which a sheet metal blank is radially drawn into a forming die by the mechanical action of a punch. Deep drawing is effective in the manufacturing of precise and intricate parts with both symmetrical and asymmetrical features that have extensive applications in our daily lives. A typical deep drawing process, which includes six steps, is schematically

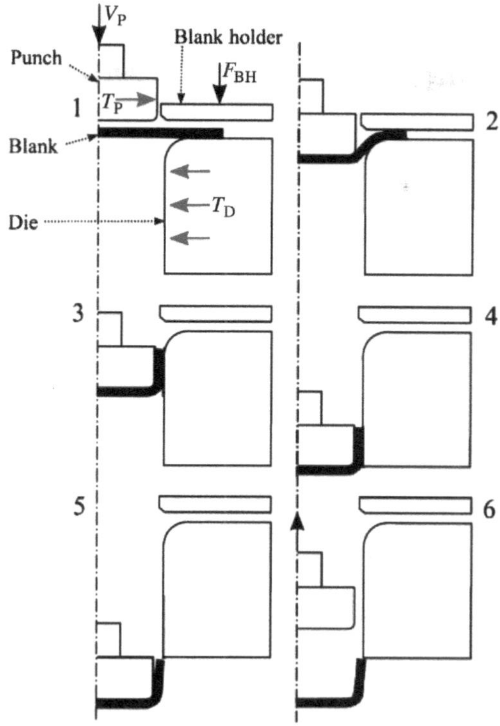

FIGURE 1.9 Deep drawing process [22].

illustrated in Fig. 1.9 [22]. In the figure, V_P is the moving speed of punch, F_{BH} is the blank holder force, and T_P and T_D are the temperatures of punch and die, respectively. Due to the ongoing trend of miniaturization in diverse applications including consumer electronics, medical devices, and telecommunication devices, the demands for micro-scaled metallic parts, e.g., micro cups, have been continuously increasing. Under this background, MDD technology has been developing rapidly. MDD is a fundamental process of microforming which has extensive applications in the forming of hollow, thin walled, cup or box like micro parts. However, due to the common problem of size effects existed in the forming of metals in microscale, research on MDD will be more comprehensive as compared to the conventional deep drawing. As a result, an acquisition of new knowledge becomes essential for the promotion of the development of MDD technology.

Tooling–workpiece interfacial lubrication condition is an important factor affecting the whole process of MDD, including drawing force, drawing ratio (DR), product surface quality, and tool life. COF decreases with increasing the amount of lubricant in both macro deep drawing and

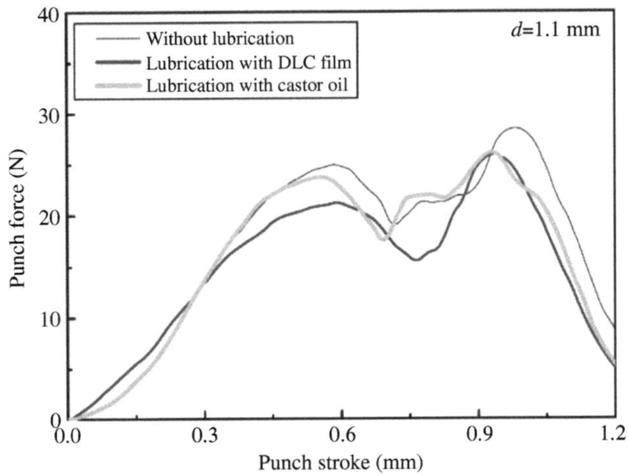

FIGURE 1.10 Punch stroke-punch force curves in MDD with punch diameter of 1.1 mm [26].

MDD processes. Differently, the decrease of COF in MDD is faster than that in macro deep drawing [23]. As liquid lubricants are not effective in lubricating in MDD process due to the tribological size effects in forming, and may also contaminate the formed parts, solid lubricants for MDD and as well as lubricant-free MDD processes have been developed. Gong et al. [24,25] found that relatively low drawing force, high limit drawing ratio (LDR) and good surface quality of the formed parts could be achieved under the lubrication conditions of polyethylene film, as compared to liquid lubricants. A number of researches have indicated that the diamond like carbon (DLC) coating of forming tools can play a positive role in decreasing the drawing force in MDD. Gong et al. [26] compared the stroke-drawing force under lubricating conditions of DLC, castor oil, and as well as that without lubricant, and found that the punch force with lubrication of castor oil was only slightly smaller than that of without lubrication due to the overflow of most of the castor oil caused by size effects on friction. However, the punch force for lubrication with DLC film could be reduced by approximately 15%, as shown in Fig. 1.10. This is because that the COF between DLC film and sheet is much lower than that without lubrication, and the COF is not affected by size effects in the MDD process. DLC coating of forming tools has an application potential in MDD in the reducing of drawing force, increasing of LDR and improving of the quality of formed micro products [27,28]. However, an important issue of cost should be considered in the application of DLC coating of forming tools for MDD process.

Blank holder force is also an important factor in MDD process. An optimal blank holder force is essential in the forming of high-quality micro products. Wrinkle and fracture will take place when the applied blank holder

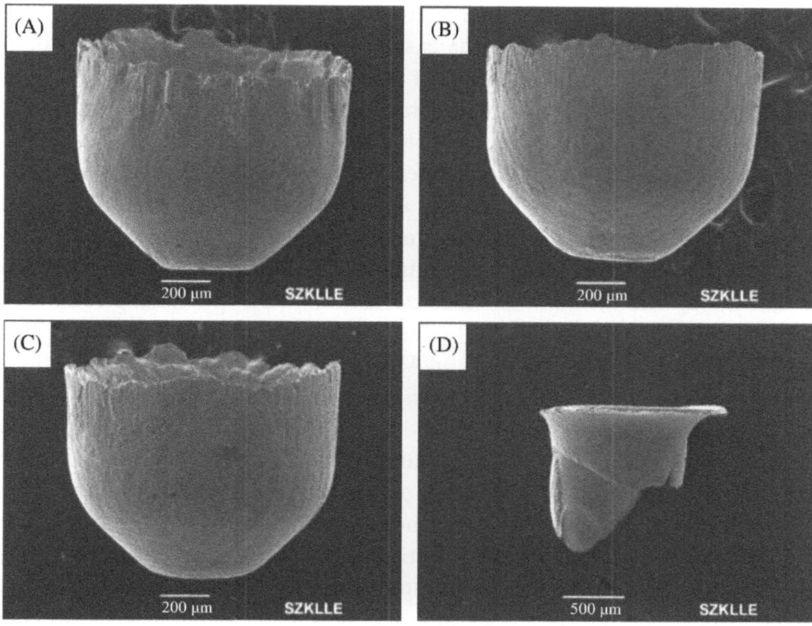

FIGURE 1.11 Formed parts in MDD under different blank holder forces: (A) 2.8 N, (B) 4.2 N, (C) 5.6 N, and (D) 7.0 N [25].

force is too small or too high for a given punch velocity [29]. For example, in the forming of copper, the rim of micro cup will be wrinkled when the blank holder force is less than 4.2 N; and when the blank holder force is higher than 5.6 N, the bottom of micro cup will be cracked, as shown in Fig. 1.11 [25]. For micro products, the surface roughness can be relatively large and has significant effects on MDD process and products' quality. Hence, surface roughness should be considered in MDD process. The work of Luo et al. [30] indicated that surface roughness in MDD process with a consideration of size effects had significant effects on the overall springback, the drawability represented by the minimum thickness and the products' quality regarding thickness evenness and shape accuracy.

In deep drawing, there is a maximum size of the blank which can be drawn out without failure. DR is defined as the ratio of the blank diameter to the punch diameter, and LDR is defined as the highest value of the ratio of the blank diameter to the punch diameter that can be drawn out without failure, which can be expressed as follows:

$$\mathrm{LDR} = \left(D_\mathrm{b}/D_\mathrm{p}\right)_{\max} \tag{1.1}$$

where D_b and D_p are the diameters of blank and punch, respectively.

The value of LDR is affected by several factors including blank diameter, punch diameter, blank thickness, microstructural characteristic of metals, blank holder force, clearance, and lubrication conditions. In MDD, a thicker foil usually has a larger LDR in the case of keeping grain size or t/d (t is the blank thickness and d is the grain size) constant [31]. Saotome et al. [32] conducted a research on the micro deep drawability of very thin sheet steels with different thickness values. Their research indicates that LDR decreases with the increase of relative punch diameter D_p/t, and an LDR of 2.2 could be achieved when D_p/t equals to 10. For a relative punch diameter of 10, bending is the dominant forming mechanism, since the effect of blank holder force on drawability has been found to be minimal [23]. In order to further improve the drawability and increase the value of LDR in MDD, processing at elevated temperatures can be adopted [33,34]. However, attention must be drawn in the use of local heating, e.g., by laser, in MDD as oxidation may occur if a local area is overheated, which can have significant effect on the quality of the formed products. In some cases, micro cups with required dimensions could not be drawn in only one stroke due to the effect of springback. As a result, redrawing, reverse redrawing, multi-step drawing, or even additional annealing process of the blank metal becomes necessary in order to form satisfactory micro products in MDD. Currently, a major challenge faced in practical MDD still exists in the achievement of as high values of LDR as possible within a limited space, which usually limits the arrangement of tooling system. Therefore, an optimal design of tooling system and processing conditions becomes necessary to enable an effective MDD process.

1.3.3 Micro Hydromechanical Deep Drawing

MHDD is a kind of sheet hydroforming method in which a counterpressure is applied to enhance material flow and achieve uniform deformation with the control of thickening and thinning behavior by deformation localization and strain allocation. MHDD is effective in the fabrication of long and complex-shaped micro products with high shape accuracy through enhancing the forming limit and improving the tribological behavior owing to the combination influences of friction holding effect, hydrodynamic lubrication effect and prebulging effect [35,36]. Fig. 1.12 shows a schematic illustration of MHDD processes and movement of forming tool [37]. As indicated in the figure, MHDD processes can be divided into four stages: ① The drawing die (blanking punch) is fixed, whereas the upper die including the drawing punch, the blank holder, the blanking die, and the bush moves downward. In this stage, the clearance between the drawing die and the bush becomes small, as a result, the die cavity is sealed and a counterpressure is generated. ② When the upper die moves further downward, the killer sheet integrated with the blank holder is fixed because of the killer pins, causing the blank holder to be simultaneously fixed. Thus, only the drawing punch, blanking die, and bush move downward during the blanking process. ③ During drawing, the drawing punch moves downward while maintaining a constant gap

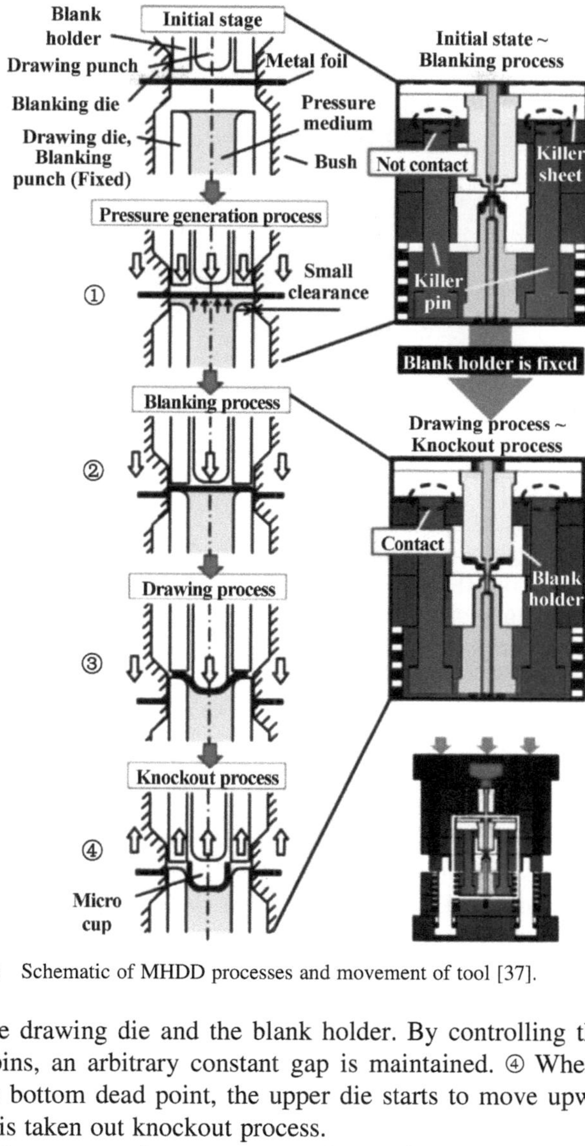

FIGURE 1.12 Schematic of MHDD processes and movement of tool [37].

between the drawing die and the blank holder. By controlling the length of the killer pins, an arbitrary constant gap is maintained. ④ When the stroke reaches the bottom dead point, the upper die starts to move upward and the drawn cup is taken out knockout process.

Fig. 1.13 compares the appearance of cups drawn by MDD and MHDD. It can be seen that wrinkling occurs frequently at the cup edges for all the tested materials in MDD. In contract, wrinkling could be suppressed when an appropriate counterpressure is applied in MHDD. In MHDD, the value of counterpressure should be controlled to be within an optimal range in order to fabricate high-quality micro products. For example, when counterpressure is increased from 4, 15, and 2 MPa to 8, 20, and 4 MPa for phosphor bronze, stainless steel and pure titanium, respectively, the cups will be fractured at

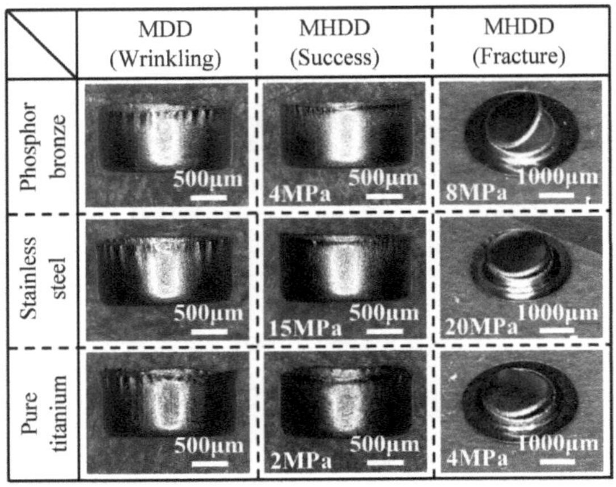

FIGURE 1.13 Appearance of cups drawn by MDD and MHDD [37].

the punch shoulder, as indicated in Fig. 1.13. For achieving high formability and fabricating high-quality micro cups without failure, fine-grained metal foils are desirable in MHDD [38].

Tribological behavior in MHDD is an important factor that may have significant effect on the whole forming process and the subsequent micro product quality. In MHDD, the effect of fluid behavior, the ratio of the punch diameter to the minimum thickness, open lubricant pocket (OLP) and closed lubricant pocket (CLP) on the tribological behavior need to be characterized in order to clarify the tribological size effects in MHDD. The work of Sato et al. [39] has indicated that MHDD can induce hydrodynamic lubrication and lubrication in OLP, which can then improve the tribological behavior in microforming process when appropriate fluid pressure is applied. MHDD has the opposite tribological behavior compared with the conventional microforming, i.e., the friction force decreases with the scaling down in MHDD, whereas it increases in the conventional microforming process. The feature size effect has a significant impact on the deformation and fluid behavior and has to be considered in the determination of the forming conditions in MHDD. The forming limit achievable in MHDD is much higher than those obtained by the conventional micro sheet forming techniques, and is effective in the fabrication of quite long micro cups with high accuracy [40]. As a promising microforming method, the research on MHDD in future can be focused on the simplification of the tooling system and optimization of the processing conditions. In order to realize further miniaturization of micro products, punchless forming technology can be developed because there is a limitation of miniaturization of target size when a mechanical punch is applied, even though its size is tiny. Moreover, development of precise positioning and control systems for MHDD is essential in order to achieve a high microforming efficiency.

1.3.4 Micro Bending

Bending, in terms of metalworking, is a major sheet metal forming process, which involves plastic deformation of workpiece over an axis to create a change in the part's geometry, in the fabrication of sheet metal products that are featured with specified geometrical shapes, e.g., V-shape, U-shape, and cylindrical shape, as illustrated in Fig. 1.14. With the development of microforming technology, bending of metals in microscale has been developing rapidly. Micro bended sheet metal products have wide applications in MEMS, e.g., micro electric contacts, micro electric switches, micro clamps, 3D profiles for mechanical and thermomechanical sensors, 3D sheet metal frames/housing for optical devices and micro sensors, etc.

Similar to other microforming methods, micro bending process is quite different from the macro one due to the common problem of size effects in the micro world. In microscale, metals cannot be considered as homogeneous owing to the limited quantities of grains involved in workpiece. The property of each grain will perform its role in micro bending process, especially, the grains in deformed region. Many factors, including the thickness of sheet metal, grain size, grain distribution, grain shape, grain orientation, and grain boundary conditions at the bend section, will have significant effects on the micro mending process and the quality of the bended products. One of the major challenging issues in the micro bending of sheet metals is springback, which may occur immediately after the release of the forming force, or occur due to the subsequent release of the residual stresses in the formed part. Springback will cause distortion of shape or instability of dimensions of the bended part in service, leading to difficulty to meet the geometry and shape accuracy requirements. Springback is affected by several factors including Young's modulus, initial yield stress, Poisson's ratio, strain hardening exponents, punch and die profile radii, clearance between the punch and the die, friction at the interface between the punch and the sheet, and blank holder force [41]. These, however, should also include considerations of the factors that may cause size effects in the evaluation of springback in micro bending.

In order to reduce springback in micro bending and improve the accuracy of micro products, optimizations of bending stroke, bending angle, tool

(A) (B) (C)

FIGURE 1.14 Schematic illustration of (A) V, (B) U, and (C) cylindrical bending techniques.

shape, punch/die clearance, etc. can be made. In addition, working at elevated temperatures can reduce the occurrence of springback in micro bending, and generally an increase in temperature decreases the springback angle. In practice, resistance heating [42] or laser heating [43,44] can be introduced in micro bending process to achieve high accuracy of the bended part's geometries. Development of other bending-related technologies, such as in-process measurement of springback and adjustment of bending angle/speed, is helpful for the control of springback in micro bending process. Moreover, numerical simulation will be beneficial for designing micro bending system and optimizing processing parameters with a purpose of minimizing the occurrence of springback in micro bending process.

1.3.5 Micro Compression

In metalworking, micro compression is a common microforming method to deform a metal along the direction of compressive loading applied to the metal. Testing of micro compression allows for a direct measurement of stress–strain behavior in volumes of metals that have microscale dimensions. In microscale, metal deformation process is characterized by high workpiece surface-to-volume ratio. With the decrease of workpiece size, problems of springback, and interfacial friction stress will be caused in micro compression process [45]. A decrease in the ratio of specimen size to grain size will induce decrease of flow stress, scatter of experimental data, and inhomogeneous deformation in micro compression [46]. When the workpiece to be compressed consists of only a few grains, the grain size, shape, and orientation have significant effect on the deformation behavior, which will lead to scatter of the obtained data and inhomogeneous deformation [47,48].

In micro compression process, the tooling–workpiece interfacial friction is high and nonuniform, and shows a dependence on the dimensions of the workpiece. The COF at the tooling–workpiece interface increases with decreasing the diameter of cylinder specimens when oil lubrication is applied in micro compression [49]. The work of Deng et al. [50] reveals that there is a real contact area at the tooling–workpiece interface, which is concentrated at the outer rim, while the CLP is located at the inner region. As the fraction of real contact area does not decrease with the decrease of specimen size, interfacial friction force will be increased. The interfacial friction in micro compression is sensitive to grain size, and the coarser the grain size, the lower the friction force will be. The decrease of friction force is related to the decrease of grain boundary strengthening effect and the increase of the fraction of surface grain to facilitate the material flow when the grain size is refined. The mechanical properties of individual grains have a significant influence on the tribological behavior. With the decrease of workpiece size, the average COF does not change much, but the distribution of COF of each random grain becomes more dispersive [51].

Using of ultra-fine grained metals in micro compression is helpful in the achieving of uniform flow of the workpiece in forming, improved surface quality, and enhanced properties of the compressed products. Fine-grained metals can be produced by either nano-powder metallurgy or severe plastic deformation. As the cost, residual porosity and health issues are concerns for the powder metallurgy process, nano-powder metallurgy is rarely used. Severe plastic deformation, which is based on inducing a very large plastic deformation without basically changing the shape of the billet, is commonly used in practice to refine the microstructure before micro compression operations. The flow stress increases, and a transition from strain hardening to strain softening in the flow stress curves appears during micro compression process as the grain size is decreased. Ultra-fine microstructure can improve the strain compatibility between adjacent grains, and can therefore improve the occurrence of roughening in micro compression process [52,53].

In microscale, the forming behavior of metals is characterized by inhomogeneous and limited formability, which can be verified by the scattering of various relevant forming parameters. To counteract such problem, working of micro compression at elevated temperatures can be adopted. A raising in temperature can reduce the flow stress, homogenize the material flow behavior and minimize the scatter of processing parameters, and as a result, the quality of compressed products can be improved. Micro compression at elevated temperatures can be achieved, e.g., with the assistance of laser heating, as schematically illustrated in Fig. 1.15. In addition, ultrasonic vibration can also be used in micro compression to reduce flow stress and decrease friction [54].

1.3.6 Other Microforming Methods

In addition to the above-mentioned microforming methods and processes, micro extrusion, micro forging, micro stamping, micro embossing, micro punching, micro blanking, micro coining, micro heading, and semisolid microforming are also attracting attentions in the forming of micro products that have a variety of applications. Currently, staged achievements in these

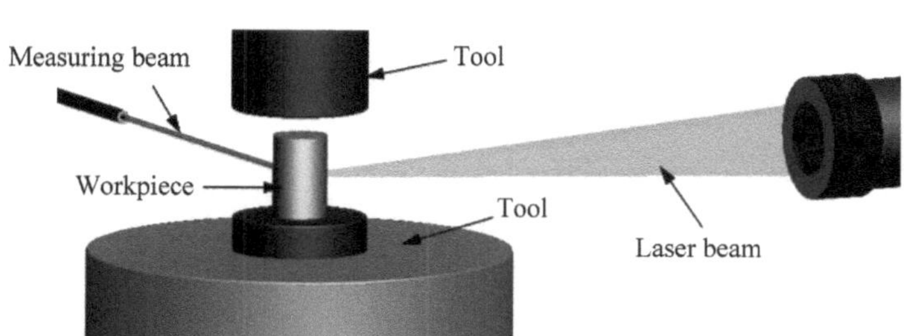

FIGURE 1.15 Schematic illustration of micro compression with assistance of laser heating.

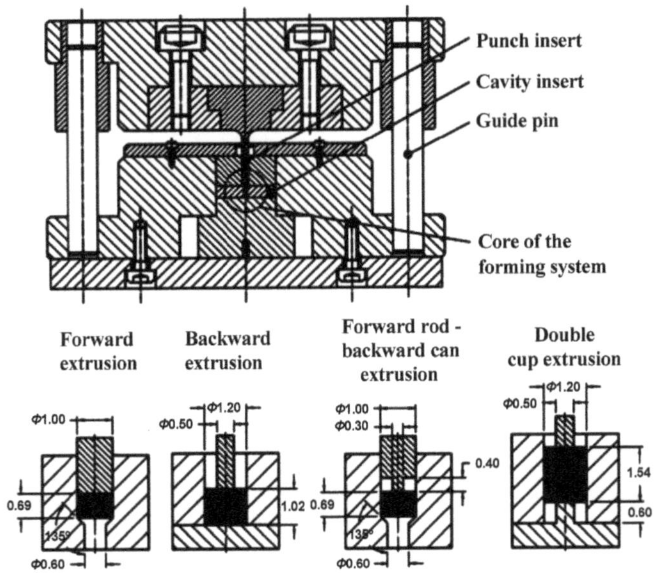

FIGURE 1.16 Schematic illustration of micro extrusion processes, die assembly and related dimensions [55].

areas have been made in modeling and experiments. For micro extrusion, e.g., four types of processing technologies, including forward extrusion, backward extrusion, forward rod-backward can extrusion, and double cup extrusion, are widely investigated in terms of the effects of specimen size, grain size, formed part feature size, tooling–workpiece interfacial friction, forming temperature, and ultrasonic vibration by either numerical or experimental methods [55–61]. Fig. 1.16 shows a schematic illustration of micro extrusion processes, and their die assembly and related dimensions. However, due to the existence of size effects in microforming, there will be an interactive effect of the interfacial friction, the plastic properties of micro-sized material, the microstructure of workpiece, the surface topology of workpiece and the workpiece size on the deformation behavior and the quality of formed micro parts. Therefore, much more efforts on comprehensive considerations should be placed in the development of these microforming methods and the optimization of relevant microforming processes.

REFERENCES

[1] F. Vollertsen, Z. Hu, H. Schulze Niehoff, C. Theiler, State of the art in the micro forming and investigations into micro deep drawing, J. Mater. Process. Technol. 151 (2004) 70–79.

[2] Y. Okazaki, N. Mishima, K. Ashida, Microfactory-concept, history, and developments, J. Manuf. Sci. Eng. 126 (2004) 837–844.

[3] Y. Okazaki, N. Mishima, K. Ashida, Microfactory and micro machine tools, Reported in The 1st Korea-Japan Conference on Positioning Technology, Daejeon, Korea, 15–17 October 2002, pp. 1–6.
[4] B. Eichenhüller, U. Engel, M. Geiger, Microforming and investigation of parameter interactions, Prod. Eng. 4 (2010) 135–140.
[5] D. Wei, H. Lu, Z. Jiang, K.I. Manabe, Surface morphology of micro stepped components in micro cross wedge rolling, Proc. Eng. 81 (2014) 1902–1908.
[6] M. Geiger, M. Kleiner, R. Eckstein, N. Tiesler, U. Engel, Microforming, CIRP Ann. Manuf. Technol. 50 (2001) 445–462.
[7] U. Engel, R. Eckstein, Microforming-from basic research to its realization, J. Mater. Process. Technol. 125–126 (2016) 35–44.
[8] Z. Pater, A. Gontarz, W. Weroński, Cross-wedge rolling by means of one flat wedge and two shaped rolls, J. Mater. Process. Technol. 177 (2006) 550–554.
[9] Z. Jiang, H. Lu, D. Wei, K.Z. Linghu, X. Zhao, X. Zhang, et al., Finite element method analysis of micro cross wedge rolling of metals, Proc. Eng. 81 (2014) 2463–2468.
[10] H.N. Lu, D.B. Wei, Z.Y. Jiang, Investigation on dimensional accuracy in micro cross wedge rolling of metals, Key Eng. Mater. 622–623 (2014) 943–948.
[11] H. Lu, A Study on the Micro Cross Wedge Rolling of Metals, University of Wollongong, 2013.
[12] M. Merklein, M. Johannes, M. Lechner, A. Kuppert, A review on tailored blanks–Production, applications and evaluation, J. Mater. Process. Technol. 214 (2014) 151–164.
[13] F. Qu, Z. Jiang, H. Lu, Effect of mesh on springback in 3D finite element analysis of flexible microrolling, J. Appl. Math. 2015 (2015) 424131-1–424131-7.
[14] F. Qu, Z. Jiang, H. Lu, Analysis of micro flexible rolling with consideration of material heterogeneity, Int. J. Mech. Sci. 105 (2016) 182–190.
[15] H.A. Kuhn, A.S. Weinstein, The minimum gauge problem in thins strip rolling, J. Lubricat. Technol. 93 (1971) 331–340.
[16] H.B. Xie, K. Manabe, T. Furushima, K. Tada, Z.Y. Jiang, An experimental and numerical investigation on micro rolling for ultra-thin strip, Int. J. Mater. Form. 9 (2016) 405–412.
[17] Z.Y. Jiang, A.K. Tieu, Contact mechanics and work roll wear in cold rolling of thin strip, Wear 263 (2007) 1447–1453.
[18] Z.Y. Jiang, D. Wei, A.K. Tieu, Analysis of cold rolling of ultra thin strip, J. Mater. Process. Technol. 209 (2009) 4584–4589.
[19] Z.Y. Jiang, S.W. Xiong, A.K. Tieu, Q. Jane Wang, Modelling of the effect of friction on cold strip rolling, J. Mater. Process. Technol. 201 (2008) 85–90.
[20] H. Xie, K.I. Manabe, T. Furushima, K. Tada, Z. Jiang, Lubrication characterisation analysis of stainless steel foil during micro rolling, Int. J. Adv. Manuf. Technol. 82 (2016) 65–73.
[21] H.B. Xie, K. Manabe, Z.Y. Jiang, A novel approach to investigate surface roughness evolution in asymmetric rolling based on three dimensional real surface, Finite Elem. Anal. Des. 74 (2013) 1–8.
[22] M. Wallmeier, E. Linvill, M. Hauptmann, J.P. Majschak, S. Östlund, Explicit FEM analysis of the deep drawing of paperboard, Mech. Mater. 89 (2015) 202–215.
[23] F. Vollertsen, Z. Hu, H. Schulze Niehoff, C. Theiler, State of the art in micro forming and investigations into micro deep drawing, J. Mater. Process. Technol. 151 (2004) 70–79.
[24] F. Gong, B. Guo, C.J. Wang, D.B. Shan, Effects of lubrication conditions on micro deep drawing, Microsyst. Technol. 16 (2010) 1741–1747.

[25] F. Gong, Z. Yang, Q. Chen, Z. Xie, D. Shu, J. Yang, Influences of lubrication conditions and blank holder force on micro deep drawing of C1100 micro conical-cylindrical cup, Precis. Eng. 42 (2015) 224–230.

[26] F. Gong, B. Guo, C. Wang, D. Shan, Micro deep drawing of micro cups by using DLC film coated blank holders and dies, Diamond Relat. Mater. 20 (2011) 196–200.

[27] Z. Hu, A. Schubnov, F. Vollertsen, Tribological behaviour of DLC-films and their application in micro deep drawing, J. Mater. Process. Technol. 212 (2012) 647–652.

[28] C.J. Wang, C.J. Wang, B. Guo, D.B. Shan, Effect of tribological behaviour of DLC film on micro-deep drawing processes, Trans. Nonferrous Met. Soc. China 24 (2014) 2877–2882.

[29] F. Vollertsen, Z. Hu, Analysis of punch velocity dependent process window in micro deep drawing, Prod. Eng. 4 (2010) 553–559.

[30] L. Luo, Z. Jiang, D. Wei, K.I. Manabe, X. Zhao, D. Wu, et al., Effects of surface roughness on micro deep drawing of circular cups with consideration of size effects, Finite Elem. Anal. Des. 111 (2016) 46–55.

[31] C.H. Chen, J.T. Gau, R.S. Lee, An experimental and analytical study on the limit drawing ratio of stainless steel 304 foils for microsheet forming, Mater. Manuf. Process. 24 (2009) 1256–1265.

[32] Y. Saotome, K. Yasuda, H. Kaga, Microdeep drawability of very thin sheet steels, J. Mater. Process. Technol. 113 (2001) 641–647.

[33] R. Erhardt, F. Schepp, D. Schmoeckel, Micro forming with local part heating by laser irradiation in transparent tools, in: Proceedings of the 7th International Conference on Sheet Metal, 27–28 September 1999, pp. 497–504.

[34] H. Tanabe, M. Yang, Design and evaluation of heat assisted microforming system, in: Proceedings of the 10th International Conference on Technology of Plasticity, 25–30 September 2011, Aachen, Germany, pp. 1020–1024.

[35] K. Nakamura, T. Nakagawa, Fracture mechanism and fracture control in deep drawing with hydraulic counter-pressure, J. Jpn. Soc. Technol. Plast. 25 (1984) 831–838.

[36] L. Lang, J. Danckert, K.B. Nielsen, Investigation into the effect of pre-bulging during hydromechanical deep drawing with uniform pressure onto the blank, Int. J. Machine Tools Manuf. 44 (2004) 649–657.

[37] H. Sato, K. Manabe, K. Ito, D. Wei, Z. Jiang, Development of servo-type micro-hydromechanical deep-drawing apparatus and micro deep-drawing experiments of circular cups, J. Mater. Process. Technol. 224 (2015) 233–239.

[38] H. Sato, K.I. Manabe, D. Kondo, D. Wei, Z. Jiang, Formability of micro sheet hydroforming of ultra-fine grained stainless steel, Proc. Eng. 81 (2014) 1463–1468.

[39] H. Sato, K.I. Manabe, D. Wei, Z. Jiang, S. Alexandrov, Tribological behavior in micro-sheet hydroforming, Tribol. Int. 97 (2016) 302–312.

[40] H. Sato, Micro Sheet Hydroforming Process Utilizing Ultra High Pressure, Metropolitan University, Tokyo, 2015.

[41] A. Albut, G. Brabie, The influence of the rolling direction of the joined steel sheets on the springback intensity in the case of Ω-shape parts made from tailor welded strips, Arch. Civil Mech. Eng. 6 (2006) 5–12.

[42] Q. Zheng, T. Aoyama, T. Shimizu, M. Yang, Experimental and numerical analysis of springback behavior under elevated temperatures in micro bending assisted by resistance bending, Proc. Eng. 81 (2014) 1481–1486.

[43] J.L. Ocaña, M. Morales, J.A. Porro, J.J. García-Ballesteros, C. Correa, Laser shock micro-forming of thin metal sheets with ns lasers, Phys. Proc. 12 (2011) 201–206.

[44] C. Pence, H. Ding, N. Shen, H. Ding, Experimental analysis of sheet metal micro-bending using a nanosecond-pulsed laser, Int. J. Adv. Manuf. Technol. 69 (2013) 319–327.

[45] W.L. Chan, M.W. Fu, J. Lu, The size effect on micro deformation behaviour in micro-scale plastic deformation, Mater. Des. 32 (2011) 198–206.

[46] H.N. Lu, D.B. Wei, Z.Y. Jiang, X.H. Liu, K. Manabe, Modelling of size effects in micro-forming process with consideration of grained heterogeneity, Comput. Mater. Sci. 77 (2013) 44–52.

[47] W.L. Chan, M.W. Fu, J. Lu, J.G. Liu, Modeling of grain size effect on micro deformation behavior in micro-forming of pure copper, Mater. Sci. Eng. A 527 (2010) 6638–6648.

[48] L.D. Cheng, C.J. Wang, C.J. Wang, B. Guo, Z.L. Wang, Size effects on plastic deformation behaviour in micro radial compression of pure copper, Trans. Nonferrous Met. Soc. China 23 (2013) 2686–2691.

[49] B. Guo, F. Gong, C.J. Wang, D.B. Shan, Flow stress and tribology size effects in scaled down cylinder compression, Trans. Nonferrous Met. Soc. China 19 (2009) s516–s520.

[50] J.H. Deng, M.W. Fu, W.L. Chan, Size effect on material surface deformation behavior in micro-forming process, Mater. Sci. Eng. A 528 (2011) 4799–4806.

[51] W. Zheng, G. Wang, G. Zhao, D. Wei, Z. Jiang, Modeling and analysis of dry friction in micro-forming of metals, Tribol. Int. 57 (2013) 202–209.

[52] J. Xu, X. Zhu, D. Shan, B. Guo, T.G. Langdon, Effect of grain size and specimen dimensions on micro-forming of high purity aluminum, Mater. Sci. Eng. A 646 (2015) 207–217.

[53] J. Xu, J. Li, X. Zhu, G. Fan, D. Shan, B. Guo, Microstructural evolution at micro/meso-scale in an ultrafine-grained pure aluminum processed by equal-channel angular pressing with subsequent annealing treatment, Mater. 8 (2015) 7447–7460.

[54] Y. Bai, M. Yang, Deformation analysis of brass in micro compression test with presence of ultrasonic vibration, Int. J. Precis. Eng. Manuf. 16 (2015) 685–691.

[55] W.L. Chan, M.W. Fu, B. Yang, Study of size effect in micro-extrusion process of pure copper, Mater. Des. 32 (2011) 3772–3782.

[56] N. Krishnan, J. Cao, K. Dohda, Study of the size effects on friction conditions in microextrusion–Part I: microextrusion experiments and analysis, J. Manuf. Sci. Eng. 129 (2007) 669–676.

[57] W.J. Kim, Y.K. Sa, Micro-extrusion of ECAP processed magnesium alloy for production of high strength magnesium micro-gears, Scr. Mater. 54 (2006) 1391–1395.

[58] A. Rosochowski, W. Presz, L. Olejnik, M. Richert, Micro-extrusion of ultra-fine grained aluminium, Int. J. Adv. Manuf. Technol. 33 (2007) 137–146.

[59] B. Eichenhueller, E. Egerer, U. Engel, Microforming at elevated temperature–forming and material behaviour, Int. J. Adv. Manuf. Technol. 33 (2007) 119–124.

[60] C. Bunget, G. Ngaile, Influence of ultrasonic vibration on micro-extrusion, Ultrasonics 51 (2011) 606–616.

[61] M. Murakawa, M. Jin, The utility of radially and ultrasonically vibrated dies in the wire drawing process, J. Mater. Process. Technol. 113 (2001) 81–86.

Chapter 2

Size Effects in Microforming

2.1 CATEGORIES OF SIZE EFFECTS

With the widespread application of metallic micro parts, their corresponding production technology-microforming has been paid close attention by many scholars. Consequently, the development of this cutting-edge technology has been greatly prompted. The popularity of microforming can be attributed to its outstanding characters, such as bulk production, high efficiency, high accuracy, short cycle, and low cost. Scaling down a certain process of conventional macroforming is an accepted strategy to investigate microforming process. Although the advance of microforming technology can benefit from the relatively mature scientific background of conventional macroforming, there is one issue that microforming cannot mechanically copy from macroforming, and that also clearly distinguish the former from the latter. This is the so-called "size effects."

In the end of the 19th century, size effects were proposed to detect and describe for upscaling purposes. However, it is understood that size effects play a significant role in miniaturization which is very frequent in microtechnology and nanotechnology nowadays. Generally, size effects can be regarded as those deviations from the expected results which are obtained when the dimension of a specimen is changed. However, Vollertsen [1] modified this statement in a more precise way: size effects are deviations from intensive or proportional extrapolated extensive values of a process which occur, when scaling the geometrical dimensions. This definition is applicable to a wide range of processes, as it not only predicts the influence that size effects could cause, but also takes the possibility that the dimensions of a wider range items, such as forces, specimen, and tools could change into consideration.

Size effects are known as unique characteristics with miniaturization, and can be categorized into three main parts in a systematic order based on the work by Vollertsen [1]. They are density size effects, shape size effects, and microstructure size effects, respectively, as illustrated in Fig. 2.1. Besides, tribological size effects are also commonly found when forming is performed in microscale.

Density size effects normally occur when the absolute value or internal value of a certain feature per volume (i.e., the density) is a constant and

Microforming Technology. DOI: http://dx.doi.org/10.1016/B978-0-12-811212-0.00002-9

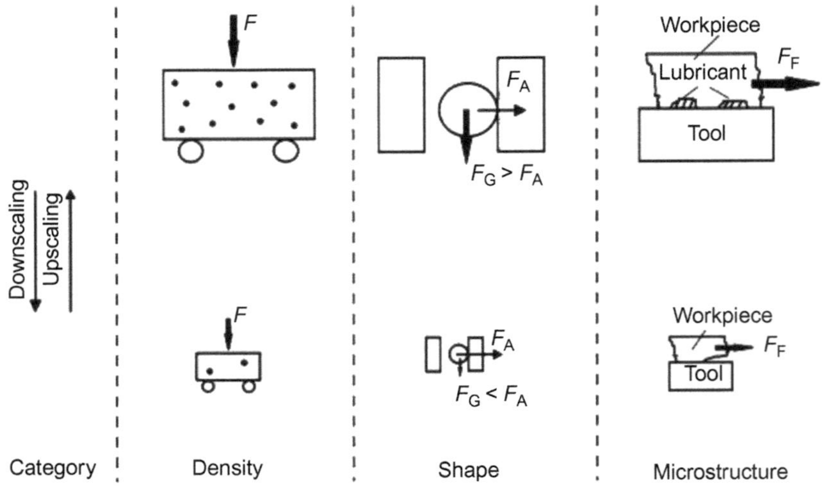

FIGURE 2.1 Categories of size effects [1].

independent on the size of an object. Besides, if the shape of an object is unchanged while its size does, the value of interest can happen to change. Density size effects are resulted due to the decrease of the density of defects with decreasing the target size. Point defects (e.g., small pores), line defects (e.g., dislocation lines), and interface areas (e.g., grain boundaries) normally exist in a material; however, these defects will be reduced or disappear when the material size is scaled down to a certain size. It causes the scale dependence of the scatter of results and material strength. Shape size effects are applicable to describe this phenomenon and relationship. Shape size effects are caused due to the change of relative dimensions among line, area, and volume. With the decrease of the target size, the ratios of the surface area and volume to the sample size decrease because the surface area of a part is proportional to the second power of its size l^2 and the volume goes with the third power l^3. For example, when the sample size becomes 1/10, the surface area and volume become 1/100 and 1/1000, respectively. Such changes of surface area and volume can be explained by *Scaling Laws*, which will be detailed in Chapter 3, Scaling Laws. Therefore, the surface effects (e.g., surface tension) are stronger than the volume effects (e.g., inertia) in microscale. Microstructure size effects contain two conditions: one is the fact that the microstructural features are not scaled down when the object size decreases, and the other is that the object size remains while the characters of microstructure (i.e., grain size) experience changes. The tribological conditions between the tool and workpiece in metal forming are of great importance for process feasibility and process quality. Tribological size effects become increasingly significant when a forming process is scaled down from the macroscale to microscale due to the high ratio of the free surface-to-volume of micro parts, and the surface topography of the specimens being held a constant.

2.2 PROBLEMS CAUSED BY SIZE EFFECTS

Size effects occur due to the fact that the ratio among all decisive features cannot be kept a constant according to the process requirements, and may cause a number of unexpected problems in microforming process in the aspects of mechanical behavior, tribology, and scatter of material behavior.

2.2.1 Size Effects on Mechanical Behavior

A number of researches have been conducted to investigate the size effects on micro deformation behavior and mechanics. When a specimen is composed of only a few grains, the grains with different sizes, shapes, and orientations are unevenly distributed in the specimen, and each grain plays a significant role in plastic deformation in microscale [2]. Ma et al. [3] studied size effects on fracture behavior in deep drawing process. Their results illustrated that limit drawing ratio (LDR) was significantly decreased with the grain size. Kals and Eckstein [4] investigated size effects in tensile tests and air bending of sheet metals by miniaturization based on similarity theory. In tensile tests, the flow stress generally decreases with miniaturizing the specimen dimensions, which is caused by the increasing share of surface grains, as shown in Fig. 2.2.

Fu and Chan [5] studied size effects on fracture behavior of annealed copper foils with different thicknesses and grain sizes by tensile tests, and

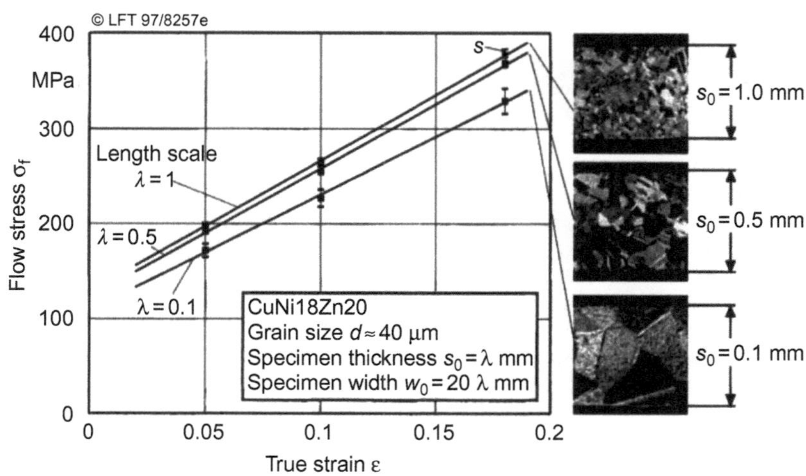

FIGURE 2.2 Flow curves of CuNi18Zn20 for different values of the length scale λ [4].

found that flow stress, fracture stress and strain, and the number of microvoids on the fracture surface decreased with decreasing the ratio (*N*) of specimen size to grain size, as shown in Fig. 2.3. Specimen thickness and grain size have significant effect on yield stress of metals in microforming. Flow stress decreases with miniaturization when the grain size of specimen is held constantly, which means the decrease of flow stress is definitely caused by miniaturization instead of different grain microstructures [6]. This tendency could be explained by the so-called surface layer model, as shown in Fig. 2.4 [7]. The moving dislocations pile up at grain boundary, but cannot pile up in the surface grains during deformation. Therefore, the grains

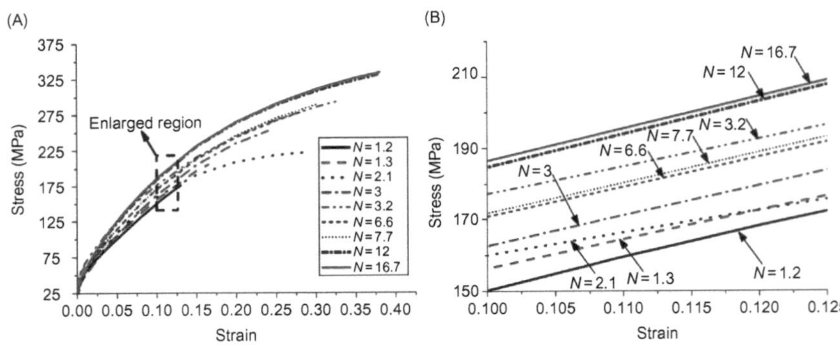

FIGURE 2.3 Tensile test results: (A) stress–strain curves and (B) enlarged region [5].

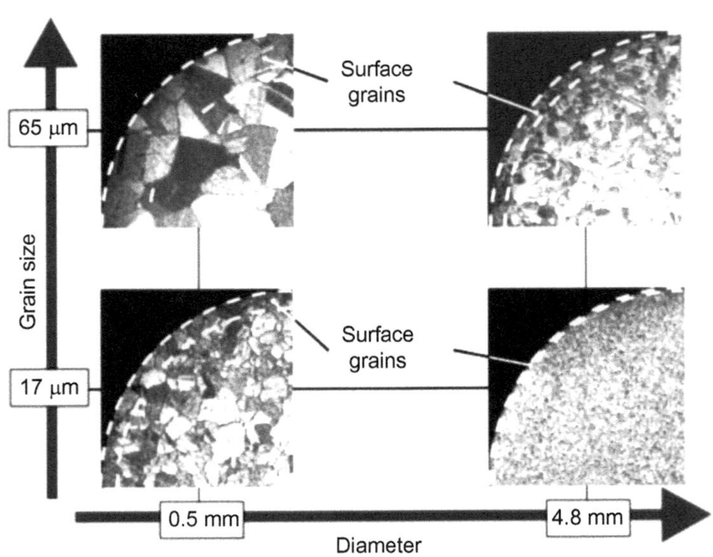

FIGURE 2.4 Change of share of surface grains with miniaturization [7].

located on the surface show less hardening and cause a low flow stress in metal foil. On the basis of tensile tests with pure copper foils with different thicknesses, Furushima et al. [8] proposed ductile fracture mechanism of metal foils as: (1) the surface roughness of metal foils increases with plastic deformation owing to free surface roughening; (2) the ratio of surface roughness to thickness increases with plastic deformation, and at the concave region formed by free surface roughening, local deformation occurs before the theoretical uniform elongation predicted by plastic instability theory; and (3) the local deformation caused by free surface roughening leads to fracture. The work of Raulea et al. [9] indicates that yield strength will decrease and increase with the decrease of the specimen thickness and grain size, respectively, which are thought to be attributed to the influence of free specimen surface, and grain-size strengthening or Hall–Petch effect, respectively. For grain size larger than the specimen thickness, yield strength increases with grain size, however, a strong increase of the variation will occur.

In the micro tensile tests of brass foils with different thicknesses and two cases of grain sizes, Li et al. [10] found that the samples with fine grains exhibited a higher flow stress than those with coarse grains, as shown in Fig. 2.5. In the figure, the numbers represent foil thickness while "Fine" and "Coarse" refer to the fine- and coarse-grained samples, respectively. They also found that the elongation of foils decreased with decreasing the foil thickness, and fracture occurred along the grain boundary in the

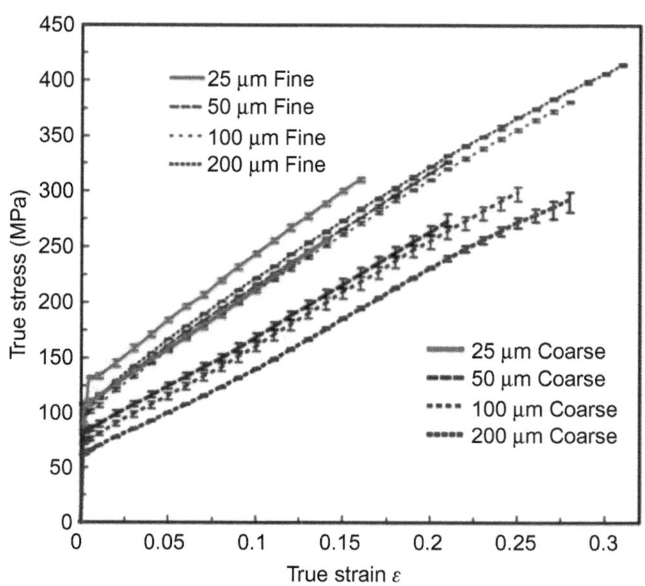

FIGURE 2.5 Tensile stress–strain curves of CuZn37 brass foils [10].

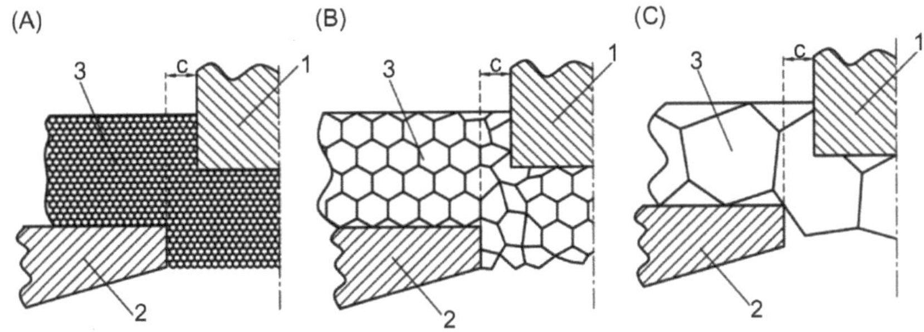

FIGURE 2.6 Size effects model of grain size in micro blanking: (A) $C/D>1$, (B) $C/D\approx 1$, and (C) $C/D<1$. 1–micro punch, 2–micro die, and 3–metal foil [11].

thickness direction when the samples were subjected to tensile loads in the plane of the sheet. For thinner foils, these fractures were penetrated across the entire foil thickness, but for thicker foils the microcracks were stopped by inner grains, resulting in longer elongation. These results indicate that higher strain gradient and more inhomogeneous deformation have occurred in thinner foils than those in thicker foils. The deformation behavior in micro blanking is related not only to the blanking clearance, but also to the grain size of metal foil. The ratio of blanking clearance (C) to grain size (D) is one of the main factors affecting micro deformation behavior in micro blanking, and the ultimate shearing strength will reach an extreme value when C/D is equal to one [11]. A model for explaining size effects in micro blanking is illustrated in Fig. 2.6 [11].

The ratio of thickness to grain size is an important parameter to describe the material flow in microscale. This ratio decreases with scaling down because the grain size does not change with decreasing the thickness. Many researches have focused this ratio on investigating the grain size effects on material properties. In micro bending, the decrease of sheet foil thickness (T) or the increase of grain size (D) will result in a big springback, and the scatter of springback angle is mainly attributed to the elastic anisotropy of surface grains and increases with the reduction of grains along the thickness direction [12,13]. Yield stress, tensile strength, and ductility decrease when T/D decreases [14].

2.2.2 Size Effects on Tribology

Tribology in metal forming processes has a significant effect on the quality of the produced part, the tool lifetime, and the process stability. The tribological behavior depends on several parameters, including the contact state between the material and tools, the contact pressure, the lubricant condition, and the surface topography. Due to high contact pressures at the tool/workpiece interface, mixed friction is predominant. Therefore, the appearance of real

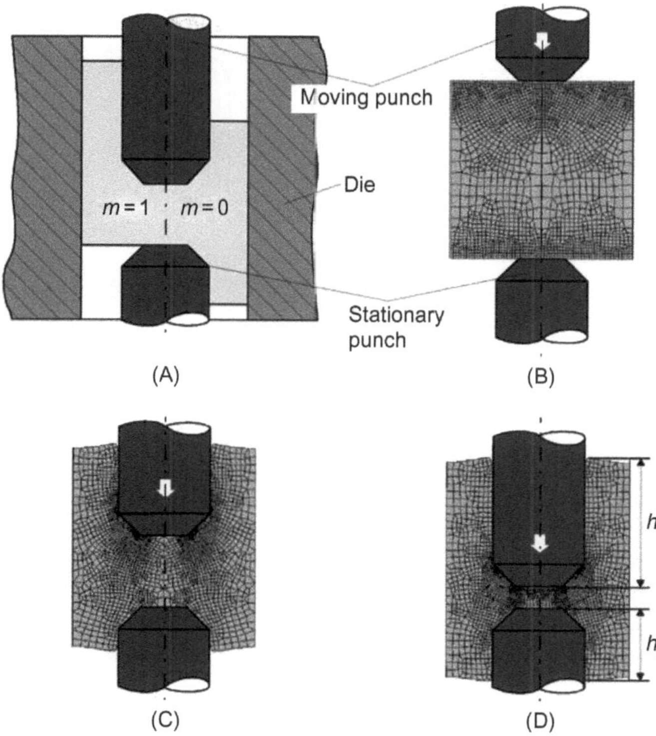

FIGURE 2.7 DCE test: (A) dependence of cup height on friction and (B, C, and D) FEM calculation on the development of the cups [7].

contact area has a major influence on the frictional conditions. Also, the normal stress and the mating materials, the topography of the workpiece and the existence of lubricant are important factors for the appearance of real contact area, thus the friction conditions.

Size effects on friction were first investigated by scaled ring-test experiments, which revealed an increasing trend of friction with miniaturization [15]. Such a finding has been confirmed by a representative study using scaled double-cup extrusion (DCE) test, as shown in Fig. 2.7 [7]. In DCE, a cylindrical specimen is placed in a die between a stationary and a moving punch of identical shape. The upper punch moves down and causes the material to form two cups with the upper cup height of h_u and lower cup height of h_l. In the theoretical case of no friction, both cups have the same height, whereas the higher friction prevents the deformation of the lower cup. The higher the friction, the more the formation of the lower cup is prevented. Therefore, the ratio of h_u/h_l is quite a sensitive measure for friction. The research of Engel and Eckstein [7] has indicated that friction increases distinctly with miniaturization, which can be reflected by an

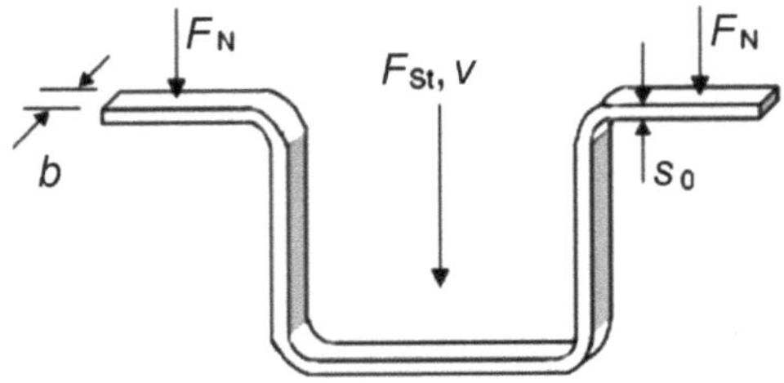

FIGURE 2.8 Principal of friction test by strip drawing [16].

increase in friction factor from 0.02 in the case of 4 mm specimen up to 0.15 and higher for the smaller specimens.

Micro sheet metal forming processes, e.g., micro deep drawing, are sensitive to the coefficient of friction (COF). The tribological size effects within sheet metal forming can be identified by strip drawing test with deflection, which is scalable and can be used for a very small process dimension [16]. The principal of friction test by strip drawing is shown in Fig. 2.8. By using such a testing method, the tangential forces, that exist while the blank is drawn into the die, can be avoided and the stress condition is simplified. This difference makes it easier to find the relationship between the punch force and the COF. In the test, the strip, under the blank holder force F_N, with a width b and a thickness s_0 is drawn by a punch with a velocity v into the die, and the punch force F_{st} can be measured. The punch force F_{st} consists of bending force at the radius of the die and punch F_B and friction force at the flange and the radius of the die F_F. For a certain workpiece bent for a certain radius the bending force can be theoretically calculated, so that the share of the friction force of the whole punch force from the strip drawing with deflection can be determined [17,18]. When changing the process dimensions, the share of friction force from the whole punch force differs: for a punch diameter of 1 mm, which is the highest; and for a punch diameter of 100 mm, which is the lowest [16].

The relation of forces during deep drawing is shown in Fig. 2.9 [19]. The following equations can be obtained to identify COF based on the punch force versus stroke curves achieved in experiments [18,19]:

$$F_{st} = 2(F_Q\cos\alpha + F_L\sin\alpha) \tag{2.1}$$

$$F_Q = \frac{M_{B(S)}}{L} \tag{2.2}$$

$$F_L = \left(2F_N\mu_1 + \frac{M_{B(A)}}{r_M + (s/2)}\right)e^{\mu_2\alpha} + \frac{M_{B(B)}}{r_M + (s/2)} \tag{2.3}$$

where F_N is the blank holder force, μ_1 is the COF on the flange, μ_2 is the COF on the die radius, $M_{B(A)}$ is the bending moment of the strip at point A,

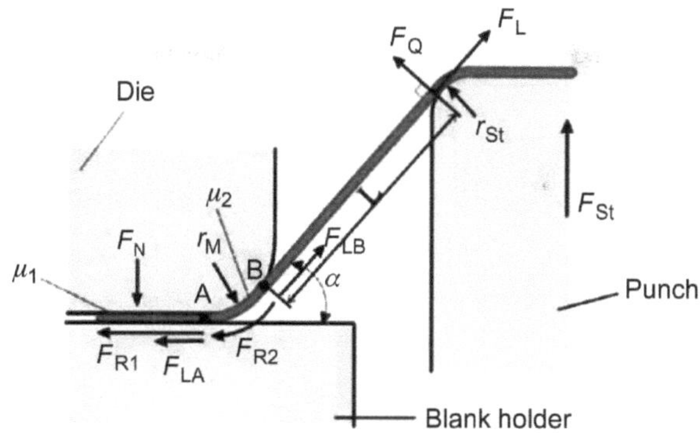

FIGURE 2.9 Analysis of the relationship of forces in strip drawing [19].

and $M_{B(B)}$ is the bending moment of the strip at point B. In this way, the COF–travel curves could be calculated from the punch force. Vollertsen et al. [19] found that the COFs on the flange (μ_1) and the die radius (μ_2) decreased and increased, respectively, with the progressing of stroke. This method provides a way for research in deep drawing tribology, especially in the field of size effects research. A study on scaled deep drawing using the same workpiece material and lubricant but with five different punch diameters shows that punch diameter has significant effect on COF. The average COF is 0.129 for punch diameter of 1 mm and 0.075 for punch diameter of 5 mm, indicating that friction increases with the process dimension decreases [20].

The size effects on lubricant friction behavior can be explained by the model of open and closed lubricant pockets (CLPs), also known as dynamic and static lubricant pockets, respectively. When a forming load is applied to a lubricated workpiece surface, the asperities (roughness peaks) start to deform plastically, thus increasing the pressure of the lubricant, which is trapped in the roughness valleys or squeezed out. In the case of roughness valleys that have a connection to the edge of the surface and cannot keep the lubricant, these are called open lubricant pockets (OLPs). With increasing normal pressure, the lubricant escapes and is not able to support or transmit the forming load. The forming load acts only on the asperities which results in a higher normal pressure, a higher degree of surface flattening and thus, a higher friction. CLPs, on the contrary, do not have a connection to the edge of the surface. The lubricant becomes trapped in those pockets and is pressurized during forming. The developing hydrostatic pressure will take a part of the external load, thus reducing the normal pressure on the asperities, which results in lower friction [7,21]. When scaling down a forming process, the contact area is reduced significantly while the size of single topography features remains approximately constant. Therefore, the ratio of CLPs is reduced drastically and the external load will be transmitted more and more

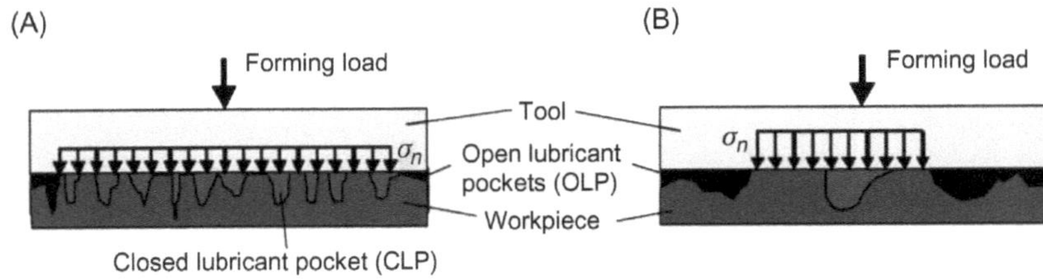

FIGURE 2.10 (A) Contact state at conventional length scale and (B) contact state in microforming processes [22].

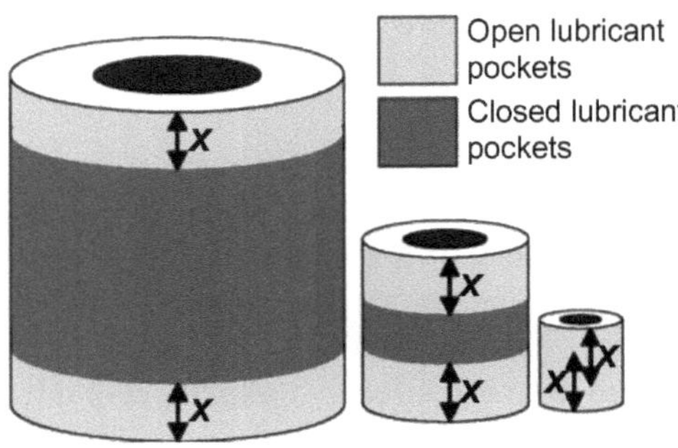

FIGURE 2.11 Effect of miniaturization on areas with open and closed lubricant pockets [7].

by the real contact area which will consequently increase, leading to an increase in friction, as shown in Fig. 2.10 [22].

When applying the lubricant pocket model to DCE test, there must be a scaling effect on the ratio of OLPs to CLPs, as shown in Fig. 2.11 [7]. It is obvious that there is a region of constant width (denoted by x), where the OLPs become effective. By reducing the specimen size, the share of OLPs increases as well as the friction factor. The ratio of OLPs to CLPs does not only increase with decreasing the specimen size, but also with decreasing the initial specimen height when the diameter is held constant. Peng et al. [23] proposed a friction model based on the OLPs and CLPs considering a function of real contact area and normal press p/σ_0 by introducing size/scale factor λ. They considered the flattening behavior under dry friction and lubricant to describe the tribological behavior in OLPs and CLPs, as shown in Fig. 2.12. Based on this model, the hydrostatic pressure reduced by enclosed lubricant in CLPs can be calculated. Their work indicates that with the increment of λ, the friction decreases until it becomes a traditional lubricant friction. On the other hand, it will increase and degenerate to dry friction with the reduction of size (size/scale effect $\lambda = 1$). Moreover, the influence of λ is greater in microscale/mesoscale ($\lambda < 10$) than that in conventional scale ($\lambda > 10$). It means that the friction of micro/meso size

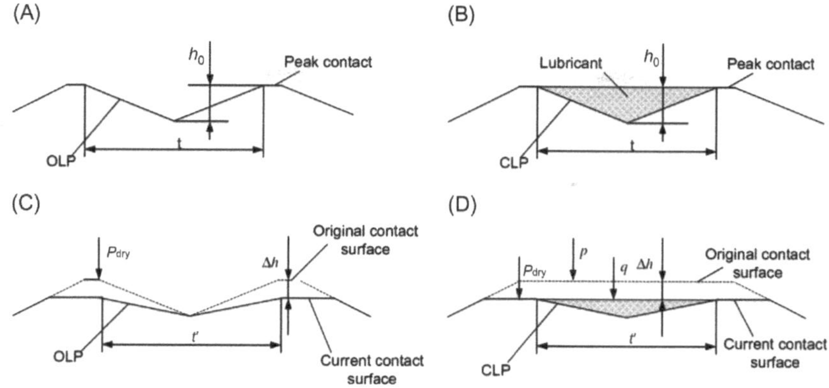

FIGURE 2.12 The deformation of materials in microscopic view: (A) initial contact surface without lubricant, (B) initial contact surface with lubricant applied, (C) current contact surface without lubricant, and (D) current contact surface with lubricant applied [23].

is between those of dry friction and conventional lubricant friction. It is the reason why size effects should be considered in microscale/mesoscale forming process, while it could be ignored in macroscale process.

2.2.3 Size Effects on Scatter of Material Behavior

In microforming, the material is not a homogeneous continuum because of the presence of only a few grains in the deformation zone. This induces a change in the material behavior resulting among others in a large scatter of forming results and geometrical accuracy of formed parts. When a microforming process is repeated under the same experimental conditions (e.g., grain size and temperature), a discrete distribution instead of a superposition of flow stress will occur due to grain heterogeneity in microscale. Fig. 2.13 shows the yield stress of aluminum obtained from bending tests [24]. It is clear that the scatter of yield strength significantly increases with decreasing the *T*/*D* ratio, especially for *T*/*D* less than one. In general, the flow stress decreases and the scatter effect increases with the reduction of specimen size in microscale deformation if the grain size keeps unchanged. When the specimen size is kept a constant, only a few grains exist in the specimen and the number of microstructural features will be decreased when the grain size is increased. As a result, the homogeneous distribution of different grains no longer exists. In this case, the characteristic of an individual grain will dominate the material behavior. For fine grains, the total grain structure can be regarded as almost homogeneous, which will yield reproducible results. However, for coarse grains, the grain structure might be more different from specimen to specimen even though mean grain size is the same, causing the increase of scatter in material

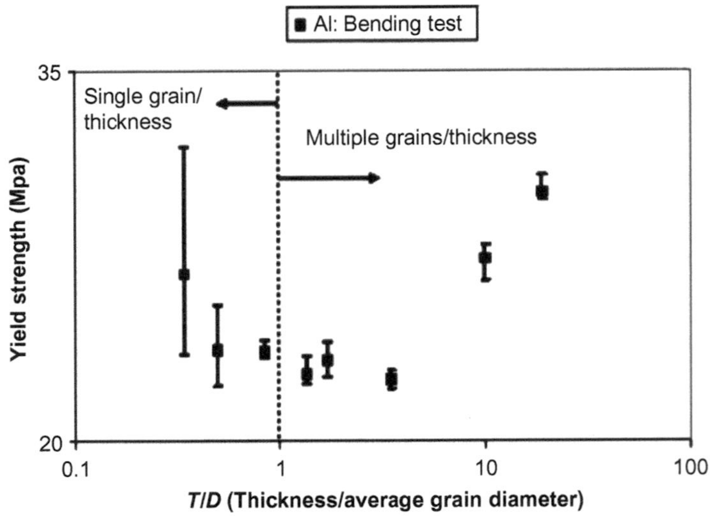

FIGURE 2.13 Yield strength of aluminum obtained from bending tests [24].

behavior in microforming [25]. Different crystallographic orientations of neighboring grains which induce inhomogeneous deformation, and the grains with different sizes, shapes, and orientations in different specimens are the main cause of scatter of the measured material properties.

In the modeling of scatter effect in microforming, different flow stress curves can be used to reflect the deformation behavior of testing material with different grain characteristics [26–28]. The scatter degree of flow stress curves could be described by a probability density function as [29]:

$$f(\sigma,\varepsilon)=\frac{1}{S(\varepsilon)\sqrt{2\pi}}\mathrm{e}^{-(1/2)[(\sigma(\varepsilon)-\sigma_m(\varepsilon))/S(\varepsilon)]^2} \tag{2.4}$$

where σ is the flow stress, ε is the strain, $\sigma(\varepsilon)$ is the flow stress at strain ε, $S(\varepsilon)$ is the standard deviation, and $\sigma_m(\varepsilon)$ is the mean flow stress. The values of $\sigma(\varepsilon)$, $S(\varepsilon)$, and $\sigma_m(\varepsilon)$ can be obtained, respectively, by:

$$\sigma(\varepsilon)=\sum_{i=1}^{n}V_i\cdot\sigma_i(\varepsilon) \tag{2.5}$$

$$S(\varepsilon)=\sqrt{\frac{\sum_{j=1}^{t}(\sigma_j(\varepsilon)-\sigma_m(\varepsilon))^2}{t-1}} \tag{2.6}$$

$$\sigma_m(\varepsilon)=\frac{\sum_{j=1}^{t}\sigma_j(\varepsilon)}{t} \tag{2.7}$$

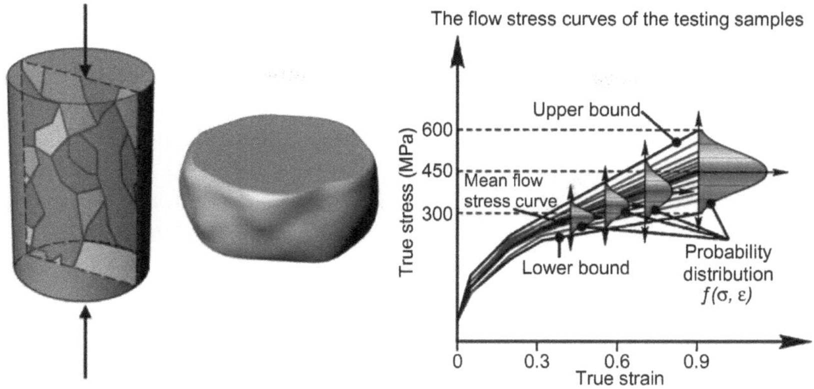

FIGURE 2.14 Schematic illustration of the modeling of the scatter effect with a normal distribution function [29].

where n is the total number of grains in the billet, V_i is the volume fraction or area fraction in 2D case of the ith grain, $\sigma_i(\varepsilon)$ is the flows stress of the ith grain and it is a function of strain considering work-hardening, $\sigma_j(\varepsilon)$ is the measured flow stress of the jth testing sample with a specific strain, and t is the total number of the testing samples. Fig. 2.14 shows a schematic illustration of the modeling of the scatter effect with a normal distribution function. To generate the upper bound of the measured flow stress curves for the testing samples, the flow stress curves of some grains must be higher than the upper bound of the measured flow stress curves of the testing samples. Similarly, the flow stress curves of some grains must be lower than the lower bound of the measured flow stress curves.

2.3 STRATEGIES FOR CONTROL OF SIZE EFFECTS

One of the approaches to compensate or reduce the influence of size effects is to elevate the temperature during the forming process. This minimizes the anisotropic material behavior resulting in a more homogeneous forming with improved reproducibility. An alternative approach is to refine the microstructure of the deformed part, which has been investigated by lots of researchers.

2.3.1 Microforming at Elevated Temperatures

2.3.1.1 Warm Forming

Mechanical properties of metals change with an increase in temperature. Therefore, the metal will react differently to the same manufacturing operation when it is performed under different temperatures, and the manufactured part may appear different properties. In industrial metal manufacturing, there are three basic temperature ranges that the metal is formed at, including cold forming, warm forming, and hot forming. Cold forming ($T < 0.35\ T_m$) is a metal forming process that is carried out at room temperature or a little above it. Hot forming ($T > 0.55T_m$) is a metal forming process that is carried out at a temperature range that is higher than the recrystallization temperature of the material being formed. Warm forming ($0.35\ T_m < T < 0.55\ T_m$) is preferred over cold forming because it will reduce the force required to perform the operation [30]. Also the amount of annealing of the material that may have been necessary for the cold-formed part may be less for warm forming. Cold forming takes advantage of strain hardening to enhance the strength of the material at the penalty of higher forming forces, while hot forming lowers the yield strength and allows simultaneous recrystallization, which controls the grain refinement. To avoid high forming temperatures and forces, warm forming is usually used as the intermediate process, which allows recovery but not recrystallization [31]. Warm forming aims to capitalize on the advantages of both hot and cold forming. As compared to hot forming process, warm forming requires higher forces due to greater material flow stress, thus making increased demand on the forming tools and equipment. From another perspective, greater part accuracy in terms of dimensions, shape, and surface finish can be achieved.

2.3.1.2 Characteristics of Microforming at Elevated Temperatures

A single grain characterized by its size, orientation, and position will significantly influence the microforming process resulting in inhomogeneous character of the material. Heat-assisted microforming is an effective process to counteract inhomogeneous deformation and improve the accuracy of formed parts. During forming at elevated temperatures, additional slip systems are activated with the effect to reduce the dependency on single grains. This improves not only the material flow but also the reproducibility of process variables like flow stress. In the micro bending of pure titanium foils, springback angle increases with decreasing the foil thickness at room temperature. Differently, springback angle decreases with decreasing the foil thickness after bending at elevated temperatures, as shown in Fig. 2.15 [32]. Due to lower flow stress of surface, the influence of surface area is thought to be the dominating factor that results in the less springback of thinner foils at elevated temperatures. A decrease in springback angle with the raising of temperature indicates a better accuracy of the parts formed at elevated temperatures. Similar results can be obtained in upsetting and backward can extrusion processes that raising of temperature influences the forming result, tending to homogenize the material flow. Further, temperature also leads to a reduced scatter of process parameters such as the force-stroke characteristic [33]. A moderate increase of forming temperature can lead to a significant reduction of the scatter problem occurred in metal microforming process, and thus can significantly improve the process stability and reliability [34].

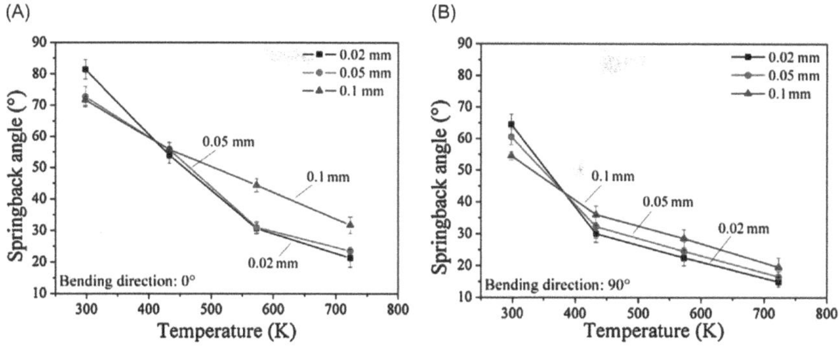

FIGURE 2.15 Springback angles measured from experiments for different thickness foils at different temperatures and bending directions of (A) 0 degree and (B) 90 degrees [32].

In the conventional metal forming under assistance of heating, the mechanical behavior of material is sensitive to temperature due to the occurrence of microstructural phenomena including grain growth, surface oxidation, and friction condition. However, when heating is applied to the microforming process, more other problems may occur. For instance, the work conducted by Wulfsberg and Terzi [35] showed that lower forming temperatures in a smaller workpiece could be expected as the surface-to-volume ratio increased, which could enhance the heat transfer to the environment with decreasing the scale factor λ, as shown in Fig. 2.16. With consideration of micro dimensional size of workpieces, the heating temperature should be controlled within a certain range in case of excessive grain growth and oxidation of metal workpieces. Warm forming is a sensible option to improve the deformability and to avoid oxidation. In practice, the upper limit of warm forming temperature is usually determined by the amount of oxidation which can be tolerated, and its lowest limit is determined by a force which can be measured using the forming machines and the material formability.

2.3.1.3 Heating Methods in Microforming

Laser heating and resistance heating are two heating methods that are commonly used in heat-assisted microforming processes. Laser radiation provides the most suitable choice for the purpose of heating the material during microforming process as it offers several advantages in comparison to other methods, including [35]: (1) the laser energy input and thus the resulting temperature in the workpiece can be easily controlled via the current of a diode laser; (2) local heating of selected areas of the workpiece is possible, allowing to limit the heating to the forming zone; (3) needed temperature gradients can be achieved by control of laser power; (4) the absorption of laser radiation allows short processing time which cannot be accomplished with heat transfer from pre-heated tools; and (5) material

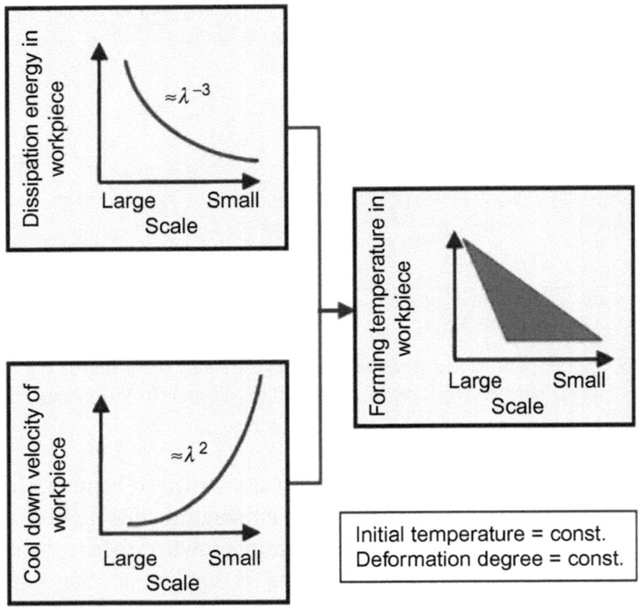

FIGURE 2.16 Influence of workpiece size on forming temperature and its distribution [35].

properties can be manipulated by controlling the down cooling time via laser power. The principal of laser-assisted microforming is schematically shown in Fig. 2.17.

One of the most two common lasers used in metal processing is the Nd: YAG laser (the other one is the CO_2) operating in the near infrared just outside the visible wavelength region [36]. The spot size of the laser can be controlled by the distance from the lens to specimen surface. The nature of laser heating is to change the internal microstructure of metal by energy absorption. The metal absorptivity of laser light is a critical factor that influences the deformation process [37]. Metal specimen is heated up to the expected temperature prior to microforming process, but the heat loss is quite significant due to the relatively large ratio of specimen surface to volume. During deformation, the laser power should be adjusted to an appropriate value to compensate the energy loss due to heat convection and radiation from specimen to atmosphere which is especially significant in microforming. Fig. 2.18 demonstrates a strategy of laser heating during microforming process. Micro specimen is first heated up to t_2, and then the deformation begins. In this way, the temperature could be kept nearly a constant because of heat compensation of laser throughout the whole microforming process (from t_2 to t_3). The actual temperature of heated specimen can be measured using an infrared thermometer.

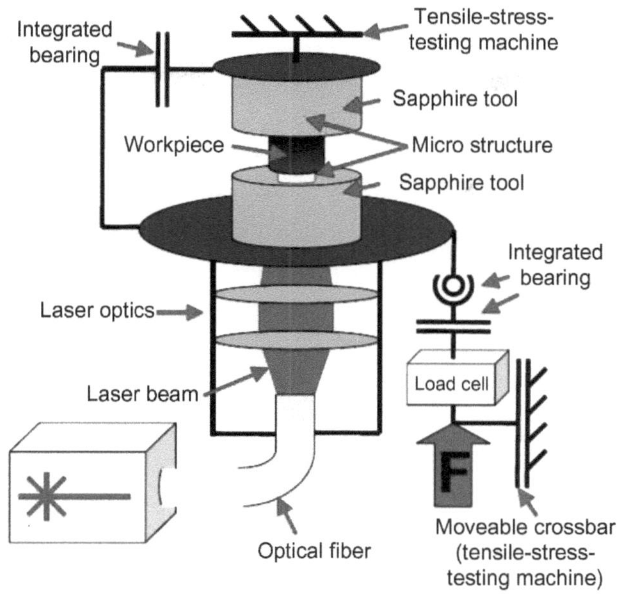

FIGURE 2.17 Principal of laser-assisted microforming [35].

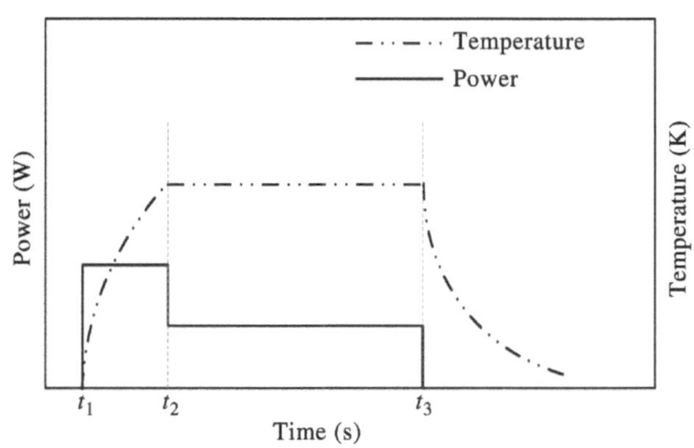

FIGURE 2.18 Laser heating strategy and specimen temperature.

Fig. 2.19 shows an example of the principal of resistance heating used in micro bending process [38]. Dies used as electrodes are connected to the power supply directly. By the contact between the blank and the die, current flows through the blank. The temperature of blank is increased by Joule heating. In order to reduce the variation in temperature of blank when punch contacts, punch can be heated up by one pair of plate heaters throughout the tests. For maintaining a constant temperature of blank, a thermo sensor can be used to measure the temperature at the center of blank.

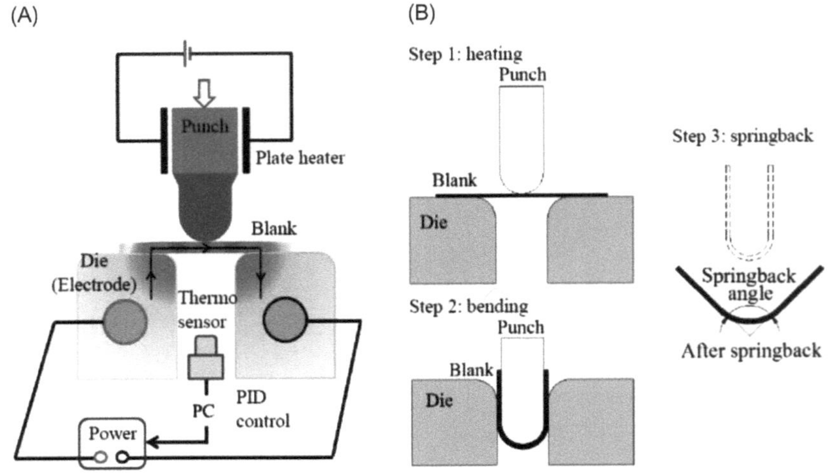

FIGURE 2.19 Illustration of micro bending process assisted by resistance heating: (A) micro bending system and (B) micro bending process [38].

2.3.2 Microstructural Refinement

An alternative approach for reducing the influence of size effects is to refine the microstructure of deformed part. Microstructure has been identified to be one of the main reasons for process scatter, as well as the uneven shape evolution in microforming processes [39,40]. A typical work on micro extrusion indicates that grain size effects on deformation load are sensitive to the friction force at tooling–workpiece interface in micro extrusion process [41]. Inhomogeneous deformation occurs in the case of material with coarse grains in micro extrusion, as shown in Fig. 2.20. This phenomenon is thought to be caused by the fact that anisotropy grain properties become significant when only a few grains flow into the microsized cavity, resulting in an irrational local deformation. A large number of slips pass through the grain boundary to accomplish the strain continuity in the case of coarse grains, and will cause the blurred grain boundary. The benefits of microstructural refinement in the reduction of size effects have been confirmed by a large number of scholars in different microforming processes [42–45].

Severe plastic deformation is a processing method by which the grain size can be reduced sufficiently to retain a high thickness to grain size ratio, and is widely used in microstructural refinement. Equal channel angular pressing (ECAP) is a severe plastic deformation processing technique that is useful in obtaining an extremely fine grain structure—often at submicron or nanometer scale—and achieving a significant improvement in strength of the material. ECAP can produce a variety of ultrafine-grained microstructures in materials depending on route, temperature, and number of passes during processing, and is commonly used in combination with microforming to reduce size effects and improve the quality of micro products [46–49].

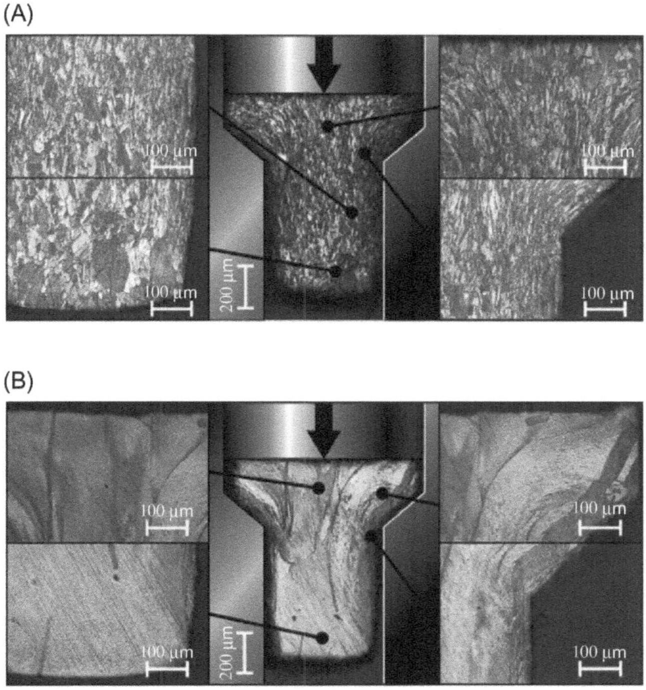

FIGURE 2.20 Microstructures of the micro-formed parts in micro forward extrusion [41]. (A) Original grain size of 20 μm and (B) original grain size of 150 μm.

In summary, size effects characterize the micro world, and have significant impact on material behavior in microforming and the quality of formed parts. In order to reduce size effects and take full advantage of microforming technology, knowledge on machine and tools design, materials science, properties evaluation, and optimization of processing parameters is needed. If appropriate strategies are adopted, microforming can be a promising method for manufacturing high-quality micro products with high production rates, excellent material utilization, and low costs.

REFERENCES

[1] F. Vollertsen, Categories of size effects, Prod. Eng. 2 (2008) 377–383.
[2] M.W. Fu, W.L. Chan, Micro-scaled Products Development via Microforming, Springer, London, 2014.
[3] X. Ma, R. Lapovok, C. Gu, A. Molotnikov, Y. Estrin, E. Pereloma, et al., Deep drawing behaviour of ultrafine grained copper: modelling and experiment, J. Mater. Sci. 44 (2009) 3807–3812.
[4] T. Kals, R. Eckstein, Miniaturization in sheet metal working, J. Mater. Process. Technol. 103 (2000) 95–101.
[5] M.W. Fu, W.L. Chan, Geometry and grain size effects on the fracture behavior of sheet metal in micro-scale plastic deformation, Mater. Des. 32 (2011) 4738–4746.

[6] M. Geiger, M. Kleiner, R. Eckstein, N. Tiesler, U. Engel, Microforming, CIRP Ann. Manufact. Technol. 50 (2001) 445–462.

[7] U. Engel, R. Eckstein, Microforming–form basic research to its realization, J. Mater. Process. Technol. 125–126 (2002) 35–44.

[8] T. Furushima, H. Tsunezaki, K.-I. Manabe, S. Alexsandrov, Ductile fracture and free surface roughening behaviors of pure copper foils for micro/meso-scale forming, Int. J. Machine Tools Manufact. 76 (2014) 34–48.

[9] L.V. Raulea, A.M. Goijaerts, L.E. Govaert, F.P.T. Baaijens, Size effects in the processing of thin metal sheets, J. Mater. Process. Technol. 115 (2001) 44–48.

[10] H.Z. Li, X.H. Dong, Y. Shen, R. Zhou, A. Diehl, H. Hagenah, et al., Analysis of microbending of CuZn37 brass foils based on strain gradient hardening models, J. Mater. Process. Technol. 212 (2012) 653–661.

[11] J. Xu, B. Guo, C. Wang, D. Shan, Blanking clearance and grain size effects on micro deformation behavior and fracture in micro-blanking of brass foil, Int. J. Machine Tools Manufact. 60 (2012) 27–34.

[12] J.G. Liu, M.W. Fu, J. Lu, W.L. Chan, Influence of size effect on the springback of sheet metal foils in micro-bending, Comput. Mater. Sci. 50 (2011) 2604–2614.

[13] J.-T. Gau, C. Principe, M. Yu, Springback behavior of brass in micro sheet forming, J. Mater. Process. Technol. 191 (2007) 7–10.

[14] R. Ben Hmida, S. Thibaud, A. Gilbin, F. Richard, Influence of the initial grain size in single point incremental forming process for thin sheets metal and microparts: experimental investigations, Mater. Des. 45 (2013) 155–165.

[15] U. Engel, A. Messner, N. Tiesler, Cold forging of microparts-effect of miniaturization on friction, in: J.L. Chenot, et al. (Eds.), Proceedings of the 1st ESAFORM Conference on Materials Forming, Sophia Antipolis, France, 1998, pp. 77–80.

[16] F. Vollertsen, Z. Hu, Tribological size effects in sheet metal forming measured by a strip drawing test, CIRP Ann. Manufact. Technol. 55 (2006) 291–294.

[17] Z. Hu, F. Vollertsen, Modelling of friction with respect to size effects, Int. J. Mater. Form. 1 (2008) 1231–1234.

[18] Z. Hu, F. Vollertsen, A new friction test method, J. Technol. Plast. 29 (2004) 1–10.

[19] F. Vollertsen, Z. Hu, H. Schulze Niehoff, C. Theiler, State of the art in micro forming and investigations into micro deep drawing, J. Mater. Process. Technol. 151 (2004) 70–79.

[20] F. Vollertsen, Z. Hu, On the drawing limit in micro deep drawing, J. Technol. Plast. 32 (2007) 1–11.

[21] U. Engel, Tribology in microforming, Wear 260 (2006) 265–273.

[22] S. Weidel, U. Engel, Characterisation of the flattening behaviour of modelled asperities, Wear 266 (2009) 596–599.

[23] L. Peng, X. Lai, H.J. Lee, J.H. Song, J. Ni, Friction behavior modeling and analysis in micro/meso scale metal forming process, Mater. Des. 31 (2010) 1953–1961.

[24] J.T. Gau, C. Principe, J. Wang, An experimental study on size effects on flow stress and formability of aluminum and brass for microforming, J. Mater. Process. Technol. 184 (2007) 42–46.

[25] M. Geiger, S. Geißdörfer, U. Engel, Mesoscopic model: advanced simulation of microforming processes, Prod. Eng. 1 (2007) 79–84.

[26] H.N. Lu, D.B. Wei, Z.Y. Jiang, X.H. Liu, K. Manabe, Modelling of size effects in microforming process with consideration of grained heterogeneity, Comput. Mater. Sci. 77 (2013) 44–52.

[27] Z. Fang, H. Lu, D. Wei, Z. Jiang, X. Zhao, X. Zhang, et al., Numerical study on springback with size effect in micro V-bending, Proc. Eng. 81 (2014) 1011–1016.

[28] Z.Y. Jiang, J. Zhao, H. Lu, D. Wei, K.I. Manabe, X. Zhao, et al., Influences of temperature and grain size on the material deformability in microforming process, Int. J. Mater. Form. (2016). Available from: http://dx.doi.org/10.1007/s12289-016-1317-4.

[29] W. Chan, M. Fu, J. Lu, J. Liu, Modeling of grain size effect on micro deformation behavior in micro-forming of pure copper, Mater. Sci. Eng. A 527 (2010) 6638–6648.

[30] F. Stachowicz, T. Trzepieciński, Warming forming of stainless steel sheet, Arch. Civil Mech. Eng. 10 (2010) 85–94.

[31] S. Fréchard, A. Redjaïmia, E. Lach, A. Lichtenberger, Mechanical behaviour of nitrogen-alloyed austenitic stainless steel hardened by warm rolling, Mater. Sci. Eng. A 445 (2006) 219–224.

[32] Q. Zhen, T. Shimizu, M. Yang, Scale effect on springback behavior of pure titanium foils in microbending at elevated temperature, J. Mater. Process. Technol. 230 (2016) 233–243.

[33] E. Egerer, U. Engel, Process characterization and material flow in microforming at elevated temperatures, J. Manufact. Process. 6 (2004) 1–6.

[34] B. Eichenhueller, E. Egerer, U. Engel, Microforming at elevated temperature – forming and material behaviour, Int. J. Adv. Manufact. Technol. 33 (2007) 119–124.

[35] J.P. Wulfsberg, M. Terzi, Investigation of laser heating in microforming applying sapphire tools, CIRP Ann. Manufact. Technol. 56 (2007) 321–326.

[36] D. Bergstrom, J. Powell, A.F. Kaplan, The absorptance of non-ferrous alloys to Nd:YAG and Nd:YLF laser light at room temperature, Appl. Opt. 48 (2007) 1290–1301.

[37] D. Bergstrom, The Absorption of Laser Light by Rough Metal Surfaces, Mid Sweden University, 2008.

[38] Q. Zheng, T. Aoyama, T. Shimizu, M. Yang, Experimental and numerical analysis of springback behavior under elevated temperatures in micro bending assisted by resistance heating, Proc. Eng. 81 (2014) 1481–1486.

[39] A. Rosochowski, W. Presz, L. Olejnik, M. Richert, Micro-extrusion of ultra-fine grained aluminium, Int. J. Adv. Manufact. Technol. 33 (2007) 137–146.

[40] S.A. Parasiz, B. Kinsey, N. Krishnan, J. Cao, M. Li, Investigation of deformation size effects during microextrusion, J. Manufact. Sci. Eng. 129 (2007) 690–697.

[41] W.L. Chan, M.W. Fu, B. Yang, Study of size effect in micro-extrusion process of pure copper, Mater. Des. 32 (2011) 3772–3782.

[42] J. Xu, X. Zhu, D. Shan, B. Guo, T.G. Langdon, Effect of grain size and specimen dimensions on micro-forming of high purity aluminium, Mater. Sci. Eng. A 646 (2015) 207–217.

[43] H. Justinger, G. Hirt, Estimation of grain size and grain orientation influence in microforming processes by Taylor factor considerations, J. Mater. Process. Technol. 209 (2009) 2111–2121.

[44] W.J. Kim, S.J. Yoo, H.K. Kim, Superplastic microforming of Mg-9Al-1Zn alloy with ultrafine-grained microstructure, Scr. Mater. 59 (2008) 599–602.

[45] W.L. Chan, M.W. Fu, Experimental studies and numerical modeling of the specimen and grain size effects on the flow stress of sheet metal in microforming, Mater. Sci. Eng. A 528 (2011) 7674–7683.

[46] A. Molotnikov, R. Lapovok, C.F. Gu, C.H.J. Davies, Y. Estrin, Size effects in micro cup drawing, Mater. Sci. Eng. A 550 (2012) 312–319.

[47] K. Durst, S. Hofmann, B. Backes, J. Mueller, M. G ken, Microimprinting of nanocrystalline metals – influence of microstructure and work hardening, J. Mater. Process. Technol. 210 (2010) 1787–1793.

[48] L.S. Tóth, R. Lapovok, A. Molotnikov, C. Gu, J.-J. Fundenberger, C.H.J. Davies, Texture evolution during micro-drawing of ultrafine grained copper, Mater. Sci. Eng. A 527 (2010) 4633–4640.

[49] A. Rosochowski, W. Presz, L. Olejnik, M. Richert, Micro-extrusion of ultra-fine grained aluminium, Int. J. Adv. Manufact. Technol. 33 (2007) 137–146.

Part II

Theory of Microforming

Chapter 3

Scaling Laws

3.1 INTRODUCTION

Miniaturization has been particularly important to many of the technological advances that have occurred over the past few decades. With the ongoing trend toward miniaturization, ever smaller mechanical, optical, and electronic devices are becoming increasingly popular. In miniaturization, the shape, size, and characteristic features of the devices must be ensured to allow them to be appropriately integrated and packaged in a successful industrial product that can meet consumer's expectations to be intelligent and multifunctional as compared to others. However, forming theories and technologies established in the macro world cannot be simply scaled down to the micro world, because it is impossible to scale down all parameters in the microforming process according to the theory of similarity due to size effects which are well known as a unique characteristic with miniaturization. Therefore, knowledge on scaling laws obtained through experience in the macro world should be well developed, so that they can be appropriately applied in the micro world with the purpose of understanding well the microforming process and manufacturing high-quality miniaturized devices.

The study of scaling laws and scaling effects was initiated by Galileo Galilei in one of his most well-known books *Two New Sciences* that was published in 1638. In this book, Galileo summarized the work he had done decades earlier on the two sciences now called kinematics and strength of materials. *Two New Sciences* begins in the Venetian Arsenal, the shipyard of the Republic of Venice, with a discussion of the scaling effects in practical construction projects, such as shipbuilding. Galileo thought that "the surface of a small solid is comparatively greater than that of a large one" because the surface goes like the square of a linear dimension, but the volume goes like the cube. As a result, the forces applied on the surfaces of objects such as viscous drag become relatively important, and bulk forces like weight become relatively less important when macro objects are scaled down. Galileo realized that large ships risk breaking when out of water under their own weight but a scaled-down model of the same ship made of the same material may exhibit much stronger behavior. He observed that animals cannot be simply scaled up because the weight increases at a third power of the size scale, and their bones must become proportionately thicker as they get

Microforming Technology. DOI: http://dx.doi.org/10.1016/B978-0-12-811212-0.00003-0

larger. Also he found a very interesting phenomenon that a thin square-shaped lamina of material denser than water can float on the water surface when the size of the lamina is small enough because when the lamina is scaled down its area and hence its weight would go down faster than the rate at which the perimeter decreases. After hundreds of years of development since Galileo's discovery, study on scaling laws has been rapidly developing due to their extensive applications in various fields, such as miniaturization, in our daily lives.

There are primarily two types of scaling laws that are applicable in miniaturization including scaling in geometry and scaling of phenomenological behavior. The first type of scaling law is dependent on the physical size of objects such as the length, surface area, and volume, while the second type of scaling law involves the scaling of various physical phenomena and properties of an object such as mechanics, dynamics, heat transfer, and electricity. All aspects of a device must be scaled in an appropriate way so that not only the geometry of the device can meet the requirement of a microsystem but also its performance is good enough to allow a successful application in the microsystem. A change in geometry is generally accompanied with a great change in physical phenomena and properties of an object if its size is sufficiently scaled down which may have significant effect on the successful design of a miniaturized system. Therefore, a comprehensive understanding of scaling laws becomes important. In this chapter, variations of both the geometry and the physical phenomena and properties of objects will be considered when scaling with a purpose of helping readers understand the scaling laws involved in miniaturization well.

3.2 SCALING IN GEOMETRY

3.2.1 Scaling of Length and Area in Two-Dimensional Geometry

Length and area are two key parameters in scaling a two-dimensional (2D) geometry. In Euclidean geometry, two types of scaling are usually considered. One type is called uniform scaling or isotropic scaling, and the other type is called nonuniform scaling or anisotropic scaling. Uniform scaling is a linear transformation that enlarges or shrinks a geometric figure by a scale factor that is same in all directions. Unless specified, "scaling" mentioned in this book means uniform scaling. Fig. 3.1 shows an example of uniform scaling of a right triangle *ABC*. Obviously, if the base *CA* is scaled down from x to $x/2$, the other two sides of *AB* and *BC* will also be scaled to half of their original lengths correspondingly, as indicated by the triangle $A'B'C'$. When it comes to area, the area of $A'B'C'$ is easily obtained to be 1/4 of the area of *ABC*. Similarly, when *CA* is scaled down to be $x/6$, the area of $A''B''C''$ will be 1/36 of the area of *ABC*. So for uniform scaling, the area of a

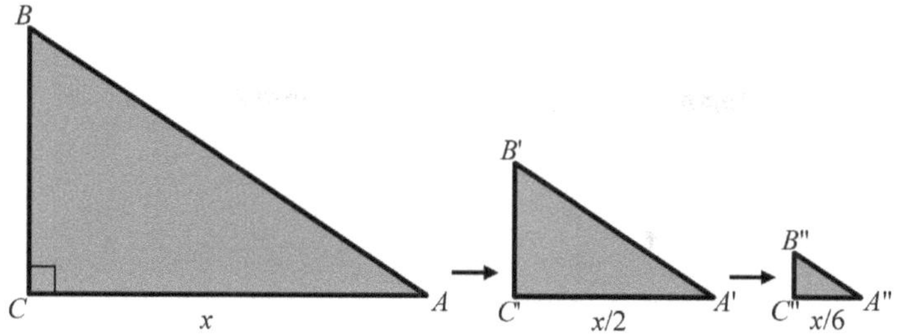

FIGURE 3.1 Uniform scaling of a two-dimensional geometric figure.

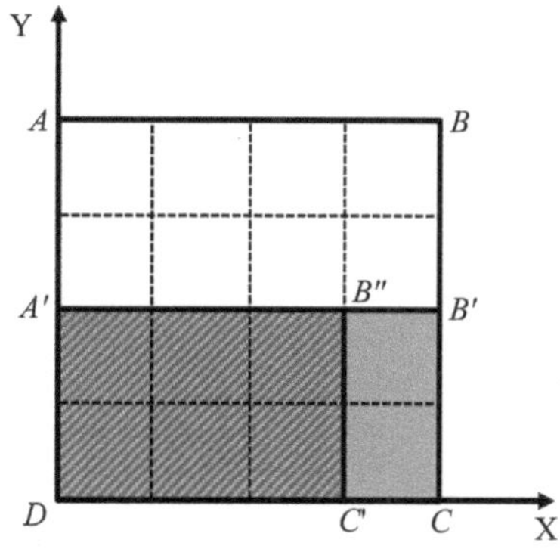

FIGURE 3.2 Nonuniform scaling of a square.

figure will change in proportion to the square of the length scale, i.e., (area ABC):(area $A'B'C'$):(area $A''B''C''$) = $1{:}(1/2)^2{:}(1/6)^2$, in which 1/2 and 1/6 are scale factors. As the scale factors are both positive but smaller than one, the scaling is sometimes also called contraction. Similarly, when the scale factor is larger than one, scaling is sometimes also called dilation or enlargement. The general rule here is the so-called "square law." If we let l be the length scale factor, then the following law can be obtained for uniform scaling:

$$\text{Area} \propto l^2 \tag{3.1}$$

Now let us see what happens if a geometric figure is scaled down nonuniformly. Nonuniform scaling is obtained when at least one of the scale factors is different from the others. In nonuniform scaling, the shape of an object can be changed, e.g., from a square into a rectangle, or into a parallelogram, if the sides of the square are not parallel to the scaling axes. If only one dimension is scaled, the area will also be scaled as a linear power of that length scale. Fig. 3.2 shows nonuniform scaling of a square. When $ABCD$ is scaled down along the Y direction only to be $A'B'CD$, the length of $A'D$ is 1/2 of the length of AD, and also the area of $A'B'CD$ is 1/2 of the area of $ABCD$. If $ABCD$ is scaled down along both the X and Y directions to be $A'B''C'D$, the length of $A'D$ is 1/2 of the length of AD, and the length of DC'

is 3/4 of the length of *DC*. When it comes to area, the area of $A'B''C'D$ can be easily obtained to be 3/8 of the area of *ABCD*. Therefore, if we let l_1 and l_2 be the length scale factors along X and Y directions, respectively, then the following law can be obtained for nonuniform scaling:

$$\text{Area} \propto (l_1 \times l_2) \tag{3.2}$$

Obviously, if l_1 equals to l_2, Eq. (3.2) expresses the same meaning as Eq. (3.1). Therefore, Eq. (3.1) is just the special form of Eq. (3.2).

3.2.2 Scaling of Surface Area and Volume in Three-Dimensional Geometry

The scaling laws for 2D geometry can be extended into three-dimensional (3D) case, as shown in Fig. 3.3 in which the length of each side of the cube is set to be $4x$. For uniform scaling, every length in the small cube is 1/2 as small as the corresponding length in the large cube. When it comes to area, the surface area of the small cube is $24x^2$ which is 1/4 of the surface area of the large cube. When it comes to volume, the volume of the small cube is $8x^3$ which is 1/8 of the volume of the large cube. The surface area contracts by a factor of 1/4 because each face of the cube is locally 2D and on each face of the cube the area is achieved by "width × height." Differently, the volume contracts by a factor of 1/8 because the small cube is 1/2 of the width, height, and length as the corresponding factors in the large cube. Therefore, if we let l be the length scale factor, then the following laws can be obtained for 3D geometry uniform scaling:

$$\text{Surface area} \propto l^2 \tag{3.3}$$

$$\text{Volume} \propto l^3 \tag{3.4}$$

Now let us consider nonuniform scaling of a 3D geometry. Let the width, height, and depth of the down-sized 3D geometry be $l_1 = 3/4$, $l_2 = 2/4$, and $l_3 = 1/4$, respectively, of the corresponding length of the large cube, as

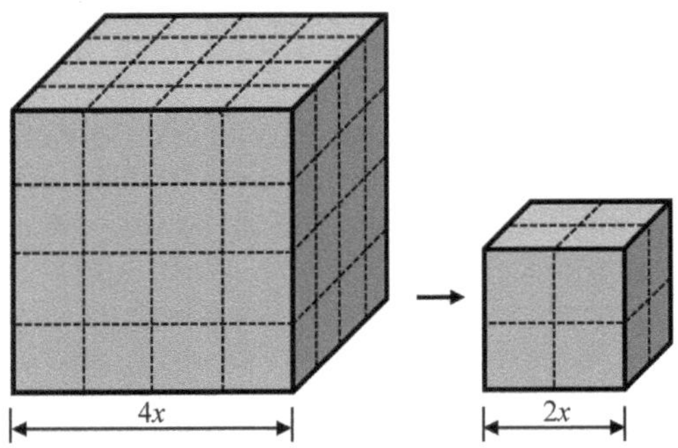

FIGURE 3.3 Uniform scaling of a cube.

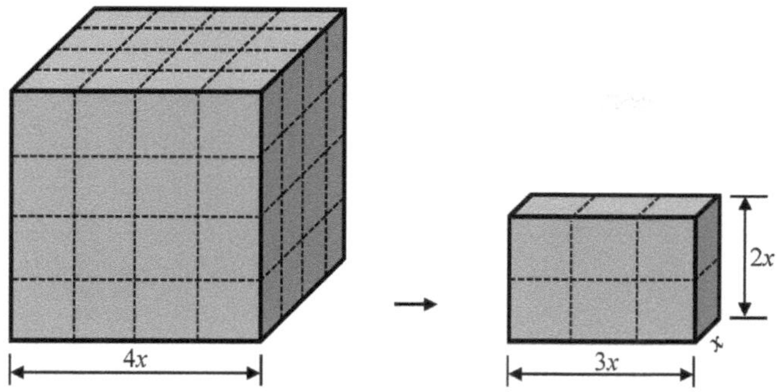

FIGURE 3.4 Nonuniform scaling of a cube.

shown in Fig. 3.4. It is easy to obtain that the surface area of the small cuboid is $22x^2$, and the volume of the small cuboid is $6x^3$, which can be expressed by the following relationships:

$$\text{Surface area} = 2 \times (l_1 \cdot l_2 + l_2 \cdot l_3 + l_1 \cdot l_3) \times (4x)^2 \tag{3.5}$$

$$\text{Volume} = (l_1 \cdot l_2 \cdot l_3) \times (4x)^3 \tag{3.6}$$

Therefore, the following laws can be obtained for 3D geometry nonuniform scaling:

$$\text{Surface area} \propto (l_1 \cdot l_2 + l_2 \cdot l_3 + l_1 \cdot l_3) \tag{3.7}$$

$$\text{Volume} \propto (l_1 \cdot l_2 \cdot l_3) \tag{3.8}$$

Obviously if $l_1 = l_2 = l_3$, Eqs. (3.7) and (3.8) express the same meanings as Eqs. (3.3) and (3.4), respectively.

The "surface area scale factor" and "volume-scale factor" relationships discussed above indicate that the volume changes faster than the surface area with the scale factor. Let us imagine scaling down the size of a stainless steel cup uniformly of 10 times, i.e., the scale factor $l = 1/10$, the surface area of the cup will be contracted $10^2 = 100$ times, while the volume of the cup will be contracted $10^3 = 1000$ times, as schematically indicated in Fig. 3.5.

3.3 SCALING IN DYNAMICS

3.3.1 Scaling in Dynamic Force

Consider two balls: one is big with a mass of 100 kg and the other is small with a mass of 1 kg, as indicated in Fig. 3.6. Which ball would experience a greater change in motion if pulled with the same force? Clearly, the smaller ball would be. This phenomenon can be described by the Newton's second law of motion. Based on Newton's second law of motion, the force acting on

FIGURE 3.5 Uniform scaling of a stainless steel cup with scale factor of 1/10.

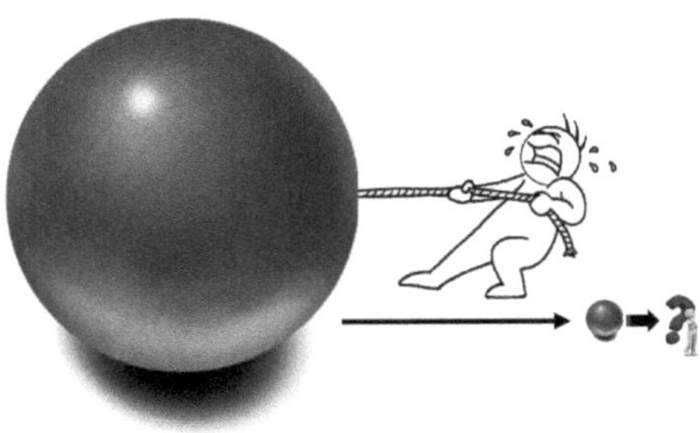

FIGURE 3.6 Difficulties in pulling two balls with different masses.

an object is equal to the mass of that object times its acceleration, which can be written in mathematical form as:

$$F = ma \tag{3.9}$$

where F is the force, m is the mass, and a is the acceleration. The law indicates that the external force causes object to accelerate, and the amount of acceleration is directly proportional to the force acting on the object and inversely proportional to the mass of the object.

After letting s be the distance that the object travels, v_0 the initial velocity, t the time taken for the travel, ρ the density of the object, V the volume of the object, and l the linear scale, the following relationships can be obtained:

$$m = \rho V \tag{3.10}$$

$$s = v_0 t + \frac{1}{2}at^2 \tag{3.11}$$

Based on Eq. (3.10), we have $m \propto l^3$ because the volume of the object $V \propto l^3$, i.e., $m \propto V$. Based on Eq. (3.11), we have:

$$a = \frac{2(s - v_0 t)}{t^2} \tag{3.12}$$

If we let the initial velocity of the object $v_0 = 0$, Eq. (3.12) can be rewritten as:

$$a = \frac{2s}{t^2} \tag{3.13}$$

Because the distance traveled s is linearly proportional to l, i.e., $s \propto l$, and then Eq. (3.9) can be expressed as:

$$F = ma = \rho V \frac{2s}{t^2} \propto l^3 \cdot l \cdot t^{-2} \tag{3.14}$$

Therefore, when scaling an object with scale factor of l, the dynamic force F acting on the object will be scaled as $l^4 t^{-2}$.

3.3.2 Scaling in Work and Power

Work refers to an activity involving a force and movement along the direction of the force. Whenever a force is applied to an object to cause the object to move, work is done by the force. Otherwise if a force is applied but the object does not move, no work is done. The physical definition of work is:

$$W = s \cdot F \cos\theta \tag{3.15}$$

where W is the work done and θ is the angle between the force and the displacement. Fig. 3.7 shows the schematic illustration of the definition of work. It is clear that if a force is applied and the object moves a distance in a direction other than the direction of the force, less work is done than if the object moves a distance in the direction of the applied force.

As discussed in Section 3.3.1, $s \propto l$ and $F \propto l^4 t^{-2}$, hence the work W will be scaled as $l^5 t^{-2}$ when scaling an object with scale factor of l.

Power (P) is the rate of doing work. Mathematically, P is calculated using the following equation:

$$P = \frac{W}{t} \tag{3.16}$$

Since $W \propto l^5 t^{-2}$, we may thus have $P \propto l^5 t^{-3}$.

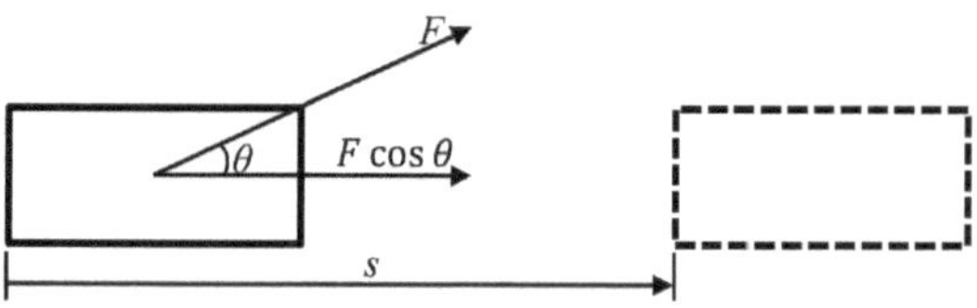

FIGURE 3.7 Schematic illustration of the definition of work.

3.3.3 Scaling in Energy

Kinetic Energy. Kinetic energy is closely related to work. The work done on a rigid object by a force applied to it equals to the change in kinetic energy of the object. Consider an object which moves a distance s with a change of velocity from v_0 to v, the following work–energy principle is:

$$W = \frac{1}{2}mv^2 - \frac{1}{2}m{v_0}^2 = E_{\mathrm{kf}} - E_{\mathrm{ki}} = \Delta E_{\mathrm{k}} \tag{3.17}$$

where E_{kf} and E_{ki} are the kinetic energies of the object at the final and initial states, respectively, and ΔE_{k} refers to the change in kinetic energy.

By letting $v_0 = 0$, i.e., the initial kinetic energy E_{ki} equals to 0, Eq. (3.17) can be expressed as:

$$W = E_{\mathrm{kf}} = E_{\mathrm{k}} \tag{3.18}$$

Therefore, kinetic energy and work done by a force have the same dimension, i.e., the kinetic energy E_{k} will be scaled as l^5t^{-2} when scaling an object with scale factor of l.

Potential Energy. Potential energy is the stored energy of an object. It is the energy by virtue of an object's position relative to other objects. The more formal definition is that potential energy is the energy difference between the energy of an object in a given position and its energy at a reference position. Potential energy is often associated with restoring forces such as a spring or the force of gravity. The action of stretching the spring or lifting the object is performed by an external force that works against the force field of the potential. This work is stored in the force field, which is said to be stored as potential energy. If the external force is removed, the force field acts on the object to perform the work as it moves the object back to the initial position, reducing the stretch of the spring or causing a body to fall. There are various types of potential energy including elastic potential energy, gravitational potential energy, electric potential energy, unclear potential energy, intermolecular potential energy, and chemical potential energy. Each of the potential energy is associated with a particular type of force.

If a force acting on an object is a function of position only, the force is also called conservative force, and it can be represented by a potential energy function which for an one-dimensional case satisfies the derivative condition of:

$$-\frac{\mathrm{d}U}{\mathrm{d}x} = F(x) \tag{3.19}$$

The integral form of the above relationship is:

$$U(x) = -\int_{x_0}^{x} F(x)\mathrm{d}x + U(x_0) \tag{3.20}$$

where U is the potential energy and x_0 and x are, respectively, the initial and final positions of an object when a force is acted on it. Thus, it is easy to obtain that the change in the potential energy is the negative of the work done by that force as:

$$\Delta U = -W \tag{3.21}$$

If the initial potential energy is set as 0, the potential energy is:

$$U = -W \tag{3.22}$$

Therefore, we can conclude that the potential energy U has the same dimension as the work done by a force, i.e., $U \propto l^5 t^{-2}$.

Mechanical Energy. Mechanical energy of a system is the sum of the potential energy which is measured by the position of the parts of the system and the kinetic energy as:

$$E_m = U + E_k \tag{3.23}$$

where E_m is the mechanical energy. It is clear that the mechanical energy also scales as $l^5 t^{-2}$ if an object is scaled with a scale factor of l.

3.4 SCALING IN MECHANICS

3.4.1 Scaling in Bending Moment

Bending is a forming process in which a force is applied to a structural element, causing it to bend at an angle and form the desired shape such as V-shape, U-shape, or channel shape along the straight axis in ductile materials, most commonly sheet metals. Bending of sheet metals is a common and vital process in manufacturing industry. In miniaturization, we need to understand well the bending properties in order to manufacture high-quality micro bent parts. Bending moment refers to the reaction induced in a structural element when an external force or moment is applied to the element that causes the element to be bent. The most common or simplest structural element subjected to bending moments is the beam. Fig. 3.8 shows two typical beams under concentrated load. Both beams have a length of L, a width of b, and a

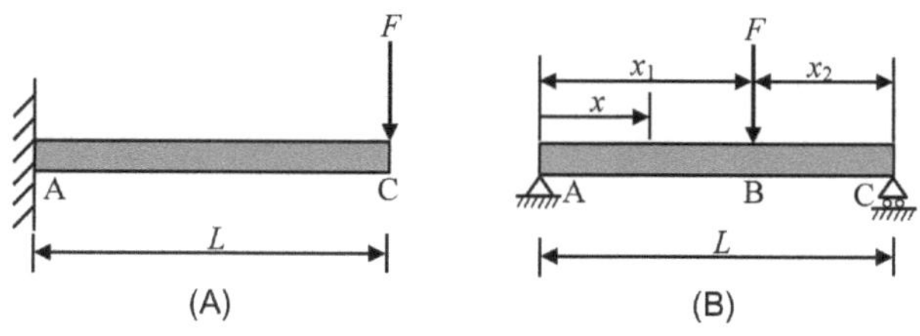

FIGURE 3.8 Concentrated load applied to (A) cantilever beam and (B) beam simply supported at ends.

thickness of h. For the cantilever beam (Fig. 3.8A), the bending moment at the fixed end "A" is the maximum and its magnitude is:

$$M_{\max} = FL \tag{3.24}$$

Similarly, for the beam supported at ends (Fig. 3.8B), the maximum bending moment is at the point of load "B," and its magnitude is:

$$M_{\max} = \frac{Fx_1x_2}{L} \tag{3.25}$$

$$M_x = \frac{Fx_2x}{L} \quad (x < x_1) \tag{3.26}$$

If $x_1 = x_2$, i.e., F is applied at the center of the beam, Eq. (3.25) can be rewritten as:

$$M_{\max} = \frac{FL}{4} \tag{3.27}$$

Next, let us consider the uniformly distributed load applied to the both beams, as schematically illustrated in Fig. 3.9. In the figure, ω refers to the load per unit length. In practice, ωL can be regarded as the self-weight of the beam with length of L. For the cantilever beam (Fig. 3.9A), the maximum bending moment at the fixed end "A" is:

$$M_{\max} = \omega L \cdot \frac{L}{2} \tag{3.28}$$

Similarly, for the beam supported at ends (Fig. 3.9B), the bending moment can be expressed as:

$$M_x = \frac{\omega \cdot x(L - x)}{2} \tag{3.29}$$

It is easy to obtain that the maximum bending moment is at the center of the beam, i.e., $x_1 = x_2$. Therefore, the maximum bending moment is:

$$M_{\max} = \omega L \cdot \frac{L}{8} \tag{3.30}$$

Based on Eqs. (3.24), (3.27), (3.28), and (3.30), it can be concluded that the bending moment M is scaled as l^4 if the beam is scaled with the scale factor of l.

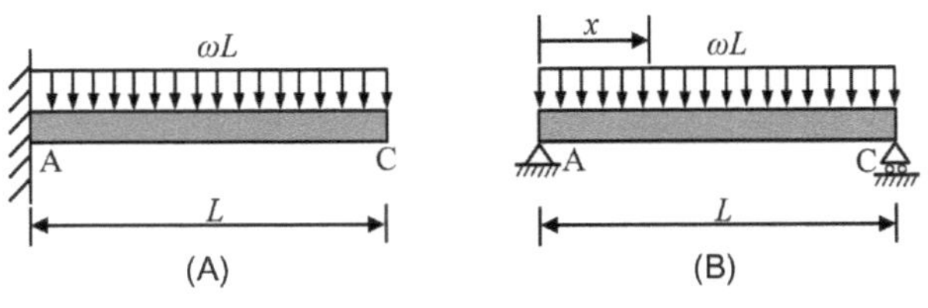

FIGURE 3.9 Uniformly distributed load applied to (A) cantilever beam and (B) beam simply supported at ends.

FIGURE 3.10 Schematic illustration of deflection of a cantilever beam under concentrated load.

3.4.2 Scaling in Deflection and Stiffness

In engineering, deflection is the degree to which a structural element is displaced under a load. It is an important factor in miniaturization as it has significant effect on the miniaturized device after microforming. Deflection may refer to an angle or a distance, as illustrated in Fig. 3.10. In the figure, δ_B and θ_B are the deflection distance and the angle of rotation at the end "B" of the beam, respectively.

For the cantilever beam under concentrated load (Fig. 3.8A), the maximum values of δ and θ are obtained at the free end "C" as:

$$\delta_{max} = \frac{FL^3}{3EI} \tag{3.31}$$

$$\theta_C = \frac{FL^2}{2EI} \tag{3.32}$$

where E is the modulus of elasticity and I is the moment of inertia.

For the beam supported at ends under concentrated load (Fig. 3.8B), if the load is at the center of the beam, the values of δ and θ are obtained as:

$$\delta_{max} = \frac{FL^3}{48EI} \tag{3.33}$$

$$\theta_A = \theta_C = \frac{FL^2}{16EI} \tag{3.34}$$

For the cantilever beam under uniformly distributed load (Fig. 3.9A), the maximum values of δ and θ are also obtained at the free end "C" as:

$$\delta_{max} = \frac{\omega L \cdot L^3}{8EI} \tag{3.35}$$

$$\theta_C = \frac{\omega L \cdot L^2}{6EI} \tag{3.36}$$

Similarly, for the beam supported at ends under uniformly distributed load (Fig. 3.9B), the maximum values of δ and θ are obtained at the center of the beam as:

$$\delta_{max} = \frac{5\omega L \cdot L^3}{384EI} \tag{3.37}$$

$$\theta_A = \theta_C = \frac{\omega L \cdot L^2}{24EI} \tag{3.38}$$

Because $I = (1/12)bh^3$, based on Eqs. (3.31), (3.33), (3.35), and (3.37), it can be concluded that the deflection distance δ is scaled as l^2, and based on Eqs. (3.32), (3.34), (3.36), and (3.38) the angle of rotation θ is scaled as l, if the beam is scaled with the scale factor of l.

Stiffness, k, is the resistance of an elastic body to deflection or deformation by an applied force, and can be expressed as:

$$k = \frac{F}{\delta} \tag{3.39}$$

Since $F \propto l^3$ and $\delta \propto l^2$, thus we have that $k \propto l$.

3.5 SCALING IN HYDRODYNAMICS

Hydrodynamics is a sub-discipline of fluid dynamics that studies the motion of liquids. It has a wide range of applications in engineering, including determination of the mass flow rate of petroleum through pipelines, optimization of propulsion efficiency, prediction of wave dynamics, and measurement of liquid metal flows. In miniaturization, hydrodynamics becomes increasingly important with the application of hydromechanical system in microforming process and the development of micro fluidics system. Knowledge on hydrodynamic characterization of the fluids helps to optimize the fluids flow and mitigate unwanted effects in the hydromechanical system and micro fluidics system.

Consider a fluid flowing through a small circular tube with a characteristic length of L and an internal radius of r, as illustrated in Fig. 3.11. For laminar flow, the pressure drop ΔP between the two ends of the tube can be obtained by using the Hagen–Poiseuille equation as:

$$\Delta P = \frac{8\mu LQ}{\pi r^4} \tag{3.40}$$

where μ is the dynamic viscosity of the fluid, Q is the volumetric flow rate, and π is the mathematic constant.

So, if the dynamic viscosity μ and the volumetric flow rate Q are kept unchanged, the pressure drop over the length of L, $\Delta P/L$, will be scaled as l^{-4}. If μ and $\Delta P/L$ are kept unchanged, Q will be scaled as l^4. This explains why it is hard to push fluid through microchannels by means of pressure drop, making conventional pressure-driven pumping difficult for use in microchannels. For example, if we decrease the radius of the tube by a factor

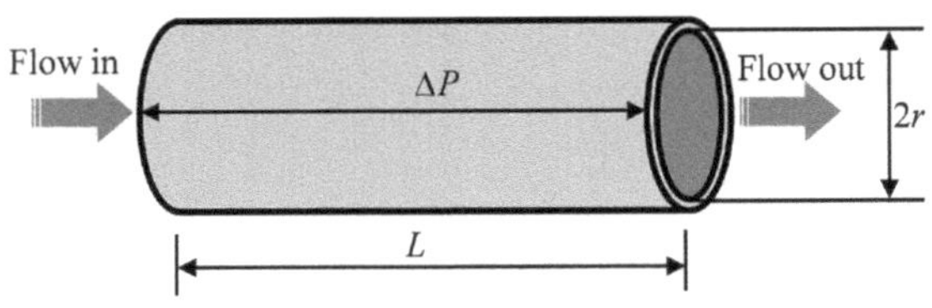

FIGURE 3.11 Fluid flow in a small circular tube.

of 10, i.e., $l = 1/10$, the pressure drop needed to pump the same amount of liquid per unit time will increase by a factor of 10^4.

It should be noted that Hagen–Poiseuille equation applies only to laminar flow. When a fluid is flowing through a tube, either laminar flow or turbulent flow may occur depending on the velocity and viscosity of the fluid. Laminar flow tends to occur at lower velocities, below a threshold at which it becomes turbulent. Turbulent flow is a less orderly flow regime that is characterized by eddies or small packets of fluid particles which result in lateral mixing. The type of flow occurring in a fluid in a tube is important in hydrodynamic problems and subsequently affects the heat and mass transfer in fluid systems. A key tool used to determine the stability and the type of a flow is the dimensionless Reynolds number (R_e) which is defined as:

$$R_e = \frac{\text{Inertia force}}{\text{Viscous force}} = \frac{\rho v L}{\mu} \tag{3.41}$$

where ρ is the density of the liquid and v is the velocity of flow.

The Reynolds number is useful because it can provide cut-off points when the flow is stable or unstable. When the Reynolds number is small, such as the fluid moves slowly or the fluid is very viscous, laminar flow generally occurs. As the Reynolds number increases, such as by increasing the flow rate of the fluid, the flow will transition from laminar to turbulent flow at a specific range of Reynolds numbers. This means that at high Reynolds number, fluid flows will be unstable. Also, if the Reynolds number is large, the effects of viscosity will be small and if it is small the effects of viscosity will be large.

Because the velocity of flow v scales as l, it can be $R_e \propto l^2$ from Eq. (3.41). Therefore, as a tube is scaled down, the Reynolds number will drop rapidly, and the effects of viscosity will become significant. At such low Reynolds number, the flow will be laminar and it will become difficult to mix fluids in miniaturized tubes, such as microchannels. If the Reynolds number is very small, e.g., much less than one, then the fluid will exhibit Stokes or creeping flow, where the viscous force of the fluid dominates the inertial force.

3.6 SCALING IN HEAT TRANSFER

Heat transfer is the change of thermal energy between physical systems, and should be considered in the development of microforming technology in which heat transfer is involved, e.g., microforming conducted at elevated temperatures. Also, heat transfer is an essential part in the design of a number of microsystems. There are fundamentally three modes of heat transfer including conduction, convection, and radiation. Heat conduction is the transfer of heat by microscopic collisions of particles and movement of electrons within a body, and takes place in all phases of matter, such as solids, liquids, gases, and plasmas. Convection is the concerted, collective movement of groups or aggregates of molecules within fluids (e.g., liquids and gases) and fluids through advection or through diffusion or as a combination of both. Radiation is the emission or transmission of energy in the form of waves or particles through space or through a material medium. In most cases, heat transfer in microsystems is in the modes of conduction and convection. However, in some special cases, e.g., warm forming with the assistance of laser heating, heat is also transferred through radiation. In this section, the scaling laws in these three heat transfer modes will be discussed.

3.6.1 Scaling in Heat Conduction

Heat conduction in homogeneous solid materials is governed by the Fourier's Law, which states that the time rate of heat transfer through a material is proportional to the negative gradient in the temperature and to the area, at right angles to that gradient, through which the heat flows. For one-dimensional heat conduction along the x-direction, it is:

$$q_x = -k\frac{dT}{dx} \tag{3.42}$$

where q_x is the heat flux along the x-direction, k is the thermal conductivity of the material, and $\mathrm{d}T/\mathrm{d}x$ is the temperature gradient along the x-direction.

The integrated form of the Fourier's Law for the rate of heat conduction per unit time, $\dot{Q}$, is expressed as follows:

$$\dot{Q} = -kA\frac{T_2 - T_1}{\Delta x} = -kA\frac{\Delta T}{\Delta x} \tag{3.43}$$

where A is the cross-sectional surface area of the material, T_1 and T_2 are, respectively, the temperatures at each end of the material, Δx is the characteristic length of the material for heat transfer, and ΔT is the temperature difference between the ends, as schematically illustrated in Fig. 3.12.

In conventional representation, thermal conductivity, k, is usually thought to be an intrinsic material property, i.e., k is often treated as a constant. In this case, the rate of heat conduction $\dot{Q}$ can be scaled as $\dot{Q} \propto l^2 \cdot l^{-1} = l$ from Eq. (3.43). It implies that a reduction in size leads to a decrease in the total heat conduction within a unit time. Such scaling law can be applied to bulk solids in macroscale. The macroscale formulation in heat conduction is based on the continuum assumption. It does not consider size effects on the

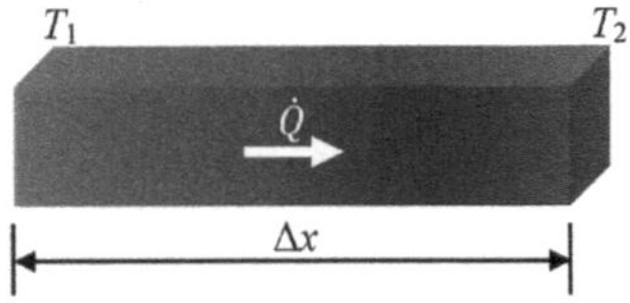

FIGURE 3.12 One-dimensional heat conduction.

heat transport, and heat is assumed to be carried by the atoms. Differently, the microscale formulation in heat conduction considers the physical mechanisms of heat transport where size effects are crucial and must be considered.

In metals, free electrons contribute to heat conduction more than the lattice vibrations. Based on the kinetic energy, the thermal conductivity can be calculated as:

$$k = \frac{1}{3} C_h v_F \lambda \tag{3.44}$$

where C_h is the specific heat per unit volume, v_F is the Fermi velocity (the average velocity of the heat carries), and λ is the mean free path. In the kinetic energy, the electrons are treated as energy carriers with a specific heat C_h, moving at the Fermi velocity v_F across a distance equal to the mean free path λ. When the mean free path is much less than the characteristic dimension of the solid ($\lambda \ll \Delta x$), the heat transport is said to be macroscopic. In this case, the Fourier's Law is still applicable and the transport is governed by a purely diffusive nature. Differently, when λ is on the order of or much greater than Δx ($\lambda \sim \Delta x$ or $\lambda \gg \Delta x$), the heat transport becomes microscopic.

As C_h is given by the Debye formula and it scales as $C_h \propto l^3$, $v_F \propto l$, and $\lambda \propto l^{-3}$; therefore, it is $k \propto l$ in microscale and even submicron scale and nanoscale.

Based on Eqs. (3.43) and (3.44), the scaling law for the rate of heat conduction $\dot{Q}$ in microscale, submicron scale, and nanoscale is as $\dot{Q} \propto k \cdot l \propto l^2$. The scaling law indicates that it is difficult to transport heat through heat conduction in a solid in microscale. For example, if we scale down a solid with scale factor of $l = 1/10$, the rate of heat conduction will be reduced by a factor of 10^2.

3.6.2 Scaling in Heat Convection

The basic relationship for heat transfer by convection can be expressed by the Newton's Law of Cooling as:

$$\dot{Q} = hA(T_a - T_b) = hA\Delta T \tag{3.45}$$

where $\dot{Q}$ is the heat transferred per unit time, A is the area of convection surface, h is the convective heat transfer coefficient, T_a is the object's surface temperature, T_b is the fluid temperature, and ΔT is the temperature difference between the surface and the fluid.

As the convective heat transfer coefficient, h, is primarily dependent on the type of media, e.g., gas or liquid, and the flow properties such as velocity, viscosity, and does not play significant role in the scaling of heat transfer, therefore, the scaling law for the heat transfer in convection is as $\dot{Q} \propto l^2$ for fluids in microscale.

3.6.3 Scaling in Heat Radiation

Heat transfer through radiation takes place in form of electromagnetic waves mainly in the infrared region. Radiation heat transfer can be described by reference to the blackbody, which is defined as a body that absorbs all radiation that falls on its surface. The radiation energy per unit time from a radiating body can be estimated by the Stefan–Boltzmann law as:

$$\dot{E} = \sigma_s A T^4 \tag{3.46}$$

where $\dot{E}$ is the radiation energy per unit time from the emitting blackbody, σ_s is the Stefan–Boltzmann constant, and A and T are the area and the temperature of the emitting blackbody, respectively.

It is readily seen from Eq. (3.46) that the radiation energy from the emitting blackbody will scale as $\dot{E} \propto l^2$ if the temperature of the blackbody is kept unchanged.

3.7 SCALING IN ELECTROMAGNETIC AND ELECTROSTATIC FORCES

3.7.1 Scaling in Electromagnetic Force

Electromagnetic force appears when a magnetic field interacts with electrically charged particles. Fig. 3.13 shows a straight and stationary wire with the length of L carrying an electric current i in a magnetic field B. Then the electromagnetic force F_{em} can be expressed as:

$$F_{\mathrm{em}} = LBi \tag{3.47}$$

The direction of the electromagnetic force generated can be determined by the left-hand rule, as indicated in Fig. 3.13. As the current i, i.e., the amount of electrons flowing through the wire, is proportional to the cross-sectional area of the wire, so it is $i \propto l^2$.

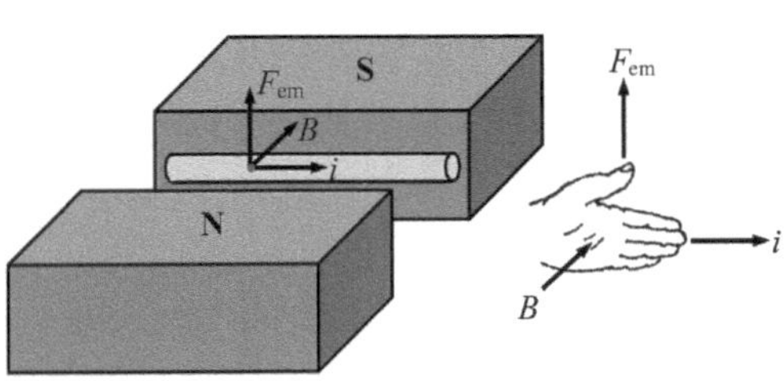

FIGURE 3.13 Determination of the direction of electromagnetic force.

Let us imagine that the magnetic field B is created by a solenoid made of a coil of wire, and then the formula for the magnetic field inside the solenoid is written as:

$$B = \frac{\mu_0 N_c i_c}{L_c} \tag{3.48}$$

where μ_0 is the permeability, N_c is the number of loops the solenoid has, i_c is the current through the solenoid producing the magnetic field, and L_c is the length of the solenoid.

Because $i_c \propto l^2$ and $L_c \propto l$, then we have $B \propto l$. Therefore, when scaling a system with the scale factor of l, the electromagnetic force F_{em} is scaled as:

$$F_{em} \propto l \cdot l \cdot l^2 = l^4 \tag{3.49}$$

3.7.2 Scaling in Electrostatic Force

Electrostatic force is also called Coulomb force, which is defined as the electrical force of repulsion or attraction induced by an electric field. Consider a capacitor composed of a pair of same-sized parallel plates with a gap of d (Fig. 3.14), the electrostatic potential energy stored in the capacitor, U, can be expressed as:

$$U = -\frac{1}{2} C V^2 \tag{3.50}$$

where C is the capacitance induced in the charged plates and V is the voltage applied to the plates.

Capacitance C is proportional to the area of the plates and can be obtained by:

$$C = \epsilon_0 \epsilon_r \frac{A}{d} \tag{3.51}$$

where A is the area of the plates and ϵ_0 and ϵ_r are, respectively, the permittivity and the relative permittivity of the dielectric medium between the plates.

Based on Eqs. (3.50) and (3.51), the following equation can be obtained:

$$U = -\frac{\epsilon_0 \epsilon_r A}{2d} V^2 \tag{3.52}$$

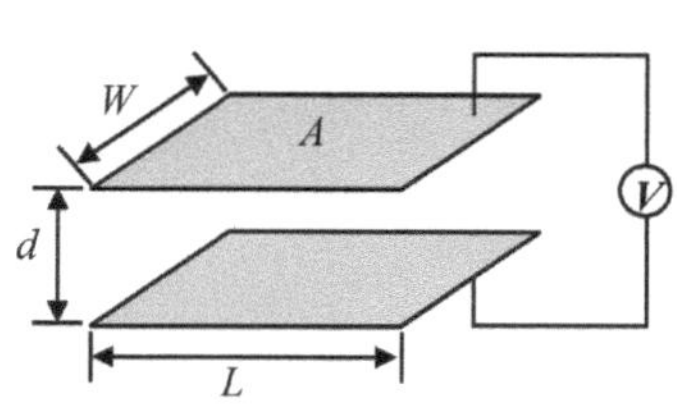

FIGURE 3.14 An electrically charged capacitor.

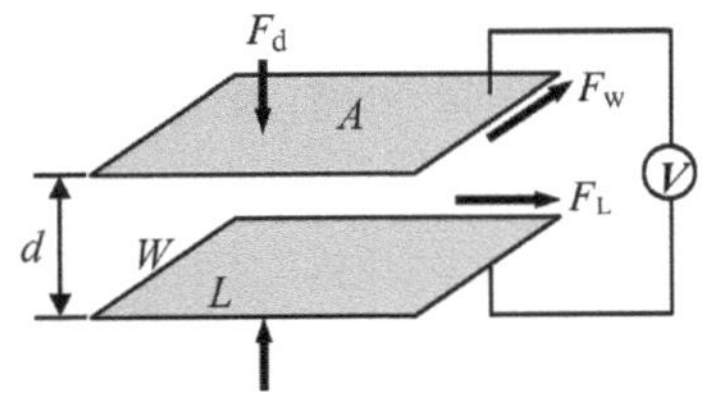

FIGURE 3.15 Electrostatic forces in a charged capacitor.

When $d > 10\,\mu m$, the breakdown voltage is taken as proportional to the gap of d. Hence the voltage V is scaled as $V \propto l$. Because both ϵ_0 and ϵ_r are constants, then $C \propto l$ from Eq. (3.51). Thus the scaling law for the electrostatic potential energy is as $U \propto l \cdot l^2 \propto l^3$ from Eq. (3.52).

Electrostatic forces can be generated in three directions in the charged parallel plates, as indicated in Fig. 3.15. The forces along different directions can be obtained by the following equations:

$$F_d = -\frac{\partial U}{\partial d} = -\frac{\epsilon_0 \epsilon_r A}{2d^2} V^2 \tag{3.53}$$

$$F_W = -\frac{\partial U}{\partial W} = -\frac{\epsilon_0 \epsilon_r L}{2d} V^2 \tag{3.54}$$

$$F_L = -\frac{\partial U}{\partial L} = -\frac{\epsilon_0 \epsilon_r W}{2d} V^2 \tag{3.55}$$

It can be readily seen from Eqs. (3.53)–(3.55) that $F_d \propto l^2$, $F_W \propto l^2$, and $F_L \propto l^2$, i.e., the electrostatic force will scale as l^2.

3.8 SCALING IN ELECTRICITY

Electricity is the set of physical phenomena associated with the presence and flow of electric charge. In miniaturization, electricity properties become important due to their involvement in a number of applications in microsystems including actuation of microsystems, thermal resistance heating, electrodynamic pumping, and electromechanical transduction.

The electric resistance of a conductor can be calculated by:

$$R = \rho \frac{L}{A} \tag{3.56}$$

where ρ, L, and A are the electrical resistivity, the length, and the cross-sectional area of the conductor, respectively.

So for a given conductor, electric resistance scales as $R \propto l \cdot l^{-2} = l^{-1}$.

For a solenoid, the inductance is given by:

$$\mathcal{L} = \frac{\mu_0 {N_c}^2 A}{L_c} \tag{3.57}$$

where the meanings of μ_0, N_c, and L_c have been given in Eq. (3.48) and A is the cross-sectional area of the solenoid. Therefore, we have $\mathcal{L} \propto l^2 \cdot l^{-1} = l$.

Electrical reactance for a capacitor is given by:

$$X_C = \frac{1}{2\pi f C} \tag{3.58}$$

where f is the signal frequency, π is the mathematic constant, and C is the capacitance. Because $C \propto l$ from Eq. (3.51), it has $X_C \propto l^{-1}$.

For a resistive circuit, the amount of power that is dissipated can be expressed as:

$$P_r = \frac{V^2}{R} \tag{3.59}$$

where V is the applied voltage and R is the electric resistance. As $V \propto l$ and $R \propto l^{-1}$, thus the dissipated power is scaled as $P_r \propto l^2 \cdot l^{-1} = l$.

Chapter 4

Strain Gradient Plasticity Theory

4.1 INTRODUCTION

Materials exhibit strong size effects when the characteristic length scale associated with nonuniform plastic deformation is in the order of microns. In microforming, size effects should be considered in order to understand well the forming process, which is essential for the design of forming tools, optimization of forming process, and quality control of produced parts. Unfortunately, conventional plasticity theories possess no internal length scale, which has limited them to predict the size dependence of material behavior in microscale. In conventional theories, stress at a point is thought to be dependent only on the strain and the strain history without considering its absolute size, and therefore, materials' property is a simple first-order quantity. Such description is applicable to the deformation process when the length scale associated with the deformation field is large and strain gradient is small. However, the stress at a point may be affected by not only the strain but also the strain gradient if the length scale associated with the deformation field is small, and in this case it becomes necessary to include the strain gradient term in the description of plasticity deformation at small length scale. Generally, strain gradient is in inverse proportion of length scale over which plastic deformation occurs, and thus the smaller the length scale, the important the strain gradient will be for the plastic deformation.

When a material is deformed, two kinds of dislocations will be generated, moved, and stored. One is called statistically stored dislocations (SSDs), which accumulate by trapping each other in a random way, and the other one is known as geometrically necessary dislocations (GNDs) which are required for compatible deformation of various parts of the material. GNDs control the work hardening of the material when their density exceeds that of the SSDs, and are related to the gradients of plastic shear in the material. They contribute to hardening by acting as individual obstacles to slip, and by creating a long-range back-stress, with wave-length equal to the particle spacing [1]. The size effects in a miniaturized material are attributed to GNDs which are associated with plastic strain gradients. Plastic strain gradients appear either because of the geometry of loading or because of the nonuniform deformation of the material. Strain gradient plasticity theory is an

Microforming Technology. DOI: http://dx.doi.org/10.1016/B978-0-12-811212-0.00004-2

extension of conventional continuum theory that takes into account the influence of strain gradient. It is based on the condition that GNDs must exist in crystal plasticity (CP) with a purpose of removing the geometric incompatibility associated with lattice curvature and elastic stretch [2].

In the past years, a number of strain gradient plasticity theories have been proposed. The existing strain gradient plasticity theories can be classified into two categories, i.e., lower-order and higher-order strain gradient plasticity theories. The lower-order strain gradient plasticity theories retain all the features of conventional J_2 flow theory but incorporate a dependence on plastic strain gradients into the incremental, or tangential, moduli. They involve only the conventional stress, and the equilibrium equations and boundary conditions are the same as those in conventional continuum theories. The theories have standard boundary conditions but introduce no higher-order stresses. The advantage of the lower-order strain gradient plasticity theories is that it is easy and straightforward to convert them into the conventional finite element codes. However, the lower-order strain gradient plasticity theories which neglect the application of the corresponding higher-order boundary conditions at interfaces and free surfaces cannot predict boundary layer effects effectively, and therefore they are usually not suitable for studying the plasticity deformation behavior in microscale. The higher-order strain gradient plasticity theories involve extra boundary conditions and additional higher-order stress quantities. For explaining the size effects encountered at the micro and submicro length scales, it is now widely recognized that gradient plasticity theories should be higher-order in the sense that it is possible to impose extra boundary conditions with respect to plastic strains or their gradients. In some special cases, however, strain gradient plasticity theories without considering higher-order conditions may also be applicable to the problems in microscale. In this chapter, higher-order strain gradient plasticity theories, including couple stress theory, phenomenological strain gradient plasticity theory, and mechanism-based strain gradient plasticity theory that are widely studied will be introduced. A conventional theory of mechanism-based strain gradient plasticity which does not involve the higher-order stress will also be introduced as this model is simple and easy for applications. The objective of this chapter is to provide readers with new thought in the solution of deformation problems in microforming.

4.2 COUPLE STRESS THEORY

The couple stress theory was developed in 1960s by a number of researchers [3–8]. Couple stress theory assumes that a surface element dS of a body may transmit both a force vector of $\mathbf{T}$dS and a couple vector of $\mathbf{q}$dS, with $\mathbf{T}$ and $\mathbf{q}$ the force traction vector and couple stress traction vector, respectively. For a body bounded by a piecewise smooth surface S, the equilibrium of forces and moments on the body can be expressed, respectively, by Eqs. (4.1) and (4.2) as [9]:

$$\int_S \mathbf{T} \mathrm{d}S = 0 \tag{4.1}$$

$$\int_S (\mathbf{r} \times \mathbf{T} + \mathbf{q}) \mathrm{d}S = 0 \tag{4.2}$$

where **r** is the radius vector from an arbitrary fixed point, and the body forces and body couples have been neglected.

In the couple stress theory, the strain energy density w per unit volume due to an arbitrary variation of displacement u is thought to be dependent upon both the symmetric strain tensor ε_{ij} and the second gradient of displacement η_{ijk} as [10]:

$$\delta w = \sigma_{ij}\delta\varepsilon_{ij} + \tau_{ijk}\delta\eta_{ijk} \tag{4.3}$$

$$\varepsilon_{ij} = \frac{1}{2}(u_{i,j} + u_{j,i}) \tag{4.3a}$$

$$\eta_{ijk} = u_{k,ij} \tag{4.3b}$$

where the symmetric Cauchy stress σ_{ij} is the work conjugate of the strain variation $\delta\varepsilon_{ij}$, and the higher-order stress τ_{ijk} is the work conjugate of the strain gradient variation $\delta\eta_{ijk}$. The higher-order stress tensor is composed of components of both couple stresses and double stresses. Based on Eq. (4.3), it is clear that the second-order symmetric stress σ can be obtained by $\sigma_{ij} = \partial w/\partial\varepsilon_{ij}$, and the third-order stress τ can be obtained by $\tau_{ijk} = \partial w/\partial\eta_{ijk}$.

Then, the work increment for a volume V is given by [3,7,8]:

$$\int_V (\delta w) dV = \int_V (\sigma_{ij}\delta\varepsilon_{ij} + \tau_{ijk}\delta\eta_{ijk}) dV \tag{4.4}$$

Eq. (4.4) can be rewritten as:

$$\begin{aligned}\int_V (\sigma_{ij}\delta\varepsilon_{ij} + \tau_{ijk}\delta\eta_{ijk}) dV = & -\int_V [\partial_i(\sigma_{ik} - \partial_j\tau_{ijk})]\delta u_k dV \\ & + \int_S [n_i(\sigma_{ik} - \partial_j\tau_{ijk})]\delta u_k dS \\ & + \int_S [n_i\tau_{ijk}(\partial_j\delta u_k) dS\end{aligned} \tag{4.5}$$

where n_i is the unit normal to the surface S of the body. The gradient $\partial_j\delta u_k$ can be resolved into a surface gradient $D_j\delta u_k$ and a normal gradient $n_j D\delta u_k$ as:

$$\partial_j\delta u_k = D_j\delta u_k + n_j D\delta u_k \tag{4.6}$$

where D_j is the surface gradient operator and D is the normal gradient operator. Both can be expressed, respectively, by Eqs. (4.7) and (4.8) as:

$$D_j \equiv (\delta_{jk} - n_j n_k)\partial_k \tag{4.7}$$

$$D \equiv n_k \partial_k \tag{4.8}$$

By substituting Eqs. (4.6)–(4.8) into Eq. (4.5), the final form for the principle of virtual work can be achieved by using Stokes' surface divergence theory [7]:

$$\int_V (\sigma_{ij}\delta\varepsilon_{ij} + \tau_{ijk}\delta\eta_{ijk})dV = \int_V (f_k\delta u_k)dV + \int_S (t_k\delta u_k)dS + \int_S r_k(D\delta u_k)dS \tag{4.9}$$

where f_k is the body force per unit volume, t_k is the surface traction on the surface S, and r_k is the double stress traction on S. These three items can be obtained, respectively, by the following expressions as:

$$f_k = -\partial_i[\sigma_{ik} - (\partial_j\tau_{ijk})] \tag{4.10}$$

$$t_k = n_i\left[\sigma_{ik} - (\partial_j\tau_{ijk})\right] + n_i n_j \tau_{ijk}(D_p n_p) - D_j(n_i\tau_{ijk}) \tag{4.11}$$

$$r_k = n_i n_j \tau_{ijk} \tag{4.12}$$

Based on Eq. (4.9), the displacement field u must satisfy three equilibrium equations given by Eq. (4.10), and six boundary conditions given by Eqs. (4.11) and (4.12).

4.3 PHENOMENOLOGICAL STRAIN GRADIENT PLASTICITY THEORY

Phenomenological strain gradient plasticity theory was first developed by Fleck and Hutchinson [9–11] and has been further developed by a number of researchers to solve problems where strain gradient effects are expected to be important [12–17]. Phenomenological strain gradient plasticity theory is based on rotation gradient which fits with the framework of couple stress theory. In this theory, material hardening is thought to be dependent upon both strain ε and curvature χ. The strain tensor ε_{ij} is dependent on the material displacement u via Eq. (4.3a), and the curvature is the spatial gradient of the material rotation θ as:

$$\chi_{ip} = \theta_{i,p} = \frac{1}{2}e_{ijk}u_{k,jp} = e_{ijk}\varepsilon_{kp,j} \tag{4.13}$$

where e_{ijk} is the permutation tensor. The strain energy density w of an incompressible solid is thought to be dependent on the second, von Mises invariant of strain as:

$$\varepsilon_e = \sqrt{\frac{2}{3}\varepsilon_{ij}\varepsilon_{ij}} \tag{4.14}$$

The density of SSDs is related to ε_e, and can be determined by:

$$\rho_S = \frac{\varepsilon_e}{\lambda_{\text{I}} b} \quad \text{(In stage I)} \tag{4.15}$$

$$\rho_S = \frac{\varepsilon_e}{\lambda_{\text{II}} b} \quad \text{(In stage II)} \tag{4.16}$$

where b is the magnitude of the Burger's vector, and λ_{I} and λ_{II} are fixed slip distances of the order of $\lambda_{\text{I}} = 1$ mm in stage I and $\lambda_{\text{II}} = 10\ \mu\text{m}$ in stage II. Both values for λ_{I} and λ_{II} are correct to within an order of magnitude for face-centered cubic (FCC) metals such as copper.

The analogous second invariant of the curvature can be expressed as:

$$\chi_e = \sqrt{\frac{2}{3}\chi_{ij}\chi_{ij}} \tag{4.17}$$

The magnitude of χ_e can be used as the scalar measure of the density of GNDs as:

$$\rho_G = \frac{\chi_e}{b} \tag{4.18}$$

Then, the simplest possible dimensionally correct relationship between the flow stress τ on the slip plane and total dislocation density can be expressed as:

$$\tau = CGb\sqrt{\rho_S + \rho_G} \tag{4.19}$$

where C is a constant and G is the shear modulus. By substituting Eqs. (4.15) or (4.16), and (4.18) into Eq. (4.19), the following equations can be obtained:

$$\tau = CG\sqrt{\frac{b\varepsilon_e}{\lambda_{\text{I}}} + b\chi_e} \quad \text{(In stage I)} \tag{4.20}$$

$$\tau = CG\sqrt{\frac{b\varepsilon_e}{\lambda_{\text{II}}} + b\chi_e} \quad \text{(In stage II)} \tag{4.21}$$

The strain energy density w is assumed to be dependent on the overall effective strain quantity ξ which has a relationship with ε_e and χ_e as:

$$\xi = \sqrt{\varepsilon_e^2 + l^2\chi_e^2} \tag{4.22}$$

where l is the material length scale required for dimensional consistency.

Define an overall stress measure Σ as the conjugate of ξ:

$$\Sigma = \frac{dw(\xi)}{d\xi} \tag{4.23}$$

In Eq. (4.23), the overall stress measure Σ is a unique function of the overall strain measure ξ. On equating the incremental work of deformation to the increase in elastic stored energy, the following equation can be obtained:

$$\delta w = \sigma_{ij}\delta\varepsilon_{ij} + \mu_{ij}\delta\chi_{ji} = s_{ij}\delta\varepsilon_{ij} + m_{ij}\delta\chi_{ji} \tag{4.24}$$

where μ is the couple stress tensor, s is the deviatoric part of the symmetric stress tensor σ, and m is the deviatoric couple stress tensor. The symmetric components s_{ij} of the deviatoric stress tensor and the deviatoric components m_{ij} of the couple stress tensor in Eq. (4.24) can be obtained by:

$$s_{ij} = \frac{\partial w}{\partial \varepsilon_{ij}} = \Sigma\frac{\partial \xi}{\partial \varepsilon_{ij}} = \frac{2}{3}\frac{\Sigma}{\xi}\varepsilon_{ij} \tag{4.25}$$

$$m_{ij} = \frac{\partial w}{\partial \chi_{ji}} = \Sigma\frac{\partial \xi}{\partial \chi_{ji}} = \frac{2}{3}l^2\frac{\Sigma}{\xi}\chi_{ji} \tag{4.26}$$

Substituting Eq. (4.25) into Eq. (4.14), and Eq. (4.26) into Eq. (4.17), the following relationship can be achieved based on Eq. (4.22) as:

$$\xi = \sqrt{\sigma_e^2 + l^{-2}m_e^2} \tag{4.27}$$

where σ_e is the usual von Mises effective stress and m_e is the analogous effective couple stress. Both can be expressed, respectively, by the following equations as:

$$\sigma_e = \sqrt{3/2s_{ij}s_{ij}} \tag{4.27a}$$

$$m_e = \sqrt{3/2m_{ij}m_{ij}} \tag{4.27b}$$

The measures of σ_e and m_e are, respectively, the conjugates of ε_e and χ_e as:

$$dw = \sigma_e d\varepsilon_e + m_e d\chi_e \tag{4.28}$$

Then a simple power law relationship between Σ and ξ can be expressed as:

$$\Sigma = \Sigma_0\xi^N = \Sigma_0(\varepsilon_e^2 + l^2\chi_e^2)^{N/2} \tag{4.29}$$

where N is the strain hardening index with values varying from 0 to 1.

Fleck and Hutchinson [10] thought that the strain energy density w for a nonlinear elastic solid could be assumed to be a nonlinear function of the combined strain quantity ξ as:

$$\xi^2 = \frac{2}{3}\varepsilon_{ij}\varepsilon_{ij} + c_1\eta_{iik}\eta_{jjk} + c_2\eta_{ijk}\eta_{ijk} + c_3\eta_{ijk}\eta_{kji} \tag{4.30}$$

where the three constants c_1, c_2, and c_3 scale the three quadratic invariants for the third-order tensor η_{ijk}. The factor of 2/3 premultiplies the $\varepsilon_{ij}\varepsilon_{ij}$ invariant, so that ξ equals to the von Mises strain invariant ε_e in the case of uniform strain. The strain gradient tensor $\boldsymbol{\eta}$ can be decomposed as:

$$\boldsymbol{\eta} = \boldsymbol{\eta}^{(1)} + \boldsymbol{\eta}^{(2)} + \boldsymbol{\eta}^{(3)} \tag{4.31}$$

where

$$\eta_{ijk}^{(1)} = \eta_{ijk}^S - \frac{1}{5}(\delta_{ij}\eta_{kpp}^S + \delta_{jk}\eta_{ipp}^S + \delta_{ki}\eta_{jpp}^S) \tag{4.31a}$$

$$\eta_{ijk}^{(2)} = \frac{1}{6}(e_{ikp}e_{jlm}\eta_{lpm} + e_{jkp}e_{ilm}\eta_{lpm} + 2\eta_{ijk} - \eta_{jki} - \eta_{kij}) \tag{4.31b}$$

$$\begin{aligned}\eta_{ijk}^{(3)} = {} & \frac{1}{6}(-e_{ikp}e_{jlm}\eta_{lpm} - e_{jkp}e_{ilm}\eta_{lpm} + 2\eta_{ijk} - \eta_{jki} - \eta_{kij}) \\ & + \frac{1}{5}(\delta_{ij}\eta_{kpp}^S + \delta_{jk}\eta_{ipp}^S + \delta_{ki}\eta_{jpp}^S)\end{aligned} \tag{4.31c}$$

where $\boldsymbol{\eta}^S$ is the fully symmetric tensor, and is expressed as:

$$\eta_{ijk}^S = \frac{1}{3}(\eta_{ijk} + \eta_{jki} + \eta_{kij}) \tag{4.32}$$

Then the combined strain quantity ξ can be rewritten in terms of ε and η as:

$$\xi^2 = \frac{2}{3}\varepsilon_{ij}\varepsilon_{ij} + l_1^2\eta_{ijk}^{(1)}\eta_{ijk}^{(1)} + l_2^2\eta_{ijk}^{(2)}\eta_{ijk}^{(2)} + l_3^2\eta_{ijk}^{(3)}\eta_{ijk}^{(3)} \tag{4.33}$$

where

$$l_1^2 = c_2 + c_3 \tag{4.33a}$$

$$l_2^2 = c_2 - \frac{1}{2}c_3 \tag{4.33b}$$

$$l_3^2 = \frac{5}{2}c_1 + c_2 - \frac{1}{4}c_3 \tag{4.33c}$$

Based on Eqs. (4.13) and (4.31a)–(4.31c), the following expressions can be obtained:

$$\chi_{ij}\chi_{ij} = \frac{3}{8}\eta_{ijk}^{(2)}\eta_{ijk}^{(2)} + \frac{5}{16}\eta_{ijk}^{(3)}\eta_{ijk}^{(3)} \tag{4.34}$$

$$\chi_{ij}\chi_{ji} = \frac{3}{8}\eta_{ijk}^{(2)}\eta_{ijk}^{(2)} - \frac{5}{16}\eta_{ijk}^{(3)}\eta_{ijk}^{(3)} \tag{4.35}$$

It is clear that $\eta_{ijk}^{(2)}$ and $\eta_{ijk}^{(3)}$ are associated with the rotational gradients, while $\eta_{ijk}^{(1)}$ is associated with the stretch gradient. Then Eq. (4.33) can be rewritten in terms of $\chi_{ij}\chi_{ij}$, $\chi_{ij}\chi_{ji}$, and $\eta_{ijk}^{(1)}\eta_{ijk}^{(1)}$ as:

$$\xi^2 = \frac{2}{3}\varepsilon_{ij}\varepsilon_{ij} + l_1^2\eta_{ijk}^{(1)}\eta_{ijk}^{(1)} + \left(\frac{4}{3}l_2^2 + \frac{8}{5}l_3^2\right)\chi_{ij}\chi_{ij} + \left(\frac{4}{3}l_2^2 - \frac{8}{5}l_3^2\right)\chi_{ij}\chi_{ji} \tag{4.36}$$

For the couple stress solid, the values of l_1, l_2, and l_3 can be determined as:

$$l_1 = 0, \quad l_2 = \frac{1}{2}l, \quad l_3 = \sqrt{\frac{5}{24}}l \tag{4.37}$$

For the more general solid which is dependent on both stretch and rotation gradients, the values of l_1, l_2, and l_3 can be determined as:

$$l_1 = l, \quad l_2 = \frac{1}{2}l, \quad l_3 = \sqrt{\frac{5}{24}}l \tag{4.38}$$

4.4 MECHANISM-BASED STRAIN GRADIENT PLASTICITY THEORY

Gao et al. [18,19] proposed a mechanism-based strain gradient plasticity theory based on a multiscale framework linking the microscale notion of SSDs and GNDs to the mesoscale notion of plastic strain and strain gradient. This theory adopts the higher-order framework laid out by Fleck and Hutchinson [10], and assumes that:

1. The microscale flow stress is governed by the dislocation motion, which obeys the law for strain gradient plasticity as:

$$\tilde{\sigma} = \sigma_Y\sqrt{f^2(\tilde{\varepsilon}) + l\eta} \tag{4.39}$$

   where $\tilde{\sigma}$ and $\tilde{\varepsilon}$ represent the stress and strain tensors, respectively, in microscale, and l is the material length which is defined by [10,11]:

$$l = 3\alpha^2\left(\frac{\mu}{\sigma_Y}\right)^2 b \tag{4.40}$$

   where μ is the shear modulus, b is the Burgers vector, and α is an empirical constant usually ranging from 0.2 to 0.5.
2. A mesoscale cell is taken as small as possible so that the variation of strain field can be approximated as linear within the cell, and yet

sufficiently large for application of the Taylor model. Higher-order strain gradients are assumed to be negligible within the mesoscale cell. The microscale and mesoscale are linked by the plastic work equality within a mesoscale cell as:

$$\int_{V_{\text{cell}}} \tilde{\sigma}'_{ij}\delta\tilde{\varepsilon}_{ij}dV = (\sigma'_{ij}\delta\varepsilon_{ij} + \tau'_{ijk}\delta\eta_{ijk})V_{\text{cell}} \tag{4.41}$$

where the stress σ, the strain ε, the higher-order stress τ, and the strain gradient η are variables in mesoscale.

3. The essential structure of conventional plasticity is preserved in microscale. Assuming dislocation slip is proportional to the resolved Schmid stress along an appropriate slip system, and then the following relationship can be obtained:

$$\frac{d\tilde{\varepsilon}_{ij}}{d\tilde{\varepsilon}} = \frac{3\tilde{\sigma}'_{ij}}{2\tilde{\sigma}_e} \tag{4.42}$$

where $d\tilde{\varepsilon}$ is the effective strain increment and $\tilde{\sigma}_e$ is the effective stress. Both can be expressed as:

$$d\tilde{\varepsilon} = \sqrt{\frac{2}{3}d\tilde{\varepsilon}_{ij}d\tilde{\varepsilon}_{ij}} \tag{4.43}$$

$$\tilde{\sigma}_e = \sqrt{\frac{2}{3}\tilde{\sigma}'_{ij}\tilde{\sigma}'_{ij}} \tag{4.44}$$

In the mechanism-based strain gradient plasticity theory, the microscale yield criterion is $\tilde{\sigma}_e = \tilde{\sigma}$ and the microscale effective strain in deformation is defined as:

$$\tilde{\varepsilon}^2 = \frac{2}{3}\tilde{\varepsilon}_{ij}\tilde{\varepsilon}_{ij} \tag{4.45}$$

$$\tilde{\varepsilon}\delta\tilde{\varepsilon} = \frac{2}{3}\tilde{\varepsilon}_{ij}\delta\tilde{\varepsilon}_{ij} \tag{4.46}$$

The three theoretical postulates described above are the base of the development of the mechanism-based strain gradient plasticity theory. Consider a cubic unit cell in mesoscale with the length of all edges equal to l_ε, which is much smaller than the intrinsic material length as defined in Eq. (4.40). When the cell is sufficiently small, the microscale strain $\tilde{\varepsilon}_{ij}$ has a relationship with the mesoscale strain ε_{ij} and the strain gradient η_{ijk} as:

$$\tilde{\varepsilon}_{ij} = \varepsilon_{ij} + \frac{1}{2}(\eta_{kij} + \eta_{kji})x_k \tag{4.47}$$

where x_k represents the local coordinates with origin at the center of the cell.

Substituting the kinematic assumption

$$\delta\tilde{\varepsilon}_{ij} = \delta\varepsilon_{ij} + \frac{1}{2}(\delta\eta_{kij} + \delta\eta_{kji})x_k \tag{4.48}$$

into Eq. (4.41), σ'_{ij} and τ'_{ijk} can be expressed by the following equations as:

$$\sigma'_{ij} = \frac{1}{V_{\text{cell}}}\int_{V_{\text{cell}}} \tilde{\sigma}'_{ij}dV \tag{4.49}$$

$$\tau'_{ijk} = \frac{1}{V_{\text{cell}}}\text{Dev}\left[\frac{1}{2}\int_{V_{\text{cell}}} \tilde{\sigma}'_{jk}x_i + \tilde{\sigma}'_{ik}x_j dV\right] \tag{4.50}$$

where Dev[...] represents the deviatoric part of [...].

Because x_k is located at the center of the cell, the following rules can be used to do the integral for the above equations:

$$\frac{1}{V_{\text{cell}}}\int_{V_{\text{cell}}} dV = 1, \quad \int_{V_{\text{cell}}} x_k dV = 0, \quad \frac{1}{V_{\text{cell}}}\int_{V_{\text{cell}}} x_k x_m dV = \frac{1}{12}l_\varepsilon^2\delta_{km} \tag{4.51}$$

Then the constitutive equations for the mechanism-based strain gradient plasticity can be expressed as:

$$\sigma'_{ij} = \frac{2\varepsilon_{ij}}{3\varepsilon}\sigma \tag{4.52}$$

$$\tau'_{ijk} = l_\varepsilon^2\left[\frac{\sigma}{\varepsilon}(\Lambda_{ijk} - \Pi_{ijk}) + \frac{\sigma_Y^2 f(\varepsilon)f'(\varepsilon)}{\sigma}\Pi_{ijk}\right] \tag{4.53}$$

where the flow stress is given in the terms of the effective strain ε and effective strain gradient as:

$$\sigma = \sigma_Y\sqrt{f^2(\varepsilon) + l\eta} \tag{4.54}$$

Λ_{ijk} and Π_{ijk} are given by:

$$\Lambda_{ijk} = \frac{1}{72}\left[2\eta_{ijk} + \eta_{kji} + \eta_{kij} - \frac{1}{4}(\delta_{ik}\eta_{ppj} + \delta_{jk}\eta_{ppi})\right] \tag{4.55}$$

$$\Pi_{ijk} = \frac{1}{54}\frac{\varepsilon_{mn}}{\varepsilon^2}\left[\varepsilon_{ik}\eta_{jmn} + \varepsilon_{jk}\eta_{imn} - \frac{1}{4}(\delta_{ik}\varepsilon_{jp} + \delta_{jk}\varepsilon_{ip})\eta_{pmn}\right] \tag{4.56}$$

The length l_ε in Eq. (4.53) is the mesoscale cell size and controls the accuracy with which the strain gradient is calculated in each cell. The value of l_ε should be sufficiently small to ensure the strain gradient in the cell is accurate. However, this cell also needs to be large enough to contain a sufficient number of dislocations to ensure the accuracy of the flow stress. Gao et al. [18] defined the cell size by considering these two conflicting requirements as:

$$l_\varepsilon = \beta L_{\text{yield}} = \beta\frac{\mu b}{\sigma_Y} \tag{4.57}$$

where L_{yield} is the mean spacing between SSDs at plastic yielding and β is a constant coefficient to be determined from experiments. In order to ensure that there are multiple dislocations within the mesoscale cell, the value of β should be greater than one. For typical metallic materials with the material properties of $\mu/\sigma_Y = 100$, $b = 0.1$ nm, and $L_{\text{yield}} = 10$ nm, the cell length l_ε is in the range of 10–100 nm, and the coefficient β is taken to be 1–10.

4.5 CONVENTIONAL THEORY OF MECHANISM-BASED STRAIN GRADIENT PLASTICITY

Huang et al. [20] proposed a conventional theory of mechanism-based strain gradient plasticity based on the Taylor dislocation model, but without involving the higher-order stress and therefore preserves the structure of conventional plasticity theories. In the constitutive law for the conventional plasticity theory, the strain rate $\dot{\varepsilon}_{ij}$ is decomposed to the elastic strain rate $\dot{\varepsilon}_{ij}^e$ and plastic strain rate $\dot{\varepsilon}_{ij}^p$ as:

$$\dot{\varepsilon}_{ij}^e = \frac{1}{2\mu}\dot{\sigma}'_{ij} + \frac{\dot{\sigma}_{kk}}{9K}\delta_{ij} \tag{4.58}$$

$$\dot{\varepsilon}_{ij}^p = \frac{3\dot{\varepsilon}^p}{2\sigma_e}\sigma'_{ij} \tag{4.59}$$

where μ and K are the shear and bulk moduli, respectively, $\dot{\sigma}'_{ij} = \dot{\sigma}_{ij} - \frac{1}{3}\dot{\sigma}_{kk}\delta_{ij}$ is the deviatoric stress rate, $\sigma_e = \sqrt{\frac{3}{2}\sigma'_{ij}\sigma'_{ij}}$ is the von Mises effective stress, and $\dot{\varepsilon}^p = \sqrt{\frac{2}{3}\dot{\varepsilon}_{ij}^p\dot{\varepsilon}_{ij}^p}$ is the equivalent plastic strain rate.

The plastic strain rate $\dot{\varepsilon}^p$ can be determined by using the power-law viscoplastic model as:

$$\dot{\varepsilon}^p = \dot{\varepsilon}\left(\frac{\sigma_e}{\sigma_Y f(\varepsilon^p)}\right)^m \tag{4.60}$$

where $\dot{\varepsilon} = \sqrt{\frac{2}{3}\dot{\varepsilon}'_{ij}\dot{\varepsilon}'_{ij}}$ with $\dot{\varepsilon}'_{ij} = \dot{\varepsilon}_{ij} - \frac{1}{3}\dot{\varepsilon}_{kk}\delta_{ij}$ being the deviatoric strain rate, and m is the rate-sensitivity exponent which usually takes a large value (≥ 20).

The equivalent plastic strain rate $\dot{\varepsilon}^p$ in Eq. (4.60) does not account for the strain gradient effect. Huang et al. [20] introduced the strain gradient effect by replacing the uniaxial flow stress $\sigma_Y f(\varepsilon^p)$ in Eq. (4.60) by the flow stress σ_{flow} based on the Taylor dislocation model as:

$$\dot{\varepsilon}^p = \dot{\varepsilon}\left(\frac{\sigma_e}{\sigma_{\text{flow}}}\right)^m \tag{4.61}$$

In the Taylor [21,22] dislocation model, the shear flow stress τ is expressed as:

$$\tau = \alpha\mu b\sqrt{\rho} \tag{4.62}$$

where μ is the shear modulus, b is the magnitude of the Burgers vector, α is an empirical coefficient around 0.3, and ρ is the dislocation density.

Consider that the dislocation density is composed of ρ_S and ρ_G, and ρ_G is related to the effective plastic strain gradient η^p which can be obtained by:

$$\rho_G = \bar{r}\frac{\eta^p}{b} \tag{4.63}$$

where $\bar{r}$ is the Nye-factor and is around 1.90 for FCC polycrystals. Then the tensile flow stress σ_{flow} which is related to the shear flow stress τ can be obtained by:

$$\sigma_{\text{flow}} = M\tau = M\alpha\mu b\sqrt{\rho} = M\alpha\mu b\sqrt{\rho_S + \bar{r}\frac{\eta^p}{b}} \tag{4.64}$$

where M is Taylor factor with value of 3.06 for FCC polycrystals.

As ρ_S can be determined by:

$$\rho_S = \left[\frac{\sigma_Y f(\varepsilon^p)}{M\alpha\mu b}\right]^2 \tag{4.65}$$

Then Eq. (4.64) can be rewritten as:

$$\sigma_{\text{flow}} = \sqrt{[\sigma_Y f(\varepsilon^p)]^2 + M^2\bar{r}\alpha^2\mu^2 b\eta^p} = \sigma_Y\sqrt{f^2(\varepsilon^p) + l\eta^p} \tag{4.66}$$

where l is the intrinsic material length in strain gradient plasticity. Because $M = 3.06$ and $\bar{r} = 1.90$, therefore:

$$l = M^2\bar{r}\alpha^2\left(\frac{\mu}{\sigma_Y}\right)^2 b = 18\alpha^2\left(\frac{\mu}{\sigma_Y}\right)^2 b \tag{4.67}$$

For typical metallic materials, the Burgers vector b is on the order of one-tenth of a nanometer, μ/σ_Y is on the order of 10^2, and α is around 0.3. Therefore, l will be on the order of microns.

Substituting Eq. (4.66) into Eq. (4.61), the following expression can be achieved:

$$\dot{\varepsilon}^p = \dot{\varepsilon}\left(\frac{\sigma_e}{\sigma_Y\sqrt{f^2(\varepsilon^p) + l\eta^p}}\right)^m \tag{4.68}$$

Substituting Eq. (4.68) into Eqs. (4.58) and (4.59), the strain rate can be expressed as:

$$\dot{\varepsilon}^p = \frac{1}{2\mu}\dot{\sigma}'_{ij} + \frac{\dot{\sigma}_{kk}}{9K}\delta_{ij} + \frac{3\dot{\varepsilon}}{2\sigma_e}\left(\frac{\sigma_e}{\sigma_Y\sqrt{f^2(\varepsilon^p) + l\eta^p}}\right)^m \sigma'_{ij} \tag{4.69}$$

Similarly, the stress rate $\dot{\sigma}_{ij}$ can be expressed as:

$$\dot{\sigma}_{ij} = K\dot{\varepsilon}_{kk}\delta_{ij} + 2\mu\left\{\dot{\varepsilon}'_{ij} - \frac{3\dot{\varepsilon}}{2\sigma_e}\left[\frac{\sigma_e}{\sigma_Y\sqrt{f^2(\varepsilon^p)+l\eta^p}}\right]^m \sigma'_{ij}\right\} \tag{4.70}$$

The effective plastic strain gradient η^p in Eqs. (4.69) and (4.70) is defined as:

$$\eta^p = \sqrt{\frac{1}{4}\eta^p_{ijk}\eta^p_{ijk}} \tag{4.71}$$

where

$$\eta^p_{ijk} = \varepsilon^p_{ik,j} + \varepsilon^p_{jk,i} - \varepsilon^p_{ij,k} \tag{4.72}$$

This theory is simple in structure and is easy for applications in solving a lot of problems in microscale. Huang et al. [20] showed that the conventional theory of mechanism-based strain gradient plasticity could give the same results with the higher-order mechanism-based strain gradient plasticity theory except within a thin boundary layer of the solid. As this theory is simple in structure and easy for applications, it has been applied by a number of researchers in the study of size effects in microscale.

REFERENCES

[1] M.F. Ashby, The deformation of plastically non-homogeneous materials, Philos. Mag. 21 (1970) 399–424.

[2] J.F. Nye, Some geometrical relations in dislocated crystals, Acta Metall. 1 (1953) 153–162.

[3] R.A. Toupin, Elastic materials with couple-stresses, Arch. Ration. Mech. Anal. 11 (1962) 385–414.

[4] R.D. Mindlin, H.F. Tiersten, Effects of couple-stresses in linear elasticity, Arch. Ration. Mech. Anal. 11 (1962) 415–448.

[5] W.T. Koiter, Couple-stresses in the theory of elasticity, I and II, Proc. K. Ned. Acad. Wet. Ser. B: Phys. Sci. 67 (1964) 17–44.

[6] R.D. Mindlin, Micro-structure in linear elasticity, Arch. Ration. Mech. Anal. 16 (1964) 51–78.

[7] R.D. Mindlin, Second gradient of strain and surface-tension in linear elasticity, Int. J. Solids Struct. 1 (1965) 417–438.

[8] R.D. Mindlin, N.N. Eshel, On first strain-gradient theories in linear elasticity, Int. J. Solids Struct. 4 (1968) 109–124.

[9] N.A. Fleck, G.M. Muller, M.F. Ashby, J.W. Hutchinson, Strain gradient plasticity: theory and experiment, Acta Metall. Mater. 42 (1994) 475–487.

[10] N.A. Fleck, J.W. Hutchinson, Strain gradient plasticity, Adv. Appl. Mech. 33 (1997) 295–361.

[11] N.A. Fleck, J.W. Hutchinson, A phenomenological theory for strain gradient effects in plasticity, J. Mech. Phys. Solids 41 (1993) 1825–1857.

[12] Y. Wei, A new finite element method for strain gradient theories and applications to fracture analyses, Eur. J. Mech. A-Solids 25 (2006) 897–913.

[13] Y. Huang, L. Zhang, T.F. Gao, K.-C. Hwang, Mixed model near-tip fields for cracks in materials with strain-gradient effects, J. Mech. Phys. Solids 45 (1997) 439–465.

[14] Z.C. Xia, J.W. Hutchinson, Crack tip fields in strain gradient plasticity, J. Mech. Phys. Solids 44 (1996) 1621–1648.

[15] B.N. Legarth, Plasticity dependents damage evolution in composites with strain-gradient effects, Int. J. Solids Struct. 63 (2015) 1–10.

[16] M.E. Gurtin, L. Anand, Thermodynamics applied to gradient theories involving the accumulated plastic strain: the theories of Aifantis and Fleck and Hutchinson and their generalization, J. Mech. Phys. Solids 57 (2009) 405–421.

[17] J.W. Hutchinson, Plasticity at the micron scale, Int. J. Solids Struct. 37 (2000) 225–238.

[18] H. Gao, Y. Huang, W.D. Nix, J.W. Hutchinson, Mechanism-based strain gradient plasticity – I. Theory, J. Mech. Phys. Solids 47 (1999) 1239–1263.

[19] Y. Huang, H. Gao, W.D. Nix, J.W. Hutchinson, Mechanism-based strain gradient plasticity – II. Analysis, J. Mech. Phys. Solids 48 (2000) 99–128.

[20] Y. Huang, S. Qu, K.C. Hwang, M. Li, H. Gao, A conventional theory of mechanism-based strain gradient plasticity, Int. J. Plast. 20 (2004) 753–782.

[21] G.I. Taylor, The mechanism of plastic deformation of crystals. Part I. Theoretical, Proc. R. Soc. Lond. A 145 (1934) 362–387.

[22] G.I. Taylor, Plastic strain in metals, J. Inst. Met. 62 (1938) 307–324.

Chapter 5

Crystal Plasticity Theory

5.1 INTRODUCTION

In conventional metal forming, the size of a workpiece is relatively very large compared with the grain size of the metal, and material deformation is usually assumed to be isotropic. However, such an assumption is inappropriate for the deformation behavior of micro devices, which is dominated by local deformations. In microforming, grain size can be similar to that of the part being formed, and as a result any region within a micro part may contain only a single-digit number of grains, which has a great difference with those in macroscale. Under such background, crystal plasticity (CP) theory, which assumes that homogenized crystallographic slip is the plastic deformation mechanism for crystalline materials, has attracted significant attention because of its ability to relate the plastic deformation of micro parts to their microstructures. Crystal plastic deformation behavior has been extensively studied from 1930s to 1960s, with achievements being related to the features such as geometry and crystallography of plastic deformation. Based on the theory of plastic deformation and strength, the concept of dislocation has been presented, and from which current dislocation theories have been developed. On the other hand, these modified dislocation theories have been promoting the development of CP theory. In essence, compared with the constitutive relationship used several decades ago, the present constitutive relationship has been improved significantly. The research work of CP can be dated back to the pioneering work of Sachs [1], Taylor [2], Bishop–Hill [3,4], Kröner [5], Orowan [6], Schmid [7], and Polanyi [8,9]. These forerunners were conscious of the tight relationship between plastic deformation and the grain structure. Their work reminded latecomers that research into crystal plastic deformation must start with crystal structure, and it should not be studied at the macroscopic level. The large strain inherent in the crystal is often accompanied by the phenomenon of strain localizing and plastic damage where the material experiencing this large strain has strongly inhomogeneous stress and strain fields. These kind of problems could not be solved analytically, but they could be handled by simulation with the finite element method (FEM).

The rapid development of computer techniques has provided significant opportunities for the use of CPFEM in microscale. CPFEM refers to a

Microforming Technology. DOI: http://dx.doi.org/10.1016/B978-0-12-811212-0.00005-4

method which directly uses the CP constitutive equations when calculating the FEM. Over the past decades, CPFEM has become a hot topic in the research field of mechanics and materials. The fundamentals of CPFEM were first brought forward by Hill and Rice [10]. Their theory is based on the Schmid rule of the crystal plastic deformation process. In 1982, Pierce et al. [11] established a simulation method and successfully used it to simulate the formation of a shear band in soft grain metal. In 1985, Asaro and Needleman [12] developed a rate dependent constitutive equation of CP and related numerical method, but since then CPFEM has become more popular in the field of materials and mechanics, and has been widely regarded as a valid tool for investigating the plastic deformation behavior of single crystals. The general method for using CPFEM is to establish a relationship between the micro polycrystal and the material, or its continuous points. At first, each grain is expressed by the constitutive equation of single-grain parameters, but then the whole deformation process is averaged and it is assumed that the single grain micro response relates to the polycrystal averaging responses. In this chapter, two currently popular CP theories including rate dependent and rate independent theories which have extensive applications in microforming will be presented. Due to the complexity of the classical crystal constitutive equations, a method for simplifying the rate dependent CP theory will be introduced in order to reduce calculation. The use of CP theory in combination with FEM in commercial software ABAQUS will be also introduced to illustrate the application method of CPFEM in simulation.

5.2 CRYSTAL PLASTICITY THEORY

5.2.1 Geometrics and Kinematics of Crystal Plastic Deformation

Mechanisms such as plastic slip and grain deformation are involved during metal forming process, but the grains can slip by becoming dislocated or deformed elastically by grain lattice rotating. The geometrics and kinematics of crystal plastic deformation were developed by Hill and Rice [10], and Asaro et al. [11,12] also described this.

During the process of single-grain deformation, the gross deformation gradient can be expressed as [5,13,14]:

$$\boldsymbol{F} = \frac{\partial x}{\partial X} = \boldsymbol{F}^{*} \cdot \boldsymbol{F}^{p} \tag{5.1}$$

where $\boldsymbol{F}$ is the gross deformation gradient, x is the position of the material point in the present configuration, X is the position of material point in the initial configuration, $\boldsymbol{F}^{*}$ is the gross deformation gradient combined by the elastic deformation (lattice distortion) and rigid rotation, and $\boldsymbol{F}^{p}$ is the displacement gradient caused by the slip. The whole deformation process of a single grain can be divided into two processes: first, the undeformed lattices move with a gradient of F^p and then the material and lattice deform and result in the gross deforming gradient F^*. After elastic deformation is unloaded, F^p is the residual deformation.

When a grain deforms and rotates, so too do the grain lattice vectors. Let $s^{(\alpha)}$ be the slip vector in the αth slip direction, then after deformation the following expression can be obtained:

$$s^{*(\alpha)} = F^* \cdot s^{(\alpha)} \tag{5.2}$$

Other vectors in the slip plane have similar transformations. After deformation, the undeformed normal vector $m^{(\alpha)}$ in the slip plane is as follows:

$$m^{*(\alpha)} = m^{(\alpha)} \cdot F^{*-1} \tag{5.3}$$

In the reference configuration, $s^{(\alpha)}$ and $m^{(\alpha)}$ are the unit vectors in the αth slip system. In general, $s^{*(\alpha)}$ and $m^{*(\alpha)}$ are not the unit vectors, but they still have orthogonal relationship of $s^{*(\alpha)} \cdot m^{*(\alpha)} = s^{(\alpha)} \cdot m^{(\alpha)} = 0$. The current velocity gradient L can be expressed as:

$$L = \frac{\partial v}{\partial x} = \frac{\partial v}{\partial X}\frac{\partial X}{\partial x} = \dot{F} \cdot F^{-1} = L^* + L^p \tag{5.4}$$

$$L^* + L^p = \dot{F}^* \cdot F^{*-1} + F^* \cdot \dot{F}^p \cdot F^{p-1} \cdot F^{*-1} \tag{5.5}$$

where v is the velocity of the material point, L^* is the elastic part of the deformation gradient, and L^p is the plastic part of the deformation gradient.

Plastic deformation of a single crystal can be expressed as:

$$F^p = I + \gamma_{(\alpha)} s_{(\alpha)} \otimes m_{(\alpha)} \tag{5.6}$$

Therefore, the polycrystal plastic deformation can be expressed as:

$$\dot{F}^p = \sum_{\alpha=1}^{n} \dot{\gamma}_{(\alpha)} s_{(\alpha)} \otimes m_{(\alpha)} \tag{5.7}$$

$$F^{p^{-1}} = I - \sum_{\alpha=1}^{n} \gamma_{(\alpha)} s_{(\alpha)} \otimes m_{(\alpha)} \tag{5.8}$$

From Eqs. (5.7) and (5.8), it can be drawn as:

$$\dot{F}^p \cdot F^{p^{-1}} = \sum_{\alpha=1}^{n} \dot{\gamma}_{(\alpha)} s_{(\alpha)} \otimes m_{(\alpha)} \tag{5.9}$$

So L^* and L^p in Eqs. (5.4) and (5.5) can be expressed as:

$$L^* = \dot{F}^* \cdot F^{*-1} \tag{5.10}$$

$$L^p = F^* \cdot \dot{F}^p \cdot F^{p^{-1}} \cdot F^{*-1} = \sum_{\alpha=1}^{n} \dot{\gamma}_{(\alpha)} s^*_{(\alpha)} \otimes m^*_{(\alpha)} \tag{5.11}$$

L^* and L^p can be obtained separately by:

$$L^* = \dot{F}^* \cdot F^{*-1} \tag{5.11}$$

$$L^p = F^* \cdot \dot{F}^p \cdot F^{p-1} \cdot F^{*-1} \tag{5.12}$$

where $\gamma_{(\alpha)}$ is the shear strain of the αth slip system and $\dot{\gamma}_{(\alpha)}$ is the shear rate of the αth slip system. Alternatively, L can be divided into two parts according to the deformation and rotating velocities, and may be written as follows:

$$L = D + \Omega \tag{5.13}$$

where D and Ω are the symmetric tensors of deforming velocity and rotating velocity, respectively. Their components can be described by the velocity v and coordinate x as:

$$D_{ij} = 1/2(\partial v_i/\partial x_j + \partial v_j/\partial x_i) \tag{5.14}$$

$$\Omega_{ij} = 1/2(\partial v_i/\partial x_j - \partial v_j/\partial x_i) \tag{5.15}$$

D and Ω can be divided into elastic part (D^*, Ω^*) and plastic part (D^p, Ω^p) as:

$$D = D^* + D^p, \quad \Omega = \Omega^* + \Omega^p \tag{5.16}$$

From Eqs. (5.4) and (5.5), it can be obtained as:

$$D^p + \Omega^p = F^* \cdot \dot{F}^p \cdot F^{p-1} \cdot F^{*-1} \tag{5.17}$$

Because the plastic deformation is caused by grain slip, Eq. (5.17) can be rewritten as:

$$D^p + \Omega^p = \sum_{\alpha=1}^{n} \dot{\gamma}^{(\alpha)} s^{*(\alpha)} m^{*(\alpha)} \tag{5.18}$$

Eq. (5.18) sums all the activated slip systems, $\dot{\gamma}^{(\alpha)}$ is the slip rate in the αth slip system, and $\dot{\gamma}^{(\alpha)}$ can be defined as:

$$\dot{F}^p F^{p^{-1}} = \sum_{\alpha=1}^{n} \dot{\gamma}^{(\alpha)} s^{(\alpha)} m^{(\alpha)} \tag{5.19}$$

For convenience, the tensors P^α and W^α of the αth slip system can be defined as:

$$P^{(\alpha)} = 1/2(s^{*(\alpha)} m^{*(\alpha)T} + m^{*(\alpha)} s^{*(\alpha)T}) \tag{5.20}$$

$$W^{(\alpha)} = 1/2(s^{*(\alpha)} m^{*(\alpha)T} - m^{*(\alpha)} s^{*(\alpha)T}) \tag{5.21}$$

Hence, the deformation and rotation tensors of Eq. (5.16) can be expressed as:

$$\boldsymbol{D}^p = \sum_{\alpha=1}^{n} \boldsymbol{P}^{(\alpha)}\dot{\gamma}^{(\alpha)} \tag{5.22}$$

$$\boldsymbol{\Omega}^p = \sum_{\alpha=1}^{n} \boldsymbol{W}^{(\alpha)}\dot{\gamma}^{(\alpha)} \tag{5.23}$$

Eqs. (5.1)–(5.23) are the basic equations of CP kinematics. These equations show the relationship between the shear rate and the rate of macroscopic deformation.

5.2.2 Rate Independent Crystal Plasticity Constitutive Equation

It is assumed that grain slip has no effect on the elastic property. According to the description of Hill and Rice [10], the elastic constitutive equation can be expressed as:

$$\overset{\nabla *}{\tau} = \tilde{\ell}{:}\boldsymbol{D}^* \tag{5.24}$$

where $\tilde{\ell}$ is the fourth-order tensor of elastic modulus and $\overset{\nabla *}{\tau}$ is the Jaumann rate of the Kirchhoff stress. The component of $\tilde{\ell}$ can be taken as the partial derivative of some free energy. The superscript * is expressed as the partial derivative for the axial stress which rotates with the lattice. $\overset{\nabla *}{\tau}$ can be defined as:

$$\overset{\nabla *}{\tau} = \dot{\tau} - \boldsymbol{\Omega}^*\cdot\tau + \tau\cdot\boldsymbol{\Omega}^* \tag{5.25}$$

where $\dot{\tau}$ is the material rate of the Kirchhoff stress which can be defined as $\rho_0/\rho\sigma$ with σ the Cauchy stress, and ρ_0 and ρ the reference and currently mass densities, respectively. In order to describe the constitutive equation of the material, the Jaumann stress rate can be defined as:

$$\overset{\nabla}{\tau} = \dot{\tau} - \boldsymbol{\Omega}\cdot\tau + \tau\cdot\boldsymbol{\Omega} \tag{5.26}$$

The difference between Eqs. (5.25) and (5.26) can be defined as:

$$\overset{\nabla *}{\tau} - \overset{\nabla}{\tau} = \sum_{\alpha=1}^{n} \beta^{(\alpha)}\dot{\gamma}^{(\alpha)} \tag{5.27}$$

where

$$\beta^{(\alpha)} = \boldsymbol{W}^{(\alpha)}\cdot\tau - \tau\cdot\boldsymbol{W}^{(\alpha)} \tag{5.28}$$

Hence, it can be obtained as:

$$\begin{aligned}\overset{\nabla}{\tau} &= \overset{\nabla^*}{\tau} - \left(\overset{\nabla^*}{\tau} - \overset{\nabla}{\tau}\right) \\ &= \tilde{\ell}:\boldsymbol{D}^* - \sum_{\alpha=1}^{n} \beta^{(\alpha)}\dot{\gamma}^{(\alpha)} \\ &= \tilde{\ell}:(\boldsymbol{D} - \boldsymbol{D}^p) - \sum_{\alpha=1}^{n} \beta^{(\alpha)}\dot{\gamma}^{(\alpha)}\end{aligned} \tag{5.29}$$

Substituting Eq. (5.22) into Eq. (5.29), then:

$$\overset{\nabla}{\tau} = \tilde{\ell}:\boldsymbol{D} - \sum_{\alpha=1}^{n} [\tilde{\ell}:\boldsymbol{P}^{(\alpha)} + \beta^{(\alpha)}]\dot{\gamma}^{(\alpha)} \tag{5.30}$$

where $\dot{\gamma}^{(\alpha)}$ in Eq. (5.30) is not determinable. It can be described by the stress and strain tensors. In order to obtain the value of $\dot{\gamma}^{(\alpha)}$, the resolving shear stress in the αth slip system can be expressed as:

$$\tau^{(\alpha)} = \boldsymbol{m}^{*(\alpha)} \cdot \tau \cdot \boldsymbol{s}^{*(\alpha)} = \boldsymbol{P}^{(\alpha)}:\tau \tag{5.31}$$

Eq. (5.31) is differentiated, then it can be expressed as:

$$\dot{\tau}^{(\alpha)} = \dot{\boldsymbol{P}}^{(\alpha)}:\tau - \boldsymbol{P}^{(\alpha)}:\dot{\tau} \tag{5.32}$$

where

$$\dot{\boldsymbol{P}} = \frac{1}{2}(\boldsymbol{s}^*\boldsymbol{m}^{*T} + \boldsymbol{m}^*\boldsymbol{s}^{*T})^* = \frac{1}{2}(\dot{\boldsymbol{s}}^*\boldsymbol{m}^{*T} + \boldsymbol{s}^*\dot{\boldsymbol{m}}^{*T} + \dot{\boldsymbol{m}}^*\boldsymbol{s}^{*T} + \boldsymbol{m}^*\dot{\boldsymbol{s}}^{*T}) \tag{5.33}$$

Firstly, $\dot{\boldsymbol{s}}^*$ and $\dot{\boldsymbol{m}}^*$ must be obtained. So Eqs. (5.2) and (5.3) can be differentiated as:

$$\begin{aligned}\dot{\boldsymbol{s}}^* &= (\boldsymbol{F}^*\boldsymbol{s}^*)^* = \dot{\boldsymbol{F}}^*\boldsymbol{s}^* \\ &= \dot{\boldsymbol{F}}^*\boldsymbol{F}^{*-1}\boldsymbol{F}^*\boldsymbol{s} \\ &= (\dot{\boldsymbol{F}}^*\boldsymbol{F}^{*-1})\boldsymbol{s}^*\end{aligned} \tag{5.34}$$

$$\dot{\boldsymbol{m}}^* = [(\boldsymbol{F}^{*-1})^T\boldsymbol{m}]^* = (\boldsymbol{F}^{*-1})^T\boldsymbol{m}$$

For

$$\begin{aligned}&(\boldsymbol{F}^*\boldsymbol{F}^{*-1})^* = 0 \\ &\Rightarrow \dot{\boldsymbol{F}}^*\boldsymbol{F}^{*-1} + \boldsymbol{F}^*\dot{\boldsymbol{F}}^{*-1} = 0 \\ &\Rightarrow \dot{\boldsymbol{F}}^{*-1} = -\boldsymbol{F}^{*-1}\dot{\boldsymbol{F}}^*\boldsymbol{F}^{*-1} \\ &\Rightarrow (\dot{\boldsymbol{F}}^{*-1})^T = -\boldsymbol{F}^{*-1}\dot{\boldsymbol{F}}^*\boldsymbol{F}^{*-1}\end{aligned}$$

So

$$\begin{aligned}\dot{\boldsymbol{m}}^* &= (\boldsymbol{F}^{*-1})^T\boldsymbol{m} \\ &= (-\boldsymbol{F}^{*-1}\dot{\boldsymbol{F}}^*\boldsymbol{F}^{*-1})\boldsymbol{m} \\ &= (\dot{\boldsymbol{F}}^*\boldsymbol{F}^{*-1})^T\boldsymbol{m}^*\end{aligned} \tag{5.35}$$

Eqs. (5.34) and (5.35) can be used in Eq. (5.33), and the result is:

$$
\begin{aligned}
\dot{\boldsymbol{P}} &= \frac{1}{2}\left[(\dot{\boldsymbol{F}}^*\boldsymbol{F}^{*-1})\boldsymbol{s}^*\boldsymbol{m}^{*T} - \boldsymbol{s}^*\boldsymbol{m}^{*T}(\dot{\boldsymbol{F}}^*\boldsymbol{F}^{*-1})\right.\\
&\quad \left. - (\dot{\boldsymbol{F}}^*\boldsymbol{F}^{*-1})^T\boldsymbol{m}^*\boldsymbol{s}^{*T} + \boldsymbol{m}^*\boldsymbol{s}^{*T}(\dot{\boldsymbol{F}}^*\boldsymbol{F}^{*-1})^T\right]\\
&= \frac{1}{2}\left[(\dot{\boldsymbol{F}}^*\boldsymbol{F}^{*-1})\boldsymbol{s}^*\boldsymbol{m}^{*T} - (\dot{\boldsymbol{F}}^*\boldsymbol{F}^{*-1})\boldsymbol{m}^*\boldsymbol{s}^{*T}\right]\\
&\quad + \frac{1}{2}\left[-\boldsymbol{m}^*\boldsymbol{s}^{*T}(\boldsymbol{F}^*\boldsymbol{F}^{*-1}) + \boldsymbol{m}^*\boldsymbol{s}^{*T}(\boldsymbol{F}^*\boldsymbol{F}^{*-1})^T\right]\\
&= \frac{1}{2}\left[(\boldsymbol{L}^*\boldsymbol{s}^*\boldsymbol{m}^{*T} - \boldsymbol{L}^{*T}\boldsymbol{m}^*\boldsymbol{s}^{*T}) + (-\boldsymbol{s}^*\boldsymbol{m}^{*T}\boldsymbol{L}^* + \boldsymbol{m}^*\boldsymbol{s}^{*T}\boldsymbol{L}^{*T})\right]\\
&= \frac{1}{4}\left[(\boldsymbol{L}^* - \boldsymbol{L}^{*T})(\boldsymbol{s}^*\boldsymbol{m}^{*T} + \boldsymbol{m}^*\boldsymbol{s}^{*T}) + (\boldsymbol{L}^* + \boldsymbol{L}^{*T})(\boldsymbol{s}^*\boldsymbol{m}^{*T} - \boldsymbol{m}^*\boldsymbol{s}^{*T})\right]\\
&\quad - \frac{1}{4}\left[(\boldsymbol{s}^*\boldsymbol{m}^{*T} + \boldsymbol{m}^*\boldsymbol{s}^{*T})(\boldsymbol{L}^* - \boldsymbol{L}^{*T}) + (\boldsymbol{s}^*\boldsymbol{m}^{*T} - \boldsymbol{m}^*\boldsymbol{s}^{*T})(\boldsymbol{L}^* + \boldsymbol{L}^{*T})\right]
\end{aligned}
\tag{5.36}
$$

$$
\Rightarrow \dot{\boldsymbol{P}}^{(\alpha)} = \boldsymbol{\Omega}^*\boldsymbol{P}^{(\alpha)} + \boldsymbol{D}^*\boldsymbol{W}^{(\alpha)} - \boldsymbol{P}^{(\alpha)}\boldsymbol{\Omega}^* - \boldsymbol{W}^{(\alpha)}\boldsymbol{D}^*
$$

Eq. (5.36) is used in Eq. (5.32), then:

$$
\begin{aligned}
\dot{\tau}^{(\alpha)} &= \dot{\boldsymbol{P}}^{(\alpha)}:\tau + \boldsymbol{P}^{(\alpha)}:\dot{\tau}\\
&= (\boldsymbol{\Omega}^*\boldsymbol{P}^{(\alpha)} + \boldsymbol{D}^*\boldsymbol{W}^{(\alpha)} - \boldsymbol{P}^{(\alpha)}\boldsymbol{\Omega}^* - \boldsymbol{W}^{(\alpha)}\boldsymbol{D}^*):\tau + \boldsymbol{P}^{(\alpha)}:\dot{\tau}
\end{aligned}
\tag{5.37}
$$

If Eq. (5.24) is used in Eq. (5.37), it can be written as:

$$
\begin{aligned}
\dot{\tau}^{(\alpha)} &= \boldsymbol{\Omega}^*\boldsymbol{P}^{(\alpha)}:\tau - \boldsymbol{P}^{(\alpha)}\boldsymbol{\Omega}^*:\tau + \boldsymbol{D}^*\boldsymbol{W}^{*(\alpha)}:\tau - \boldsymbol{W}^{*(\alpha)}\boldsymbol{D}^*:\tau\\
&\quad + \boldsymbol{P}^{(\alpha)}:\left(\overset{\nabla *}{\tau} - \boldsymbol{\Omega}^*\tau + \tau\boldsymbol{\Omega}^*\right)\\
&= \boldsymbol{D}^*\boldsymbol{W}^{(\alpha)}:\tau - \boldsymbol{W}^{(\alpha)}\boldsymbol{D}^*:\tau + \boldsymbol{P}:\overset{\nabla *}{\tau}
\end{aligned}
$$

($AB:C = A:CB^T = B:A^TC$, τ is symmetric tensor.)

$$
\begin{aligned}
&= \boldsymbol{D}^*:\boldsymbol{W}^{(\alpha)}\tau - \boldsymbol{D}^*:\tau\boldsymbol{W}^{(\alpha)} + \boldsymbol{P}^{(\alpha)}:\overset{\nabla *}{\tau}\\
&= \boldsymbol{D}^*:\beta + \boldsymbol{P}^{(\alpha)}:\tilde{\ell}:\boldsymbol{D}^*\\
&= (\boldsymbol{P}^{(\alpha)}:\tilde{\ell} + \beta):\boldsymbol{D}^*
\end{aligned}
\tag{5.38}
$$

$$
= \boldsymbol{m}^*\cdot\left(\overset{\nabla *}{\tau} - \boldsymbol{D}^*\cdot\tau + \tau\cdot\boldsymbol{D}^*\right)\cdot\boldsymbol{s}^{*T}
\tag{5.39}
$$

According to the theory of dislocation, dislocation is determined by the value of the shearing stress where, after $\tau^{(\alpha)}$ equals to the critical value τ_0, dislocation in the αth slip system is activated and the slip direction of dislocation is up to the direction of τ_0.

After grain yields, the resolving shearing stress increases as the grains harden. The increment of stress is determined by the increment of slipping strain and the coefficient of hardening. It can be expressed as:

$$\sum_{\beta=1}^{n} h_{\alpha\beta}\dot{\gamma}^{(\beta)} \tag{5.40}$$

where $h_{\alpha\beta}$ expresses the effect of slip in the βth slip system on the resolved shear stress in the αth slip system. So in the activated slip system it can be obtained as:

$$\dot{\tau}^{(\alpha)} = \sum_{\beta=1}^{n} h_{\alpha\beta}\dot{\gamma}^{(\beta)} \tag{5.41}$$

It can also be obtained from Eq. (5.38) as:

$$\begin{aligned}\tau^{(\alpha)} &= (\boldsymbol{P}^{(\alpha)}:\tilde{\ell} + \beta):\boldsymbol{D}^{*}\\ &= (\boldsymbol{P}^{(\alpha)}:\tilde{\ell} + \beta)(\boldsymbol{D} - \boldsymbol{D}^{p})\end{aligned} \tag{5.42}$$

If Eq. (5.39) is used in Eq. (5.41), it can be obtained as:

$$\dot{\tau}^{(\alpha)} = (\boldsymbol{P}^{(\alpha)}:\tilde{\ell} + \beta):\boldsymbol{D} - \sum_{\beta}(\boldsymbol{P}^{(\alpha)}:\tilde{\ell}:\boldsymbol{P}^{(\beta)} + \beta^{(\alpha)}:\boldsymbol{P}^{(\beta)})\dot{\gamma}^{(\beta)} \tag{5.43}$$

For the mobile slip system, the result is as:

$$(\boldsymbol{P}^{(\alpha)}:\tilde{\ell} + \beta^{(\alpha)}):\boldsymbol{D} = \sum_{\beta=1}^{n} \boldsymbol{N}_{\alpha\beta}\dot{\gamma}^{(\beta)} \tag{5.44}$$

The matrix $\boldsymbol{N}_{\alpha\beta}$ in Eq. (5.44) can be expressed as:

$$\boldsymbol{N}_{\alpha\beta} = h_{\alpha\beta} + \boldsymbol{P}^{(\alpha)}:\tilde{\ell}:\boldsymbol{P}^{(\beta)} + \beta:\boldsymbol{P}^{(\beta)} \tag{5.45}$$

From Eq. (5.44), if the matrix $\boldsymbol{N}_{\alpha\beta}$ is reversible, the slip strain rate $\dot{\gamma}^{(\beta)}$ can be obtained. If the hardening coefficients are mutually independent, this condition can always be met. Therefore, the slip rate can be obtained from Eq. (5.44) as:

$$\dot{\gamma}^{(\alpha)} = (\boldsymbol{N}^{-1})_{\alpha\beta}(\boldsymbol{P}^{(\beta)}:\tilde{\ell} + \beta^{(\beta)}):\boldsymbol{D} \tag{5.46}$$

If Eq. (5.46) is used in Eq. (5.32), the constitutive equation of material can be obtained as:

$$\overset{\nabla}{\tau} = \boldsymbol{C}:\boldsymbol{D} \tag{5.47}$$

where $\boldsymbol{C}$ is the rigid matrix, and can be written as:

$$\boldsymbol{C} = \tilde{\ell} - \sum_{\alpha}\sum_{\beta}(\boldsymbol{P}^{(\alpha)}:\tilde{\ell} + \beta^{(\alpha)})(\boldsymbol{N}^{-1})_{\alpha\beta}(\boldsymbol{P}^{(\beta)}:\tilde{\ell} + \beta^{(\beta)}) \tag{5.48}$$

5.2.3 Rate Dependent Crystal Plasticity Constitutive Equation

The obvious drawback in the rate independent CP constitutive equations is that it cannot make the shear rate unique and its hardening coefficient is complicated. In order to overcome these problems, Pierce et al. [15] and Asaro [16] developed a rate dependent CP constitutive theory.

This theory considers that deformation has no effect on the elastic property of grain, and therefore the Jaumann rate of the Kirchhoff stress based on the intermediate state can be expressed as:

$$\overset{\nabla^*}{\tau} = \tilde{\ell}:\boldsymbol{D}^* \tag{5.49}$$

It can be obtained from a similar deduction in Section 5.2.2:

$$\overset{\nabla}{\tau} = \tilde{\ell}:\boldsymbol{D} - \sum_{\alpha} \boldsymbol{R}^{(\alpha)}\dot{\gamma}^{(\alpha)} \tag{5.50}$$

where $\boldsymbol{R}^{(\alpha)}$ is expressed as:

$$\boldsymbol{R}^{(\alpha)} = \tilde{\ell}:\boldsymbol{P}^{(\alpha)} + \beta^{(\alpha)} \tag{5.51}$$

A feature of this rate dependent constitutive equation is that the shearing rate is determined by the current state and stress since it is independent on the stress rate and deforming path. This is the fundamental difference from the rate independent constitutive equation. There is an exponential function relationship between the slip rate and resolving shear stress which is shown in the following equation:

$$\dot{\gamma}^{(\alpha)} = \dot{a}^{(\alpha)} \left[\frac{\tau^{(\alpha)}}{g^{(\alpha)}}\right] \left[\frac{\tau^{(\alpha)}}{g^{(\alpha)}}\right]^{(1/m)-1} \tag{5.52}$$

where m is the stress sensitive parameter of the slip system. If the slip rate is always $\dot{a}$, the resolving shearing stress $\tau^{(\alpha)}$ equals $g^{(\alpha)}$. In fact, equation $\tau^{(\alpha)} = g^{(\alpha)}$ can be taken as a reference state and $\dot{a}$ is the slip rate in the reference state. $g^{(\alpha)}$ represents the ability of the grain to harden. However, a practical expression of $g^{(\alpha)}$ is probably complicated, so in order to simplify the simulation, a simplified function of $g^{(\alpha)}$ is used where it is only a function of the gross slip deformation.

$$g^{(\alpha)} = g^{(\alpha)}(\gamma) \tag{5.53}$$

where γ is the gross slip deformation which can be determined by:

$$\gamma = \sum_{\alpha} \left|\gamma^{(\alpha)}\right| \tag{5.54}$$

The hardening feature of material can be embodied by the function of $g^{(\alpha)}$. When $\gamma = 0$, the initial value of $g^{(\alpha)}$ is τ_0. The variation of $g^{(\alpha)}$ can be expressed as:

$$\dot{g}^{(\alpha)} = \sum_{\beta} h_{\alpha\beta}\left|\dot{\gamma}^{(\beta)}\right| \tag{5.55}$$

where $h_{\alpha\beta}$ is the function of γ.

In the rate dependent constitutive equation, there is no obvious yield point in metal, but if the resolving stress in the slip system is not zero, plastic shearing continues. However, because the stress sensitive parameter m is small ($m < 0.02$), the shear rate is minimal when the resolving shear stress is less than τ_0. Therefore, τ_0 can be taken as the yield resolving shear stress.

The hardening modulus matrix $h_{\alpha\beta}$ includes not only the effect of self-hardening (hardening caused by slip in the self-slip system), but also the effect of latent hardening (hardening caused by slip of the else slip systems). It can be expressed as:

$$h_{\alpha\beta} = qh + (1 - q)h\delta_{\alpha\beta} \tag{5.56}$$

where q is the ratio of self-hardening and latent hardening, and in general $1 < q < 1.4$. In order to make the $h_{\alpha\beta}$ symmetric, it can also be expressed as:

$$h_{\alpha\beta} = qh + (1 - q)h\delta_{\alpha\beta} + \frac{1}{2}(\beta^{(\alpha)}{:}\boldsymbol{P}^{(\beta)} - \beta^{(\beta)}{:}\boldsymbol{P}^{(\alpha)}) \tag{5.57}$$

The hardening of the slip system can be fitted by the hyperbolic curve, which is expressed by the following equation:

$$h(\gamma) = h_0 \sec h^2\left(\frac{h_0\gamma}{\tau_s - \tau_0}\right) \tag{5.58}$$

5.3 SIMPLIFICATION OF RATE DEPENDENT CRYSTAL PLASTICITY THEORY

The CP model is the base for CPFEM modeling. Different CP models have different applications in FE modeling. Classical crystal constitutive equations are expressed by the objective stress rate and deformation rate. However, it is complicated to use these constitutive equations in numerical simulations. So, in order to reduce the calculation process, a simplification method is introduced in the rate dependent CPFEM simulation.

5.3.1 Decomposition of the Crystal Plastic Deformation Gradient

The kinematics of crystal plastic deformation is described in Section 5.2. According to the decomposition of deformation gradient, it is very difficult to obtain a rotation matrix of the present configuration. Therefore, a polar decomposition of the elastic deformation gradient (right polar decomposition) shall be carried out to solve the rotation matrix [17]. During this process, the elastic deformation gradient $\boldsymbol{F}^*$ is polarized and decomposed to the right stretching gradient U^* and rotation gradient R^* which is expressed as [18,19]:

$$F^* = U^* R^* \tag{5.59}$$

5.3.2 Elastic Constitutive Equation

The crystal elastic constitutive equation can be expressed as:

$$T^{(1)} = \mathcal{L}:E^{(1)} \tag{5.60}$$

where $T^{(1)}$ is the second Piola–Kirchhoff stress tensor, $\mathcal{L}$ is the fourth-order tensor, and $E^{(1)}$ is the Green's elastic strain tensor. $T^{(1)}$ can be expressed as follows:

$$T^{(1)} = F^{*-1}[\det(F^*)\sigma](F^*)^{-T} \tag{5.61}$$

where σ is the Cauchy stress.

$E^{(1)}$ can be expressed as:

$$E^{(1)} = \frac{1}{2}(F^{*T} \cdot F^* - I) \tag{5.62}$$

5.3.3 Flow Rule of Plastic Deformation

Flow rule of plastic deformation gradient F^p is expressed as:

$$\dot{F}^p F^{p^{-1}} = \sum_{\alpha=1}^{n} \dot{\gamma}_{(\alpha)} s_{(\alpha)} \otimes m_{(\alpha)} \tag{5.63}$$

where $\dot{\gamma}_{(\alpha)}$ is the plastic shear rate of the αth slip system.

5.3.4 Equation of Kinematics

The relationship between shear rate $\dot{\gamma}_{(\alpha)}$ and resolving shear stress $\tau_{(\alpha)}$ of the αth slip system can be expressed by a power exponent equation [11,15,20–23]. The power exponent equation is expressed as:

$$\dot{\gamma}_{(\alpha)} = \dot{\gamma}_0 \mathrm{sign}(\tau_{(\alpha)}) \left| \frac{\tau_{(\alpha)}}{s_{(\alpha)}} \right|^{1/m} \tag{5.64}$$

where $s_{(\alpha)}$ is the slip resistance of the αth slip system and has the following relationship:

$$\tau_{(\alpha)} \approx T^{(1)} \cdot s_{(\alpha)} \otimes m_{(\alpha)} \tag{5.65}$$

5.3.5 Hardening Law

For cubic metal, the hardening equation of the slip system can be simplified as [24]:

$$\dot{s}_{(\alpha)} = \sum_{\beta=1}^{n} h_{\alpha\beta}\left|\dot{\gamma}_{(\beta)}\right| \tag{5.66}$$

The hardening matrix of the slip system α led by the slip system β is expressed as [25–27]:

$$h_{\alpha\beta} = h_0\left[q_l + (1-q_l)\delta_{\alpha\beta}\right]\left\{1-\frac{s_{(\beta)}}{s_s}\right\}^a \tag{5.67}$$

where q_l is the latent hardening parameter, $\delta_{\alpha\beta}$ is the Kronecker symbol, and h_0, a, and s_s are material parameters.

5.3.6 Models of Polycrystal Homogenization

5.3.6.1 *Taylor Averaging Procedure*

The essential assumptions in the Taylor type polycrystal model are that all the grains have an equal volume and the local deformation gradient in each grain is homogeneous and identical to the macroscopic deformation gradient at the continuum material point level. The local deformation gradient can be expressed as [20,28]:

$$\boldsymbol{F} = \overline{\boldsymbol{F}} \tag{5.68}$$

where $\boldsymbol{F}$ is the local deformation gradient in each grain and $\overline{\boldsymbol{F}}$ is the macroscopic deformation gradient at the continuum material point level.

The stress response at each macroscopic continuum material point is given by the average volume of the multitude of microscopic single-crystalline grains comprising the material point. Then, with $\sigma^{(k)}$ denoting the Cauchy stress in the kth crystal, these assumptions lead to:

$$\overline{\sigma} = \frac{1}{N}\sum_{k=1}^{N}\sigma^{(k)} \tag{5.69}$$

where $\overline{\sigma}$ is the volume averaged stress and N is the total number of grains comprising the material point. The influences of grain size, shape, and distribution are not considered in this equation.

5.3.6.2 *Finite Element Averaging Procedure*

In this approach, the FE calculations have been used to make a transition from the response of a single grain (or a region within a grain) to the response of a polycrystalline aggregate. It is assumed that each element represents one grain, and is assigned an orientation as the initial texture [20,24,28]. So each grain is modeled to allow for nonuniform deformation between and within the grains, and both the equilibrium and compatibility are satisfied in the weak FE sense.

5.4 NUMERICAL INTEGRATION OF RATE DEPENDENT CRYSTAL PLASTICITY THEORY

Currently, researchers have already brought forward many methods of explicit and implicit integrations. The above-mentioned methods mainly include the rate dependent constitutive equation and rate independent constitutive equation. In this section, the Total Lagrangian implicit integration procedure is used to integrate the CP constitutive model, and the numerical integration of the crystal elastic-plastic constitutive equations is carried out for the application of CP theory in FEM modeling.

5.4.1 Total Lagrangian Formulation

There are two choices to solve the crystal elastic-plastic constitutive equations. One is the Total Lagrange formulation and the other is the Updated Lagrange formulation. In the Total Lagrange formulation, all the initial variables ($t = 0$) will be taken as the reference configuration, but in the Updated Lagrange formulation, all the variables of time t will be taken as a reference configuration.

The advantages of using the Lagrange formulation in the CP model are as follows: (1) Optional solution can be applied to solve the grain orientation of the special time step, but not the orientations of each time step. The time for calculation will be reduced significantly. (2) Application of fully implicit integration is available. (3) Interfaces can be supplied for the application of other deformation mechanisms (such as twins) in the CP model.

5.4.2 Fully Implicit Integration Procedure

In order to reduce the consumption of time, Kalidindi [20,29] brought forward a fully implicit numerical procedure for CP constitutive equations. The assumptions are as follows:

(1) Initial configurations ($s_{(a)}$, $m_{(a)}$) are given; (2) values of ${}^{t}\boldsymbol{F}$, ${}^{t}\boldsymbol{F}^{p}$, and ${}^{t}s_{(\alpha)}$ are also given; and (3) ${}^{\tau}\boldsymbol{F}$ of time $\tau = t + \Delta t$ can be estimated.

According to the flow rule in Eq. (5.63), $\boldsymbol{F}^{p}$ of time $t + \Delta t$ can be obtained as:

$$
{}^{\tau}\boldsymbol{F}^{p} \doteq \left\{ \boldsymbol{I} + \sum_{\alpha=1}^{n} {}^{\tau}\Delta\gamma_{(\alpha)} s_{(\alpha)} \otimes \boldsymbol{m}_{(\alpha)} \right\} {}^{t}\boldsymbol{F}^{p} \tag{5.70}
$$

where ${}^{\tau}\Delta\gamma_{(\alpha)}$ is obtained as:

$$ {}^{\tau}\Delta\gamma_{(\alpha)} = \dot{\gamma}_{(\alpha)}({}^{\tau}\tau_{(\alpha)}, {}^{\tau}s_{(\alpha)})\Delta t \tag{5.71} $$

${}^{\tau}\Delta\gamma_{(\alpha)}$ can be expressed as:

$$ {}^{\tau}\Delta\gamma_{(\alpha)} = \dot{\gamma}_{(\alpha)}(\tau_{(\alpha)}({}^{\tau}\boldsymbol{T}^{(1)}), {}^{\tau}s_{(\alpha)})\Delta t \tag{5.72} $$

Simultaneously,

$$ {}^{\tau}\boldsymbol{F}^{p^{-1}} = {}^{t}\boldsymbol{F}^{p^{-1}}\left\{\boldsymbol{I} - \sum_{\alpha=1}^{n} {}^{\tau}\Delta\gamma_{(\alpha)}\boldsymbol{s}_{(\alpha)} \otimes \boldsymbol{m}_{(\alpha)}\right\} \tag{5.73} $$

At the time $t + \Delta t$, the stress can be obtained as:

$$ {}^{\tau}\boldsymbol{T}^{(1)} = \boldsymbol{\mathcal{L}}:\left[\frac{1}{2}\left\{{}^{\tau}\boldsymbol{F}^{p^{-T}}\,{}^{\tau}\boldsymbol{F}^{T\,\tau}\boldsymbol{F}\,{}^{\tau}\boldsymbol{F}^{p^{-1}} - \boldsymbol{I}\right\}\right] \tag{5.74} $$

if

$$ \boldsymbol{X} = {}^{\tau}\boldsymbol{F}^{p^{-T}}\,{}^{\tau}\boldsymbol{F}^{T\,\tau}\boldsymbol{F}\,{}^{\tau}\boldsymbol{F}^{p^{-1}} \tag{5.75} $$

If Eq. (5.73) is used in Eq. (5.75), it can be obtained as:

$$ \begin{aligned} \boldsymbol{X} = {} & {}^{t}\boldsymbol{F}^{p^{-T}}\,{}^{\tau}\boldsymbol{F}^{\mathrm{T}\,\tau}\boldsymbol{F}\,{}^{t}\boldsymbol{F}^{p^{-1}} \\ & - \sum_{\alpha=1}^{n} {}^{\tau}\Delta\gamma(s_{(\alpha)} \otimes \boldsymbol{m}_{(\alpha)})^{T\,t}\boldsymbol{F}^{p^{-T}}\,{}^{\tau}\boldsymbol{F}^{\mathrm{T}\,t}\boldsymbol{F}^{p^{-1}} \\ & - {}^{t}\boldsymbol{F}^{p^{-T}}\,{}^{\tau}\boldsymbol{F}^{\mathrm{T}\,\tau}\boldsymbol{F}\,{}^{t}\boldsymbol{F}^{p^{-1}}\sum_{\alpha=1}^{n} {}^{\tau}\Delta\gamma(s_{(\alpha)} \otimes \boldsymbol{m}_{(\alpha)}) + O(\Delta\gamma^{\alpha^2}) \end{aligned} \tag{5.76} $$

If $O(\Delta\gamma^{\alpha^2})$ is omitted, then Eq. (5.76) can be expressed as:

$$ {}^{\tau}\boldsymbol{F}^{p^{-T}}\,{}^{\tau}\boldsymbol{F}^{T\,\tau}\boldsymbol{F}\,{}^{\tau}\boldsymbol{F}^{p^{-1}} \doteq \boldsymbol{A} - \sum_{\alpha=1}^{n} {}^{\tau}\Delta\gamma\boldsymbol{B}_{(\alpha)} \tag{5.77} $$

$$ \boldsymbol{A} = {}^{t}\boldsymbol{F}^{p^{-T}}\,{}^{\tau}\boldsymbol{F}^{T\,\tau}\boldsymbol{F}\,{}^{t}\boldsymbol{F}^{p^{-1}} \tag{5.78} $$

$$ \boldsymbol{B}_{(\alpha)} = \boldsymbol{A}(s_{(\alpha)} \otimes \boldsymbol{m}_{(\alpha)}) + (s_{(\alpha)} \otimes \boldsymbol{m}_{(\alpha)})^{T}\boldsymbol{A} \tag{5.79} $$

Finally, Eq. (5.77) is used in Eq. (5.74) as:

$$ {}^{\tau}\boldsymbol{T}^{(1)} \doteq \boldsymbol{T}^{(1)tr} - \sum_{\alpha=1}^{n} \Delta\gamma[\tau_{(\alpha)}({}^{\tau}\boldsymbol{T}^{(1)}, {}^{\tau}s_{(\alpha)})]\boldsymbol{C}_{(\alpha)} \tag{5.80} $$

Among the equations,

$$ \boldsymbol{T}^{(1)tr} = \boldsymbol{\mathcal{L}}:[(1/2)\{\boldsymbol{A} - \boldsymbol{I}\}] \tag{5.81} $$

$$ \boldsymbol{C}_{(\alpha)} = \boldsymbol{\mathcal{L}}:[(1/2)\boldsymbol{B}_{(\alpha)}] \tag{5.82} $$

where $\boldsymbol{T}^{(1)tr}$, $\boldsymbol{A}$, $\boldsymbol{B}_{(\alpha)}$, and $\boldsymbol{C}_{(\alpha)}$ can be obtained by the given tensors, and $\Delta\gamma_{(\alpha)}$ is the function of unknown tensors ${}^{\tau}\boldsymbol{T}^{(1)}$ and ${}^{\tau}s_{(\alpha)}$.

Further, fully implicit integration is used in the equation of slip resistance $s_{(\alpha)}$ as:

$$ {}^{\tau}s_{(\alpha)} = {}^{t}s_{(\alpha)} + \sum_{\beta=1}^{n} h_{\alpha\beta}({}^{\tau}s_{(\beta)})\left|\Delta\gamma_{(\beta)}(\tau_{(\alpha)}({}^{\tau}\boldsymbol{T}^{(1)}, {}^{\tau}s_{(\beta)}))\right| \tag{5.83} $$

Eqs. (5.80) and (5.83) are the nonlinear equations of ${}^{\tau}\boldsymbol{T}^{(1)}$ and ${}^{\tau}s_{(\alpha)}$. Eq. (5.83) can be solved by two step iterations. In the first step iteration, ${}^{\tau}s_{(\alpha)}$ can be estimated and ${}^{\tau}\boldsymbol{T}^{(1)}$ will be solved by Eq. (5.80). The Newton iterative method has been used in the calculation. The equation is shown as:

$$ {}^{\tau}\boldsymbol{T}^{(1)}_{n+1} = {}^{\tau}\boldsymbol{T}^{(1)}_{n} - \mathcal{J}_n^{-1}[\boldsymbol{G}_n] \tag{5.84} $$

where

$$ \mathcal{J}_n = \boldsymbol{I} + \sum_{\alpha=1}^{n}\left\{\frac{\partial}{\partial\tau_{(\alpha)}}\Delta\gamma_{(\alpha)}(\tau_{(\alpha)}({}^{\tau}\boldsymbol{T}^{(1)}_{n}, {}^{\tau}s_{(\alpha)k}))\right\}\boldsymbol{C}_{(\alpha)} \otimes (\boldsymbol{s}_{(\alpha)} \otimes \boldsymbol{m}_{(\alpha)}) \tag{5.85} $$

where tensors with subscripts of n and $n+1$ refer to the estimated tensors after n and $n+1$ step iterations, respectively, and tensor $s_{(\alpha)k}$ means the value of ${}^{\tau}s_{(\alpha)}$ after k times updating.

The increment of iteration can be expressed as:

$$ \Delta\boldsymbol{T}^{(1)} = {}^{\tau}\boldsymbol{T}^{(1)}_{n+1} - {}^{\tau}\boldsymbol{T}^{(1)}_{n} \tag{5.86} $$

if

$$ \left|\Delta\boldsymbol{T}^{(1)}_{ij}\right| < \eta s_0 \tag{5.87} $$

The first step iteration will be finished, otherwise:

$$ {}^{\tau}\boldsymbol{T}^{(1)}_{n+1} = {}^{\tau}\boldsymbol{T}^{(1)}_{n} + \eta s_0 \mathrm{sign}\{\Delta\boldsymbol{T}^{(1)}\} \tag{5.88} $$

where η is a constant.

After convergence of ${}^{\tau}\boldsymbol{T}^{(1)}$, the second step iteration of ${}^{\tau}s_{(\alpha)}$ will start as:

$$ {}^{\tau}s_{(\alpha)k+1} = {}^{t}s_{(\alpha)} + \sum_{\beta=1}^{n} h_{\alpha\beta}({}^{\tau}s_{(\beta)k})\left|\Delta\gamma_{(\beta)}(\tau_{(\alpha)}({}^{\tau}\boldsymbol{T}^{(1)}_{n+1}, {}^{\tau}s_{(\beta)k}))\right| \tag{5.89} $$

The initial values of ${}^{\tau}\boldsymbol{T}^{(1)}$ and ${}^{\tau}s_{(\alpha)}$ are, respectively, ${}^{t}\boldsymbol{T}^{(1)}$ and ${}^{t}s_{(\alpha)}$ at time t. Criteria of convergence is as follows:

$$ \mathrm{Max}\{{}^{\tau}\boldsymbol{T}^{(1)}\} < 10^{-4}, \quad \mathrm{Max}\{{}^{\tau}s_{(\alpha)}\} < 10^{-3} \tag{5.90} $$

where ${}^{\tau}\Delta\gamma_{(\alpha)}$ can be obtained by a combination of Eqs. (5.64), (5.66), and (5.72). Then ${}^{\tau}\boldsymbol{F}^{p}$and ${}^{\tau}\boldsymbol{F}^{p^{-1}}$ can be obtained by Eq. (5.70), ${}^{\tau}\boldsymbol{F}^{*}$ can be obtained from Eq. (5.1), ${}^{\tau}\boldsymbol{E}^{(1)}$ can be obtained by Eq. (5.62), and ${}^{\tau}\sigma$ can be updated by Eq. (5.61).

5.5 CALCULATION OF GRAIN ORIENTATION

If the initial grain orientation can be expressed by a matrix $\boldsymbol{g}$, the grain orientation in the specimen coordinate can be expressed as [30–33]:

$$\boldsymbol{b}^{g} = \boldsymbol{g}\boldsymbol{b}^{c} \tag{5.91}$$

where $\boldsymbol{b}^{c}$ is the grain orientation in grain coordinates and $\boldsymbol{b}^{g}$ is the same vector of specimen coordinates. At time τ, the vector of the specimen can be expressed as:

$${}^{\tau}\boldsymbol{b}^{g} = \boldsymbol{R}^{*}\boldsymbol{b}^{g} = \boldsymbol{R}^{*}\boldsymbol{g}\boldsymbol{b}^{c} \tag{5.92}$$

Therefore, the grain orientation matrix is $\boldsymbol{R}^{*}\boldsymbol{g}$, by the new grain orientation matrix, and three Euler Angles can be obtained.

5.6 CRYSTAL PLASTICITY FINITE ELEMENT METHOD IN ABAQUS

ABAQUS is software used for engineering modeling which is widely applied as advanced general FE software. It can solve problems, ranging from simplified linear analysis to complicated nonlinear simulations. ABAQUS provides a lot of subroutines which allow users to define their own model with special functions. The user subroutine User-defined Material Mechanical Behaviour (UMAT) is an interface for the FORTRAN program by which a user can define a material model that is different from the ABAQUS Program library [34,35]. ABAQUS does not provide a CP constitutive model, so a CP constitutive model can be established by programming the ABAQUS/Standard UMAT.

5.6.1 UMAT of ABAQUS

UMAT exchanges the data with ABAQUS by an interface with the main ABAQUS solving program. In the input file, "USER MATERIAL" expresses the property of defined material. Owing to the data exchange and some shared variables, UMAT has a certain rule that is shown as:

```
SUBROUTINE UMAT (STRESS, STATEV, DDSDDE, SSE, SPD, SCD,
1 RPL, DDSDDT, DRPLDE, DRPLDT,
2 TRAN, DSTRAN, TIME, DTIME, TEMP, DTEMP, PREDEF, DPRED, CMNAME,
3 NDI, NSHR, NTENS, NSTATV, PROPS, NPROPS, COORDS, DROT, PNEWDT,
4 CELENT, DFGRDO, DFGRD1, NOEL, NPT, LAYER, KSPT, KSTEP, KINC)
C
INCLUDE'ABA_PARAM.INC'
```

```
C
CHARACTER*80 CMNAME
DIMENSION STRESS (NTENS), STATEV (NSTATV),
1 DDSDDE (NTENS, NTENS), DDSDDT (NTENS), DRPLDE (NTENS),
2 STRAN (NTENS), DSTRAN (NTENS), TIME (2), PREDEF (1), DPRED (1),
3 PROPS (NPROPS), COORDS (3), DROT (3, 3), DFGRDO (3, 3), DFGRD1 (3,
3)
user coding to define DDSDDE, STRESS, STATEV, SSE, SPD, SCD
and, if necessary, RPL, DDSDDT, DRPLDE, DRPLDT, PNEWDT
RETURN
END
```

This interface has two functions, one is to make sure the stress state of the material is at the current increment, and calculate all kinds of variables of the material state. The other is to calculate the Jacobian matrix of current material configuration. The Jacobian matrix is the matrix of all first-order partial derivatives of a vector-valued function. Let $\boldsymbol{f}:\mathbb{R}^m \rightarrow \mathbb{R}^n$ be a differentiable function which takes as input the vector $x \in \mathbb{R}^m$ and produces as output the vector $\boldsymbol{f}(\mathbf{x}) \in \mathbb{R}^n$, then the Jacobian matrix $\boldsymbol{J}$ of $\boldsymbol{f}$ is a $n \times m$ matrix, which is defined and arranged as follows:

$$\boldsymbol{J} = \frac{\mathrm{d}f}{\mathrm{d}\mathbf{x}} = \left[\frac{\partial f}{\partial x_1} \cdots \frac{\partial f}{\partial x_m}\right] = \begin{bmatrix} \frac{\partial f_1}{\partial x_1} & \frac{\partial f_1}{\partial x_2} & \cdots & \frac{\partial f_1}{\partial x_m} \\ \frac{\partial f_2}{\partial x_1} & \frac{\partial f_2}{\partial x_2} & \cdots & \frac{\partial f_2}{\partial x_m} \\ \vdots & \vdots & \ddots & \vdots \\ \frac{\partial f_n}{\partial x_1} & \frac{\partial f_n}{\partial x_2} & \cdots & \frac{\partial f_n}{\partial x_m} \end{bmatrix} \tag{5.93}$$

If $n=1$, $\boldsymbol{f}$ is a scalar field and the Jacobian matrix is reduced to a row vector of partial derivatives of $\boldsymbol{f}$, i.e., the gradient of $\boldsymbol{f}$. If $n=m$, then $\boldsymbol{f}$ is a function from $\mathbb{R}^m$ to itself and the Jacobian matrix is a square matrix, and in this case its determinant is the Jacobian determinant of $\boldsymbol{f}$.

The Jacobian determinant at a given point gives important information about the behavior of $\boldsymbol{f}$ near that point. If the Jacobian determinant at a point p is non zero, the continuously differentiable function $\boldsymbol{f}$ is invertible near $p \in \mathbb{R}^m$. If the Jacobian determinant at point p is positive, then $\boldsymbol{f}$ preserves orientation near p; and if it is negative, $\boldsymbol{f}$ reverses orientation.

The implementation of UMAT in ABAQUS requires the computation of the Jacobian matrix $\boldsymbol{J}$ defined as:

$$\boldsymbol{J} = \frac{\partial \Delta \sigma}{\partial \Delta \varepsilon} \tag{5.94}$$

where $\Delta\sigma$ and $\Delta\varepsilon$ are the increments in the stress and strain tensors under a given time increment, respectively. The Jacobian matrix plays an important role in the rate of convergence of the solution to the global equilibrium equations, but has no effect on the accuracy of the solution.

UMAT has no special requirements for the material constitutive model so it is quite suitable for theoretical study. The stability and rate of convergence depends on the material constitutive model. In the UMAT, there are many matrices such as DDSDDE, DDSDDT, and DRPLDE. which express the stress and strain state, etc. Different tensors have elements with different freedoms so when the UMAT code is compiled, the sorts of element should be considered. Some explanations of variables are given as follows:

DDSDDE (NTENS, NTENS) is the Jacobian matrix ($\partial\Delta\sigma/\partial\Delta\varepsilon$) of constitutive equations, with $\Delta\sigma$ the stress increment at the current step incremental step, and $\Delta\varepsilon$ the strain increment. DDSDDE (*I*, *J*) expresses the variance ratio of the *I*th stress component interacted by the *J*th strain component. It is a symmetric matrix.

STRESS (NTENS) is an array of the stress tensor which includes NDI normal stress components and NSHR shear stress components. At the beginning of incremental step calculation, the matrix of stress tensors is transferred to the UMAT by an interface between UMAT and the main program. When the calculation of incremental steps has finished, the matrix of stress tensors will be updated by UMAT. In the finite-strain problem of rigid rotation, the stress tensors will be rotated before calling the incremental step. In the UMAT, the stress tensors are the Cauchy stresses (true stress).

STATEV (NSTATEV) is used to store the array of state variables related to the solution. They are also transferred to UMAT at the beginning of incremental step calculation. The data can be updated in the subroutine USDFLD or UEXPAN, and then transferred to UMAT. The dimension of state variables can be defined by the "*DEPVAR."

PROPS (NPROPS) are the array of material constants. NPROPS is the number of materials. It is the same as the definition of "CONSTRANTS" in the subroutine "*USER MATERIAL." The data of PROPS (NPROPS) is the same as the data of the subroutine "*USER MATERIAL." At each incremental step, SSE, SPD, SCD can separately define the elastic strain energy, the plastic dissipative energy, and the creep dissipative energy.

PNEWDT is the ratio of the suggested time increment and current time increment. The step increment can be controlled by this variable. If the value of PNEWDT is less than 1, the current iteration will stop and select a smaller step size for the solution. The new step is the value of PNEWDT $\times$ DTIME. If the value of PNEWDT is larger than 1, the iteration of the current incremental step will converge. The size of the next loading step will increase, and the new step size is the value of PNEWDT $\times$ DTIME. The value of PNEWDT will be the minimal value of all the iterations. This is for the automatic step mode. If the mode is not activated there will be no response when the value of PNEWDT is larger than 1, and the calculation process will stop when the value of PNEWDT is smaller than 1.

STRAN (NTENS) is the array of total strain tensors at the beginning of the iteration of the incremental step. The rigid rotation has been taken into consideration.

DSTRAN (NTENS) is the array of strain increment. It only includes the mechanical strain increment, but no thermal strain increment.

TIME (2) is the total time when the incremental step iteration starts.

DTIME is the time increment of the incremental step.

NDI is the number of normal stress or strain components.

NSHR is the number of shear stress or shear strain components.

NTENS is the number of the total stress or strain components, NTENS = NDI + NSHR.

NSTATV is the number of variables of material definition.

NPROPS is the number of material constants.

COORDS is the array of current material configuration.

DROT (3, 3) is the matrix of rotation increment.

DFGRD0 (3, 3) is the deformation gradient tensor when the incremental step iteration starts.

DFGRD1 (3, 3) is the deformation gradient when the incremental step iteration finishes.

NOEL is the number of elements.

NPT is the number of integration points.

KSTEP is the number of loading steps.

KINC is the number of incremental steps.

5.6.2 Flowchart of Crystal Plasticity Finite Element Method in ABAQUS

On the basis of a numerical integration algorithm, the CP constitutive model can be written into UMAT and used in the FE software ABAQUS. The flowchart of the UMAT code used in ABAQUS is shown in Fig. 5.1.

When UMAT is first used, the increment step will be initialized and the data of grain orientation and slip system will be input. At a starting time t of the incremental step, the ABAQUS main program will input UMAT with the initial grain orientation matrix, time increment, and all the updating variables calculated from the last step. Then the variables ${}^{\tau}s_k^{\alpha}$ and ${}^{\tau}\boldsymbol{T}_k^{(1)}$ of time τ will be iterated and solved. The values of variables ${}^{\tau}\boldsymbol{F}_k^p, {}^{\tau}\boldsymbol{\sigma}_k, {}^{\tau}\Delta\gamma_k^{\alpha}$, and Jacobian matrix ${}^{\tau}\boldsymbol{J}_k$ can be obtained by the related equations. If the Jacobian matrix converges, the values of ${}^{\tau}s_{\alpha}$, ${}^{\tau}\boldsymbol{T}^{(1)}$, ${}^{\tau}\boldsymbol{F}^p$, ${}^{\tau}\boldsymbol{\sigma}$, ${}^{\tau}\Delta\gamma^{\alpha}$, and ${}^{\tau}\boldsymbol{J}$ will be updated, and then the iteration of the next increment will be calculated. If the Jacobian matrix does not converge, the time increment will be estimated and iterated.

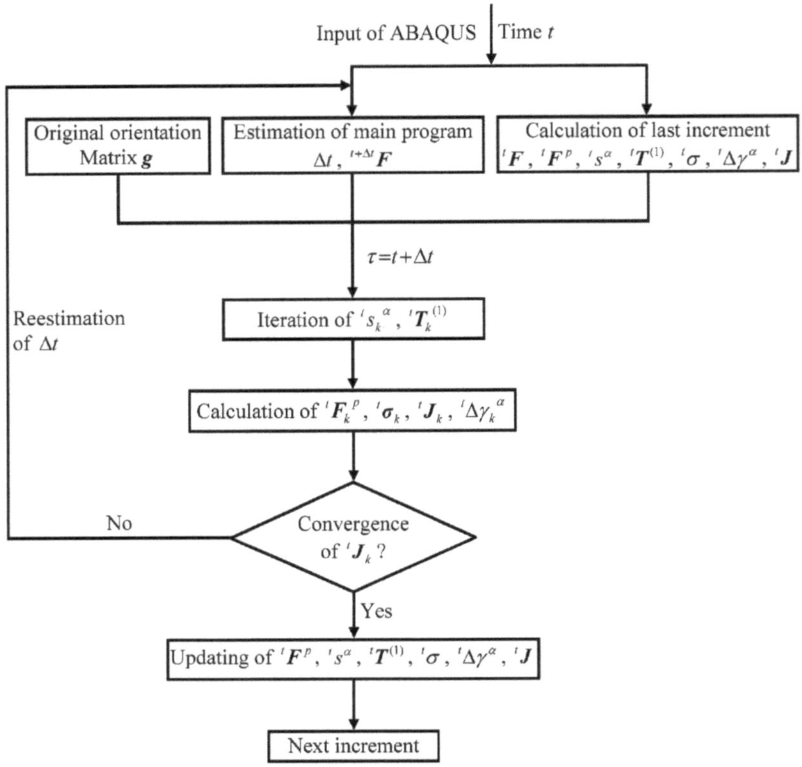

FIGURE 5.1 Flowchart of the UMAT code used in ABAQUS.

The Newton–Raphson iteration method is used in ABAQUS to solve the nonlinear FE equilibrium equation. It is as:

$$\sum \left(\int \boldsymbol{B}^{T}\boldsymbol{J}\boldsymbol{B}dV \right) \boldsymbol{C}^{k} = \boldsymbol{P}(t+\Delta t) - \sum \int \boldsymbol{B}^{T}[\sigma(t) + \Delta\sigma(\Delta u^{k})]dV \quad (5.95)$$

$$\Delta u^{k+1} = \Delta u^{k} + \boldsymbol{C}^{k} \quad (5.96)$$

where $\boldsymbol{B}$ is the transfer matrix from the dislocation increment to the strain increment, $\Delta\varepsilon = \boldsymbol{B}\Delta u$. The stress increment is a function of the strain increment and a function of the dislocation increment. After k times' iteration, the corrected displacement is obtained. When the Newton–Raphson iterating method is used to solve the FE equations, the Jacobian matrix ${}^{t+\Delta t}\boldsymbol{J} = {}^{t+\Delta t}(\partial\Delta\sigma/\partial\Delta\varepsilon)$ (Matrix DDSDDE) will be updated. If the Jacobian matrix cannot be expressed explicitly, the rigid matrix can replace it. In the solving process, the time increment and deformation gradient of time t is input into UMAT by the ABAQUS program. The increment of stress tensors will be updated, and the stress tensor at time $t + \Delta t$ will be obtained. If the Jacobian matrix can be expressed implicitly, the Matrix DDSDDE needs to be updated to improve the rate of convergence.

REFERENCES

[1] G. Sachs, Zur Ableitung einer Fliessbedingung, Z. Ver. Dtsch. Ingen. 72 (1928) 734–736.
[2] G.I. Taylor, Plastic strain in metals, J. Inst. Met. 62 (1938) 307–324.
[3] J.F.W. Bishop, R. Hill, A theory of the plastic distortion of a polycrystalline aggregate under combined stresses, Philos. Mag. 42 (1951) 414–427.
[4] J.F.W. Bishop, R. Hill, A theory of the plastic distortion of a polycrystalline aggregate under combined stresses, Philos. Mag. 42 (1951) 1298–1307.
[5] E. Kröner, On the plastic deformation of polycrystals, Acta Metall. 9 (1961) 155–161.
[6] E. Orowan, Dislocations and mechanical properties, in: M. Cohen (Ed.), Dislocations in Metals, The American Institute of Mining and Metallurgy Engineering, New York, 1954.
[7] E. Schmid, W. Boas, Plasticity of Crystals: With Special Reference to Metals, Chapman & Hall, London, 1968.
[8] J.C. Polanyi, N. Sathyamurthy, Location of energy barriers. VIII. Reagent → product energy conversion on surfaces with sudden or gradual late-barriers, Chem. Phys. 37 (1979) 259–264.
[9] J.C. Polanyi, N. Sathyamurthy, Location of energy barriers. VII. Sudden and gradual late-energy-barriers, Chem. Phys. 33 (1978) 287–303.
[10] R. Hill, J.R. Rice, Constitutive analysis of elastic-plastic crystals at arbitrary strain, J. Mech. Phys. Solids 20 (1972) 401–413.
[11] D. Pierce, R.J. Asaro, A. Needleman, An analysis of nonuniform and localized deformation in ductile single crystals, Acta Metall. 30 (1982) 1087–1119.
[12] R.J. Asaro, A. Needleman, Texture development and strain hardening in rate dependent polycrystals, Acta Metall. 33 (1985) 923–953.
[13] E. Kröner, Continuum theory of defects, in: R. Balian, M. Kléman, J.P. Poirer (Eds.), Physics of Defects, North-Holland Publishing Company, Amsterdam, 1981.
[14] E. Kröner, Allgemeine Kontinuumstheorie der Versetzungen und Eigenspannungen, Arch. Ration. Mech. Anal. 4 (1959) 273–334.
[15] D. Pierce, R.J. Asaro, A. Needleman, Material rate dependence and localized deformation in crystalline solids, Acta Metall. 31 (1983) 1951–1976.
[16] R.J. Asaro, Crystal plasticity, J. Appl. Mech. 50 (1983) 921–934.
[17] S. Nemat-Nasser, Plasticity: A Treatise on Finite Deformation of Heterogeneous Inelastic Materials, Cambridge University Press, Cambridge, 2004.
[18] S.C. Hunters, Mechanics of Continuous Media, second ed., Ellis Horwood Ltd., Chichester, 1983.
[19] P. Haupt, Continuum Mechanics and Theory of Materials, Springer, Berlin and New York, 2000.
[20] S.R. Kalidindi, C.A. Bronkhorst, L. Anand, Crystallographic texture evolution in bulk deformation processing of FCC metals, J. Mech. Phys. Solids 40 (1992) 537–569.
[21] J.R. Rice, Inelastic constitutive relations for solids: an internal variable theory and its application to metal plasticity, J. Mech. Phys. Solids 19 (1971) 433–455.
[22] J.W. Hutchinson, Bounds and self-consistent estimates for creep of polycrystalline materials, Proc. Roy. Soc. A 348 (1976) 1001–1127.
[23] J. Hirsch, K. Lucke, M. Hatherly, Overview no. 76: mechanism of deformation and development of rolling textures in polycrystalline f.c.c. metals-I. Description of rolling texture development in homogeneous CuZn alloys, Acta Metall. 36 (1988) 2863–2882.
[24] S.R. Kalidindi, L. Anand, Macroscopic shape change and evolution of crystallographic texture in pre-textured FCC metals, J. Mech. Phys. Solids 42 (1994) 459–490.

[25] U.F. Kocks, The relation between polycrystal deformation and single crystal deformation, Metall. Trans. 1 (1970) 1121–1143.
[26] R.J. Asaro, J.R. Rice, Strain localization in ductile single crystals, J. Mech. Phys. Solids 25 (1977) 309–338.
[27] J.W. Hutchinson, Elastic-plastic behaviour of polycrystalline metals and composites, Proc. Roy. Soc. A 319 (1970) 247–272.
[28] V. Bachu, S.R. Kalidindi, On the accuracy of the predictions of texture evolution by the finite element technique for fcc polycrystals, Mater. Sci. Eng. A 257 (1998) 108–117.
[29] S.R. Kalidindi, L. Anand, An approximate procedure for predicting the evolution of crystallographic texture in bulk deformation processing of fcc metals, Int. J. Mech. Sci. 34 (1992) 309–332.
[30] V. Randle, O. Engler, Introduction to Texture Analysis: Macrotexture, Microtexture and Orientation Mapping, CRC Press, Boca Raton, 2000.
[31] H.J. Bunge, Texture Analysis in Materials Science: Mathematical Methods, Butterworths, London and Boston, 1982.
[32] U.F. Kocks, Texture and Anisotropy: Preferred Orientations in Polycrystals and their Effect on Materials Properties, Cambridge University Press, Cambridge, 1998.
[33] G.J. Davies, Texture and the Properties of Materials, Metals Society, London, 1976.
[34] ABAQUS/Standard: User's Manual, Karlsson & Sorensen, Inc, Hibbitt, 2002.
[35] Simulink, Abaqus GUI Toolkit Reference Manual, USA, 2008.

Part III

Simulation of Microforming Process

Chapter 6

Simulation Models in Microforming

6.1 INTRODUCTION

When scaling down the geometrical dimensions of a material from conventional scale to microscale, the so-called size effects will appear in forming. In microforming process, the characteristic length of inhomogenous plastic deformation is in microscale, and the material behavior is significantly influenced by size effects. The stress is not only related to strain but also to the strain gradient, as discussed in Chapter 4, Strain Gradient Plasticity Theory. Size effects cannot be conveyed by the classical theory of continuum plastic mechanics which is scale-independent. Classical continuum mechanics considers material continua as simple point-continua with points having three displacement-degrees of freedom, and the response of a material to the displacement of its points is characterized by a symmetric Cauchy stress tensor presupposing that the transmission of loads through surface elements is uniquely determined by a force vector, neglecting couples. Such a model may be insufficient for the description of certain physical phenomena. Nonclassical behavior due to size effects has been extensively observed in the forming of miniaturized parts. Also, it is still difficult to investigate microforming using quantum mechanics or atomistic simulation. Under this background, simulation models that consider size effects and account for the heterogeneity of deformation in microscale have been developing between the classical plastic mechanics and quantum mechanics. In this chapter, simulation models, including surface layer model, mesoscopic model, and Voronoi model, that are popular in the investigation of microforming process will be introduced. The purpose of this chapter is to familiarize readers with the research methodology of simulation which may benefit the analysis of material deformation behavior and design of forming tools and processes in microforming.

6.2 SURFACE LAYER MODEL

The basic idea of surface layer model is that the geometry of a workpiece is subdivided into two areas, i.e., surface and inner areas, in which grains distribution is different, as illustrated in Fig. 6.1. In the free surface area, the

Microforming Technology. DOI: http://dx.doi.org/10.1016/B978-0-12-811212-0.00006-6

grains are less restricted due to the slight grain boundary influence, and it is assumed that the behavior is similar to a crystal. Differently, the inner area is taken as polycrystal. In surface layer model, the flow stress of material is contributed by two parts: the stress of the surface layer which is determined by the single crystal model, and the stress of the inner layer which is controlled by the polycrystal model.

The single crystal deformation is characterized by the critical resolved shear stress, τ_C, and follows the Schmid law as:

$$\tau_C = m\sigma \tag{6.1}$$

where σ is the tensile stress and m $(0 < m \leq 1/2)$ is known as the Schmid factor which can be determined by:

$$m = \cos\phi \cos\lambda$$

where ϕ is the angle between the tensile direction and the direction normal to slip plane, and λ is the angle between the tensile direction and the slip direction, as illustrated in Fig. 6.2.

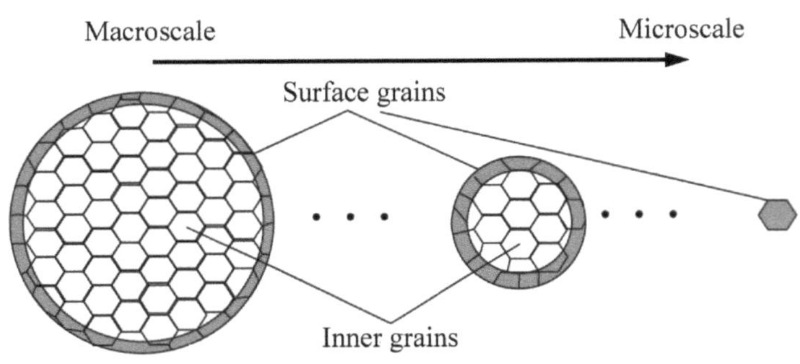

FIGURE 6.1 Grains distribution in a material section with the decrease of the feature scale [1].

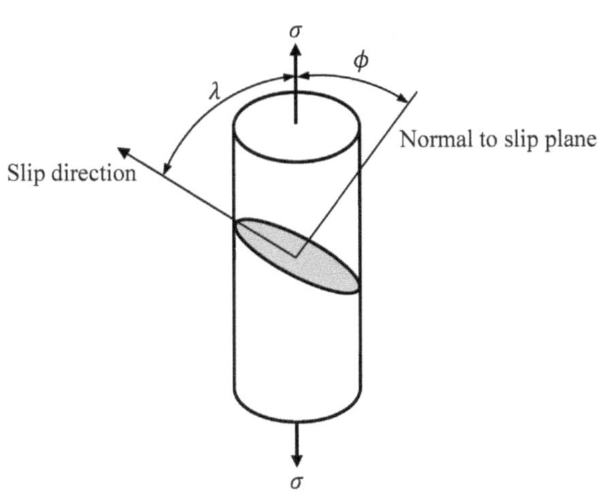

FIGURE 6.2 Determination of the Schmid factor.

Based on the assumption of surface layer model, the stress of the surface grains, $\sigma_s(\varepsilon)$, can be expressed as:

$$\sigma_s(\varepsilon) = \frac{1}{m}\tau_R(\varepsilon) \tag{6.2}$$

where τ_R is the resolved shear stress.

For the inner area where the polycrystal model applies, the stress of grains, $\sigma_i(\varepsilon)$, can be obtained by [2]:

$$\sigma_i(\varepsilon) = \sigma_0(\varepsilon) + \frac{k(\varepsilon)}{\sqrt{D}} \tag{6.3}$$

where $\sigma_0(\varepsilon)$ and $k(\varepsilon)$ are constants under a given strain (ε) and D is the grain size. $\sigma_0(\varepsilon)$ means the friction stress required to move individual dislocations in microyielded slip band pile-ups confined to isolated grains, whereas $k(\varepsilon)$ is the locally intensified stress needed to propagate general yielding across the polycrystal grain boundaries. The value of $\sigma_0(\varepsilon)$ obtained from Hall–Petch analysis has a relationship with the resolved shear stress of a single crystal as [3]:

$$\sigma_0(\varepsilon) = M\tau_R(\varepsilon) \tag{6.4}$$

where M is the orientation factor to account for slips on deformation systems. M is the average value of all the different orientations of the grains constituting a whole specimen, and thus, may be considered as the flow stress of a single crystal oriented for multi-slip. By combining Eqs. (6.3) and (6.4), the polycrystal flow stress can be rewritten as the following form:

$$\sigma_i(\varepsilon) = M\tau_R(\varepsilon) + \frac{k(\varepsilon)}{\sqrt{D}} \tag{6.5}$$

For FCC crystals, M equals to 3.06 and 2.23 for the Taylor and Sachs models, respectively. The Taylor model is an upper-bound model which requires at least five active slip systems. It assumes that the strains to be the same in all grains and equal to the macroscopic strain. The Sachs model is a lower-bound model assumes equal stress in all grains, and requires only one active slip system [4].

Surface layer model can be used to explain the decrease in flow stress with miniaturization. The dislocation movements in the grains located at the surface area have less constraint compared with grains inside the material, and dislocations moving through the grains during deformation pile-up at grain boundaries but not at the free surface. Therefore, the surface grain shows less hardening and lower resistance against deformation, since the ratio of the free surface grain to the internal grain increases with miniaturization, which would result in a decrease in flow stress. Therefore, the change of volume fraction of surface grains has a great effect on flow stress. The flow stress of materials in microscale can be expressed as:

$$\sigma_{\text{micro}}(\varepsilon) = \frac{N_s \sigma_s(\varepsilon) + N_i \sigma_i(\varepsilon)}{N} \tag{6.6}$$

where N_s and N_i are the numbers of surface and inner grains, respectively, and N is the total number of grains.

By substituting Eqs. (6.3) and (6.5) into Eq. (6.6), the following expression can be obtained:

$$\sigma_{\text{micro}}(\varepsilon) = \frac{N_s \frac{1}{m} \tau_R(\varepsilon) + N_i \left[M \tau_R(\varepsilon) + \frac{k(\varepsilon)}{\sqrt{D}} \right]}{N} \tag{6.7}$$

If we let η be the size factor, and it has the following expression:

$$\eta = \frac{N_s}{N} \tag{6.8}$$

Then, Eq. (6.7) can be rewritten as:

$$\sigma_{\text{micro}}(\varepsilon) = \eta \frac{1}{m} \tau_R(\varepsilon) + (1 - \eta) \left[M \tau_R(\varepsilon) + \frac{k(\varepsilon)}{\sqrt{D}} \right] \tag{6.9}$$

If the size factor $\eta = 0$, Eq. (6.9) expresses the flow stress of polycrystal. While for size factor $\eta = 1$, Eq. (6.9) expresses the flow stress of crystal.

Eq. (6.9) can be divided into two different terms: size-dependent model and size-independent model. Then the surface layer model can be expressed as:

$$\begin{cases} \sigma_{\text{micro}}(\varepsilon) = \sigma_{\text{dep}} + \sigma_{\text{ind}} \\ \sigma_{\text{dep}} = \eta \left[\frac{1}{m} \tau_R(\varepsilon) - M \tau_R(\varepsilon) - \frac{k(\varepsilon)}{\sqrt{D}} \right] \\ \sigma_{\text{ind}} = M \tau_R(\varepsilon) + \frac{k(\varepsilon)}{\sqrt{D}} \end{cases} \tag{6.10}$$

where σ_{dep} and σ_{ind} are the stress values contributed by the size-dependent and size-independent parts, respectively.

6.3 MESOSCOPIC MODEL

In surface layer model, the transition between the surface layer and the inner layer is thought to be discontinuous, i.e., the behavior of grains in the surface area is assumed to be similar to single crystal behavior, while the behavior of grains in the inner area is described by that of polycrystal. Surface layer model can reach a good agreement between simulation and experimental results, however, it is unable to consider the occurrence of scatter of the forming results and as well as the local forming phenomenon during microforming process. Mesoscopic model is based on the idea of surface layer model, in which material is subdivided into single grains with each characterized by individual properties, depending on its size and position within the material [5]. As the material structure used in mesoscopic model is similar to the real one, it is possible to study the local forming behavior controlled by the local grain structure.

According to the description of Geißdörfer et al. [5,6] and Geiger et al. [7], the materials plastic response to external forces follows the theory of Hall–Petch [8,9] and Ashby [10], and is mainly controlled by the grain size, the position of the grain within the material, and surrounding of the grain. The shear stress τ has a relationship with grain size as:

$$\tau = \tau_0 + \frac{K}{\sqrt{D}} \tag{6.11}$$

where τ_0 is the critical shear stress, K is the Hall–Petch factor, and D is the grain size.

If the effect of neighboring grains on the mechanical properties of an individual grain is taken into account in the forming behavior of few grains, the shear stress τ can be expressed as:

$$\tau = \tau_0 + \frac{m_{1,2}\tau_i\sqrt{\delta}}{\sqrt{D}} \tag{6.12}$$

where $m_{1,2}$ represents a transformation matrix for the different sliding systems in the adjoining grains, τ_0 is the critical shear stress in the considered grain, and δ is the distance between the pile-up source and the dislocation. If dislocations are assumed to pile-up until the stress concentration at the grain boundary exceeds the stress τ_i, dislocation sources will be activated in neighboring grains. Therefore, the Hall–Petch factor for an individual grain K_i differs from the integral Hall–Petch factor K, i.e., K_i is valid only for an individual grain.

The Hall–Petch factor K_i is given by:

$$K_i = m_{1,2}\tau_i\sqrt{\delta} \tag{6.13}$$

The relationship between K_i and K can be expressed as:

$$K_i = K\xi_i \tag{6.14}$$

where ξ_i is a pile-up factor which describes the amount of dislocation pile-up at the grain boundary region of an individual grain, and is defined as:

$$\xi_i = \frac{\overline{\sigma}_{\mathrm{nG}}}{\sigma_{\mathrm{G}}} = \frac{1}{\sigma_{\mathrm{G}}}\frac{1}{\alpha_{\mathrm{G}}}\sum \alpha_{\mathrm{c}}\sigma_{\mathrm{nG}} \tag{6.15}$$

where σ_{G} is the yield stress of an individual grain and $\overline{\sigma}_{\mathrm{nG}}$ is the average value of yield stress σ_{nG} of the neighboring grains related to the contact length α_c. In case of 3D studies, α_c/α_G represents the fraction of the real contact area of the grains. As dislocation pile-up in the free surface area is impossible, the pile-up factor there is set to be zero, and it will increase with an increase of the distance of grains from the surface to the inner. This explains why the flow stress of the grains in or close to the free surface area is lower than that of the grains in the inner area of a specimen in microforming.

Substituting Eqs. (6.13) and (6.14) into Eq. (6.12), then the shear stress of an individual grain can be rewritten as:

$$\tau = \tau_0 + \frac{K\xi_i}{\sqrt{D}} \tag{6.16}$$

The mesoscopic model can be enhanced based on the theory of Meyers and Ashworth [11] with consideration of dislocation pile-up in the region of a grain boundary. An individual grain is subdivided into two regions, i.e., the grain boundary region and the inner grain region, and the integral yield stress σ_f of an individual grain is expressed as:

$$\sigma_f = \alpha_{GB}\sigma_{f,GB} + \alpha_I\sigma_{f,I} \tag{6.17}$$

where α_{GB} is the volume fraction of grain boundary, $\sigma_{f,GB}$ is the flow stress at grain boundary due to the occurrence of dislocation pile-up, α_I is the volume fraction of inner grain, and $\sigma_{f,I}$ is the flow stress of inner grain. If the individual grain is assumed to be a sphere with a diameter of D and with thickness of a grain boundary layer t, then the values of α_{GB} and α_I can be obtained as follows:

$$\alpha_{GB} = \frac{\frac{4}{3}\pi\left[\left(\frac{D}{2}\right)^3 - \left(\frac{D}{2} - t\right)^3\right]}{\frac{4}{3}\pi\left(\frac{D}{2}\right)^3} = 6\frac{t}{D} - 12\left(\frac{t}{D}\right)^2 + 8\left(\frac{t}{D}\right)^3 \tag{6.18}$$

$$\alpha_I = \frac{\frac{4}{3}\pi\left(\frac{D}{2} - t\right)^3}{\frac{4}{3}\pi\left(\frac{D}{2}\right)^3} = 1 - \alpha_{GB} \tag{6.19}$$

Substituting Eqs. (6.18) and (6.19) into Eq. (6.17), then:

$$\sigma_f = \sigma_{f,I} + \left[6\frac{t}{D} - 12\left(\frac{t}{D}\right)^2 + 8\left(\frac{t}{D}\right)^3\right]\left(\sigma_{f,GB} - \sigma_{f,I}\right) \tag{6.20}$$

As the flow stress of $\sigma_{f,GB}$ is higher than that of $\sigma_{f,I}$ due to dislocation pile-up at grain boundary, it can be assumed that a stress gradient exists from the grain boundary into the grain itself. Therefore, a higher value of α_{GB} will induce a higher yield stress of the material. Let's consider an extreme condition of $D = 2t$, i.e., the whole material is composed of only grain boundary layer, then $\alpha_{GB} = 1$, $\alpha_I = 0$, and in this case, $\sigma_f = \sigma_{f,GB}$.

In many cases, if a section through a material composed of spherical grains with equal diameter is taken, the interactions may not be all the diametric. As indicated in Fig. 6.3, the cross sections of s_1, s_2, and s_3 may yield different values of the diameter and the grain boundary thickness. For correcting the values of t and D, Meyers and Ashworth [11] recommended to use the average values in calculation. When t is substantially smaller than D, the average value of grain boundary thickness $\bar{t}$ is approximately equal to $1.8t$. Margolin and Stanescu [12] used the expression of:

$$\bar{t} = 1.571t \tag{6.21}$$

The average grain size $\overline{D}$ can be obtained by [11]:

$$\overline{D} = \frac{\pi}{4}D \tag{6.22}$$

Calculating the average volume fractions of grain boundary and inner grain per cross-section by assuming an average grain boundary thickness $\bar{t}$ and an average grain size $\overline{D}$, Eq. (6.20) can be rewritten as:

$$\sigma_f = \sigma_{f,I} + \left[6\frac{\bar{t}}{\overline{D}} - 12\left(\frac{\bar{t}}{\overline{D}}\right)^2 + 8\left(\frac{\bar{t}}{\overline{D}}\right)^3\right](\sigma_{f,GB} - \sigma_{f,I}) \tag{6.23}$$

Substituting Eqs. (6.21) and (6.22) into Eq. (6.23), the following expression can be obtained:

$$\sigma_f = \sigma_{f,I} + \left[12\frac{t}{D} - 48\left(\frac{t}{D}\right)^2 + 64\left(\frac{t}{D}\right)^3\right](\sigma_{f,GB} - \sigma_{f,I}) \tag{6.24}$$

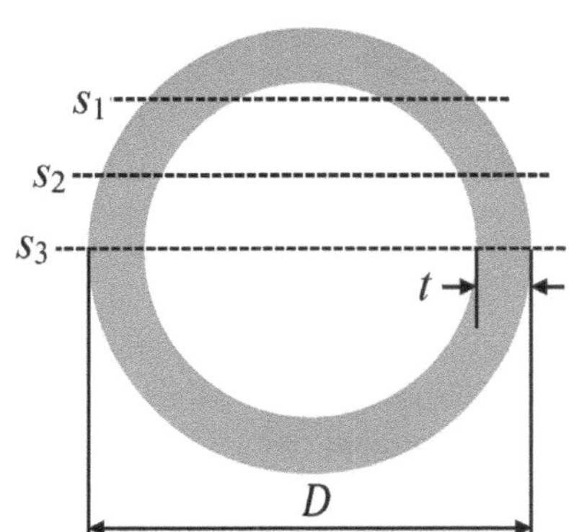

FIGURE 6.3 Schematic illustration of sections of s_1, s_2, and s_3 that yield different values for the diameter and the grain boundary thickness.

When D is greater than $2t$, considering the variation of t as follows:

$$t = kD^{1/2} \tag{6.25}$$

where k is a parameter. Physically, Eq. (6.25) indicates that t increases rapidly with grain size when the grain size is small; and at large grain size, the dependence becomes nonsignificant. In the work done by Geiger et al. [7], t is assumed to have a relationship with D as: $t = 0.133D^{0.7}$ in the investigation of microforming process. Substituting Eq. (6.25) into Eq. (6.24), it has:

$$\sigma_{\text{f}} = \sigma_{\text{f,I}} + \left[12k\frac{1}{\sqrt{D}} - 48k^2\left(\frac{1}{\sqrt{D}}\right)^2 + 64k^3\left(\frac{1}{\sqrt{D}}\right)^3\right]\left(\sigma_{\text{f,GB}} - \sigma_{\text{f,I}}\right) \tag{6.26}$$

6.4 VORONOI MODEL

As the geometry of a specimen approaches the dimension of grain size, only a small amount of grains locate in the deformation area or the force bearing area of the specimen where size effects of grains become significant. This feature leads to grains with different sizes, shapes, and orientations, which will result in scatter of the measured mechanical properties. In order to numerically investigate the material heterogeneity in microforming, Voronoi model, also known as Dirichlet tessellation or Voronoi diagram, is introduced in this section. Voronoi model has been used to conduct research by generating the like polycrystalline material structure, and it can subdivide an area or space into convex polygons or cells that fill in the space without overlap.

The idea of Voronoi model can be dated back to the 17th century when a book entitled *Principia Philosophiae* in Latin was published by Descartes [13] in 1644. An English version of this book was published in 1982 [14]. In this book, Descartes claims that the solar system consists of vortices. His illustrations show a decomposition of space into convex regions, with each consisting of matter revolving round one of the fixed stars, as shown in Fig. 6.4. Even though Descartes did not explicitly define the extension of these regions, the underlying idea is clear. This concept has independently emerged, and proven useful, in various fields of science. Different names particular to the respective fields, including medial axis transform in biology and physiology, Wigner–Seitz zones in chemistry and physics, domains of action in crystallography, and Thiessen polygons in meteorology and geography, have been used. The mathematicians Dirichlet [15] and Voronoi [16], who used this concept to study quadratic forms where the sites were integer lattice points and the influence was measured by the Euclidean distance, were the first scholars to formally introduce this concept.

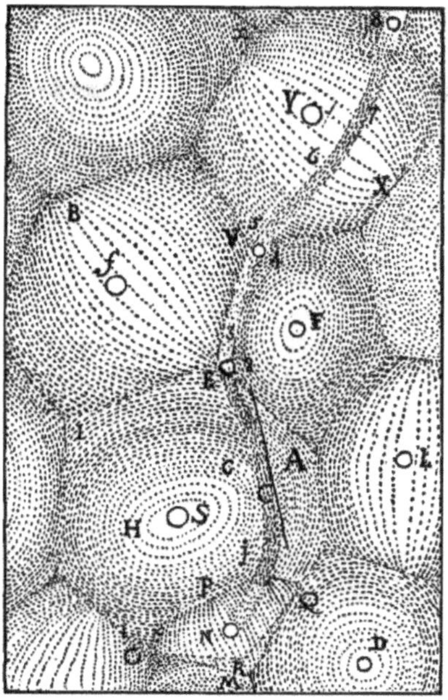

FIGURE 6.4 Descartes' decomposition of space into vortices [13].

The work by Dirichlet [15] and Voronoi [16] was the origin of Voronoi model being widely used in a number of fields with a purpose of solving numerous scientific problems.

6.4.1 Definitions

S is denoted in the plane by a set of n ($n > 2$) point sites p, q, ... for points $p = (x_p, y_p)$ and $r = (x_r, y_r)$, let $d(p, r) = \sqrt{(x_p - x_r)^2 + (y_p - y_r)^2}$ denote their Euclidean distance. The term $\overline{pq}$ is used to denote the line segment from p to q.

For p, $q \in S$, let

$$B(p, q) = r|d(p, r) = d(q, r) \tag{6.27}$$

be the bisector of p and q. $B(p, q)$ is the perpendicular line through the center of the line segment $\overline{pq}$. It separates the half plane

$$D(p, q) = r|d(p, r) < d(q, r) \tag{6.28}$$

containing p from the half plane $D(p, q)$ containing q. It is:

$$VR(p, S) = \bigcap_{q \in S, q \neq p} D(p, q) \tag{6.29}$$

the Voronoi region of p with respect to S. Then, the Voronoi diagram of S is defined as follows:

$$V(S) = \bigcup_{p,q \in S, q \neq p} \overline{VR(p,S)} \cap \overline{VR(q,S)} \tag{6.30}$$

By definition, each Voronoi region $VR(p,S)$ is the intersection of $n-1$ open-half plane containing the site p. Therefore, $VR(p,S)$ is open and convex. Different Voronoi regions are disjointed, as shown in Fig. 6.5.

The common boundary of two Voronoi regions belongs to $V(S)$ and is called a Voronoi edge, if it contains more than one point. If the Voronoi edge e borders the regions of p and q then $e \in B(p,q)$ holds. The endpoints of Voronoi edges are called Voronoi vertices which belong to the common boundary of three or more Voronoi regions.

There is an intuitive way of looking at the Voronoi diagram $V(S)$. Let x be an arbitrary point in the plane. Let's center a circle, C, at x and let its radius grow, from 0 onwards. At some stage the expanding circle will, for the first time, hit one or more sites of S. Then three different cases will occur [17]:

Case 1: If the circle C expands from x and hits exactly one site, p, then x belongs to $VR(p,S)$. If C hits exactly two sites p and q, then x is an interior point of a Voronoi edge separating the regions of p and q. If C hits three or more sites simultaneously, then x is a Voronoi vertex adjacent to those regions whose sites have been hit.

Case 2: A point p of S lies on the convex hull of S if its Voronoi region $VR(p,S)$ is unbounded.

Case 3: The Voronoi diagram $V(S)$ has many edges and vertices. The average number of edges in the boundary of a Voronoi region is less than six.

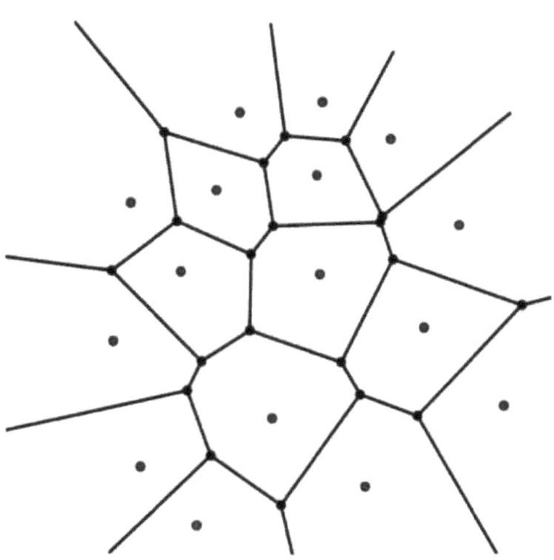

FIGURE 6.5 A 2D Voronoi diagram of 16 points in the Euclidean plane.

In general, a triangulation of S is a planar graph with the vertex set S and straight line edges, which is maximal in the sense that no further straight line edge can be added without crossing other edges. Each triangulation of S contains the edges of the convex hull of S. Its bounded faces are triangles, due to maximality. Let's call a subset of edges of a triangulation a tessellation of S if it contains the edges of the convex hull, and if each point of S has at least two adjacent edges. Consequently, two points of S are joined by a Delaunay edge if their Voronoi regions are edge-adjacent. Therefore, $DT(S)$ is the graph-theoretical dual of $V(S)$, realized by straight line edges. Fig. 6.6 gives an example that the Voronoi diagram $V(S)$ is drawn by solid lines and $DT(S)$ by dash lines.

In order to help readers understand the use of Voronoi model in practice, a comprehensive introduction of implementation methods of both 2D and 3D Voronoi diagrams in commercial FE software ABAQUS/CAE will be stated. The implementation is conducted with the assistance of other two sets of software: MATLAB and Python, and the related functions are expressed.

6.4.2 Implementation of Two-Dimensional Voronoi Diagram in ABAQUS/CAE

For implementing a 2D Voronoi diagram into ABAQUS preprocessing, the first job is to plot the 2D Voronoi diagram in MATLAB. The algorithm for generating Voronoi diagram is sound in MATLAB, and the syntax chosen to generate a Voronoi diagram is *voronoi*(x, y) which plots the bounded cells of the Voronoi diagram for the points (x, y). For obtaining the topology of the Voronoi diagram, e.g., the vertices for each Voronoi cell, the syntax $[V, C] =$ *voronoin*$([x, y])$ is used. For any point p, it has at least one neighborhood which belongs to the p's Voronoi cell, which means that any

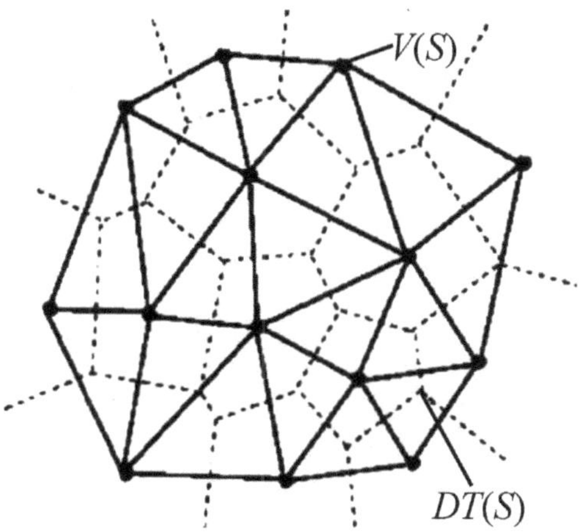

FIGURE 6.6 Voronoi diagram and Delaunay tessellation.

domain $V(p)$ cannot be null. This declares that for a Voronoi diagram from n unique points:

- $V(S)$ contains n domains exactly, and parts of these domains are infinite.
- Every vertex in Voronoi diagram goes through at least three circle centers of seeds, and these circles do not contain any other seeds.
- Seeds are decisive to the position and size of Voronoi cells.

For recrystallized metal microstructure, the grain shape can be considered as equiaxed which is consistent with the topology structure of Voronoi cells. Therefore, the key issue for successfully generating grain-like Voronoi diagram in MATLAB is seeding. Here the seeding method is stated in the following.

1. Assuming that a Voronoi diagram consists of $i \times j$ $(i, j \geq 1)$ Voronoi cells with average "grain" size D. So the length of the plane can be calculated as $D \times i$, and its width is $D \times j$ where $i, j \geq 1$.
2. Dividing the plane into $i \times j$ equal rectangles, and seeding at the center of each rectangle. The coordinates of the seeds can be expressed as:

$$x_{(i,j)} = \frac{D}{2} + (j-1) \times D \tag{6.31}$$

$$y_{(i,j)} = \frac{D}{2} + (i-1) \times D \tag{6.32}$$

3. Taking r which is named as shape control factor as an arbitrary value between -1 and 1, the expression of r is:

$$r = 2 \times \text{rand}(1,1) - 1 \tag{6.33}$$

where $\text{rand}(1,1)$ is a MATLAB syntax distributing pseudorandom number valued between 0 and 1. So the new coordinates of the seeds can be expressed as:

$$x_{(i+1,j)} = x_{(i,j)} + r \times \frac{D}{2} \tag{6.34}$$

$$y_{(i,j+1)} = y_{(i,j)} + r \times \frac{D}{2} \tag{6.35}$$

Then in this way the seeds' coordinates can be obtained.

The key point in seeding process is that changing the coordinates of seeds to change the shape of Voronoi cells by assigning different r values. Otherwise, the seeds will be just distributed in the center of each rectangle, and grain shape-like Voronoi cells cannot be generated. Fig. 6.7 shows the Voronoi diagram plotting in MATLAB, when r is assigned with different value ranges.

It can be seen that, when r values from -1 to 0 and from 0 to 1, the plots are similar to normal rectangles and the shape factor does not take much effect. However, when the value of r is extended from -1 to 1, it increases the irregularity of the Voronoi diagram which can be clearly visualized in MATLAB plotting.

After plotting 2D Voronoi diagram in MATLAB, the coordinates of every vertex in each Voronoi cells can be obtained. Two groups of information are essential for copying the Voronoi diagram in MATLAB. One is the coordinates of all the vertices in the diagram and the other is the indices of connecting each vertex for every Voronoi cell. More specifically, each Voronoi cell can contain several vertices, and this topology information can be found when the syntax [V, C] = *voronoin*([x, y]) is adopted. V delivers the Voronoi vertex information, and is an m-by-n array of the m-Voronoi vertices in n-dimensional space, and each row corresponds to a Voronoi vertex. C is a vector array where each element contains the indices into V of the vertices of the corresponding Voronoi cell. Fig. 6.8 illustrates a Voronoi cell

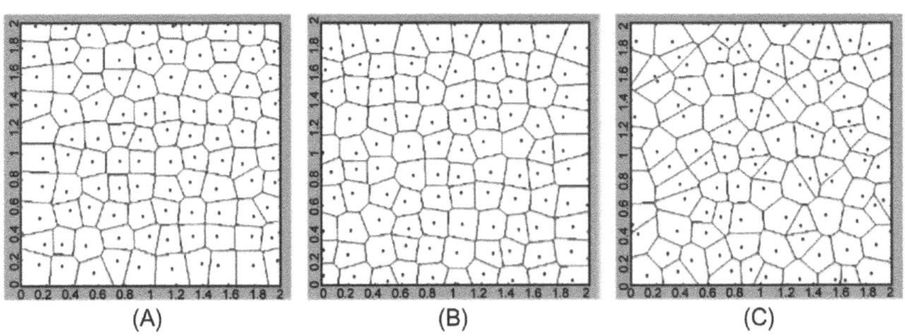

FIGURE 6.7 2D Voronoi diagram under different r values: (A) $-1<r<0$, (B) $0<r<1$, and (C) $-1<r<1$.

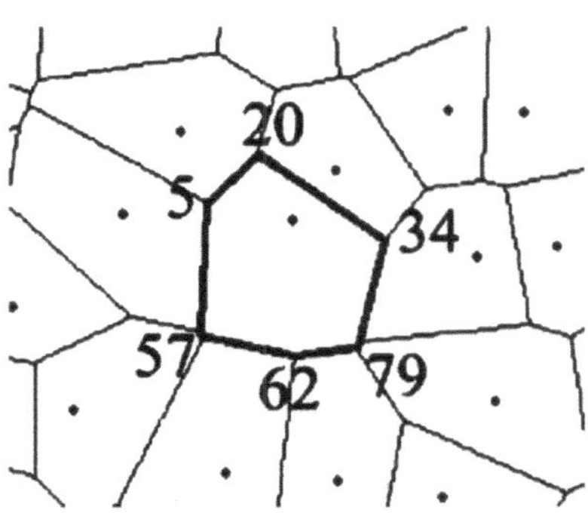

FIGURE 6.8 2D Voronoi cell with indices.

marked with the indices of vertices, and it can only be plotted successfully when the vertices are connected in a clockwise or anticlockwise order. Otherwise, e.g., if vertex 20 was connected with vertex 79 instead of 5 or 34, the cell can be an error.

Therefore, when programming Python script language, it is significant to confirm that each vertex should be linked in the given orders provided by MATLAB. In this book, Python script files are generated through MATLAB programming, and a function has been created in MATLAB to finish this task. In this function, all the Python script statements should follow Python's syntax and semantics since it is going to be run as a script file in ABAQUS/CAE.

The technique used here is that loops have been applied to record all the vertices coordinates in the diagram, and to every two vertices which can be connected through lines, they are renumbered to fit in the loops. The programming of the loops in MATLAB is as follows:

```
for i = 1:size(LineX_Coords,2)
  LinePts(i,1) = LineX_Coords(1,i);
  LinePts(i,2) = LineY_Coords(1,i);
  LinePts(i,3) = LineX_Coords(2,i);
  LinePts(i,4) = LineY_Coords(2,i)
end
```

This is a loop for renumbering the coordinates of connectable vertices stored from *V*.

```
for i = 1:size(LinePts,1);
fprintf(fa_prg,'%s\n', ['mySketch.Line(point1 = (',num2str(LinePts
(i,1)),',',num2str(LinePts(i,2)),'),point2 = (',num2str(LinePts
(i,3)),',',num2str(LinePts(i,4)),'))']);
end
```

This is a loop for programming Python script language to connect the vertices according to the indices in *C*.

Moreover, when all the edges in Voronoi diagram are drawn, it is necessary to draw a rectangle to include all the Voronoi cells and to limit the boundary of the specimen because the side Voronoi cells contain edges with infinite length. Last but not the least, every Voronoi cell should be partitioned into a face which ensures that Voronoi cells are not simply frames but faces with boundaries. If Voronoi cells are just frames connected from edges, mechanical properties cannot be assigned into Voronoi tessellations in ABAQUS preprocessing.

Other Python statements for acting as ABAQUS/CAE script language can be found in Python-related reference books, so these contents will be skipped here. Fig. 6.9 shows 2D Voronoi diagrams with different size Voronoi cells (grain sizes) in the same specimen dimension.

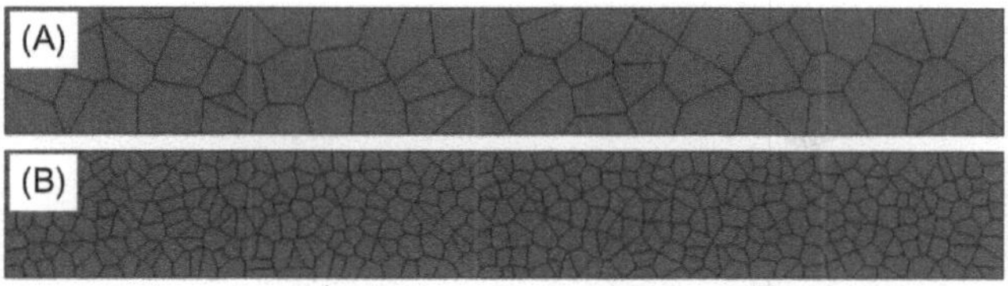

FIGURE 6.9 Different quantity Voronoi cells in a fixed specimen dimension: (A) 66 grains and (B) 375 grains.

6.4.3 Implementation of Three-Dimensional Voronoi Diagram in ABAQUS/CAE

The general flow path for 3D Voronoi diagram implementation in ABAQUS/CAE is evolved from that for 2D one. Therefore, it is necessary to point out the differences between these two methods. On the surface, although one more dimension is added, and one more parameter needs to be considered, some details for 3D Voronoi diagram are completely distinctive to the method for its 2D counterpart.

First, a 3D Voronoi tessellation is generated by means of programming in MATLAB. As 2D Voronoi diagram can be plotted according to the build-in function *voronoi*(x, y) in MATLAB itself, the programming code for 3D counterpart is evolved and modified based on this function, and can make the size of Voronoi cells controllable. For 3D Voronoi diagram implementation, *voronoi*(x, y) is no longer applicable since it is for 2D plotting. Instead, $[V, C] =$ *voronoin*($[x, y, z]$) is firstly used to obtain the coordinates of 3D Voronoi cells' vertices and the indices of connecting every vertex for each cell. x, y, z are the coordinates of seeds in X, Y, Z directions, respectively, and the distance between two adjacent seeds is the grain size of physical specimen. In visualization, another function *convhulln*(*Verts*) is used. *Verts* represent the whole coordinates of the vertices in 3D Voronoi diagram. To 3D visualization, this function returns to the indices of the *Verts* points that comprise the facets of the convex hull of these points, so it actually computes the vertices of the facets that make up the Voronoi cell. Then *patch* function is applied to generate the figure in MATLAB plotting. Fig. 6.10 displays an example of $1 \times 1 \times 1$ mm^3 cubic 3D Voronoi diagram with 0.1 mm Voronoi cell size (grain size) plotted in MATLAB.

Second, extracting the coordinate values of the entire vertex from the 3D Voronoi tessellation plot, and then stores all these data into a Python file. This file is also generated by means of MATLAB programming, and the codes, like 2D Voronoi diagram implementation, should be consistent with Python's syntax and semantics because this file will work as a script file in ABAQUS/CAE. However, more than the 2D counterpart, the vertices should be connected to edges in the given indices, and the edges are

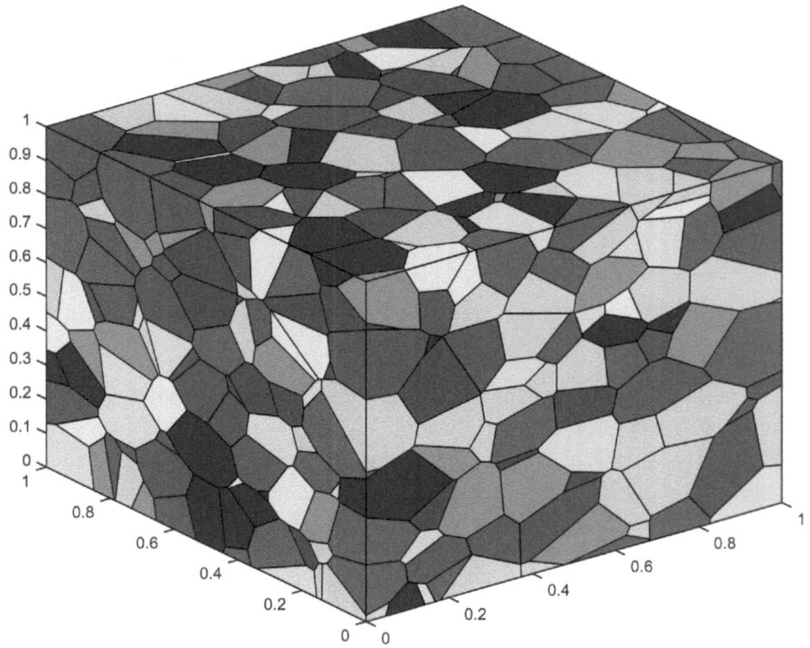

FIGURE 6.10 3D Voronoi diagram in MATLAB.

connected to make up facets. Last, these facets are combined together to generate the Voronoi cells while facets can directly represent Voronoi cells in 2D plotting.

Finally, entering GUI of ABAQUS/CAE, and clicking "file" then choosing "run script." From the popped-up window, the generated Python file which is suffixed by "py." can be traced and selected according to the route which had been set up in the second step. Python programming language runs as scripting language that can be embedded in ABAQUS/CAE to achieve less time consuming, less repetitive, and more flexible operation. Instead of human operation in preprocessing, Python scripts can promptly and compactly execute commands in ABAQUS/CAE through typing given statements and expressions. Especially for the multiple Voronoi tessellations implementation, Python has a natural advantage when inputting a large number of coordinates values and connecting them repetitively. Fig. 6.11 illustrates how ABAQUS scripting interface commands interact with the ABAQUS/CAE kernel.

If there are syntax mistakes, then debugging until the file can be fully interpreted by ABAQUS/CAE which delivers that Voronoi diagram can be completely duplicated from the 3D plot in MATLAB. Voronoi tessellations

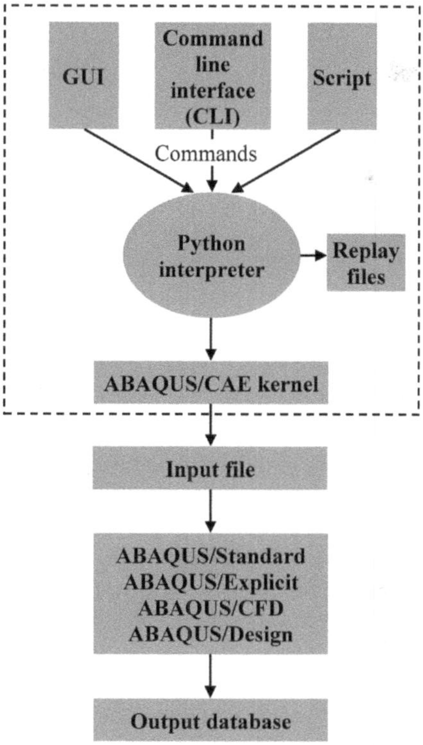

FIGURE 6.11 ABAQUS scripting interface commands and ABAQUS/CAE.

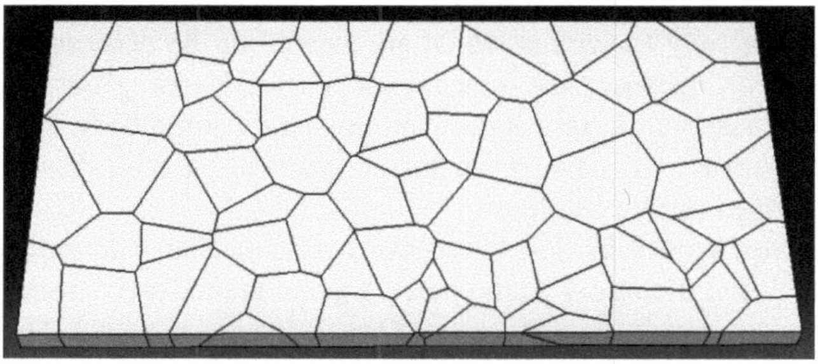

FIGURE 6.12 An example of specimen after Voronoi treatment in ABAQUS/CAE.

are especially suitable in ABAQUS/CAE to compose experimental specimens (to realize polycrystalline aggregate) in the microforming FE models. Fig. 6.12 illustrates an example of specimen after Voronoi treatment (the ratio of specimen thickness T to grain diameter D is less than one) which is computationally divided into multiple grains and each single grain contains grain interior and grain boundary.

After implementation of Voronoi diagram in ABAQUS/CAE, the virtual specimens, which are made up of Voronoi diagram, are equipped with a certain number of grain-like structures. The grain interior and grain boundary can be modeled as a single body. This is because stress incompatibility at grain boundary will disappear when grain interior is undergone plastic deformation. Nevertheless, in microforming simulation which is featured with inhomogeneity and heterogeneity, no matter 2D or 3D Voronoi diagram, each grain involved in the specimen needs to be employed with a particular property to make it inhomogeneous to each other.

It is acknowledged that different orientations, sizes, and shapes of different grains have a momentous impact on inhomogeneous deformation behavior and scatter of material's mechanical properties. At macrolevel, the interest of mechanical property tests, e.g., tensile test, is the mean stress–strain curve, which is normally obtained by averaging three testing results. Generally, these three testing results are not repeatable, and it is accepted that this scatter of experimental results is caused by errors coming from measurement, observation, and operation. However, in microforming, for any grain involved in a polycrystalline sample, it has a unique stress–strain curve which may be higher or lower than or approximately the same as the sample's averaged stress–strain curve. This proposes two different points of views on the scatter for testing results. In macroforming, the scatter comes from human errors, while in microforming, the scatter comes from and represents the grain heterogeneity itself. Theoretically, different grains in one sample actually have different mechanical properties, but this feature will normally be ignored when considering macroscopic conventional processing. However, it should never be neglected in microforming since each grain's mechanical properties can influence uniquely and greatly on deformation behavior, especially for grains which are located in the deformation zone of specimens. For instance, the mechanical properties of grains in bending deformation zone will have a significant impact on springback. However, the grain inside areas will have much less impact on it since these grains are hardly deformed during bending.

Here comes a problem: how the range of grain mechanical properties can be confirmed? It is difficult to find the exact range of grain's mechanical properties for a certain specimen. However, a close answer can be acquired. The probability density function can be used to illustrate how to obtain discrete grain's mechanical properties from measured material's mechanical properties [18]:

$$f(\sigma,\varepsilon) = \frac{1}{S(\varepsilon)\sqrt{2\pi}} e^{-\frac{1}{2}\left[\frac{\sigma(\varepsilon)-\sigma_m(\varepsilon)}{S(\varepsilon)}\right]^2} \tag{6.36}$$

where the standard deviation $S(\varepsilon)$ is described as:

$$S(\varepsilon) = \sqrt{\frac{\sum_{j=1}^{t}\left[\sigma_j(\epsilon) - \sigma_m(\epsilon)\right]^2}{t-1}} \tag{6.37}$$

and the mean flow stress $\sigma_m(\varepsilon)$ is expressed as:

$$\sigma_m(\varepsilon) = \frac{\sum_{j=1}^{t} \sigma_j(\varepsilon)}{t} \tag{6.38}$$

where t is the total number of tensile samples and $\sigma_j(\varepsilon)$ is the jth tensile testing result.

Based on tensile testing results, the mean grain's mechanical properties, the upper and the lower bounds of grain's mechanical properties can be calculated through Eqs. (6.37) and (6.38). Finally, different mechanical properties of grains can be obtained, and the scatter of grain's mechanical properties can be displayed. In FE simulation, these distinctive properties of grains can be assigned into Voronoi tessellations with normal distribution probability by Python scripting. Fig. 6.13 shows a specimen which contains grains assigned with different mechanical properties displayed by different colors.

Fig. 6.13 indicates that each Voronoi tessellation can be equipped with different and unique mechanical properties in the model, and each Voronoi tessellation will have influence on deformation behavior significantly. In this way, with the combination of grain's mechanical properties with Voronoi tessellations, the grain heterogeneity can be achieved in ABAQUS/CAE.

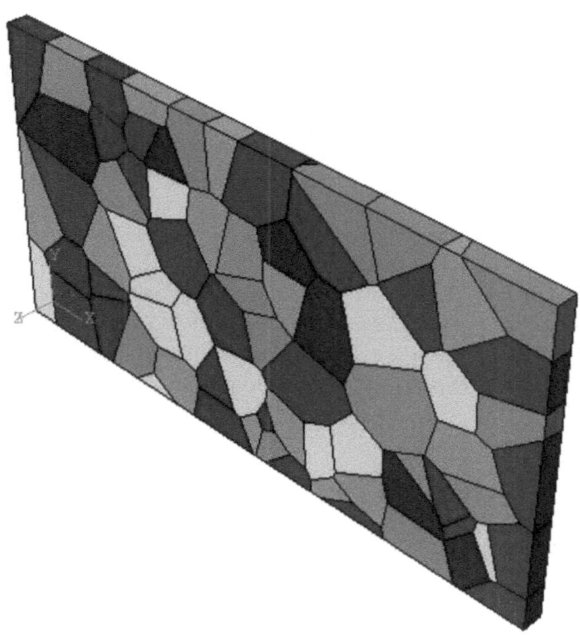

FIGURE 6.13 A Voronoi-treated specimen with different grain mechanical properties (each color represents one grain mechanical property).

REFERENCES

[1] X. Lai, L. Peng, P. Hu, S. Lan, J. Ni, Material behavior modelling in micro/meso-scale forming process with considering size/scale effects, Comput. Mater. Sci. 43 (2008) 1003–1009.

[2] R.W. Armstrong, I. Codd, R.M. Douthwaite, N.J. Petch, Plastic deformation of polycrystalline aggregates, Philos. Mag. 7 (73) (1962) 45–58.

[3] R.W. Armstrong, The Yield and Flow Stress Dependence on Polycrystal Grain Size, Applied Science Publishers, London, 1983.

[4] G.Y. Kim, J. Ni, M. Koç, Modeling of the size effects on the behavior of metals in microscale deformation processes, Trans. ASME 129 (2007) 470–476.

[5] S. Geißdörfer, U. Engel, M. Geiger, FE-simulation of microforming processes applying a mesoscopic model, Int. J. Mach. Tools Manufact. 46 (2006) 1222–1226.

[6] S. Geißdörfer, U. Engel, Mesoscopic model-simulation of size effects in microforming, in: J. Kusiak et al. (Eds.), Proceedings of the 10th International Conference on Metal Forming, Kraków, Poland, 19–22 September 2004, pp. 699–703.

[7] M. Geiger, S. Geißdörfer, U. Engel, Mesoscopic model: advanced simulation of microforming processes, Prod. Eng. Res. Dev. 1 (2007) 79–84.

[8] E.O. Hall, The deformation and ageing of mild steel: III discussion of results, Proc. Phys. Soc. Sect. B 64 (1951) 747–753.

[9] N.J. Petch, The cleavage strength of polycrystals, J. Iron Steel Inst. 173 (1953) 25–28.

[10] M.F. Ashby, The deformation of plastically non-homogeneous materials, Philos. Mag. 21 (1970) 399–424.

[11] M.A. Meyers, E. Ashworth, A model for the effect of grain size on the yield stress of metals, Philos. Mag. A 46 (1982) 737–759.

[12] H. Margolin, M.S. Stanescu, Polycrystalline strengthening, Acta Metall. 23 (1975) 1411–1418.

[13] R. Descartes, Principia Philosophiae, Apud Ludovicum Elzevirium, 1644.

[14] R. Descartes, in: V.R. Miller, R.P. Miller (Eds.), Principles of Philosophy, Springer, 1982.

[15] G.L. Dirichlet, Über die Reduction der positiven quadratischen Formen mit drei unbestimmten ganzen Zahlen, J. Reine Angew. Math. 40 (1850) 209–227.

[16] G. Voronoi, Nouvelles applications des paramètres continus à la théorie des formes quadratiques. Deuxième mémoire. Recherches sur les parallélloèdres primitifs, J. Reine Angew. Math. 134 (1908) 198–287.

[17] F. Aurenhammer, R. Klein, Voronoi diagrams, in: J.-R. Sack, J. Urrutia (Eds.), Handbook of Computational Geometry, Elsevier, 2000, pp. 201–290.

[18] W. Chan, M. Fu, J. Lu, J. Liu, Modeling of grain size effect on micro deformation behavior in micro-forming of pure copper, Mater. Sci. Eng. A 527 (2010) 6638–6648.

Chapter 7

Simulation of Micro Cross Wedge Rolling

7.1 INTRODUCTION

Cross wedge rolling (CWR) is one of the rolling technologies. One important characteristic of CWR is that the CWR process is highly automated [1,2] where the feeding, rolling, and cutting end of the billet can be accomplished in a single pass, resulting in higher productivity and higher material utilization than traditional machining, forging, and casting processes. Micro technology is gaining an increasing relevance due to the growth of the use of miniature metallic parts. Many microforming methods have been developed by previous researchers, e.g., micro upsetting, deep drawing, stamping, bending, and blanking. However, there exists a variety of challenges during scaling down the conventional forming process to be applied in the microscale [3]. Taking the advantage of CWR, MCWR has been discussed for the first time in this chapter [4,5]. Besides the similar issues existing in CWR, MCWR has to confront with the common problems in other micro manufacturing processes such as size effects.

Many explorations on the issue of size effects were conducted using both experimental and numerical methods and the corresponding constitutive models were developed as well. Hall [6] and Petch [7] studied the effect of grain size on the yield stress and found a linear relationship between the yield stress and the reciprocal of the square root of grain size. Miyazaki et al. [8] investigated the effect of specimen size on the flow stress of rod specimen of polycrystalline Cu-Al alloy and an affect zone model is proposed to explain why the flow stress decreases with the dimensional reduction of specimen. Leu [9] established a flow stress model as a function of the ratio of the sheet thickness to grain size based on the affect zone model, the pile-up theory of dislocation and Hall–Petch relationship.

Geiger et al. [1] proposed the surface layer model, which divided the specimen into two portions, inner and surface layer portions, to explain the reduction of flow stress in metal microforming process. A symmetrical influence of the sample size was proposed by Kals et al. [10] by defining a quantitative value for the share of surface grains. A mixed model was developed by combining the modified Hall–Petch relationship and surface layer model [11].

Microforming Technology. DOI: http://dx.doi.org/10.1016/B978-0-12-811212-0.00007-8

Based on the composite and surface layer models, Liu [12] developed a constitutive model considering the grain boundary strain strengthening. The grained heterogeneity was taken into account to investigate the decrease of flow stress and scatter of grained behavior due to different grain sizes, shapes, and orientations [13]. The grained heterogeneity was identified using nanoindentation test to simulate the micro cross wedge rolling (MCWR) process and an inhomogeneity coefficient has been introduced [14].

The Voronoi tessellation was introduced to generate polycrystalline aggregate using probability theory. This method subdivides a space tessellation into Voronoi polyhedral, which is similar to morphogenetic process of nucleation and growth from random seeds [15,16]. The concept of the Voronoi polygon has been used extensively in material science, especially for modeling random microstructure of grain aggregates [17,18]. Zhang and Tong [19] presented the numerical computation of 3D as a general investigation tool that featured an aggregate computation including element type and geometrical representation of grain boundary. Based on Voronoi tessellation, log-normal, and gamma distribution were used to describe the distribution of grain size in polycrystalline by Zhu et al. [20] and Dalla Torre et al. [21].

In this chapter, a mixed material model based on modified Hall–Petch relationship, surface layer model, and grained heterogeneity is proposed. The 3D aggregate of polycrystalline is represented by a Voronoi tessellation. The effect of grain size on flow stress is an important aspect of polycrystalline metal plastic deformation. The simulation of MCWR has been conducted with the consideration of size effects from grain size and feature size. The influence of temperature on the surface asperity in MCWR is discussed. The development of a more effective microforming method can produce the micro stepped shafts or cylindrical components with complicated cross sections. MCWR has been proposed for the first time, which can solve the problems of size effects in terms of materials, tools, machines, and equipment.

7.2 SIMULATION PROCEDURE OF MCWR

Fundamentals have been developed to build up a finite element (FE) model considering the occurrence of size effects at microscale using the ANSYS/LS-DYNA program. The newly developed material model is implemented considering grained heterogeneity. As shown in Fig. 7.1, a cylindrical workpiece of $\varnothing$ $0.8 \times 1.2\ \mathrm{mm}^2$ and two forming tools are meshed eight-noded structural solid element 164 with six degrees of freedom per node, translation and rotation in x, y, and z directions. One point integration and viscous hourglass control were employed for all the elements in the model to reduce computational time and ensure stability during large deformation. For each grain size, 10 different polycrystalline aggregates of workpiece were generated stochastically by the algorithm of 3D Voronoi tessellation. The simulation was performed by applying equal and opposite velocities to forming

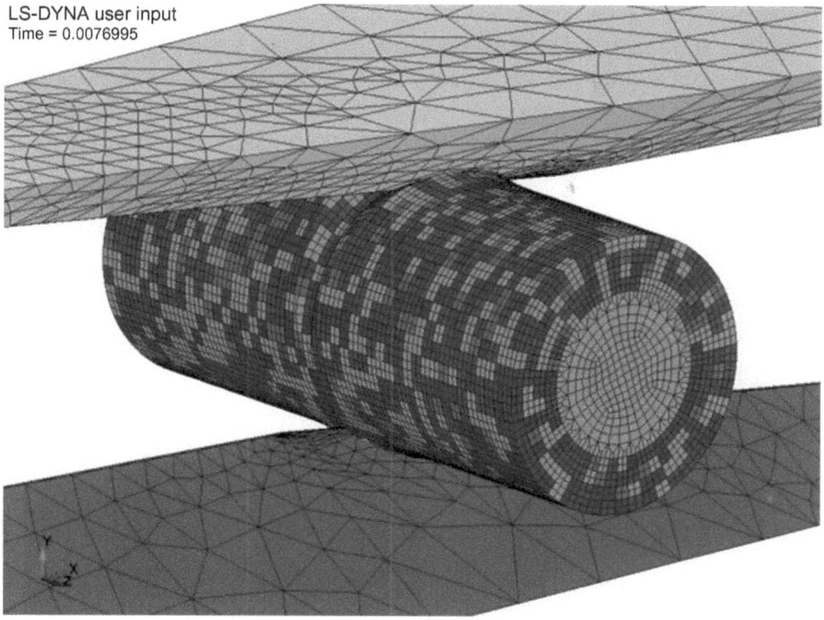

FIGURE 7.1 FE model in grained heterogeneities of workpiece.

tools in the horizontal (x) direction. In whole process, the workpiece is left unconstrained and the tools are held in vertical (y) direction and out-of-plane (z) direction [22].

In meshing process, space was discretized and described with a 3D digital image composed of voxels (pixels in 3D). According to the characteristics of Voronoi diagram, a distance function was introduced and a segmentation algorithm associates each voxel to the nearest nucleus. According to the experimental conditions, the applied method of MCWR modeling has been developed on the grounds of the following assumptions:

1. grains involved in the workpiece can be approximately described by Voronoi tessellation;
2. the micro workpiece is heated up to the expected temperature within a negligible short time;
3. the heat generated by plastic deformation is ignored within the workpiece;
4. an even distribution of constant temperature within the workpiece during compression; and
5. a constant coefficient of friction on the material–tool contact surface.

The workpiece and tools were defined to be Aluminum 1100 H16 alloy and hardened carbon steel 1045, respectively. The mechanical properties of studied materials are listed in Table 7.1. In this table, c is the inhomogeneity coefficient following the statistical distribution as shown in Fig. 7.2.

Numerical simulation was conducted to investigate the effect of grain orientations on the distribution of stress and strain, and dimensional accuracy in the MCWR process of metals. The influence of grain size and position were taken into account by subdividing the material into different regions, each characterized by individual properties. In such a case of modeling where the grain profile for simulation is idealized as cubic solid, it was possible to simulate local forming behavior controlled by grain size. A bilinear kinematic hardening constitutive plasticity model (Fig. 7.3) is defined for workpiece. In the initial stage, the inhomogeneities of material due to dimension miniaturization is realized to attribute individual properties to each grain by assumption that all grains have the uniform Young's modulus and individual tangent modulus.

TABLE 7.1 Mechanical Properties of Materials

Materials	Density (mg/m^3)	Young's Modulus (GPa)	Poisson Ratio	Yield Strength (MPa)	Tangent Modulus (MPa)
A1	2.710	69	0.33	138	202 × c
Tool steel	7.8	207	0.292	/	/

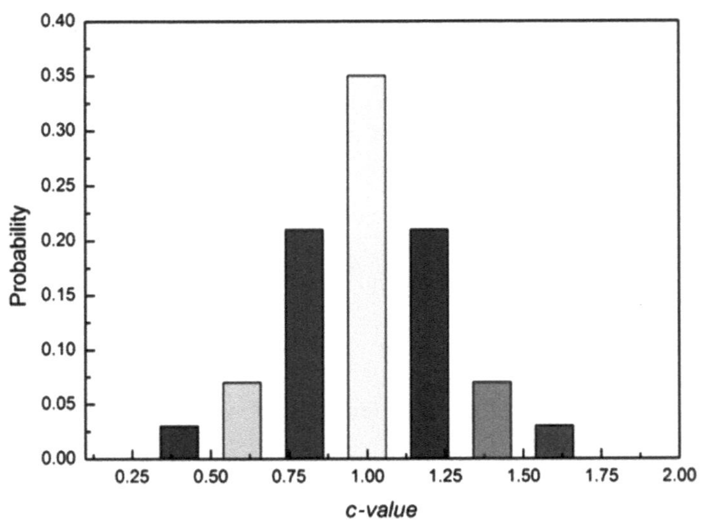

FIGURE 7.2 Statistical distribution of c-value.

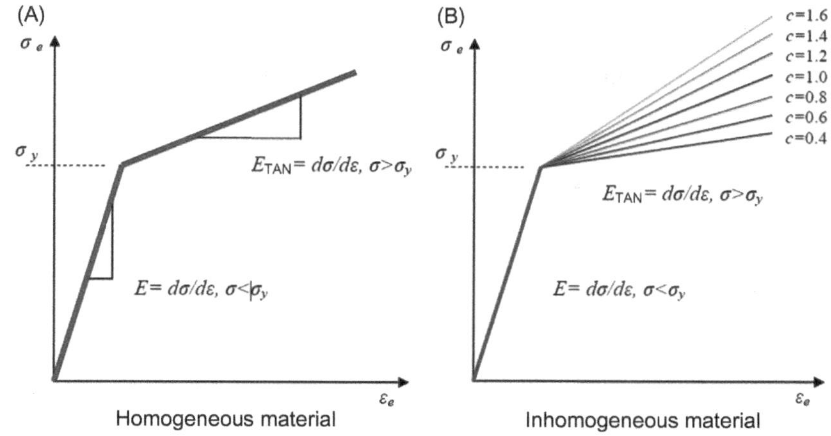

FIGURE 7.3 Material model of bilinear kinematic hardening: (A) homogeneous material and (B) inhomogeneous material.

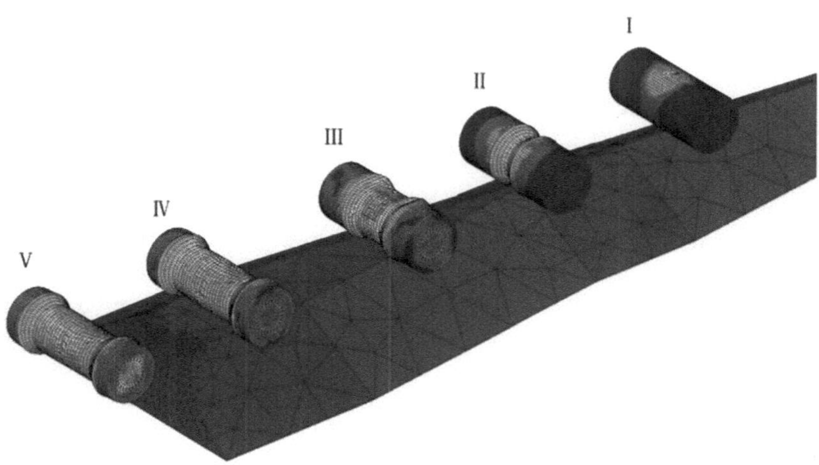

FIGURE 7.4 Process of forging shape during rolling.

$$\begin{cases} E = d\sigma/d\varepsilon, & \sigma < \sigma_y \\ c^{*}E_{TAN} = d\sigma/d\varepsilon, & \sigma < \sigma_y \end{cases} \tag{7.1}$$

where E and E_{TAN} are Young's modulus and tangent modulus of homogeneous material, respectively, and the variation of c-value represents material heterogeneity.

The diameter of cylindrical workpiece is less than 2 mm to ensure that there are two dimensions at least in the submillimeter category. The grain size ranges from 25 to 100 μm. Due to FEM application, it is possible to analyze proper change of workpiece shape during the rolling process. In Fig. 7.4, progression of step shaft during rolling is illustrated where cross sections of five rolling stages have been schematically shown: (I) knifing zone, (II) guiding zone, (III) stretching zone, (IV) sizing zone, and (V) unloading zone. This figure shows that wedges cut into workpiece in its middle part. Then the V-shaped groove is generated along the entire circumference after the guiding zone. Subsequently, in the stretching zone the reduction of cross section is developed to the required width. Finally the geometrical dimensions of workpiece are refined during the sizing zone.

It is also available to detect precisely evolution of stress and strain states during the deformation process with the assistance of FE simulation. In this part of workpiece, typical for CWR process laminar distribution of strain was also obtained. Numerical values of this strain lower with moving from layers toward the central region of workpiece. Area reduction, forming angles, and spreading angles are the primary parameters in the CWR process. These parameters play an important role in predicting the likelihood of interfacial slip and internal defects. The FE model built in the following section is used to obtain the optimum values of those parameters.

7.3 SIMULATION RESULTS

7.3.1 Forming Angle α and Stretching Angle β

The geometrical parameters of α and β are of significance in CWR process, in which can determine the final properties of products. Fig. 7.5 illustrates the variation of maximum stresses in core part of workpiece under a series of combinations (α, β) with the area reduction 36%. Despite irregular tendencies that the stresses in transverse and axial direction follow, the maximum effective stress falls obviously with a decrease of α and β. It is indicated that the stress value in the angles region around points with smaller spreading angles is smaller and the optimum parameters should locate in that region in consideration of core effective stresses that are vital inducements of significance of internal void.

The decrease of β value results in elongation of tools, which makes the specimen undergo more repeating loads to achieve final profile. Consequently, not only does it consume more space and time but also harm the qualities of products for too small spreading angles. However, β value exceeding some critical angle prevents the metal at surface of workpiece from rolling out, so the accumulated metal in transverse direction causes large tensile stress that induce internal defects (Fig. 7.6A).

If forming angle α is too small to generate enough force to push material out in axial direction, tensile stress will increase due to material accumulation at surface. However, if it is too large, the rising force in axial direction can cause necking (Fig. 7.6B).

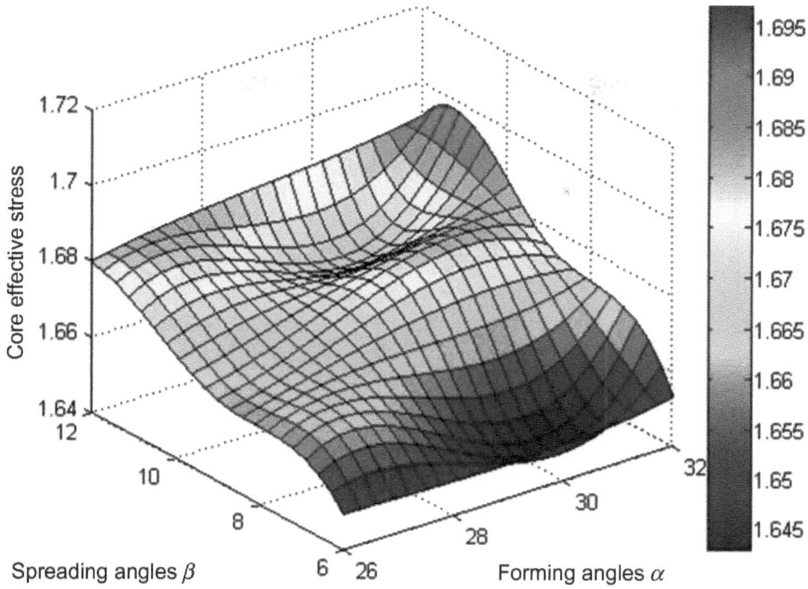

FIGURE 7.5 Tendency of core stress on variation of area reduction of workpiece using the tools with a forming angle 28 degrees and spreading angle 8 degrees.

Based on the analysis above, the combination of (28, 8 degrees) should be the optimum parameters considering time, space, and product qualities. On the basis of numerical simulation results, the optimum processing parameters are obtained, the area reduction of specimen cross section is 36% and the optimum combinations of forming and spreading angles locate around (28, 8 degrees).

7.3.2 Area Reduction ΔA

Simulations with different area reductions were conducted when α and β are 28 and 8 degrees, respectively. The numerical results demonstrate that the stresses in the core part of workpiece change sensitively with the area reduction. As shown in Fig. 7.7, stresses in axial and transverse direction goes down with an increase of area reduction ΔA while the effective stress goes up. When ΔA is quite small, the deformation occurs only at the surface of workpiece which incurs accumulation of dislocation in core part. When ΔA raises from 20% to 35% approximately, the increasing tendency of material flow in transverse direction predominates over that in axial direction and the value of stress in axial direction slides down slightly until the critical ΔA. However, once ΔA exceeds critical value approximately around 35%, stress in axial direction rises remarkably due to the increase of tensile force and deduction of cross-section area of deformation part actually reduction in size.

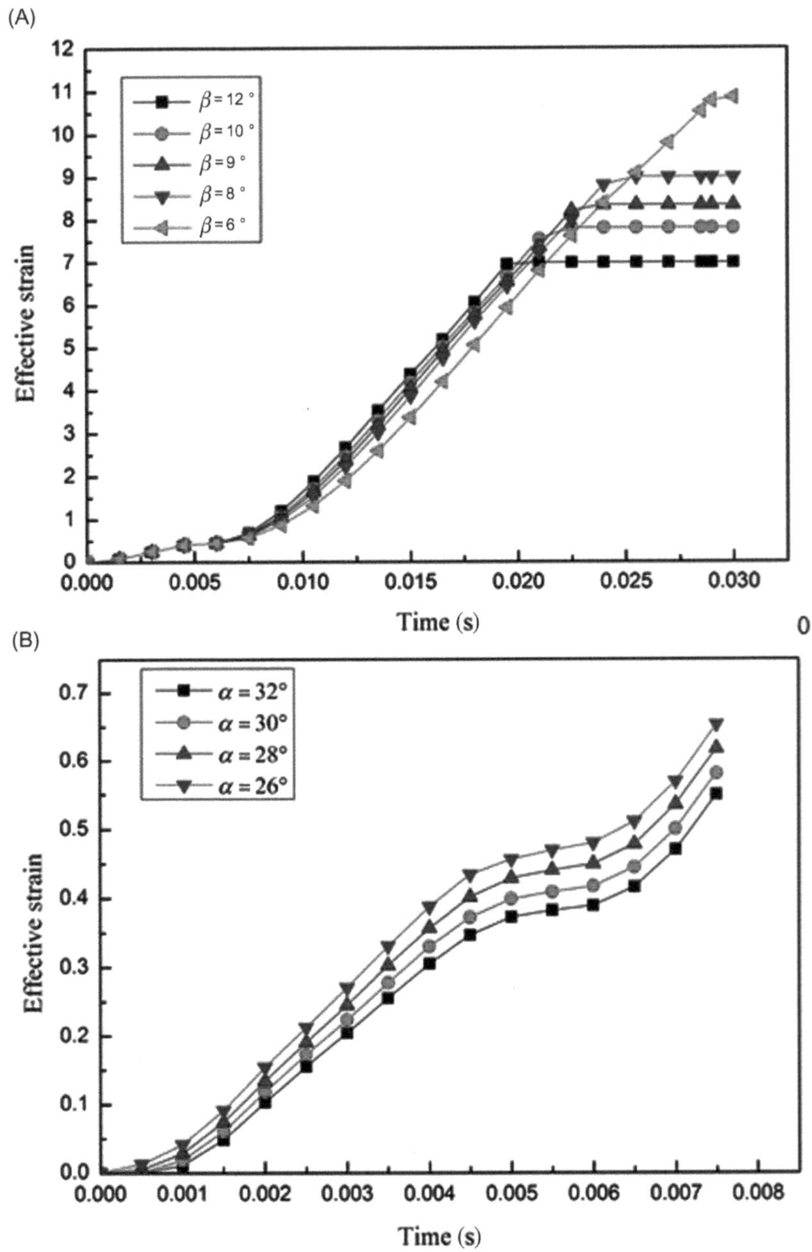

FIGURE 7.6 Influences of forming angle α and spreading angle β on effective stress in core region: (A) spreading angle effect and (B) forming angle effect.

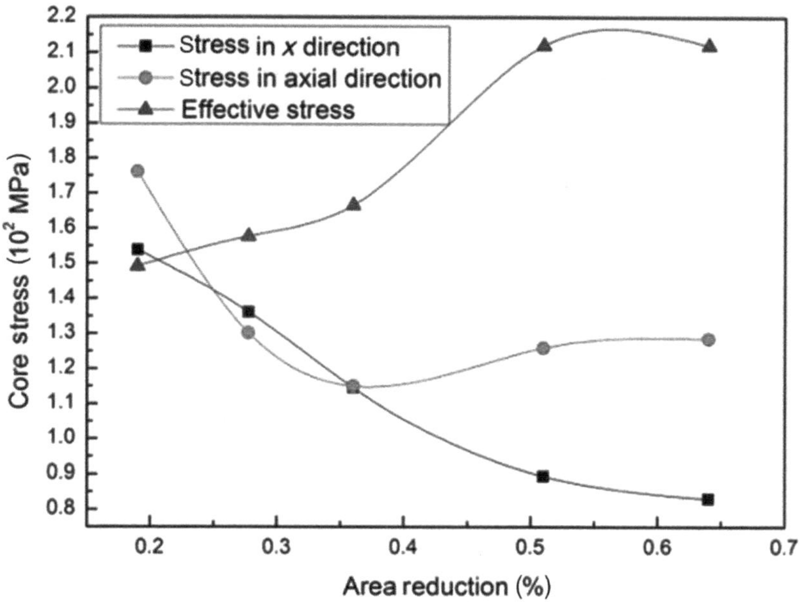

FIGURE 7.7 Effective strain status (A) effective strain with $\alpha = 28$ degrees and, (B) effective strain with $\beta = 8$ degrees in knifing zone and guiding zone.

7.3.3 Strain and Stress During MCWR

The strain and stress evolution during forming process may indicate the final product quality according to characteristics of metal forming [23]. In order to predict the mechanical property of MCWR product, the strain and stress development was numerically studied in this section. Laminar cricoid distribution of strain is typical in conventional CWR, where the material properties are regarded as homogeneous. This regular distribution is easier to predict the probability of material failure. The material behavior does not obey the conventional constitutive relation, so the novel material model should be applied to simulate MCWR. The effects of grain heterogeneities caused by grain size distribution and deformability anisotropy of crystals on the prediction of mechanical properties and defect pregnancy were investigated with assistance of FE method. In this numerical study, the forming and stretching angles are 28 and 8 degrees. The coefficients of friction at interface between tools and workpiece are 0.15, 0.3, and 0.45, respectively. The grain sizes of workpiece are assumed to be 6, 50, 120, and 240 μm with two kinds of grain size distributions. Considering the statistical precision simulation were repeated for 10 times in each grain size dimension. The materials of workpiece and tools are pure aluminum and tool steel, respectively.

Statistical log-normal density probability theories have been proved to be two typical distributions of measured grain diameters [24]. A log-normal distribution function $Lgn(D, M, S)$ of the variable D (with D positive and being the grain diameter) is defined so that the variable $x = \ln(D)$ follows a normal distribution. M and S are the mean value and standard deviation, respectively, of the variable x. Thus, the probability density function for the log-normal distribution is

$$P(D|M,S) = \frac{1}{S\sqrt{2\pi D}}\exp - \frac{(\ln(d)-M)^2}{2S^2} \tag{7.2}$$

and the cumulative distribution function should be

$$Cdf(D|M,S) = \frac{1}{2}\left(1 + \operatorname{erf}\frac{(\ln(d)-M)^2}{2S^2}\right) \tag{7.3}$$

The nth raw moment of the distribution writes

$$\mu_n = \exp\left(nM + \frac{1}{2}n^2S^2\right) \tag{7.4}$$

The mean μ, the variance Var, and the skewness α of the log-normal distribution are, respectively

$$\mu = \exp\left(M + \frac{S^2}{2}\right) \tag{7.5}$$

$$\text{Var} = \exp(S^2 + 2M)\cdot{}^*(\exp(S^2) - 1) \tag{7.6}$$

$$\alpha = (\exp(S^2)-1)^{1/2}\cdot(2 + \exp(S^2)) \tag{7.7}$$

Using the statistic tools of MATLAB 2012, a set of 2 log-normal distributions with different couples (M, S) has been generated. Table 7.2 gathers the statistical discrete diameters of 25 μm. The values of μ, Var, and α are derived from Eqs. (7.5)–(7.7). Regarding discrete distributions, the mean grain diameter D_{mean} and parameters characterizing the grain size descriptions: absolute range ($D_{min} - D_{max}$), standard deviation (SD), and mean absolute (MAD) are given in Table 7.2.

Here, the 3D Voronoi tessellation was utilized to represent the metal polycrystalline structure. These grain diameter distributions finally were converted into the distributions of equivalent Voronoi cell diameter. The geometrical

TABLE 7.2 Statistic Data Characterizing the Log-Normal Distribution

Set	M	S	μ	Var	α	D_{min}	D_{max}	D_{mean}	Rang	SD	MAD
1	1.782	0.140	19.94	7.87	0.42	12.2	32.2	20	20.0	2.80	2.22
2	2.860	0.510	19.89	117.48	1.80	3.0	103.2	20	100.2	10.81	8.0

models were imported into FE simulation software—ANSYS/LS-DYNA. For analyzing the influences of distribution of strain and material heterogeneity on the surface asperity of workpiece, FEM simulations with uniform material properties and inhomogeneous properties were developed, respectively. The distribution of stress and strain and the surface status of workpiece in both cases were compared.

Laminar cricoid distribution of strain is typical in conventional CWR, where the material properties are regarded as homogeneous. This phenomenon also occurs in MCWR if the material of billet is homogeneous (Fig. 7.8A). In microscale, the influence of grained heterogeneity on the metal deformability and strain distribution cannot be neglected. As shown in Fig. 7.8B–D, the continuous laminar distribution of strain in the workpiece has been disturbed due to the inhomogeneous mechanical properties. The location of the maximum strain and stress cannot be determined easily as that in the conventional CWR process.

Since the distribution of grain size in inhomogeneous materials provides heterogeneity, it appears fundamental to get an accurate description of effect of grain size on the local interaction and behavior, and also, a relevant mathematics description of grain size statistics. The different microstructure features of polycrystalline aggregates (such as different grain sizes and plastic anisotropies of the crystals) lead to the development of intergranular interactions which are sources of the internal stress, accommodation, and texture changes. When interaction becomes significant, material flow is modified by internal stresses and plastic accommodation occurs. For materials with complex local behavior, the overall behavior does have the same reaction as the local one in macroscale. However, the micromechanical heterogeneity of metal polycrystalline which is mainly the result of different grains to grain interactions is a crucial issue.

According to the generation mechanism of workpiece interior failure, the stress and strain evolution in the central region should be focused. Therefore, the disturbance of the grained heterogeneity is an important issue because it may cause the misdirection to predict and avoid failure generation. In order to investigate the influence of material inhomogeneity caused by

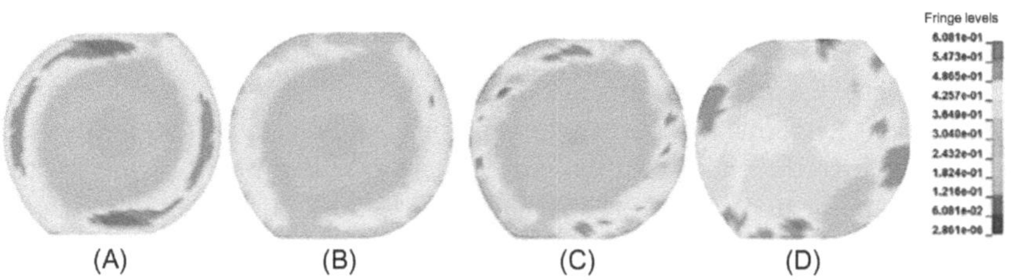

FIGURE 7.8 Distribution of effective strain of the medial section in axial direction (A) uniform material properties and (B, C, D) inhomogeneous material properties with grain sizes of 6, 40, and 120 μm, respectively.

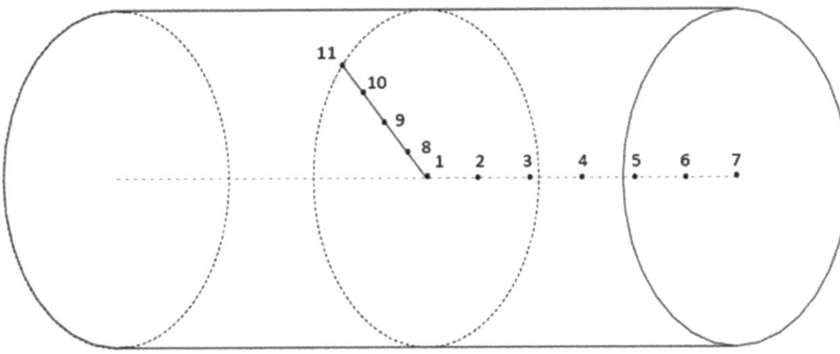

FIGURE 7.9 The points selected to capture the variations of stress and strain.

miniaturization, 11 spots within the cylindrical workpiece are selected as shown in Fig. 7.9. Points 1–7 distribute evenly on the semi axis and the Points 1, 8, 9, 10, and 11 spread evenly on the radius of the workpiece middle circular cross section. The evolution of stresses and strains in vicinity of these spots are captured during the whole MCWR process.

In the conventional CWR process, there is tendency that the internal defect may occur under some conditions. This has been regarded as a special type of Manesmann effect which is the most common defect that limits the usage of the CWR technique. The central portion of a billet becomes significantly weakened due to the Manesmann effect when it is rotated and compressed. This weakness ultimately leads to the generation of internal voids. The large tensile stresses initiate the opening of the micro and local shear determines the level of aggregation and the size of void within the workpiece. The slip deformation occurs due to the tensile stresses in the direction of 45 degrees angle to the tensile stresses. With the development of deformation process, voids merge into larger fractures. Therefore, the Manesmann effect still exists in MCWR and may play a significant role in the micro product quality. Moreover, the grained heterogeneity complicates the analysis of strain and stress evolutions that is the mercury of the material failure within workpiece.

Figs. 7.10 and 7.11 demonstrate the stresses in *X*, *Y*, *Z* directions and von Mises effective stresses within the homogenous and inhomogeneous workpieces, respectively. In Fig. 7.10, the material located near spots 1, 8, and 9 undergo repeated hydrostatic pressures, which indicates that there are a high probability of internal-void formation and propagation. In contrast, it is difficult to identify the exact position where there is a likelihood of the internal void development because the distribution of stress fluctuates dramatically within the deformed region of workpiece. As it can be seen in Fig. 7.11, the variation of v-m effective stresses at spots 10 and 11 seem not so regular as

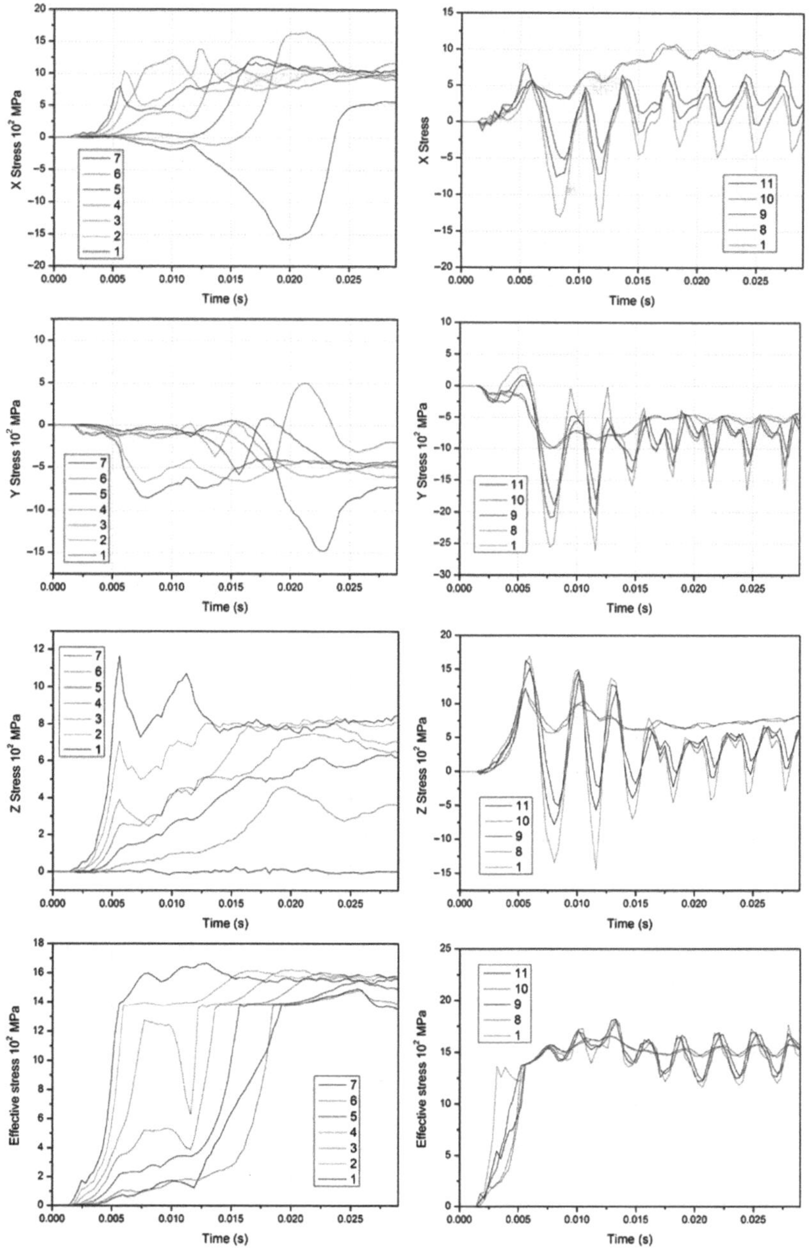

FIGURE 7.10 Evolution of stresses at the select points within the homogeneous workpiece.

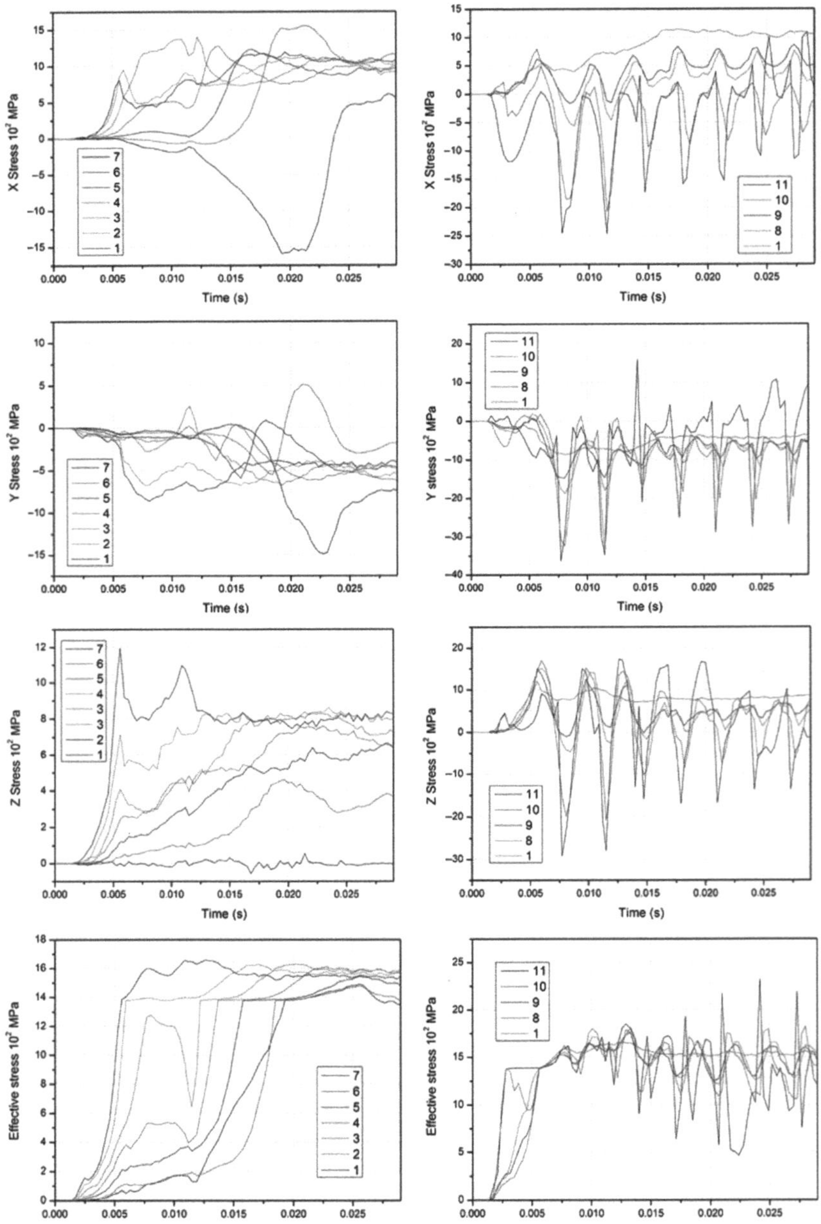

FIGURE 7.11 Evolution of stresses within workpiece with grain size of 120 μm in consideration of grain heterogeneities.

that in Fig. 7.10 when the material is considered as homogeneous. That is because the deformability of individual grain affect the stress distribution resulting in poor strain compatibility. However, the influence of grained heterogeneity on the stress distribution near the axil is less significant according to the comparison between the stress evolutions at spots 1–7 in Figs. 7.10 and 7.11.

The numerical results also show the grained heterogeneity affects the strain distributions. As described in Fig. 7.12, although there is a small difference between the effective plastic strain evolutions in axial direction, the influence of grained heterogeneity is significant on the effective plastic strain distribution in radius direction.

Not only the mean grain size but also the width of the grain size range is important to the material deformed behavior. The grain size distribution is another factor that contributes to the influence of grained heterogeneity. The effect of the scatter of grain size with the workpiece on the stress distribution has been studied in order to find out the optimum material state for the MCWR deformation. Two sets of polycrystalline aggregates with meaning grain size 20 μm were generated. The grain size distributions of these two

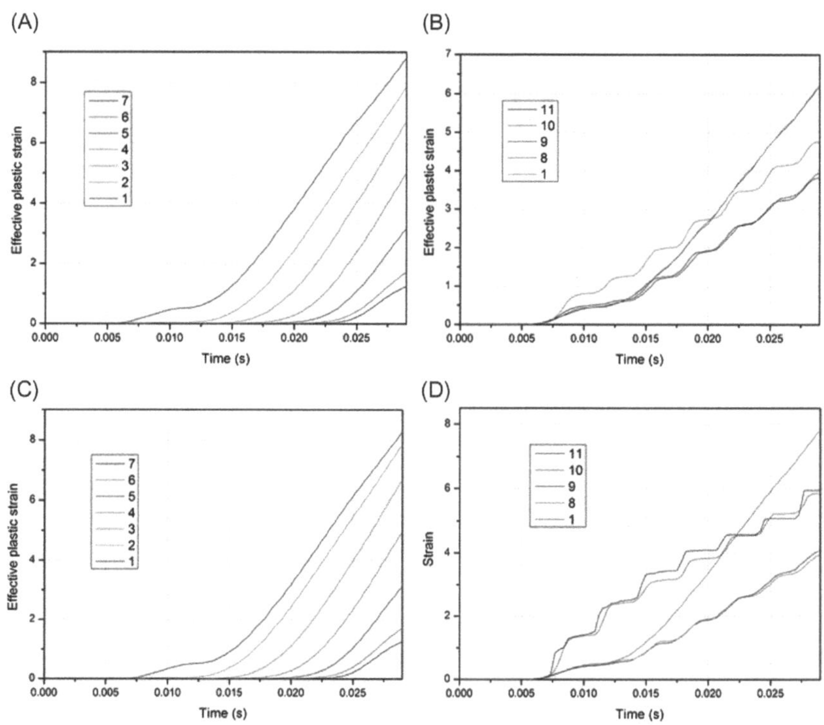

FIGURE 7.12 Evolution of stresses at the select points (A, B) simulation with homogeneous material and (C, D) simulation with inhomogeneous material.

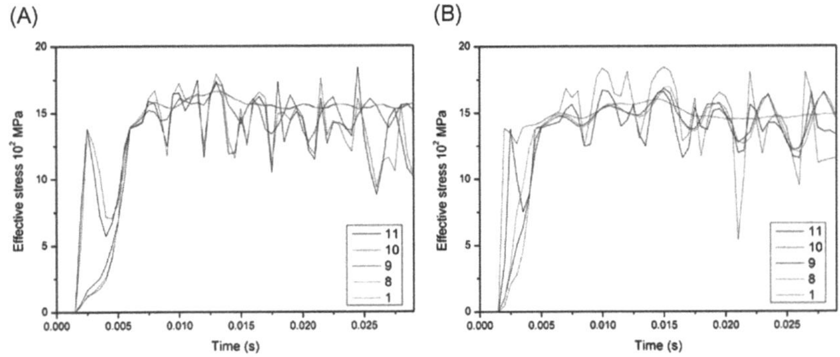

FIGURE 7.13 The stress evolutions of spots in radius direction of workpieces with mean grain size 20 μm: (A) variance of 7.87 and (B) variance of 117.48.

sets follow log-normal distribution and the variances are 7.87 and 117.48, respectively. When the scatters of grain size distributions increase, the influence of grained heterogeneity becomes obvious. Fig. 7.13 demonstrates, as to the workpiece with grain size distribution variance of 7.87, the evolution of effective stresses is more similar to that of homogenous workpiece in contrast to the workpiece with grain size distribution variance of 117.48. Consequently, in the practical production, the metal polycrystalline should be treated to obtain the blank material with more homogeneous grain size distribution, which is an effective way to lessen grained heterogeneity.

7.3.4 Size Effects in MCWR Process

The relationship between the specimen geometrical dimension and material grain size is taken into account to investigate the size effect during deformation by the ratio of the specimen size (g) to grain size (s) as shown in Eq. (7.8).

$$\lambda = \frac{\text{Specimen radius}(g)}{\text{Grain size}(s)} \tag{7.8}$$

As shown in Fig. 7.14A, an FE model of MCWR process of pure copper has been established taking into account the size effects using the novel developed model. It is used to predict and analyze the distribution of strain and stress and generation of defects. The geometrical dimension of the cylindrical specimen is $\varnothing\ 0.8 \times 0.8\ \text{mm}^2$ and the area reduction of deformed part is 43.75%. The specimen model involves 1313×40 solid elements and other parameters are the same as that in micro compression. The equal and opposite velocities were applied on two wedges in x direction, and the two wedges are fixed in both y and z directions.

A characteristic of CWR is the laminar cricoid distribution of strain when the material properties are regarded as homogeneous. This phenomenon also

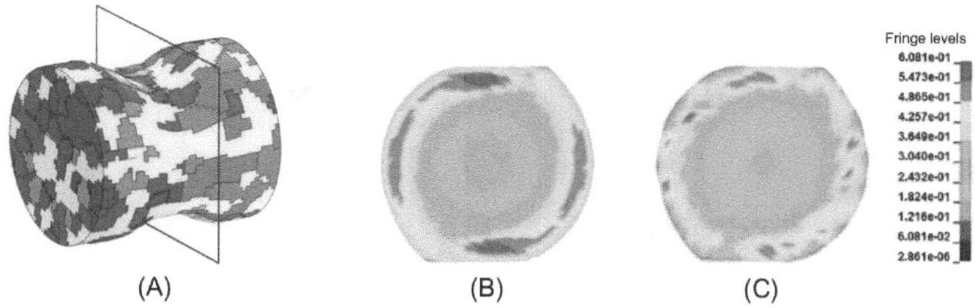

FIGURE 7.14 Distribution of effective strain of the medial section in axial direction: (A) FE model of MCWR in proceeding, (B) uniform material model, and (C) inhomogeneous material model [4].

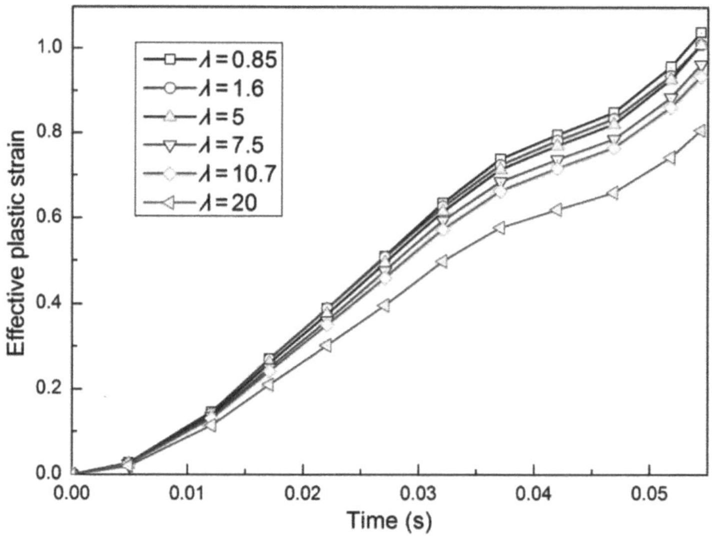

FIGURE 7.15 Effective strain in knifing and guiding zones [4].

occurs in MCWR if the material of billet is homogeneous (Fig. 7.14B). In microscale, the influence of grained heterogeneity on the metal deformability and strain distribution cannot be neglected. The cross sections shown in Fig. 7.14 located in the middle of the cylindrical specimen with $\lambda = 3$. As shown in Fig. 7.14C, the continuous laminar distribution of strain in the workpiece has been disturbed due to the inhomogeneous mechanical properties. The location of maximum strain and stress cannot be determined easily as that in the CWR process without considering the size effects. Fig. 7.15 shows the curves of the effective strain at vicinity of specimen core region in knifing and guiding zones. The strain falls down with the increase of λ value, which means there is more deformation around the core region of specimen with larger grain size.

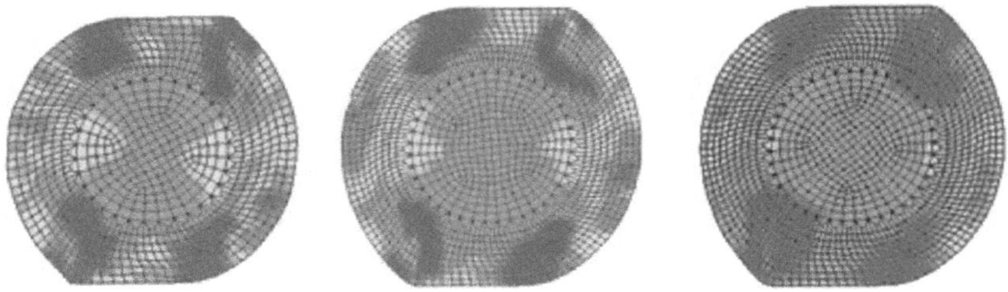

FIGURE 7.16 von Mises stress distribution on middle cross section of workpieces with different grain sizes: (A) 248, (B) 40, and (C) 6 μm [25].

The grain size effects make the strain and stress distribution become more complicated. The cross sections shown in Fig. 7.16 located in the middle of the cylindrical billets when the deformation of billets is in the guiding stage. As shown in Fig. 7.16, with an increase of grain size, the continuous laminar distribution of strain in the workpiece has been more significantly disturbed due to inhomogeneous mechanical properties. Because the material at macroscale is regarded as homogeneous, the location of maximum strain and stress cannot be determined easily as that in the conventional CWR process.

When the grain size is 40 μm, the disturbance of stress distribution is much smaller than that of the workpiece with grain size of 248 μm. Moreover, the workpiece with grain size of 6 μm shows a similar distribution of stress. Fig. 7.17 illustrates the stress distribution evolution of the workpiece with grain size of 40 μm in each rolling stage. It can be seen that the effects of grain size on the stress in radial direction are more obvious than that in axial direction because the geometry dimension in axial direction is larger than that of radial direction.

7.3.5 Influence of Temperature on the Surface Asperity in MCWR

The influences of temperature on flow behavior are even more pronounced in micro metal forming, because so-called "size effects" cause inhomogeneity in terms of the material flow behavior and inaccuracy in the shape of the items produced. The latter was focused in this paper. The MCWR tests as well as FE simulations were implemented at the temperatures of 300, 450, and 600 K.

A material model of pure copper was developed based on dislocation mechanics theory and Hall–Petch relation. It is expressed as:

$$\sigma_u = \sigma_0 + 0.478\varepsilon^{-0.1}d^{-1/2} + 808.\sqrt{1-\exp(-10\varepsilon)}\left[1+10^{-4}T\ln\left(\frac{\dot{\varepsilon}}{\dot{\varepsilon_0}}\right)\right]^{0.45} \tag{7.9}$$

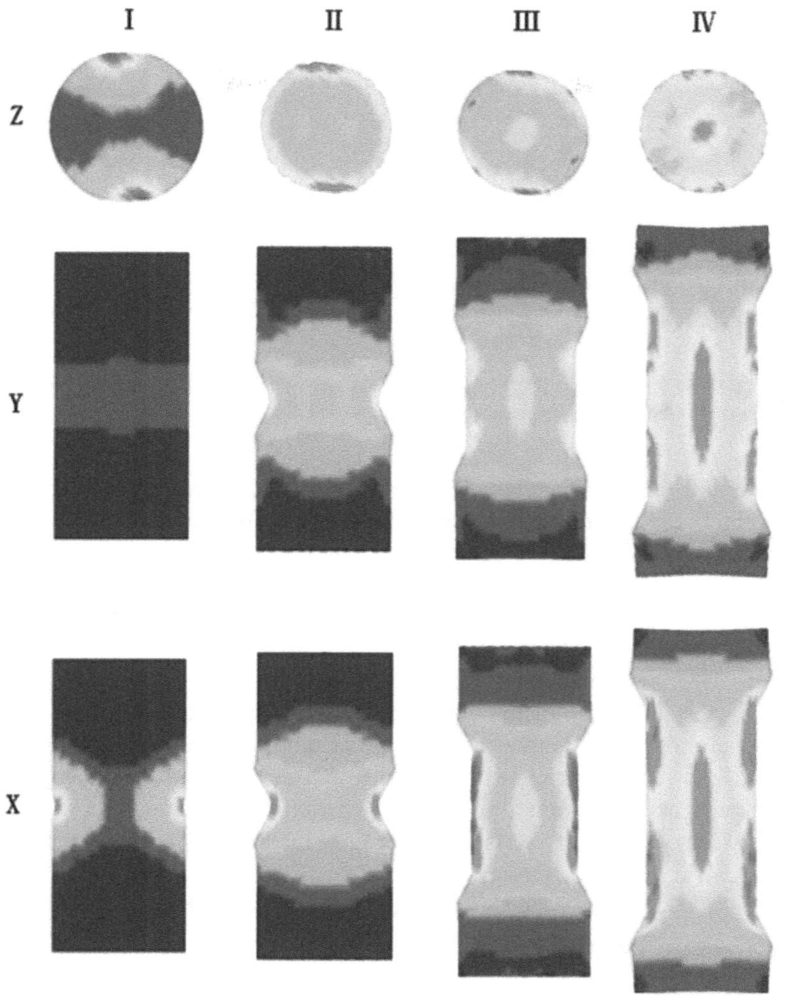

FIGURE 7.17 Evolution of von Mises stress distribution in longitudinal and cross sections of cylindrical workpiece [25].

where σ_u is the stress, ε and $\dot{\varepsilon}$ are the strain and strain rate, respectively, T is the deforming temperature, and d is the grain size. In this simulation, $\sigma_0 = 15$ MPa, $\dot{\varepsilon}_0 = 1.76 \times 10^8\ \mathrm{s}^{-1}$, and $\dot{\varepsilon} = 0.065\ \mathrm{s}^{-1}$. A series of flow stress curves can be obtained at different temperatures.

The experimental and simulation results obtained were analyzed in terms of the surface quality. In order to measure the surface asperity quantitatively, 12 generatrices uniformly spreading on the side surface of cylindrical specimen were determined as illustrated in Fig. 7.18. After deformation, surface becomes rough comparing with the original sample surface, which can be

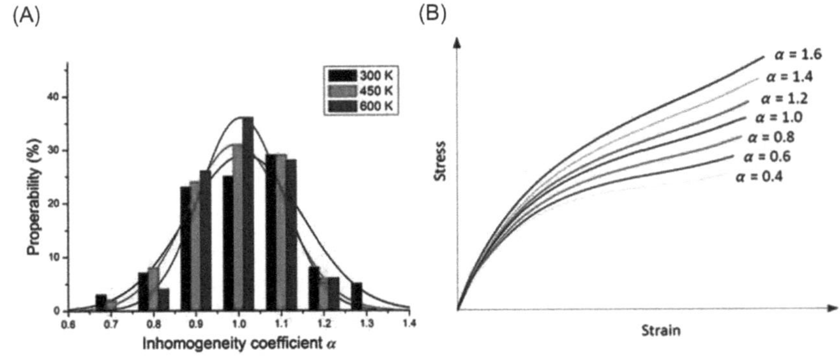

FIGURE 7.18 (A) Statistic distribution of inhomogeneity coefficient influenced by temperature and (B) strain–stress curve varies along with inhomogeneity coefficients [26].

measured by the distance between the nodes at the generatrice and the original center, and the average values of the distance obtained from the four generatrices were calculated. Assuming the deformation behaviors of grains inside the sample are homogenous, side surface of the compressed sample is perfectly smooth. One generatrice of this assumed compressed sample is used as a reference. The results from the homogeneous model and inhomogeneous model are compared.

The absolute value of the difference between the values by adopting conventional model with homogeneous material property and the model considering grained heterogeneity is defined as H_{sa}. The mean value of the absolute differences obtained from all of the 12 generatrices is defined as H_{sa} and has been used to assess the surface roughening quantitatively. The experimental and numerical results were calculated and analyzed to obtain H_{sa}.

Fig. 7.19 illustrates the influence of temperature on H_{sa}. It is noticed that there was a significant decrease of H_{sa} with increasing temperature for copper specimen with grain size of 235 μm. H_{sa} declined linearly from 1.8 to 1.1 when temperature increased from room temperature to 600 K as illustrated in numerical results and the experimental data show the same tendency. During the plastic deformation of metals, dislocations move and accumulate, which has to overcome the obstacles spreading among the crystal structure. For the typical face-centered cubic (FCC) metal, copper, the primary short-range barriers are other dislocations which intersect the slip plane and impede the motion of gliding dislocation. The increasing temperature of deformed part can activate the constrained slip system resisted by short-range obstacles, which homogenize grained deformability of the metal polycrystalline.

This constitutive model was capable of modeling the FCC material responses in microscale at certain range of temperature. The experimental and numerical results illustrate a homogeneous phenomenon of deformability due to elevated temperature, which was analyzed based on surface asperity.

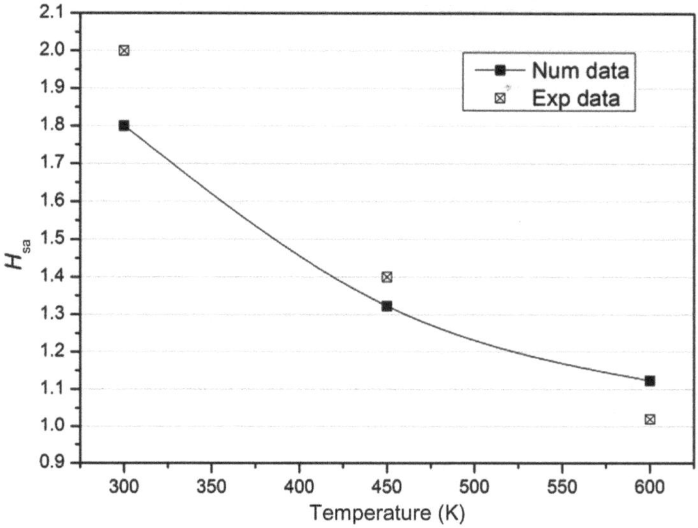

FIGURE 7.19 Influence of temperature on H_{sa} [26].

With the thermal activation, more slip systems were activated, leading to improvements in the edge quality of the deformed workpieces in MCWR.

In summary, CWR technique has been introduced into microscale for the first time due to its unique advantages, but it is inevitable to confront with the common issue in microforming process—size effects. A constitutive model considering grain boundary strengthening effect, surface layer model, and grained heterogeneity has been proposed. A set of 3D aggregates of polycrystalline with different grain sizes were generated using Voronoi tessellation. The scatter of grained deformation behavior has been investigated through flow stress and side surface asperity of cylindrical specimen. The results show that the scatter of deformation behavior becomes significant as the grain size increases while specimen size kept constant. The size effects on the material flow behavior during MCWR were also numerically investigated. The strain and stress status within workpiece during MCWR were studied with assistance of FE simulation and the numerical results reflect the distribution and evolution of strain and stress become irregular when the grained heterogeneity is regarded in material model. The numerical simulation results also show the contribution of grain size distributions to the grained heterogeneity. It is applicable to predict MCWR using the proposed FE simulation models.

REFERENCES

[1] M. Geiger, M. Kleiner, R. Eckstein, N. Tiesler, U. Engel, Microforming, Ann. CIRP 50 (2001) 445–462.

[2] Q. Li, M. Lovell, Cross wedge rolling failure mechanisms and industrial application, Int. J. Adv. Manufact. Technol. 37 (2008) 265–278.

[3] J. Jeswiet, M. Geiger, U. Engel, M. Kleiner, M. Schikorra, J. Duflou, et al., Metal forming progress since 2000, CIRP J. Manufact. Sci. Technol. 1 (2008) 2–17.

[4] H.N. Lu, D.B. Wei, Z.Y. Jiang, X.H. Liu, K. Manabe, Modelling of size effects in microforming process with consideration of grained heterogeneity, Comput. Mater. Sci. 77 (2013) 44–52.

[5] Z.Y. Jiang, H.N. Lu, K.Z. Linghu, X.M. Zhao, X.M. Zhang, D. Wu, Finite element method analysis of micro cross wedge rolling of metals, Proc. Eng. 81 (C) (2014) 2463–2468.

[6] E.O. Hall, Deformation and ageing of mild steel, Phys. Soc. Proc. 64 (1951) 747–753.

[7] N.J. Petch, Cleavage strength of polycrystals, Iron Steel Inst. 174 (1953) 25–28.

[8] S. Miyazaki, K. Shibata, H. Fujita, Effect of specimen thickness on mechanical properties of polycrystalline aggregates with various grain sizes, Acta Metall. 27 (1979) 855–862.

[9] D.K. Leu, Modeling of size effect on tensile flow stress of sheet metal in microforming, ASME J. Manufact. Sci. Eng. 131 (2009) 0110021–0110028.

[10] R. Kals, H.J. Pucher, F. Volertesen, Effects of specimen size and geometry in metal forming, in: S.J. Hashmii (Ed.), Proceedings of the 2nd International Conference on Advances in Materials and Processing Technologies, vol. 3, 1995, pp. 1288–1297.

[11] X.M. Lai, L.F. Peng, P. Hu, S.H. Lan, J. Ni, Material behavior modelling in micro/meso-scale forming process with considering size/scale effects, Comput. Mater. Sci. 43 (2008) 1003–1009.

[12] J.G. Liu, M.W. Fu, W.L. Chan, A constitutive model for modelling of the deformation behavior in microforming with a consideration of grain boundary strengthening, Comput. Mater. Sci. 55 (2012) 85–94.

[13] W.L. Chan, M.W. Fu, J. Lu, J.G. Liu, Modeling of grain size on micro deformation behaviour in micro-forming of pure copper, Mater. Sci. Eng. A 527 (2010) 6638–6648.

[14] H.N. Lu, D.B. Wei, Z.Y. Jiang, K. Manabe, T. Furushima, Steel Res. Int. (2011) 962–967.

[15] G.F. Voronoi, Nouvelles applications des parameters continus à la théorie des formes quadratiques, J. Reine Angew. Math. 134 (1908) 198–287.

[16] G.F. Voronoi, Propriétés des formes quadratiques positives parfaits, J. Reine Angew. Math. 136 (1909) 67–181.

[17] O. Diard, S. Leclercq, G. Rousselier, G. Cailetaud, Evaluation of finite element based analysis of 3D multicrystalline aggregates plasticity: application to crystal plasticity model identification and the study of stress and strain fields near grain boundaries, Int. J. Plast. 21 (2005) 691–722.

[18] A.G. Crocker, P.E.J. Flewitt, G.E. Smith, Computational modeling of fracture in polycrystalline materials, Int. Mater. Rev. 50 (2005) 99–124.

[19] N. Zhang, W. Tong, An experimental study on grain deformation and interactions in an Al-0.5% Mg multicrystal, Int. J. Plast. 20 (2004) 523–542.

[20] B. Zhu, R.J. Asaro, P. Krysl, K. Zhang, J.R. Weertman, Effects of grain size distribution on the mechanical response of nanocrystalline metals: part II, Acta Mater. 54 (2006) 3307–3320.

[21] F. Dalla Torre, H. Van Swygenhoven, M. Victoria, Nanocrystalline electrodeposited Ni: microstructure and tensile properties, Acta Mater. 50 (2002) 3957–3970.

[22] H.N. Lu, A Study on the Micro Cross Wedge Rolling of Metals, Doctoral dissertation, University of Wollongong, Australia, 2013.

[23] F. Vollertsen, Categories of size effects, Prod. Eng. 2 (2008) 377–383.

[24] R.Z. Valiew, R.K. Islamgaliev, I.V. Alexandrov, Bulk nanostructured materials from severe plastic deformation, Progr. Mater. Sci. 45 (2000) 103–189.

[25] Z.Y. Jiang, H.N. Lu, D.B. Wei, K.Z. Linghu, X.M. Zhao, X.M. Zhang, et al., Finite element method analysis of micro cross wedge rolling of metals, Proc. Eng. 81 (2014) 2463–2468.

[26] H.N. Lu, D.B. Wei, Z.Y. Jiang, D. Wu, X.M. Zhao, Study on the Influence of Temperature on the Surface Asperity in Micro Cross Wedge Rolling, AIP Publishing, 2013, pp. 1032–1037.

Chapter 8

Simulation of Micro Flexible Rolling

8.1 INTRODUCTION

Flexible rolling, a novel forming method for the production of strips with longitudinal thickness variation, has been rapidly developed and gradually used on a large industrial scale, such as automotive, construction, and energy industries. In recent years to meet the rising demand for weight-saving and cost-efficient products, the roll gap may be adjusted to achieve a variable thickness along the metal sheet's length during the rolling process [1]. In the meantime, numerous research activities have been conducted to obtain a better understanding and make further development of this forming technology. Engler et al. [2] presented the process optimization and control of material properties in the flexible rolling of aluminum alloy sheet based on lab-scale experiments, which elucidated the range of properties that could be realized through flexible rolling while maintaining reproducible constant material properties. Huang et al. [3] proposed a flexible rolling simulation model on the basis of Deform-3D and verified the feasibility of flexible rolling by comparison between simulation curve and target profile curve. They later discussed the influence of different process parameters on the quality of flexibly rolled products and optimized these parameters to obtain better results via orthogonal trial [4]. Kopp et al. [5] reviewed the flexible rolling process launched at the Institute of Metal Forming and the company Mubea. With a controlled online adjustment of the roll gap, any thickness transitions could be selected and adapted exactly to the demanded load of the product in use. Zhang and Tan [6] derived a mathematical formula to control the vertical motion of the roll in flexible rolling, which proved effective by simulation outcomes, and they also found that the vertical velocity of the roll was significantly impacted by the length of the transition zone of flexibly rolled blank and the diameter and rotation speed of the roll. Liu et al. [7–9] set up formulas for calculating parameters, such as rolling force, forward slip, and horizontal velocity in flexible rolling, and performed finite element simulations and experiments to further verify the theoretical calculations. This technology had proved feasible and would become a new growth point in rolling area.

Nowadays it is promising to transfer the knowledge of flexible rolling from the macro to micro world due to the current trend toward miniaturization

Microforming Technology. DOI: http://dx.doi.org/10.1016/B978-0-12-811212-0.00008-X

of systems and devices, particularly in the field of electronics production. When the workpiece thickness is scaled down to the micrometer range with other two workpiece-dimensions decreased to the millimeter range in micro flexible rolling process, the so-called size effects that characterize the micro world have to be taken into account. The size effects have also been revealed through the flow behavior of the material obtained from the tensile test [10–12], in which a decrease in flow stress as well as integral flow curve was observed with a decrease in specimen thickness. With decreasing specimen thickness, the share of surface grains increases. As the surface grains are less constrained than grains inside the material, the increasing number of surface grains leads to less hardening and lower resistance against deformation of the whole specimen, which results in lower flow stress and flow curve [13–15]. On account of the dwindling specimen size, the influence of each grain on the micro-scaled deformation is enhanced. The grains with various sizes, orientations, and shapes are distributed in the specimen nonuniformly, giving a rise to inhomogeneous material behavior and the scatter of experimental results [16,17]. Consequently in the finite element simulation of microforming process, Voronoi tessellation has been widely applied to model the microstructure of grain aggregate, and then the material heterogeneity caused by different grain properties can be reflected via assigning a certain property to each Voronoi polyhedron [18–21].

In this chapter, a mathematical model is set up to explicate the flexible rolling phases with thicker and thinner thicknesses respectively. According to the geometric parameters analyzed in the mathematical model, the 3D micro flexible rolling finite element analysis (FEA) model is built, where initial workpiece thicknesses of 100, 250, and 500 μm with reduction changing from 20% to 50% are respectively taken into account. Voronoi polyhedrons with varying material properties classified by micro hardness testing are introduced to embody the material inhomogeneity of the workpiece. Five types of grain sizes are respectively employed in the workpiece to ascertain the scatter of the springback in thickness direction, and a model in accordance with the FEA results is deduced to quantitatively predict the springback. The surface asperity deformation of workpiece with different reductions has been analyzed. Four kinds of initial workpiece surface roughness are separately applied to the workpiece, and how the coefficient of friction (COF), rolling speed, and roll gap adjusting speed influence the final surface roughness are evaluated for each sort of initial surface roughness.

8.2 SIMULATION PROCEDURE

8.2.1 Theory Analysis of Flexible Rolling Process

8.2.1.1 Angle of Bite

During the micro flexible rolling process, the exit position of the workpiece is not at the central line of the rolls because of the existence of tilt angle of

$$r_x = \Delta - \delta_x = (\Delta - \delta_0) + 2v \cdot t \cdot \tan\theta \tag{8.7}$$

where Δ is the initial workpiece thickness, and likewise, the reduction at arbitrary location x in the rolling phase with thinner thickness (as shown in Fig. 8.4) can be expressed as

$$r_x = \Delta - \delta_x = (\Delta - \delta_0) - 2v \cdot t \cdot \tan\theta \tag{8.8}$$

It can be seen from Eqs. (8.7) and (8.8) that the reduction r_x is a function of speed v, time t, and tilt angle θ, and it only changes in view of whatever t takes on when the other two variables remain a constant during micro flexible rolling process.

8.2.1.3 Length of Arc of Contact

The length of arc of contact is defined to be the horizontal projection length of the arc of contact between the rolls and the rolled workpiece [24]. In conformity with this definition, the length of arc of contact in the rolling phase with thicker thickness is indicated by line *DE*, as shown in Fig. 8.5. Using geometrical relationship

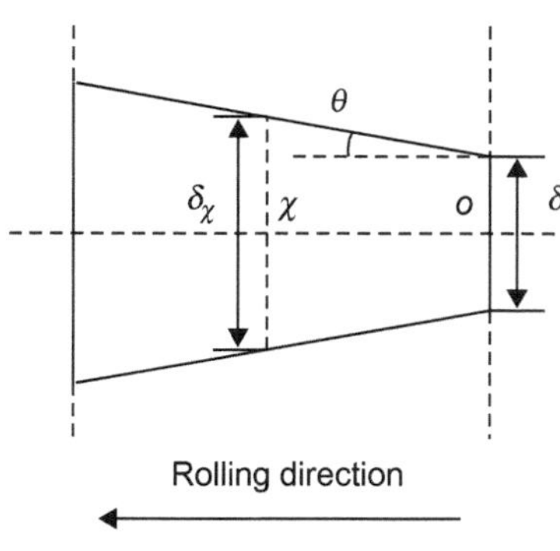

FIGURE 8.4 Variation of workpiece thickness in the rolling phase with thinner thickness.

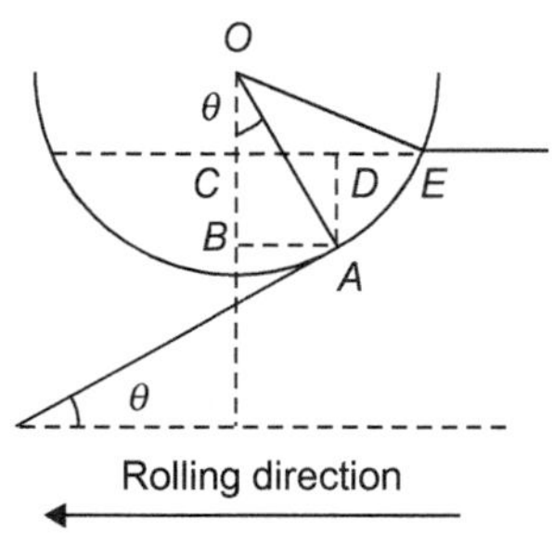

FIGURE 8.5 Contact situation in the rolling phase with thicker thickness.

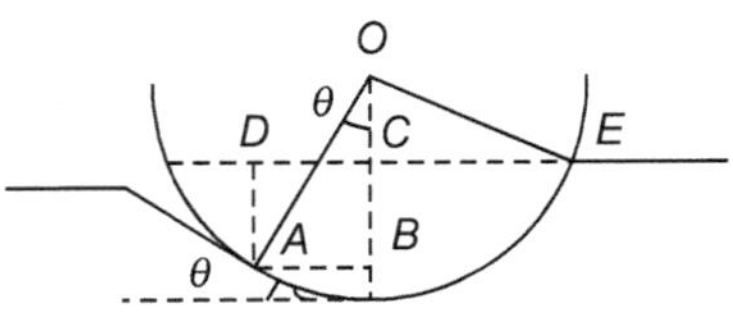

FIGURE 8.6 Contact situations in the rolling phase with thinner thickness.

$$CE = \sqrt{OE^2 - OC^2} = \sqrt{OE^2 - (OB - BC)^2}$$
$$= \sqrt{R^2 - \left(R\cos\theta - \frac{r}{2}\right)^2} = \sqrt{(R\sin\theta)^2 + R\cdot r\cdot\cos\theta - \frac{r^2}{4}} \tag{8.9}$$

where R is the roll radius and r is the reduction. Then

$$DE = CE - CD = \sqrt{(R\sin\theta)^2 + R\cdot r\cdot\cos\theta - \frac{r^2}{4}} - R\sin\theta \tag{8.10}$$

As is shown in Fig. 8.6, the length of arc of contact in the rolling phase with thinner thickness can be calculated in the same way

$$DE = CE + CD = \sqrt{(R\sin\theta)^2 + R\cdot r\cdot\cos\theta - \frac{r^2}{4}} + R\sin\theta \tag{8.11}$$

From Figs. 8.5 and 8.6 it can be seen that with respect to an equal reduction, the length of arc of contact gets shorter in the rolling phase with thicker thickness while it becomes longer in the rolling phase with thinner thickness compared with that in the rolling phase for an invariable thickness, which thereby brings about the change of corresponding deformation area in micro flexible rolling process.

8.2.2 Numerical Simulation of Micro Flexible Rolling Process

8.2.2.1 Solution Method

The elastic–plastic behavior of the material in the micro flexible rolling process depends on the history of loading and deformation, so the solution to this nonlinear problem can be developed by a series of increments [25]. The applied load is first divided into a number of increments, and the displacement is discretized into successive segments conformable to load increments. Assuming the load $\boldsymbol{P}_m$ and displacement $\boldsymbol{u}_m$ at the end of mth increment step to be known, the updated load $\boldsymbol{P}_{m+1}$ and displacement $\boldsymbol{u}_{m+1}$ can be attained by adding incremental load and displacement to $\boldsymbol{P}_m$ and $\boldsymbol{u}_m$ separately.

The convergence of the solution can be ensured by providing the small enough load increments.

The Newton–Raphson iterations are subsequently performed within an increment step to acquire the equilibrium condition between the externally applied load and the internal structural resistance [26]. The $(n+1)$th iteration within the $(m+1)$th increment step can be described as

$$\boldsymbol{f}_{m+1}^{(n+1)} = \boldsymbol{D}\left(\boldsymbol{u}_{m+1}^{(n+1)}\right) - \boldsymbol{P}_{m+1} = \boldsymbol{D}\left(\boldsymbol{u}_{m+1}^{(n)}\right) - \boldsymbol{P}_{m+1} + \left(\boldsymbol{K}_T^n\right)_{m+1}\boldsymbol{C}_m^n = 0 \quad (8.12)$$

where $(\boldsymbol{K}_T^n)_{m+1}$ is the nth correction of $(\boldsymbol{K}_T)_{m+1}$ and equals $\boldsymbol{K}_T(\boldsymbol{u}_{m+1}^{(n)})$. Further, the nth correction of $\boldsymbol{C}_m$ may be calculated as

$$\boldsymbol{C}_m^n = \left(\boldsymbol{K}_T^n\right)_{m+1}^{-1}\left(\boldsymbol{P}_{m+1} - \boldsymbol{D}\left(\boldsymbol{u}_{m+1}^{(n)}\right)\right) \quad (8.13)$$

As a consequence, the improved value of $\boldsymbol{u}_{m+1}$ at the $(n+1)$th increment step can be achieved by [27]

$$\boldsymbol{u}_{m+1}^{(n+1)} = \boldsymbol{u}_{m+1}^{(n)} + \boldsymbol{C}_m^n \quad (8.14)$$

Let $\boldsymbol{u}_{m+1}^0 = \boldsymbol{u}_m$ at the beginning of the iterative procedure, and then the same procedure will be repeated till the satisfactory convergence is reached under this increment step.

8.2.2.2 Flow Rule

When the material goes into the plastic deformation, the components of the plastic strain increment may be defined by

$$d\varepsilon_{ij}^p = d\lambda\frac{\partial Q}{\partial \sigma_{ij}} \quad (8.15)$$

where dλ is a positive proportionality constant, concerned with the hardening rule, and Q is the plastic potential function that characterizes the relationship between stress state and plastic strain in general [28,29].

For the strain hardening material, the yield surface expands with continuous loading and the plastic potential function is assumed to have the same form as the current yield function. Eq. (8.15) is afterwards termed associated and can be rewritten as

$$d\varepsilon_{ij}^p = d\lambda\frac{\partial F}{\partial \sigma_{ij}} \quad (8.16)$$

where $\frac{\partial F}{\partial \sigma_{ij}}$ is normal to the yield surface $F = 0$ in the stress space [30].

8.2.2.3 Material Heterogeneity

The numerical modeling of the workpiece has been implemented based on 3D Voronoi tessellation so as to look into the effect of material inhomogeneity aroused by divergent grain properties on the deformation manner in micro flexible rolling. Voronoi tessellation can be interpreted as q distinguishing regions V_k regarding an open set Ω and q disparate sites o_k [31]

$$V_k = \{\omega \in \Omega | d(\omega, o_k) < d(\omega, o_l) k, l = 0, 1, \ldots, q-1, k \neq l\} \tag{8.17}$$

where d is a function of distance.

The workpiece was first split into several cubes. Next, the sites were laid out in these cubes. Suppose that the center of a cube is (x, y, z), the site will stand $\eta a/2$ away from the central point, in which a is the edge length of the cube and η is a random number between -1 and 1 [32]. As a result, the exact position of the site ought to be ($x + \eta a/2$, $y + \eta a/2$, $z + \eta a/2$).

After sites were distributed in the workpiece, Voronoi polyhedrons were created by means of computational software MATLAB, which take those sites as their mass centers. This geometrical feature was plotted by applying the 3D Voronoi function in MATLAB itself, and all the topology information of the generated Voronoi tessellation was extracted by MATLAB programming which produced a Python file at the same time. The Python script was then imported to ABAQUS/CAE to accomplish the setup with a large quantity of grains in ABAQUS/CAE. Fig. 8.7 shows the Voronoi tessellation in the 250 μm thick workpiece, which comprises 671 grains with average grain size of 250 μm [33]. With an alteration in a value, the average grain size may be adjusted. When the average grain size decreases, the number of grains through workpiece thickness increases, which brings about more grains boundaries to hinder the dislocation glide along the slip planes, and thus leads to an enhanced constraint for grains to rotate into their preferred orientations, making the workpiece harder to deform, and vice versa.

Since the Voronoi tessellation is just geometrical feature, each grain needs to be endowed with some specific property to procure a heterogeneous workpiece, in as much as even if for an invariable average grain size different dislocation densities, grain shapes, and grain orientations all suggest

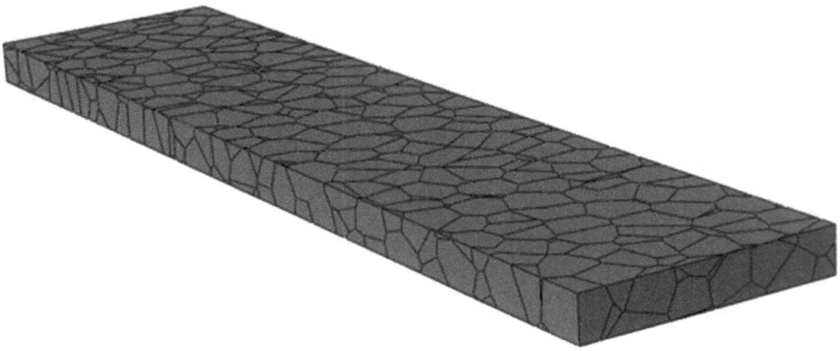

FIGURE 8.7 Voronoi tessellated workpiece with 671 grains in it [33].

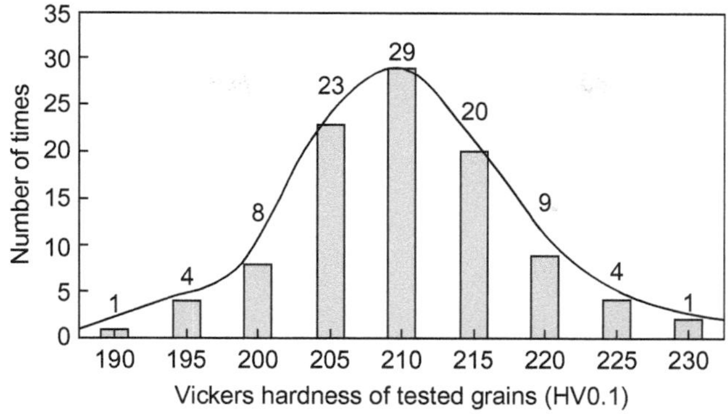

FIGURE 8.8 Vickers hardness distributions in the 304 stainless steel micro workpiece [33].

differing formabilities of workpieces. In this study, micro hardness testing has been used to figure out the ability of grain to resist plastic deformation, which provides a relative idea as well as an effectual estimation of grain properties. The micro hardness testing was conducted at 100 random locations on a 304 stainless steel sample, and the results are exposed in Fig. 8.8.

As can be noticed from Fig. 8.8, nine varieties of Vickers hardness values representing nine sorts of grain properties have been found in the tests, which approach the normal distribution. Moreover, as stainless steels as well as a number of other materials do not have a clear demarcation between elastic and plastic regimes and yield strengths are conventionally defined by 0.2% offset strain, so the variation of yield stress with regard to this tiny elastic zone is small. For this reason together with the outcomes acquired from the micro hardness testing, the bilinear isotropic hardening material model has been chosen for the workpiece, considering the material inhomogeneity by presuming that all grains share the same Young's modulus but possess dissimilar tangent modulus [34]. As are exhibited in Fig. 8.9, nine classes of heterogeneity coefficients are in use to alter the tangent modulus to obtain nine categories of mechanical properties.

In Fig. 8.9, the Young's modulus E is 192 GPa, initial yield stress σ_{s0} 205 MPa, and tangent modulus E_{TAN} is 1630 MPa for 304 stainless steel. All these mechanical properties were attributed to the grains in the workpiece via Python file, as displayed in Fig. 8.10, and the colors of the grains correspond to those utilized in Fig. 8.11, where ξ values are subject to the normal distribution.

8.2.3 Simulation of Springback

The FEA model was halved along the symmetry plane, i.e., the middle surface of the workpiece to the thickness direction, to save the computational time, as shown in Fig. 8.12.

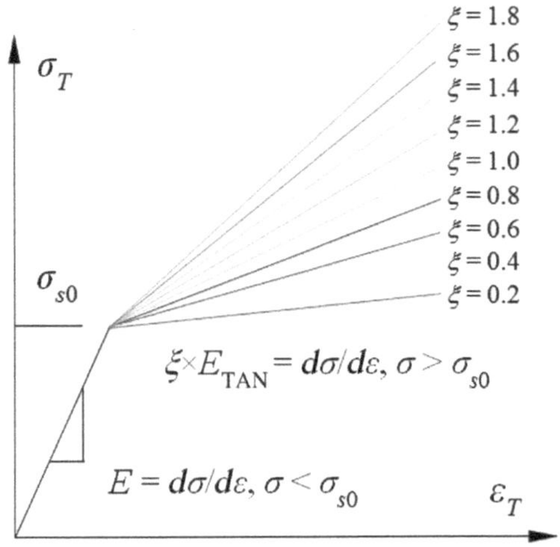

FIGURE 8.9 Bilinear isotropic hardening material model with consideration given to material inhomogeneity [33].

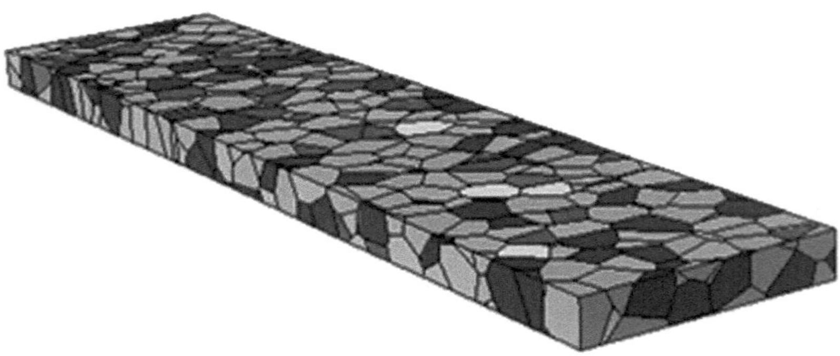

FIGURE 8.10 Grains equipped with distinct mechanical properties [33].

The workpiece was finely meshed with 10-node modified quadratic tetrahedron elements (C3D10M) by utilizing mesh size of 0.05 mm to guarantee a good convergence. To avoid penetration into the workpiece, 4-node 3D bilinear rigid quadrilateral elements (R3D4) with a coarser element size of 0.25 mm were exercised for the roll, which has been idealized as a rigid cylindrical surface in view of two major concerns, whereof one is that the workpiece can be exactly rolled to the predefined reductions as there is no deformation of the roll at the contact location, thus the influence of roll deformation on the thickness change of the workpiece can be neglected, namely there is no need to define a further reduction for the compensation of elastic deformation of the roll, and the workpiece springback from the predefined reductions after rolling; and the other is that the motion of the rigid roll is

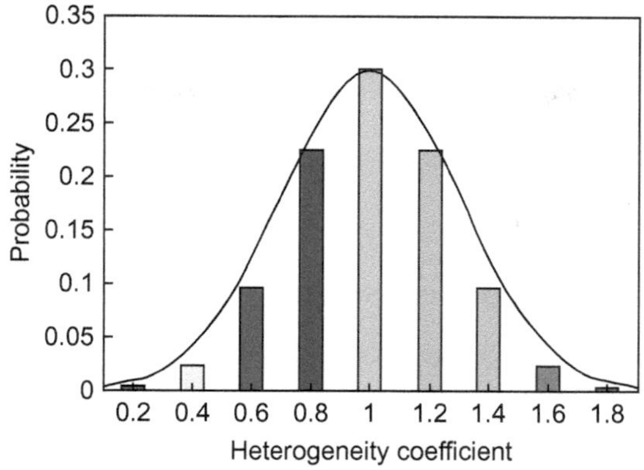

FIGURE 8.11 Statistical distribution of ξ values.

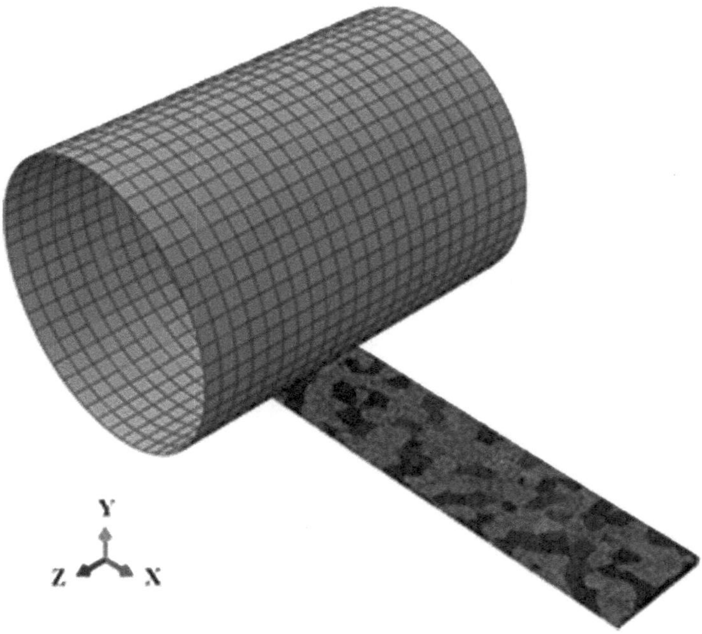

FIGURE 8.12 Halved 3D micro flexible rolling finite element model [33].

governed completely by the motion of a single reference point with no more than six degrees of freedom, which is able to boost the computational efficiency and save the computational time.

The aforementioned symmetry plane which allows only a translational degree of freedom in X direction was endued with an initial speed to propel the workpiece forward. The speed was withdrawn after the workpiece made a full contact with the roll, which kept moving forward taking advantage of the frictional force between the two instances. The COF between the roll and workpiece was selected as 0.15 [35]. A translational degree of freedom in Y direction, combined with the rotational degree of freedom about Z direction, was bestowed on the geometrical center of the roll to realize the micro flexible rolling.

8.2.4 Modeling of Surface Roughness Deformation

The roll was regarded as a rigid body, adopting R3D4 elements with uniform mesh size of 0.05 mm, in order to eliminate the effect of its elastic deformation on the deformation pattern of the workpiece asperities. Whereas the workpiece was set as an elastoplastic body and meshed with C3D10M elements, as these elements provide a better approximation of the waveform on the workpiece surface, as exposed in Fig. 8.13. The element size of the workpiece matrix was selected to be 0.02 mm, which moderately decreased to a much finer order of 0.005 mm for the workpiece surface layer, so that penetration can be avoided in the zone of contact between the roll and the workpiece.

The actual rough surface of the workpiece is approximated as an ideal rough surface in this investigation, which comprises a series of convex and concave surfaces of identical radii of curvature along rolling direction, as shown in Fig. 8.14 [36].

Consider the workpiece segment of length L, which is divided into n subsegments by $(n-1)$ points evenly spaced on this workpiece segment. The central line average roughness R_a may be calculated as:

$$R_a = \frac{1}{n}\sum_{i=1}^{n} |y_i| \tag{8.18}$$

where y_i is the distance between the curve and the x-axis which represents the central line. The R_a value can be computed with sufficient accuracy if n is large enough.

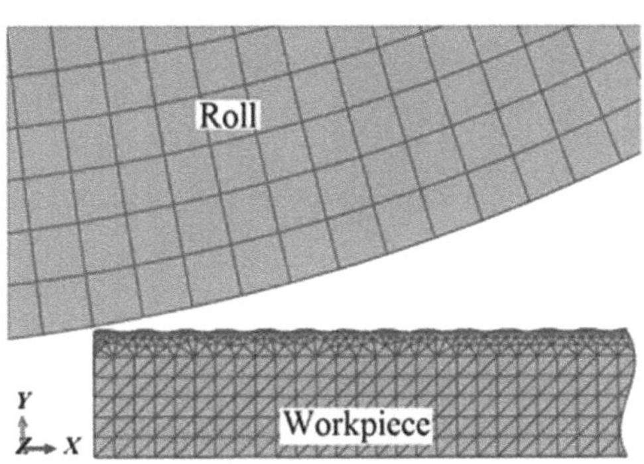

FIGURE 8.13 Mesh distributions of the micro flexible rolling model [36].

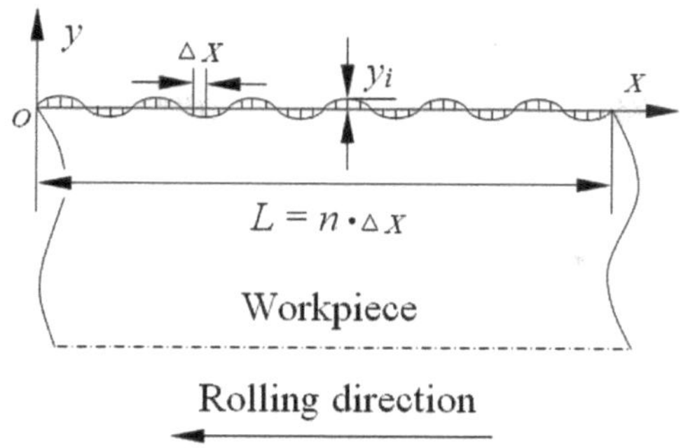

FIGURE 8.14 Idealized surface roughness model for the workpiece [36].

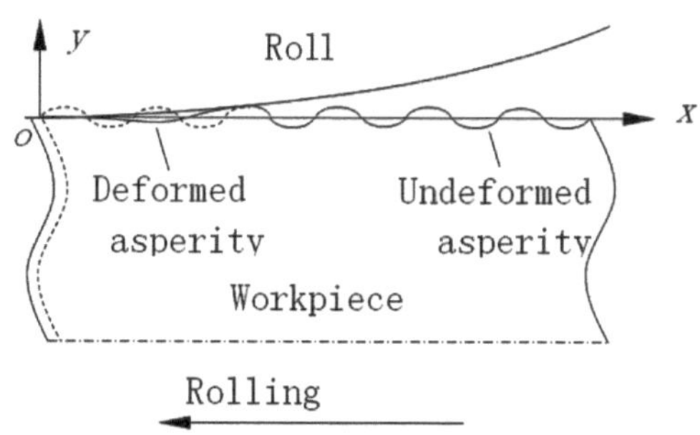

FIGURE 8.15 Surface asperity deformation in micro flexible rolling [36].

Fig. 8.15 illustrates the surface asperity deformation mechanism during micro flexible rolling process. In the interaction between the roll and the workpiece, the roll surface is assumed to be smooth and rigid, and the workpiece asperities are flattened on account of the combined action of normal and tangential contact forces, namely both convex and concave surfaces get flatter as their radii of curvature become larger, as depicted in Fig. 8.15. The final surface roughness profile depends on all deformations of the asperities in mixed characteristic, i.e., the asperities may be stretched or compressed in disparate directions, and the surface roughness transfer behavior can be quantitatively estimated by the difference in R_a value before and after flexible rolling.

8.3 SIMULATION RESULTS

8.3.1 Springback Analysis

The stress–strain distribution on the profile for the halved 250 μm thick workpiece consisting of grains with average grain size of 250 μm is illustrated in Fig. 8.16. It can be found that the stress–strain distribution is

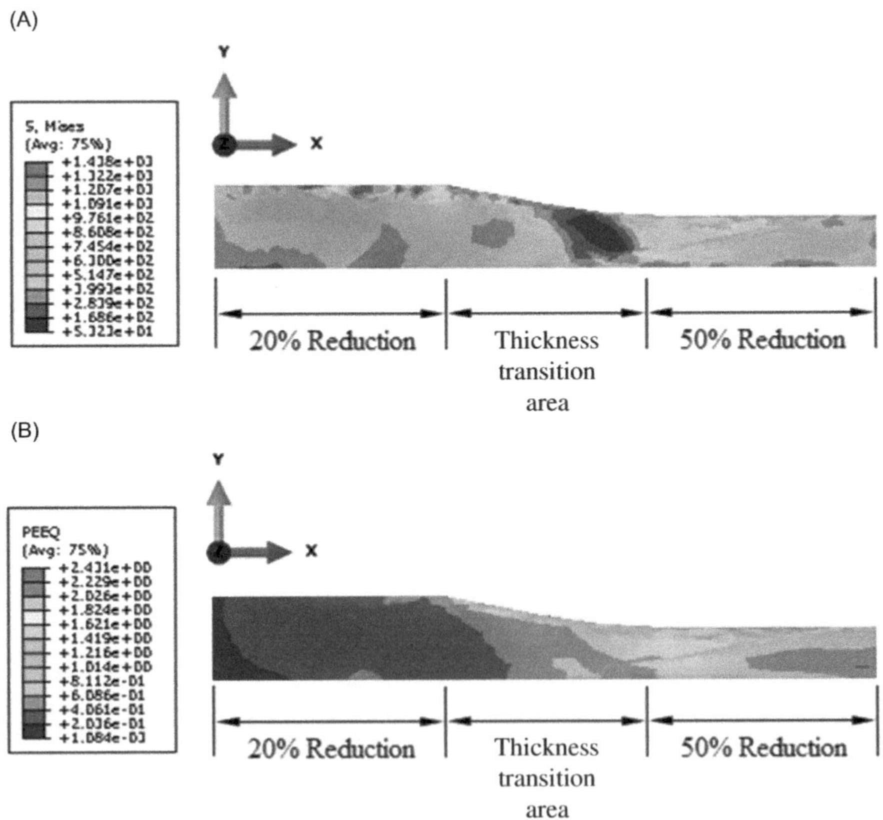

FIGURE 8.16 Stress–strain distribution on the profile after springback: (A) von Mises stress distribution and (B) equivalent plastic strain distribution [33].

nonuniform owing to the fact that not all grains undergo identical plastic deformation and some grains are still in elastic strain while others have entered the plastic regime in the whole rolling process.

This study has been focused on the springback in thickness direction, which is deemed a critical indicator to judge the forming quality and can be worked out through the formula

$$s_x = \delta_x^{'} + r_x - \Delta \tag{8.19}$$

where $\delta_x^{'}$ is the average thickness of the workpiece after springback, namely the average distance between the upper and lower surface nodes of the workpiece after the whole simulation procedure is completed. In regard to the simulated rolling phase with thinner thickness, the average springback has been evaluated within the extended zones of initial reduction of 20% and final reduction of 50% separately. Utilizing the COF of 0.15 in conjunction with Eq. (8.5), the maximum tilt angle θ has been fixed as 5.58, 8.95, and 11.93 degrees correspondingly for workpieces with initial thicknesses of 500, 250, and 100 μm, respectively, to complete the rolling phase with thinner thickness. Tables 8.1–8.3 list the length of thickness transition area, respectively, in respect of different initial workpiece thicknesses under an identical tilt angle of 5.00 degrees. As can be observed from these tables, (1) the length of thickness transition area decreases with a decrease in initial workpiece thickness, (2) the discrepancy between calculated

TABLE 8.1 Length of Thickness Transition Area for 500 μm Thick Workpiece ($\theta = 5.00$ degrees)

Average grain size (μm)	250	100	50	10	1
Measured length of thickness transition area (mm)	0.857	0.855	0.861	0.857	0.859
Calculated length of thickness transition area (mm)	0.860	0.860	0.860	0.860	0.860

TABLE 8.2 Length of Thickness Transition Area for 250 μm Thick Workpiece ($\theta = 5.00$ degrees)

Average grain size (μm)	250	100	50	10	1
Measured length of thickness transition area (mm)	0.422	0.419	0.417	0.420	0.417
Calculated length of thickness transition area (mm)	0.420	0.420	0.420	0.420	0.420

TABLE 8.3 Length of Thickness Transition Area for 100 μm Thick Workpiece ($\theta = 5.00$ degrees)

Average grain size (μm)	250	100	50	10	1
Measured length of thickness transition area (mm)	0.180	0.179	0.180	0.179	0.181
Calculated length of thickness transition area (mm)	0.180	0.180	0.180	0.180	0.180

values and simulation values shrinks as the workpiece becomes thinner, and (3) the average grain size has a minor influence on the length of thickness transition area.

8.3.1.1 *Influence of Tension on Average Springback*

Front and back tensions of 25, 50, 75, and 100 MPa were individually applied to workpieces with initial thicknesses of 100, 250, and 500 μm in accordance with the following formula [23]

$$\overline{\sigma}_z = (0.1 \sim 0.6)\sigma_{s0} \tag{8.20}$$

where $\overline{\sigma}_z$ is the average unit tension.

Fig. 8.17 presents the effect of tension on the average springback obtained from the reference models in which all grains have ξ value equal to 1.0. It is found that the average springback declines moderately as the front and back tensions increase from 0 to 100 MPa in increments of 25 MPa regardless of the initial workpiece thickness, as well as the preset reduction amount. It is also seen that for a thicker workpiece, more front and back tensions play a more crucial role in eliminating the average springback despite the change of reduction. This is because front and back tensions are capable of ameliorating metal flow and relaxing residual stresses, the thickness precision of rolled workpiece can thereupon be improved with application of these tensions.

For the sake of investigation of springback in a free state, all work detailed in the following section has been executed under condition of zero front and back tensions.

8.3.1.2 *Scatter of Springback for Initial Workpiece Thickness of 250 μm*

Grains with the average sizes of 1, 10, 50, 100, and 250 μm were successively shaped in the 250 μm thick workpiece. Fig. 8.18 shows the relationship between the springback and the grain size by way of 10 representative groups of simulation results when the reduction is 20% and 50% individually, where the dashed lines stand for the average springback.

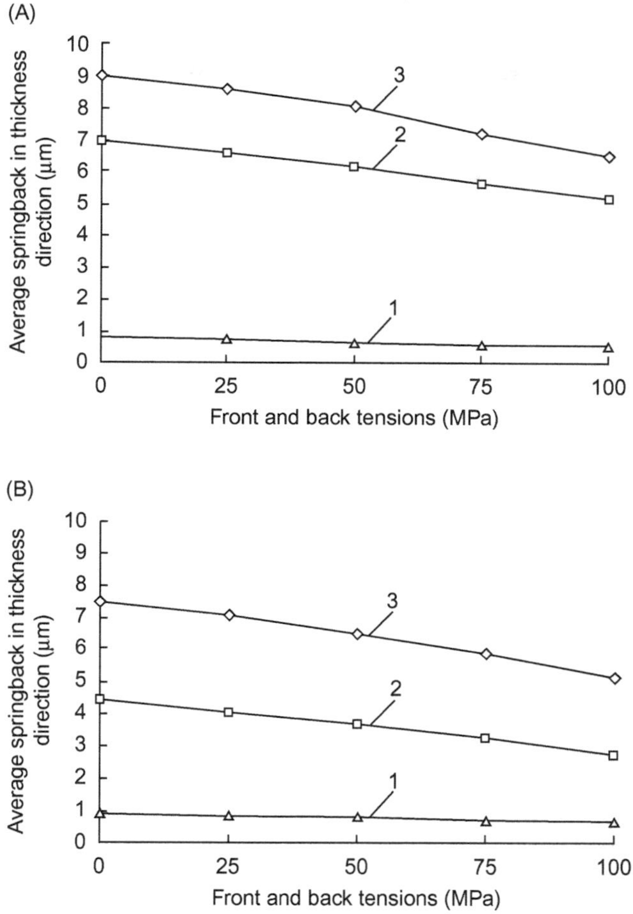

FIGURE 8.17 Relationship between average springback in thickness direction and front and back tensions for initial workpiece thicknesses of 100, 250, and 500 μm: (A) 20% reduction and (B) 50% reduction [33].

As can be seen in Fig. 8.18, a scatter of springback exists for each grain size thanks to the random association of Voronoi polyhedrons with mechanical properties. If grains with ξ value smaller than 1.0 exceed those with ξ value bigger than 1.0 in thickness direction, which makes the average heterogeneity coefficient less than 1.0, consequently the elastic springback is smaller after unloading from the same plastic strain value. However, if grains with ξ value bigger than 1.0 exceed those with ξ value smaller than 1.0 in thickness direction, which makes the average heterogeneity coefficient greater than 1.0, the elastic springback is therefore larger after unloading from the same plastic strain value, as demonstrated in Fig. 8.19.

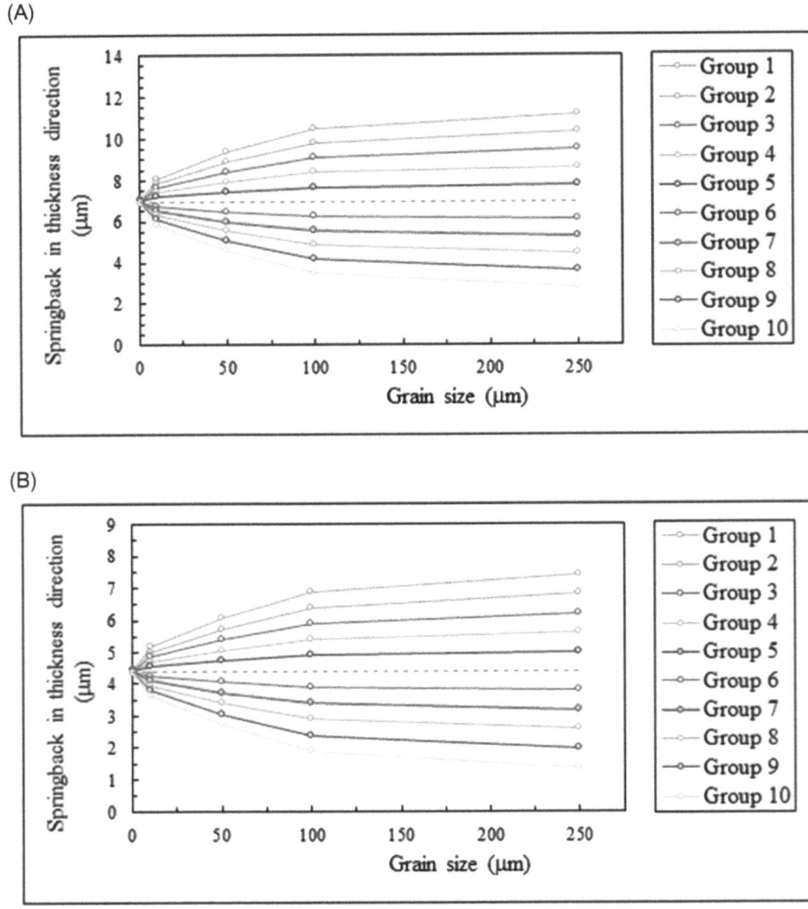

FIGURE 8.18 Springback in thickness direction versus grain size for initial workpiece thickness of 250 μm: (A) 20% reduction and (B) 50% reduction.

For both reductions, the largest scatter of springback occurs at the grain size of 250 μm as there is almost one grain in the thickness direction, so that the springback amount mainly depends on the single grain's mechanical property. When the number of grains increases in the thickness direction owing to the decrease of grain size, the effect of each single grain on the springback is diminished, making the scatter of springback more compact and gradually converge to that from the reference model.

8.3.1.3 Comparison of Scatter of Springback for Different Initial Workpiece Thicknesses

Workpieces with initial thicknesses of 100 and 500 μm were also modeled choosing the grain size in alignment with that for the 250 μm thick workpiece

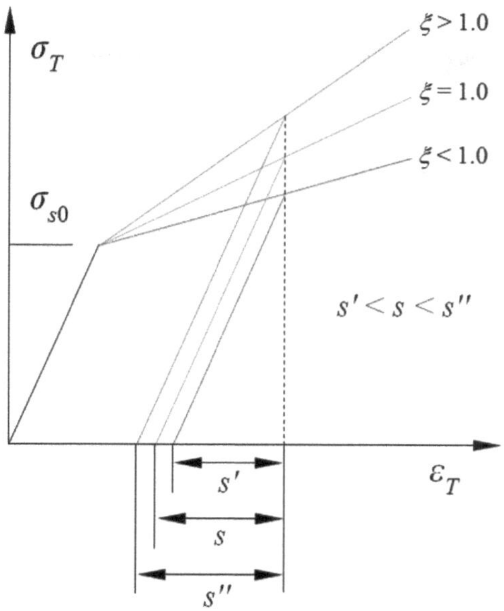

FIGURE 8.19 Elastic springback for different ξ values.

and the simulation outcomes are separately depicted in Figs. 8.20 and 8.21. As can be noted by comparing Figs. 8.18, 8.20, and 8.21, for either reduction the scatter of springback in thickness direction expands at each grain size when the initial workpiece thickness lessens. This is because the grain number drops in the thickness direction due to the decrease of workpiece thickness for a certain grain size, the effect of each single grain on the springback becomes more prominent, which makes the scatter of springback larger. This being the case a relatively significant narrowing of the scatter of springback only happens when the grain size shrinks to 50 μm for the 100 μm thick workpiece, whereas for the 500 μm thick workpiece, the scatter of springback has exhibited convergence even at the maximum grain size of 250 μm.

Meanwhile, for each thickness the curves for both reductions indicate similar trends and the difference of the scatter of springback between both reductions keeps within 10.5% for each grain size in line with the simulation data, which can be referred to the qualitative prediction of the scatter of springback for other reductions.

8.3.1.4 Quantitative Prediction of Springback in Thickness Direction

Seeing that springback is related with many essential factors, such as material properties, work hardening exponent, yield strength, workpiece

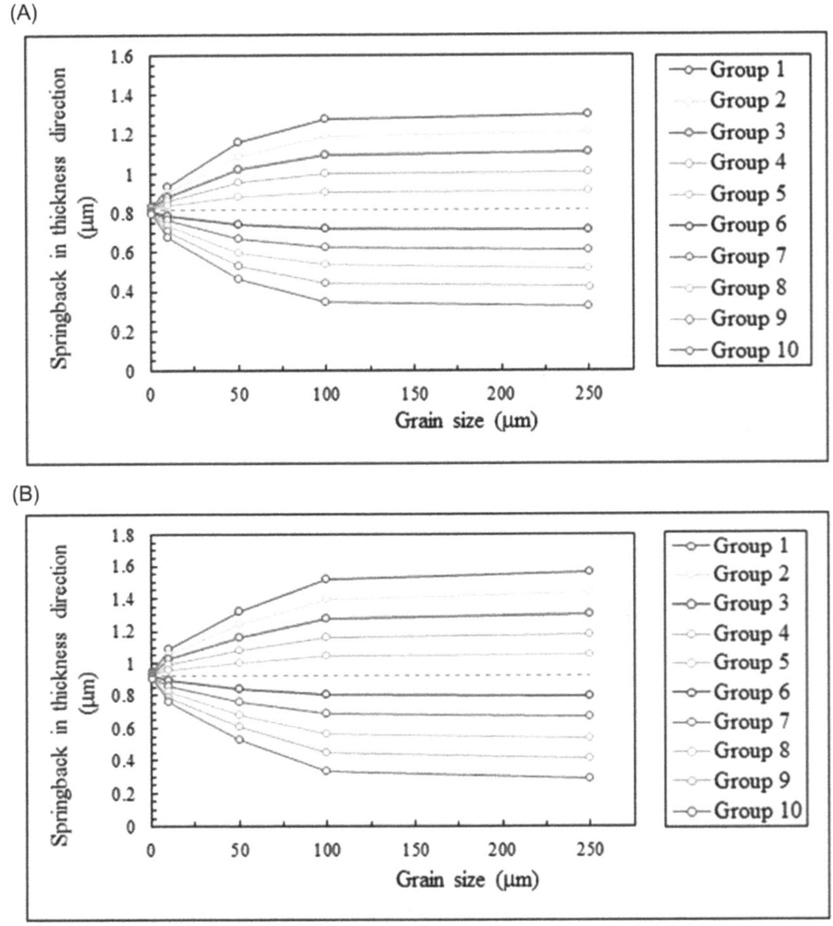

FIGURE 8.20 Springback in thickness direction versus grain size for initial workpiece thickness of 100 μm: (A) 20% reduction and (B) 50% reduction.

thickness, grain size, reduction amount, and bending ratio, during the metal forming processes including rolling, bending, and deep drawing, multiple linear regression that infers the relationship between two or more independent variables and one dependent variable can be deployed in constructing a model to quantitatively predict the springback, which is based on design of experiment, i.e., experimenter chooses n sets of values of $x_1, x_2, \ldots, x_p$ and observes the response y on n value sets with the purpose to inquire into the impact of variables $x_1, x_2, \ldots, x_p$ on the response variable y.

In the regression model with regard to the springback prediction in micro flexible rolling, four independent variables x_1–x_4 are taken as factors affecting the thickness directional springback y, which are predefined reduction

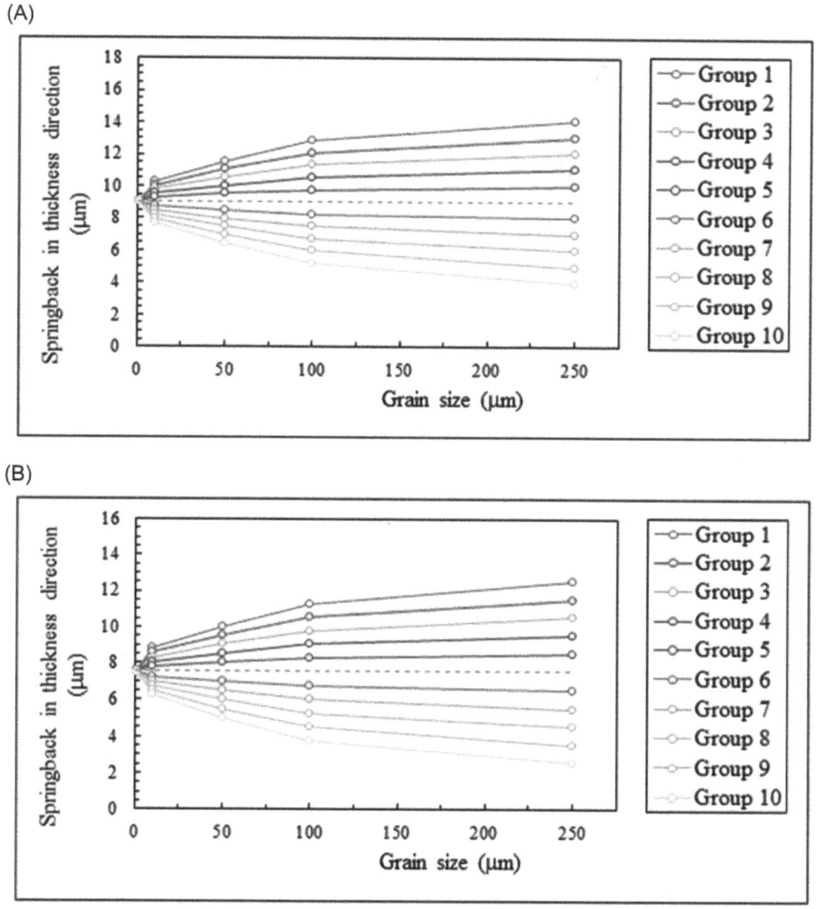

FIGURE 8.21 Springback in thickness direction versus grain size for initial workpiece thickness of 500 μm: (A) 20% reduction and (B) 50% reduction.

amount, heterogeneity coefficient, the ratio of initial yield stress to Young's modulus and the ratio of initial workpiece thickness to average grain size, respectively. Given n observations, the model is formally written as

$$\begin{cases} y_i = \beta_0 + \beta_1 x_{i1} + \beta_2 x_{i2} + \beta_3 x_{i3} + \beta_4 x_{i4} \\ \quad + \varepsilon_i, i = 1, 2, \ldots, n \\ \varepsilon_i \sim N(0, \sigma^2) \end{cases} \tag{8.21}$$

where β_0 is the intercept, $\beta_1 - \beta_4$ are the partial coefficients, which are estimated applying the least-squares method, and ε_i is the deviation of ith observed value y_i from ith fitted value $\hat{y}_i$, which obeys the normal distribution with mean 0 and variance σ^2 calculated by

$$\sigma^2 = \frac{\sum \varepsilon_i^2}{n-p-1} \tag{8.22}$$

where p is the number of independent variables.

So for the set of simulation data with 270 groups of values of x_1–x_4 and y, the model is finally fitted as

$$\begin{cases} \hat{y}_i = 4.11505 - 4.40316x_{i1} \\ \quad +1.96573x_{i2} + 0.00412x_{i3} \\ \quad +0.0062x_{i4} + \varepsilon_i, i = 1, 2, \ldots, 270 \\ \varepsilon_i \sim N(0, 10.26) \end{cases} \tag{8.23}$$

After that springback prediction can be carried out within the range of values of x_1–x_4 in the dataset wielded for model-fitting by interpolation. For instance, by accessing the dataset the first observation of values of x_{11}, x_{12}, x_{13}, and x_{14} is 0.2, 1.8, 0.001, and 100 individually, and the deviation ε_1 is -6.55932 calculated by the regression model. By substituting $x_{11} = 0.2$, $x_{12} = 1.8$, $x_{13} = 0.001$, $x_{14} = 100$, and $\varepsilon_1 = -6.55932$ into Eq. (8.23), the forecasted thickness directional springback $\hat{y}_1$ is computed to be 0.833 μm. The springback predictions for the other observations of values of x_1, x_2, x_3, and x_4 can be generated in the same manner.

Furthermore, simulations using different kinds of materials have been performed for the verification of the results obtained from the deduced regression model. The mechanical properties and simulation parameters for each type of material are given in Table 8.4, while the values of x_1, x_2, x_3, x_4 and the springback comparisons for each type of material are shown in Table 8.5, where the error is determined as follows:

TABLE 8.4 Mechanical Properties and Simulation Parameters for Different Types of Materials

Material	Young's Modulus (GPa)	Yield Stress (MPa)	Tensile Stress (MPa)	Elongation (%)	Workpiece Thickness (μm)	Average Grain Size (μm)
Low-carbon steel 1020	210	200	380	25	250	250
Aluminum alloy 1060	75	76	83	16	250	250
Copper alloy C85200	117	110	241	25	250	250

TABLE 8.5 Values of x_1, x_2, x_3, x_4 and Springback Comparisons for Different Types of Materials

Material	x_1	x_2	x_3	x_4	Predicted Springback (μm)	Measured Springback (μm)	Error
Low-carbon steel 1020	0.2	1.0	0.00095	1	6.972	6.805	2.40%
	0.5				4.393	4.298	2.16%
Aluminum alloy 1060	0.2	1.0	0.00101	1	6.972	6.603	5.29%
	0.5				4.393	4.157	5.37%
Copper alloy C85200	0.2	1.0	0.00094	1	6.972	6.718	3.64%
	0.5				4.393	4.230	3.71%

$$\text{Error} = \frac{|\text{Predicted springback} - \text{Measured springback}|}{\text{Predicted springback}} \times 100\% \tag{8.24}$$

As is seen from the last column of Table 8.5, for low-carbon steel 1020 the error in springback prediction is less than 3%, and for aluminum alloy 1060 and copper alloy C85200, the errors are around 5.33% and 3.68%, respectively. The results show that the developed regression model can provide sufficient accuracy in thickness directional springback prediction in micro flexible rolling, even though it slightly overestimates the springback for these materials because of their lower strength and greater ductility that promote them to springback less in thickness direction.

8.3.2 Surface Roughness Evolution

Initial surface roughness of 1, 2, 5, and 10 μm, in conjunction with dissimilar rolling parameters, was successively adopted in the simulation to examine their interactive effects on the flattening process of surface asperities. Fig. 8.22 presents the changes of workpiece surface morphology after flexible rolling, where initial surface roughness is 1 μm, the COF 0.1, rolling speed 20 mm/s and roll gap adjusting speed 2.5 mm/s, respectively. It is observed that the surface asperities are largely flattened and the final surface profile approaches a smooth line.

In addition, a variety of mesh sizes were individually selected for the surface layer of the workpiece in order to optimize the mesh size, which yields satisfactory computational accuracy with least running time. Table 8.6 exhibits the influence of mesh size on the computational time and accuracy with other simulation parameters identical to those for Fig. 8.22. As can be seen in Table 8.6, for either reduction the final surface roughness decreases sharply as

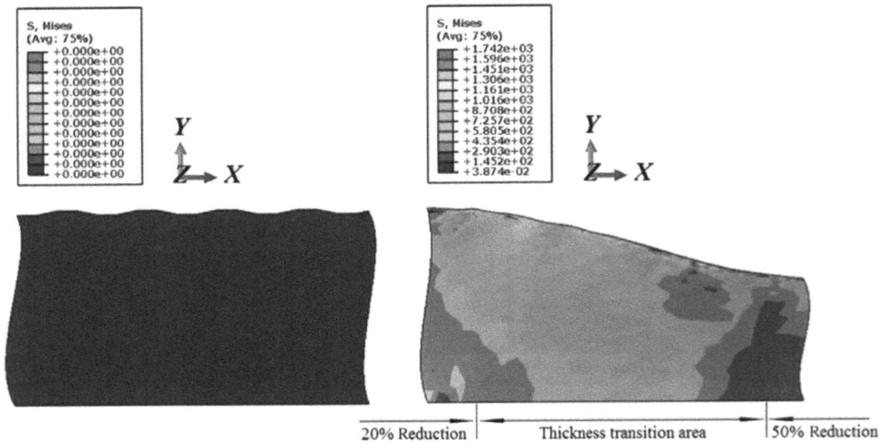

FIGURE 8.22 Comparison of workpiece surface morphology before and after flexible rolling [36].

TABLE 8.6 Influence of Mesh Size on Computational Time and Accuracy

Mesh size (μm)	20	10	5	2.5
Number of meshes	8483	12,986	30,383	140,285
Computational time (h)	14	21	34.5	226
Final surface roughness for 20% reduction (μm)	0.684	0.639	0.603	0.601
Final surface roughness for 50% reduction (μm)	0.597	0.562	0.548	0.547

the mesh size descends to 5 μm, and then makes a very slight decrease as the mesh size keeps continually dropping to 2.5 μm. Although the mesh size of 2.5 μm seems to be able to obtain more accurate results than other mesh sizes, the computational time soars to 226 h which is nearly seven times longer than that for the mesh size of 5 μm. Based on this evidence, the mesh size of 5 μm has been used in the surface layer of the workpiece in the simulation.

8.3.2.1 Effect of the COF

As the COF usually ranges from 0.1 to 0.3 in cold rolling without lubrication, typical values of 0.1, 0.2, and 0.3 have been chosen for micro flexible rolling. Fig. 8.23 demonstrates the relationship between the final surface roughness and the COF for different initial surface roughness with regard to the reductions of 20% and 50%.

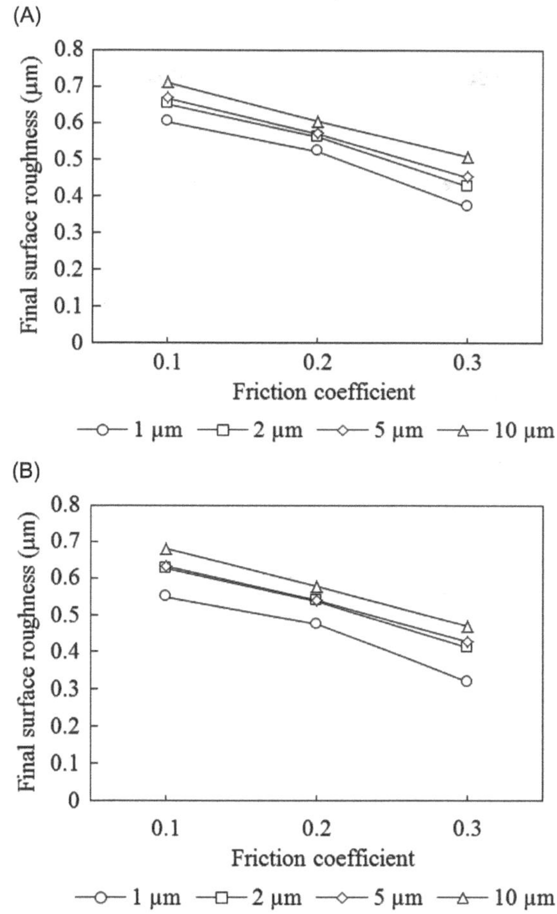

FIGURE 8.23 Final surface roughness versus COF for different initial surface roughness: (A) 20% reduction and (B) 50% reduction.

It can be seen from Fig. 8.23 that for both reductions, the final surface roughness of the workpiece decreases significantly with an increase of the COF for each level of initial surface roughness. This is mainly because the friction formulation between the roll and the workpiece employed the penalty method to place constraint to the workpiece. The constraint grows stronger as the COF turns out to be greater, which leads to a more powerful prevention of workpiece deformation but a quicker occurrence of surface asperity flattening. Therefore an increasing COF has an accelerating effect on the surface asperity flattening process.

Table 8.7 shows the comparison between the simulation and experimental results on the effect of the COF on final surface roughness. The experimental data are approximate values which were originated from Ref. [37], where the

TABLE 8.7 Comparison Between Simulation and Experiment

Friction Coefficient	Final Surface Roughness (μm)	
	Simulation	Experimental
20% Reduction		
0.1	0.352	0.245
0.3	0.273	0.238
Difference (%)	22.44	2.86
50% Reduction		
0.1	0.226	0.101
0.3	0.164	0.075
Difference (%)	27.43	25.74

initial surface roughness was very close to the value of 0.75 μm, hence simulations with initial surface roughness of 0.75 μm have been performed and the results are utilized in Table 8.7. It is seen that the simulation results are in agreement with the experimental ones, namely the final surface roughness can be further reduced by increasing the COF. Another point to be noticed is that the increase in reduction gives rise to the decrease of the final surface roughness, which is rather obviously reflected in the experimental data.

The positive effect of the COF on the flattening of surface asperities can also get convinced by Bay et al.'s work [38,39], where they argued that the real area of contact increased as the surface asperity flattening process was enhanced by the large friction factor. The exemplary simulation results with initial surface roughness of 10 μm are displayed in Table 8.8, together with the experimental data appearing in Ref. [39]. It can be seen from Table 8.8 that the final surface roughness may be decreased with increasing the coefficient/factor of friction, which results in a bigger real area of contact irrespective of the reduction or normal pressure.

8.3.2.2 Influence of Rolling Speed

The influence of rolling speed on the final surface roughness is shown in Fig. 8.24, where the COF is 0.1 and roll gap adjusting speed is 2.5 mm/s for all cases. As is noticed in Fig. 8.24, the increase in rolling speed can promote the process of surface asperity flattening and acquire a rolled surface with lower surface roughness, no matter what the original surface roughness of the workpiece and the reduction are. This being the case a higher rolling speed means a higher strain rate, which speeds up the progressive shearing process especially in the surface zone, and the applied stress may rise up to a

TABLE 8.8 Effect of COF/factor on Surface Asperity Flattening: (A) Simulation and (B) Experiment

(A)

Friction Coefficient	Final Surface Roughness (μm)	
	20% Reduction	50% Reduction
0.1	0.711	0.680
0.3	0.507	0.471

(B)

Friction Factor	Real Area of Contact	
	1 Unit of Normal Pressure	2 Units of Normal Pressure
0	0.405	0.750
0.3 ~ 0.4	0.495	0.850

higher level in the surface area under this condition, which eventually results in a larger amount of surface deformation of the workpiece [40].

8.3.2.3 Effect of Roll Gap Adjusting Speed

Fig. 8.25 shows the relationship between the final surface roughness and the roll gap adjusting speed with respect to the same COF and rolling speed for all instances. As can be noted in Fig. 8.25, the final surface roughness changes little with increasing roll gap adjusting speed, in as much as the variation of roll gap adjusting speed alters only the strain rate in the thickness transition area. In other words, the deformation history within the zones of 20% and 50% reductions hardly changes even though the roll gap adjusting speed varies in different cases. Consequently the roll gap adjusting speed plays an insignificant role in the surface asperity flattening process in the rolling phase with an invariable thickness.

8.3.2.4 Effect of Reduction and Initial Surface Roughness

It can be found out from Figs. 8.23–8.25 that the final surface roughness decreases with reduction and also reduces considerably as the value of the initial surface roughness falls down. The reason for this phenomenon is that the bigger reduction causes the asperity to be more depressed, thus the contact area between the roll and workpiece surface increases, making the whole surface layer deform more evenly and the rolled surface smoother. According to the same reasoning, the asperities with smaller initial roughness are easier

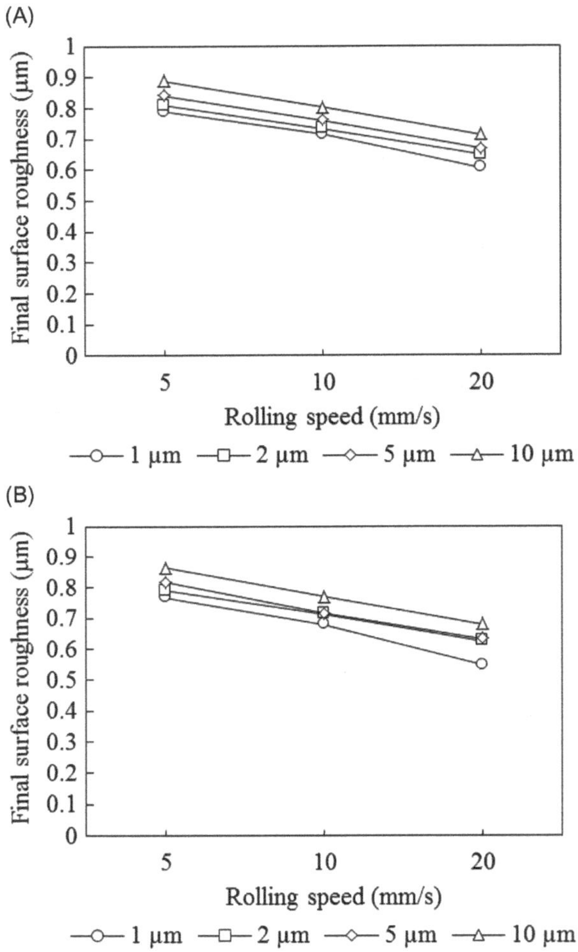

FIGURE 8.24 Final surface roughness versus rolling speed for different initial surface roughness: (A) 20% reduction and (B) 50% reduction [36].

to be flattened and attain a larger contact area during the rolling process, which brings about better surface finish of the rolled workpiece.

In summary, this research numerically enquires into the springback of the workpiece along thickness direction in micro flexible rolling process taking material heterogeneity into account. A 3D finite element model is created to simulate micro flexible rolling process on the basis of mathematical model of the forming process, which is verified by the simulation with reduction changing from 20% to 50%. The 3D Voronoi structure is embedded in the workpiece to geometrically model the grain morphology, and nine generations of grain properties are normally distributed in Voronoi cells according

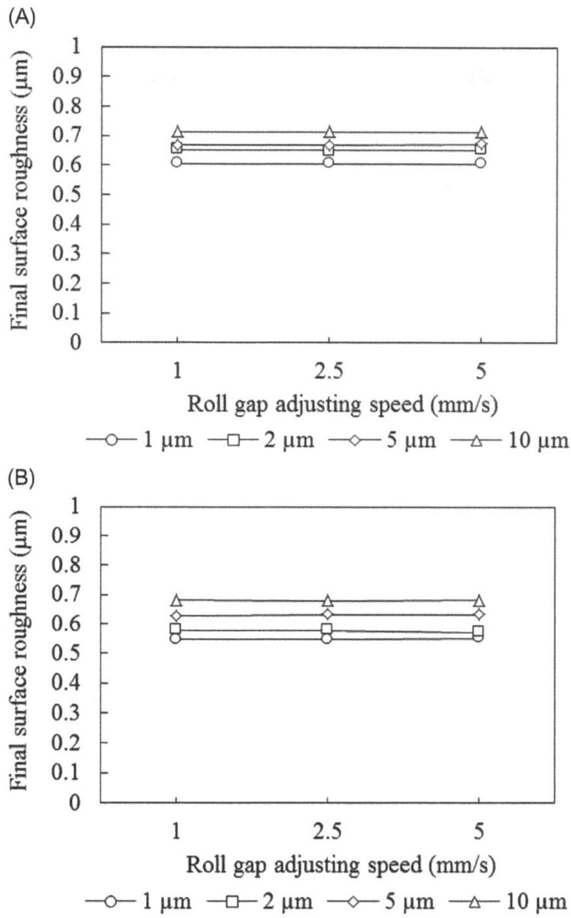

FIGURE 8.25 Final surface roughness versus roll gap adjusting speed for different initial surface roughness: (A) 20% reduction and (B) 50% reduction [36].

to the grain hardness distribution discovered via micro hardness testing to crystallize the material heterogeneity. The portrayed scatter diagrams can qualitatively forecast the trend and the amplitude of springback in thickness direction while a regression model is developed to undertake quantitative prediction of springback in thickness direction inside the limits of values of the foregoing factors that govern the springback. Lower surface roughness can be achieved in the rolling phase with thinner thickness or smaller surface asperity under constant pass reduction. The increase of the COF accompanies an obvious acceleration of the surface asperity flattening process. The final surface roughness in the rolling phase with a uniform thickness decreases as the rolling speed increases. A higher roll gap adjusting speed leads to a lower level of final surface roughness in the thickness transition area.

REFERENCES

[1] A.E. Tekkaya, W. Homberg, 60 Excellent Inventions in Metal Forming, Springer, Berlin Heidelberg, 2015.

[2] O. Engler, C. Schäfer, H.J. Brinkman, J. Brecht, P. Beiter, K. Nijhof, Flexible rolling of aluminium alloy sheet – process optimization and control of materials properties, J. Mater. Process. Technol. 229 (2016) 139–148.

[3] C. Huang, Y. Gan, J.T. Du, C.Z. Chen, Q.J. Chen, Shape rolling simulation of tailor rolled blanks based on deform-3D, Appl. Mech. Mater. 101–102 (2012) 897–900.

[4] J.T. Du, C. Huang, C.Z. Chen, Q.J. Chen, Optimization to control parameters on rolling of tailor rolling blanks by orthogonal test, Appl. Mech. Mater. 80–81 (2011) 546–550.

[5] R. Kopp, C. Wiedner, A. Meyer, Flexibly rolled sheet metal and its use in sheet metal forming, Sheet Metal 2005 – 11th International Conference, 2005, pp. 81–92.

[6] Y. Zhang, J. Tan, Numerical simulation and vertical motion control of rolls for variable gauge rolling, J. Iron Steel Res. Int. 22 (8) (2015) 703–708.

[7] X.H. Liu, Prospects for variable gauge rolling: technology, theory and application, J. Iron Steel Res. Int. 18 (1) (2011) 1–7.

[8] G.J. Zhang, X.H. Liu, X.L. Hu, Y. Zhi, Horizontal velocity of variable gauge rolling: theory and finite elements simulation, J. Iron Steel Res. Int. 20 (10) (2013) 10–16.

[9] G.J. Zhang, X.H. Liu, Research on roll force for variable gauge rolling, Adv. Mater. Res. 418–420 (1) (2012) 1232–1236.

[10] T.A. Kals, R. Eckstein, Miniaturization in sheet metal working, J. Mater. Process. Technol. 103 (1) (2000) 95–101.

[11] S. Miyazaki, H. Fujita, H. Hiraoka, Effect of specimen size on the flow stress of rod specimens of polycrystalline CuAl alloy, Scr. Metall. 13 (6) (1979) 447–449.

[12] L.V. Raulea, A.M. Goijaerts, L.E. Govaert, F.P.T. Baaijens, Size effects in the processing of thin metal sheets, J. Mater. Process. Technol. 115 (1) (2001) 44–48.

[13] U. Engel, R. Eckstein, Microforming – from basic research to its realization, J. Mater. Process. Technol. 125 (2002) 35–44.

[14] J.G. Liu, M.W. Fu, J. Lu, W.L. Chan, Influence of size effect on the springback of sheet metal foils in micro-bending, Comput. Mater. Sci. 50 (9) (2011) 2604–2614.

[15] M. Geiger, M. Kleiner, R. Eckstein, N. Tiesler, U. Engel, Microforming, CIRP Ann. Manufact. Technol. 50 (2) (2001) 445–462.

[16] H.N. Lu, D.B. Wei, Z.Y. Jiang, X.H. Liu, K. Manabe, Modelling of size effects in microforming process with consideration of grained heterogeneity, Comput. Mater. Sci. 77 (2013) 44–52.

[17] W.L. Chan, M.W. Fu, J. Lu, J.G. Liu, Modeling of grain size effect on micro deformation behavior in micro-forming of pure copper, Mater. Sci. Eng.: A 527 (2010) 6638–6648.

[18] Z. Fang, Z.Y. Jiang, D.B. Wei, Modeling of grained heterogeneity with Voronoi tessellation in microforming process, Appl. Mech. Mater. 553 (2014) 66–70.

[19] L. Luo, Z.Y. Jiang, D.B. Wei, X.F. He, Study on micro hydroforming of metals, Adv. Mater. Res. 887–888 (2014) 1133–1138.

[20] L. Luo, Z.Y. Jiang, H.N. Lu, D.B. Wei, K.Z. Linghu, X.M. Zhao, et al., Optimisation of size-controllable centroidal voronoi tessellation for FEM simulation of micro forming processes, Proc. Eng. 81 (2014) 2409–2414.

[21] Z. Fang, Z.Y. Jiang, D.B. Wei, X.H. Liu, Study on springback in micro V-bending with consideration of grain heterogeneity, Int. J. Adv. Manufact. Technol. 78 (5–8) (2015) 1075–1085.

[22] T.T. Wang, K.M. Qi, Plastic Forming of Metals: Principles and Technology of Rolling, second ed., Metallurgical Industry Press, Beijing, 2001.
[23] Y.H. Pang, Plastic Forming of Metals, Northwestern Polytechnical University Press, Xi'an, 2005.
[24] S. Hashmi, G.F. Batalha, C.J.V. Tyne, B.S. Yilbas, Comprehensive Materials Processing, Elsevier, Amsterdam, 2014.
[25] Abaqus 6.11 Tutorials, Dassault Systèmes, 2011.
[26] N. Yidris, The post-buckled coupled mode interaction behaviour of thin-walled members in compression using finite element simulation, PhD Thesis, Loughborough University, England, 2012.
[27] F.K. Kong, Reinforced Concrete Deep Beams, Blackie, Glasgow, 1990.
[28] O.C. Zienkiewicz, The Finite Element Method in Engineering Science, second ed., McGraw-Hill Book Company, London, 1971.
[29] A.F. Bower, Applied Mechanics of Solids, CRC Press, Boca Raton, 2009.
[30] W.F. Chen, D.J. Han, Plasticity for Structural Engineers, Springer-Verlag, New York, 1988.
[31] S. Valette, J.M. Chassery, Approximated centroidal voronoi diagrams for uniform polygonal mesh coarsening, Comput. Graph. Forum 23 (3) (2004) 381–389.
[32] H.N. Lu, D.B. Wei, Z.Y. Jiang, Application of centroidal voronoi diagram in numerical model of microforming process, Adv. Mater. Res. 476–478 (2012) 508–512.
[33] F.J. Qu, Z.Y. Jiang, H.N. Lu, Analysis of micro flexible rolling with consideration of material heterogeneity, Int. J. Mech. Sci. 105 (2016) 182–190.
[34] H.N. Lu, D.B. Wei, Z.Y. Jiang, K. Manabe, T. Furushima, Study on size effect in micro cross wedge rolling, Steel Res. Int., Spec. Edn. (2011) 962–967.
[35] H.B. Xie, K. Manabe, T. Furushima, K. Tada, Z.Y. Jiang, Deformation characterization of micro rolling for stainless steel foil, Proc. Eng. 81 (2014) 179–184.
[36] F.J. Qu, H.B. Xie and Z.Y. Jiang, Finite element method analysis of surface roughness transfer in micro flexible rolling, MATEC Web of Conferences, NUMIFORM 2016, the 12th International Conference on Numerical Methods in Industrial Forming Processes, University of Technology of Troyes-France 4–7 July 2016.
[37] H.J. Li, A study of surface roughness in the metal forming process, PhD Thesis, University of Wollongong, Australia, 2012.
[38] T. Wanheim, N. Bay, A.S. Petersen, A theoretically determined model for friction in metal working processes, Wear 28 (1974) 251–258.
[39] N. Bay, T. Wanheim, Real area of contact and friction stress at high pressure sliding contact, Wear 38 (1976) 201–209.
[40] H.J. Li, Z.Y. Jiang, D.B. Wei, Study on effect of strain rate on 3D surface asperity flattening in uniaxial planar compression by crystal plasticity finite element modelling, Wear 301 (2013) 11–18.

Chapter 9

Simulation of Micro Ultrathin Strip Rolling

9.1 INTRODUCTION

The utilization of metallic foil has been expanded in recent years with sophistication of electrical machinery and the parts for automobiles, mobile phone, medical devices, etc. For example, stainless steel foil has been used for metallic substrate of automotive catalytic converter for two decades, because the property of stainless foil as the honeycombs substrate shows a good performance in exhaust gas resistance and heat resistance. In such usages, ultrathin foil of 50 μm or less is strongly hoped for as to achieve higher performance. The stainless steel foil manufactured by rolling method is regarded as a hard material because of high-flow stress and low ductility. Such ultrathin foils were often produced under low productivity with lots of rolling passes and the rolling techniques that enable more efficient production have been expected strongly. Therefore, it is necessary to clarify the characteristic of ultrathin foil rolling.

Fleck and Johnson [1] developed a theoretical framework for a new theory of cold rolling thin metallic foil. The theory predicts that plastic reduction occurs near entry and near exit of the roll bite, separated by a central region where the strip does not suffer reduction and does not slip relative to the work rolls. Yuen et al. [2] proposed the modeling of the mechanics of deformation in flat rolling. Depending on the rolling region, models of different complexities are required to predict accurately the strip and roll deformation in the roll gap. Zhang [3] presented the mechanism of cold rolling thin foil with the aid of the slab method in conjunction with an incremental analysis. The effects of initial thickness and yield stress of foil materials are discussed in detail. Jiang et al. [4] reported mechanics of roll edge contact in cold rolling of thin strip. Simulation of cold rolling of thin strip due to roll edge contact with oil lubrication was performed successfully using a developed influence function method. Langlands and McElwain [5] proposed a modified Hertzian foil rolling model with approximations based on perturbation methods. This enables good approximations to the roll torque and rolling force to be obtained in a small fraction of the time required for the full iterative solution.

Microforming Technology. DOI: http://dx.doi.org/10.1016/B978-0-12-811212-0.00009-1

Tribology plays a significant role for process feasibility and quality, especially in microforming where the influence of the tribological condition is very essential due to an extensive increase in friction while scaling processes from macro to micro. Asperity flattening in surface roughness affects the friction behavior at the interface and the transfer of surface roughness during rolling process [6–8]. As advanced rolling technology, asymmetrical rolling has been applied in foil production. Hwang and Tzou [9] carried out analytical and experimental study on asymmetrical sheet rolling. As reduction and frictional factor increase, the relative bonding length increases. In the miniaturization of thin strip rolling, the proportion of surface roughness of the materials to total thickness becomes larger with a decrease of strip thickness, which should be considered as nonuniformity of thickness.

In this chapter, foil rolling characteristic has been discussed and testified by 2D finite element method (FEM) simulation. Numerical investigations have been performed on ultrathin strips by micro rolling to study and to evaluate the influence of the specimen size and rolling parameters. Based on the actual strip surface, a novel finite element model of surface roughness has been developed and the characterization of asymmetrical thin strip rolling has also been identified. The comparison examines the evolution of real contact area, surface roughness, and contact pressure distribution at the interface between rolls and the rolled thin strip under various rolling conditions. It is the first attempt to investigate the surface roughness evolution in asymmetrical rolling based on real surface features.

9.2 SIMULATION PROCEDURE

9.2.1 Theory Analysis of Ultrathin Strip Rolling

In case of thin strip or foil rolling, deformation may be considered fairly homogeneous across the thickness of the foil. In foil rolling, rolling pressure is several times of the flow stress of the rolled material. When these earlier models with the simplifying assumption that the rolls remained circular in profile were applied to the rolling of thin foil, where the roll is far from circular, convergence problems arose. Fleck and Johnson assumed the roll profile included a central flattened region when the rolls deformed [1]. The pressure was taken to be near Hertzian and the deviation of the roll profile from flat was incorporated by using a modified mattress model.

The new theory assumes plane deformations with no lateral spread of the strip. Since the strip is much thinner than the elastic deformation of the rolls, the elastic and contained plastic compression of the strip in the z-direction can be ignored. Most of the theory is taken directly from the work by Fleck and Johnson [1]. Note that in this model, the strip (foil) thickness is considered to be small compared to the length of the roll gap and its transverse width. There is no variation of stress in vertical direction while the stress varies as the strip moves along the length of the roll gap. The modified Hertzian model was developed by Fleck and Johnson to simulate the rolling of thin foil [1].

The inlet thickness and outlet thickness of the strip are denoted by b_0 and b_2, respectively. At the contact surface between rolls and the rolled material, initially circular roll of radius R adapts a noncircular form and there is a sticking zone with a certain length. A small slab element of the strip with thickness of b and at a distance of x from the center of rolls is shown in Fig. 9.1A.

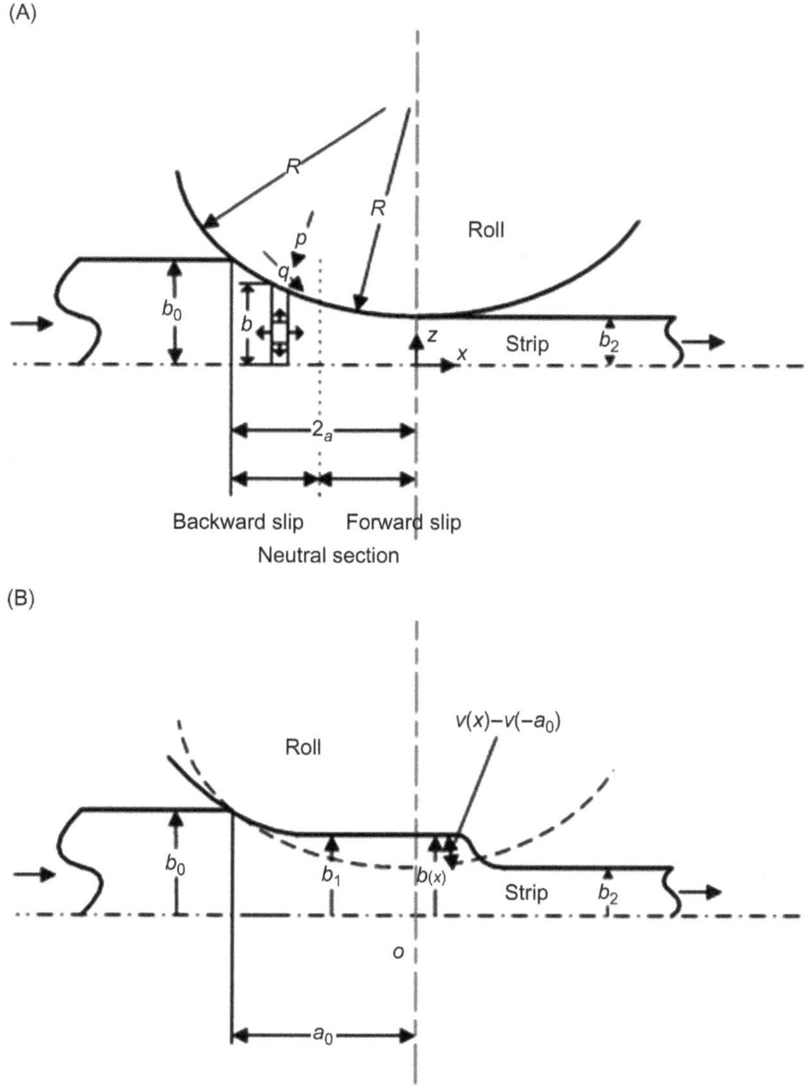

FIGURE 9.1 Definition of nip geometry: (A) slab element method and (B) sketch of deformed roll.

Given that $\sigma_z = -p$, since $db/dx \ll 1$, while horizontal equilibrium leads to,

$$b\frac{d\sigma}{dx} + (\sigma + p)\frac{db}{dx} + 2q = 0 \tag{9.1}$$

where x is the rolling distance, b is the strip thickness, σ is the tensile stress in the rolling direction, averaged through the thickness of the strip, p is the interface pressure, and q is the shear stress, where σ, p, and q are, respectively, the longitudinal stress, normal pressure, and interfacial shear of the element. The equilibrium equation for a small slab element will be used throughout the bite.

An exact treatment of the elastic deformation of the rolls is difficult since the elastic displacements (normal and tangential) of a surface point in the nip are integral functions of the pressure and traction distributions throughout the nip. For thin foil, the plastic compression of the foil is generally small compared with the roll elastic compression, so that the overall pressure distribution is approximately Hertzian (elliptical).

The variation of strip thickness t through the bite is given by

$$b = b_c + 2b(x) \tag{9.2}$$

where $b_c = b_0 - (a_0^2 - x^2)/R$ is the strip thickness variation with an initial strip thickness b_0 corresponding to the circular arc of the undeformed roll, as illustrated in Fig. 9.1B. Here a_0 is the value of x at the entry to the roll bite. $b(x)$ is the elastic displacement of the roll in the vertical z direction at a point x due to the applied normal pressure distribution p, measured relative to the displacement at the entry position. Using the half-space solution described by Johnson, $b(x)$ is given by [1]

$$b(x) = -\frac{2}{\pi E_R^{\varphi}}\int_{-a0}^{a2} p(x)\ln\left|\frac{s + xa}{s - x}\right| ds + \frac{(1 - 2\nu_R)(1 + \nu_R)}{E_R}\left(\int_{-a0}^{x} q(x')dx'\right) \tag{9.3}$$

where a_2 is the value of x at the exit of the roll bite and $E_R^{\varphi} = E_R/(1 - \nu_R^2)$ is the plane strain Young's modulus of the roll, with ν_R the Poisson's ratio of the roll. $p(x')$ and $q(x')$ are the normal pressure distribution and shear stress, respectively, at a point (x').

An increase of the contact pressure results in some roll deformation. With decreasing inlet thickness, there is a substantial nearly flat sticking region in the middle of the roll bite. A flat central region, a nearly Hertzian pressure distribution and a pressure spike just after the neutral zone. In this analysis, the strip equilibrium equation is still valid, but a simpler treatment is made to the roll deformation. The pressure distribution on the roll is assumed to be given by that from a Hertzian contact, modified by the contribution due to the local deviation of the roll profile from the Hertzian contact:

$$p(x) = p_0\sqrt{1 - \left(\frac{x - x_a}{x_a}\right)^2} - \frac{1}{2}K\Delta b \tag{9.4}$$

where p_0 is the maximum Hertzian contact pressure, Δh is the reduction in strip thickness (from its initial value h_1), and

$$K = \frac{8}{3\pi}\left\{\frac{E_r}{x_a(1 - \nu_r^2)}\right\} \tag{9.5}$$

where p_0 is the maximum pressure and a_0 is the distance from the contact at entry to the center-line of the rolls. Eq. (9.5) relates the reduction in strip thickness, hence the roll deformation, to the local roll pressure. The first term of Eq. (9.4) is the Hertzian pressure distribution whilst the second term represents a modification of the Hertzian pressure distribution associated with the change of strip thickness, Δb, at the point x. Note that the pressure, $p(x)$, is less than Hertzian when the strip reduction, Δb, is nonzero. A key feature of foil rolling is the fact that the elastic deformation of the work roll is strongly coupled to the plastic flow in the nip. Contact length is longer in foil rolling than that in thickness strip rolling.

The strip cannot expand perpendicular to the direction of the highest pressure, i.e., toward the roll surfaces and expansion of the strip is in contradiction to the plastic flow rule. The relation between the relative surface velocities of the roll and strip must be consistent with the direction of the frictional stress. The contact length is split into a series of zones, according to whether the strip is plastic or elastic and whether there is slip between the roll and strip as shown in Fig. 9.2: elastic loading zone, plastic reduction, central flattened zone, plastic expansion zone, plastic reduction, and elastic recovery zone. The deformation of the elastic rolls is assumed to be governed by the matrix foundation model [1,3].

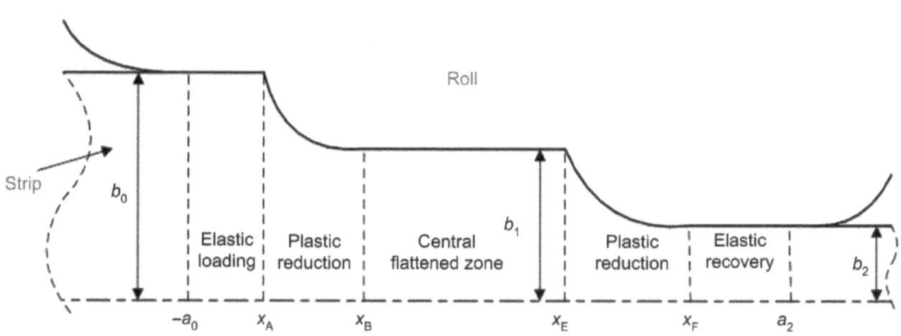

FIGURE 9.2 Schematic of the zones in the roll gap: elastic loading, plastic reduction, central flattened zone, plastic reduction, and elastic recovery.

Elastic loading: this zone starts from the entry and ends when the strip yields $\sigma_x + p = Y_{s.}$ The strip slips backward relative to the roll.

Plastic reduction: this starts from the point where the strip starts to yield and ends when the strip profile becomes horizontal. The condition $\sigma_x + p = Y_s$ holds throughout this zone. The strip slips backward relative to the roll.

Central flattened zone: this zone starts when the strip profile becomes flat. Here $-\sigma_s$ increases faster than p such that $\sigma_x + p < Y_s$. Eventually $-\sigma_x > p$ and the yield criterion $\sigma_x + p = -Y_s$ is satisfied at which the zone ends. The strip slips backward relative to the roll in this zone.

Plastic expansion zone: this zone starts when the yield criterion is satisfied and $\sigma_x + p = -Y_s$ holds throughout the zone. This strip increases in thickness in this zone, and this zone ends when the horizontal stress equals to that from zone E. The strip slips backward relative to the roll in the zone and the neural point could be at the end point.

Plastic reduction: the condition $\sigma_x + p = Y_s$ holds throughout the zone. It starts when the horizontal stress equals to that from the zone D and ends when the strip profile becomes horizontal. The neutral point could exist in this zone so that part of the strip slips backward and the other part slips forward relative to the roll in the zone.

Elastic recovery: this zone starts the strip profile and becomes horizontal and ends when the pressure and the horizontal stress drop to zero. The strip slips forward relative to the roll in the zone.

Fig. 9.3 shows the roll shape and contact pressure for different initial thickness under various rolling conditions. In case a (initial thickness: 0.20 mm), the deformation in the work rolls is small and the rolling deformation characteristic is similar to conversional strip rolling. Plastic reduction takes place throughout the roll bite, and slip occurs and changes direction at the neutral section where the pressure peaks. In case b (initial thickness: 0.10 mm), the increased sharp peak in pressure is bit different from the circular deformed arc in the form of a noticeable indentation of the roll profile. In case c (initial thickness: 0.05 mm), the peak pressure makes the rolls to deform locally to be a concave shape during rolling, and plastic reduction of the strip in such a no-slip zone is effectively uniform. With decreasing strip thickness, the length of the no-slip neutral zone increases and plastic reduction is confined to short zones at entry and exit. To flatten a segment of an elastic cylindrical roll requires the semielliptical pressure distribution of Hertzian. A characteristic feature of the pressure distribution for thin strip is the sharp pressure peak which occurs just near the end of the neutral zone. It is associated with the rather sharp reentrant corner in the roll deformation at the onset of plastic reduction at exit. The pressure rises steadily through the no-slip zone, the stresses in the strip in that zone are at yield status and the material is in a state of contained plastic flow. If the pressure falls in the on-slip zone, elastic unloading of the strip could occur.

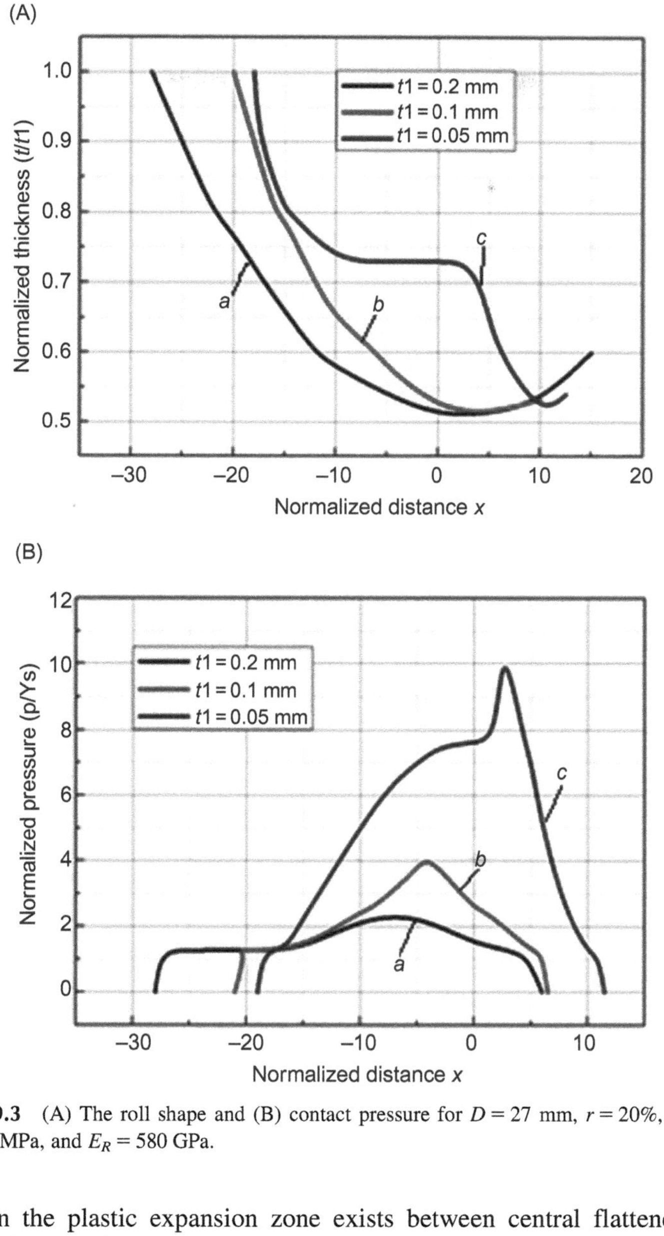

FIGURE 9.3 (A) The roll shape and (B) contact pressure for $D = 27$ mm, $r = 20\%$, $\mu = 0.12$, $Y_s = 1250$ MPa, and $E_R = 580$ GPa.

When the plastic expansion zone exists between central flattened zone and plastic reduction zone during the rolling stage, the material of the strip at the exit side of this zone moves slower than the material at the entry side of the zone in order to preserve the material continuity. The speed of the points on the roll surface is relatively uniform. Thus, the neutral point cannot be in this zone or at the interface between this zone and plastic reduction zone. The speed of the material in plastic reduction zone increases along the

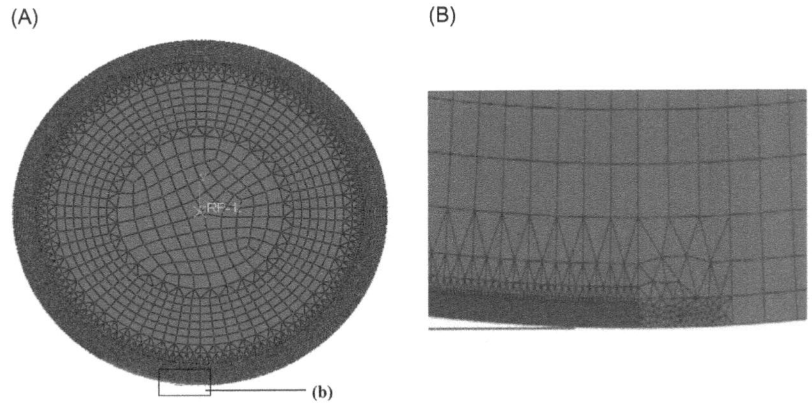

FIGURE 9.4 Mesh generation of the model [10]: (A) meshing work roll and foil and (B) transition between coarse and fine region.

rolling direction and at some point, it matched the speed of the roll surface, so the neural point could be confirmed at this zone.

9.2.2 Two-Dimensional Foil Rolling Modeling

Upper work roll is modeled as an elastic body whereas upper half foil is modeled as an elastoplastic material. Considering that foil is very thin in thickness, an extremely fine discretization of the work roll sleeve near the contact region is essential to obtain reliable data. A value of 100 μm has been chosen for work roll element length at contact surface. The corresponding foil surface element is <50 μm. The friction between the foil surface and work roll surface is modeled by utilizing the coulomb friction law with a coefficient of friction (COF) of 0.05. Explicit dynamic algorithms are selected in this case. The mesh generation of model is shown in Fig. 9.4A [10]. Transition between coarse and fine meshed region is given in Fig. 9.4B.

Considering that the foil width is far greater than its thickness and geometric symmetry, an upper half model of 2D plane strain is set up. Parameters of the model are shown in Table 9.1.

9.2.3 Surface Roughness Modeling in FEM

In order to simulate the rolling process accurately, appropriate models for material behavior across a large range of size scales are necessary. Fig. 9.5 shows the measured surface roughness of the steel strip obtained from measurement system: 3D Laser Scanning Microscope, VK-X100, where the rolling direction and rolling trace have also been indicated. The characterization of height distribution is significant along the rolling direction [11]. It can be seen that the surface roughness profile seems in random, and the maximum

TABLE 9.1 Model Parameters

Roll	Work roll diameter (mm)	260
	Rolling speed (mm/s)	7000
	Roll Young's modulus (MPa)	2,06,000
	Poisson ratio	0.3
	Density (tonne mm^3)	7.8×10^{-9}
Foil	Entry foil thickness (mm)	0.1
	Reduction (%)	50
	Young's modulus (MPa)	68,000
	Density (tonne mm^3)	2.7×10^{-9}
	Poisson ratio	0.34

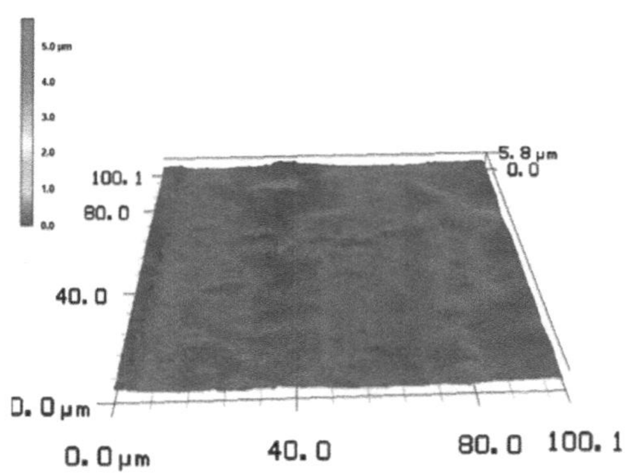

FIGURE 9.5 Measured three-dimensional surface roughness.

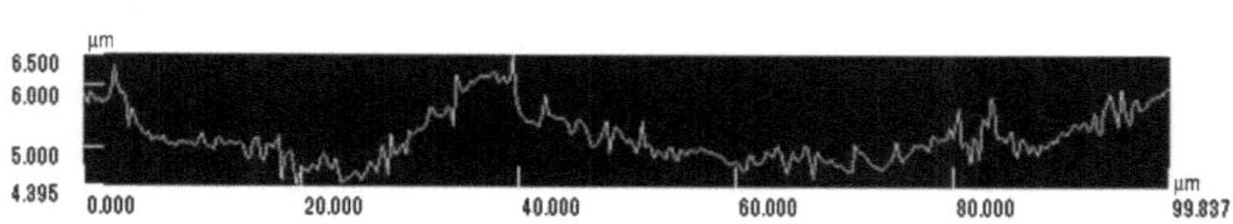

FIGURE 9.6 Two-dimensional surface profile.

relative height is around 5.8 μm. The surface height distribution of the rolled strip material is obtained from the surface profiles of the rough surfaces (Fig. 9.5). The two-dimensional surface profile is shown in Fig. 9.6.

The rolling simulation is carried out with the explicit dynamic finite element code, LS-DYNA. The SOLID164 element type has been used (a three-dimensional 8 NODE solid element) for the simulation. Refined meshing is employed around the surface roughness of the rolled strip to improve the simulation accuracy and further surface roughness calculation. A continuous function is required to eliminate the need of integrating discrete functions during the numerical procedure. Fourier series makes it possible to describe nonsmooth asymmetrical distribution functions from which the accuracy of the evaluation depends on the number of expansions used. However this kind of function still cannot represent the real surface feature.

In this chapter, the surface roughness model based on the real measured value of surface roughness has been set up through the ANSYS Parametric Design Language (APDL) as shown in Fig. 9.7A. In order to simulate the rolling process accurately, appropriate model for surface roughness across a large range of size scales is necessary. Key points are firstly created at the base of the block and then bottom-up solid modeling approach is applied to create the block volume. Fig. 9.7B shows the fine-meshing of surface asperities of rolled material.

The rolled strip was assumed to be isotropic elastic–plastic, while the model of the work rolls was assumed to be linear-elastic material. Table 9.2 shows the mechanical properties of the work roll and rolled strip used in the simulation. The static COF and the kinetic COF between the strip and roll were assumed to be 0.35 and 0.25, respectively. The result of simulation was

FIGURE 9.7 Modeling surface roughness based on real surface roughness: (A) surface roughness modeling and (B) fine-meshing of surface asperities [11].

TABLE 9.2 Properties of the Roll and the Rolled Strip in Simulation

Items	Roll	Steel Strip
Material model	Elastic	Elastoplastic
Density (kg/m^3)	7.85×10^3	7.85×10^3
Young's modulus (MPa)	2.1×10^5 (SKD 11)/5.8×10^5 (cemented carbide)	2.1×10^5
Poisson's ratio	0.3	0.3

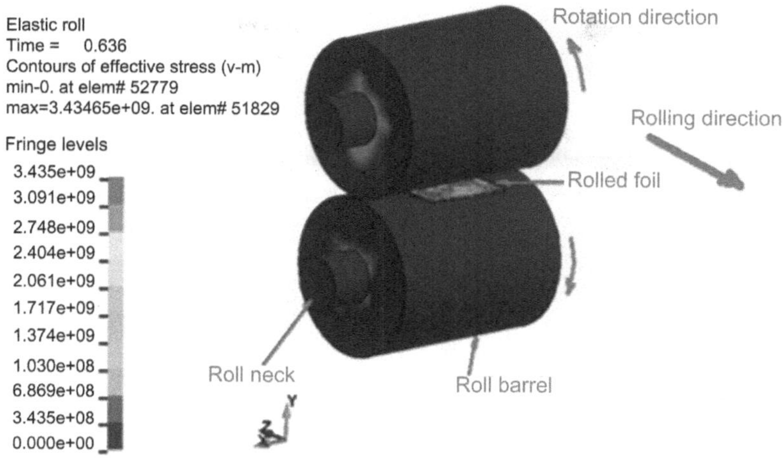

FIGURE 9.8 Effective stress distributions.

evaluated on the basis of rolling speed ratio between the upper and lower rolls, surface roughness and strain distribution of the rolled strip.

In the process simulation, the roll rotates with a fixed angular velocity. There is a predefined initial velocity of the strip and once in the rolling bite zone, the strip is pulled forward due to the friction of the rolls. A surface-to-surface contact condition is applied between the rotating roll and the thin strip. The rolled strip is assumed to be isotropic elastoplastic. Considering the elastic deformation, the work roll is assumed to be linear-elastic material and roll neck is rigid material. The roll rotation is driven by roll neck. Fig. 9.8 shows the effective stress distribution in the simulation model for 2-high rolling process. Y direction is the normal direction and $-Z$ direction is the rolling direction. The result of simulation is evaluated on the basis of rolling force distribution and surface characterization of the rolled strip.

9.3 SIMULATION RESULTS

9.3.1 Two-Dimensional Foil Rolling Simulation

9.3.1.1 Contact Pressure and Roll Contour in Roll Bite

By using parameters in Table 9.1, the von Mises stress distribution inside the foil and work roll in steady state is demonstrated in Fig. 9.9.

The profile contact surface and contact pressure are shown in Fig. 9.10. It should be emphasized that the roll profile in contact area with foil is consistent with the contact foil surface because of coupling relationship. A neutral zone of finite length, namely a central flattened region exists in roll bite in which the contact pressure reaches peak and no slip between the foil and the roll.

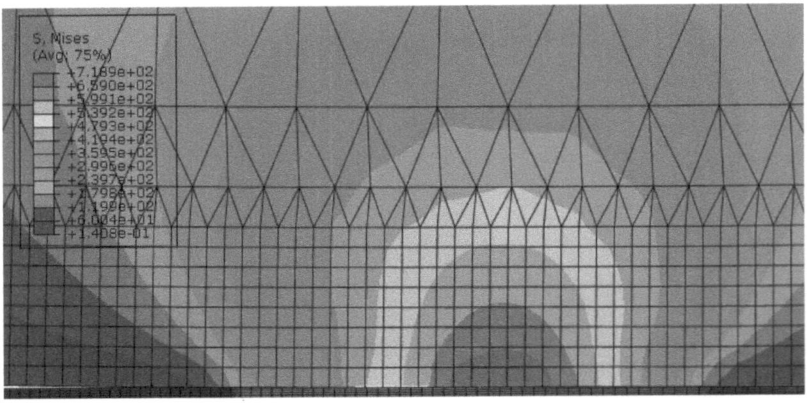

FIGURE 9.9 von Mises stress distributions inside the foil and work roll in steady state.

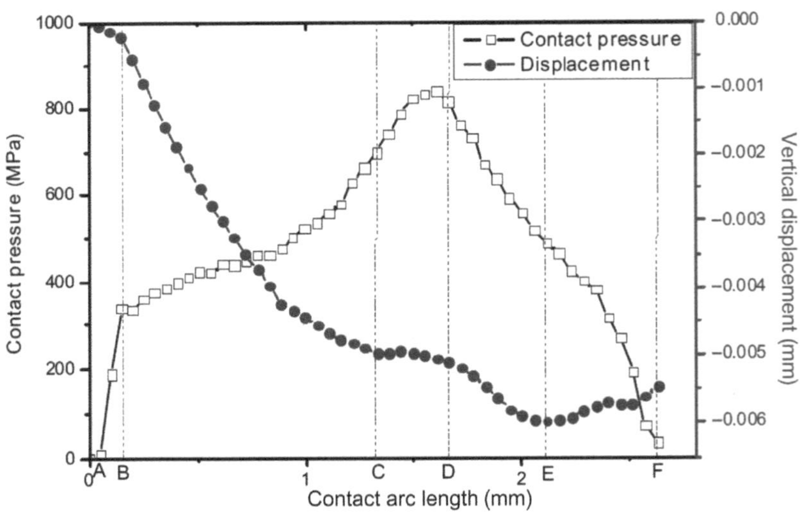

FIGURE 9.10 Distributions of contact surface and vertical foil displacement.

The simulation results are in the similar trend with the ones from Fleck and Johnson [1] assumptions about foil rolling. However, entry elastic zone and exit elastic zone are included in this model. In the whole contact arc length, it consists of entry elastic zone (arc length AB), entry plastic zone (arc length BC), a central flattened zone (arc length CD), exit plastic zone (arc length DE), and exit elastic zone (arc length EF). It is clearly seen from Fig. 9.10, entry and exit elastic zones are the obvious proportion parts of the roll bite.

9.3.1.2 *Effect of Entry Thickness*

In order to investigate the effect of entry foil thickness on behavior in roll bite, a set of four groups from (a) entry thickness 0.2 mm to (d) 0.04 mm are

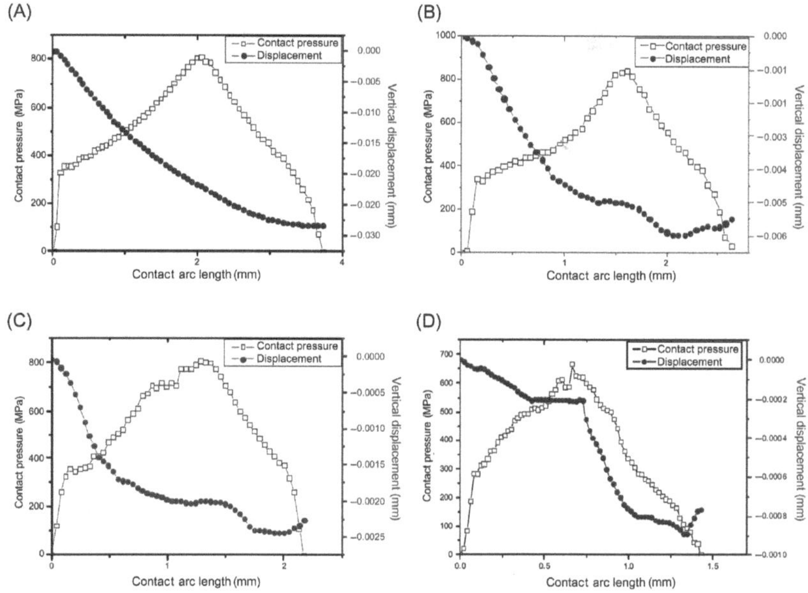

FIGURE 9.11 Effect of entry thickness on distribution of contact pressure and vertical displacement: (A) entry thickness 0.2 mm, (B) entry thickness 0.1 mm, (C) entry thickness 0.08 mm, and (D) entry thickness 0.04 mm.

presented in Fig. 9.11. Other rolling parameters, such as reduction rate, rolling speed, and COF are kept constant. In Fig. 9.11A, the roll profile at contact surface is approximate to be circular arc. Slip occurs through the whole roll bite except at the neutral section where the contact pressure reaches peak, which satisfies Bland and Ford rolling theory. In Fig. 9.11B, the roll profile at contact surface is to be noncircular and contact pressure becomes sharper. A neutral zone exists in roll bite where no slip occurs. In Fig. 9.11C and D, with the decrease of entry thickness, the proportion of neutral zone for the whole roll bite increases. Entry plastic reduction occurs in shorter zone whereas the exit plastic reduction occurs in relative longer zone.

9.3.1.3 Effect of COF

The friction between the work roll and the foil is a key parameter in cold rolling process. The values of the COF may change significantly at different stages of rolling and also has significant effect on the distribution of contact force and work roll flattening. With rolling parameters shown in Table 9.1 and other two coefficients of friction 0.03 and 0.07, distributions of contact pressure and vertical displacement are shown in Fig. 9.12.

In Fig. 9.12, it is shown that COF has great influenced on contact pressure and vertical displacement. With the increase of the COF, contact arc length reduces and contact pressure peak moves to entry side. When the COF is 0.03, the contact arc is nearly circular, however, the noncircular with neutral

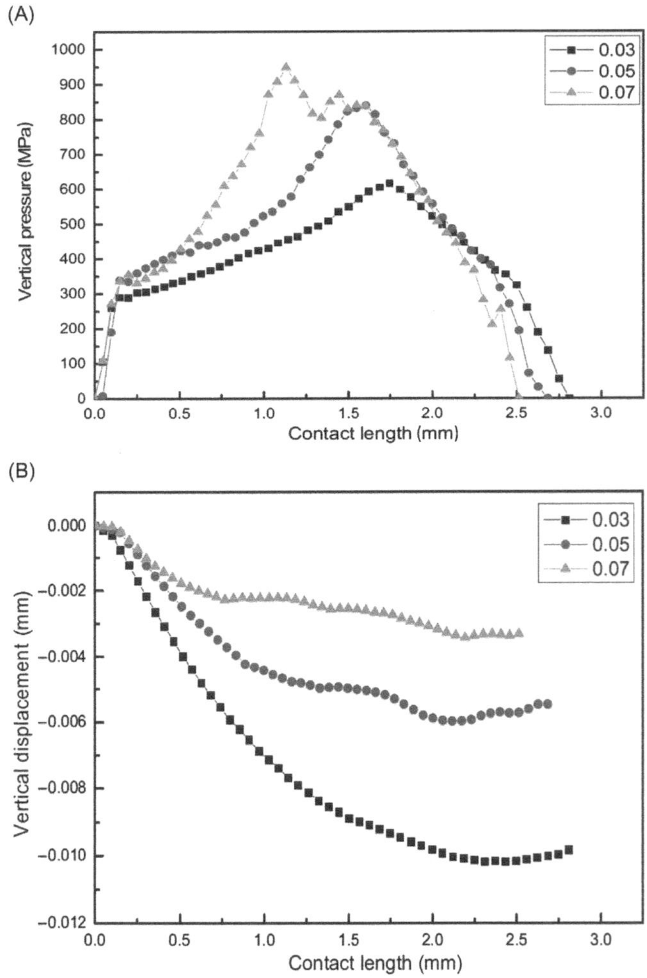

FIGURE 9.12 Effect of COF in roll bite: (A) contact pressure distribution and (B) vertical displacement distribution.

zone appears under 0.05 and 0.07 coefficients. Vertical displacement reduces obviously whereas the neutral zone increases when the COF increases.

9.3.1.4 *Effect of Reduction Rate in Roll Bite*

Rolling parameters shown in Table 9.1 and other two different reductions rates, 10% and 30% are applied, respectively, in the simulation.

Fig. 9.13A shows the effect of reduction rate on distribution of contact pressure. It can be seen that the contact arc length becomes longer and pressure distribution hill gets sharper as the reduction rate increases. Fig. 9.13B,

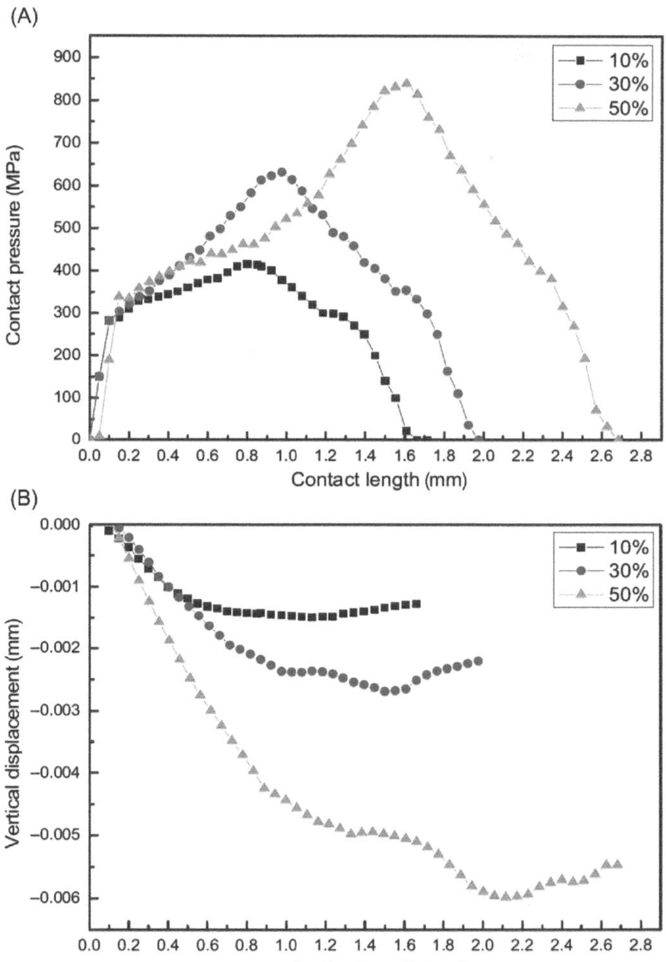

FIGURE 9.13 Effect of pass reduction in roll bite: (A) Contact pressure distribution and (B) vertical displacement distribution.

when the reduction rate is 10%, the contact arc profile is nearly circular. It changes to be noncircular with 30% and 50% reduction rates. Length of neutral zone at 50% is greater than that of 30%.

9.3.2 Simulation in Surface Roughness Evolution

Analytical studies show that the real contact area between the rolled strip and the roll occurs at isolated points where the asperities came together. This constitutes a very small fraction of the apparent area for flat surfaces. With

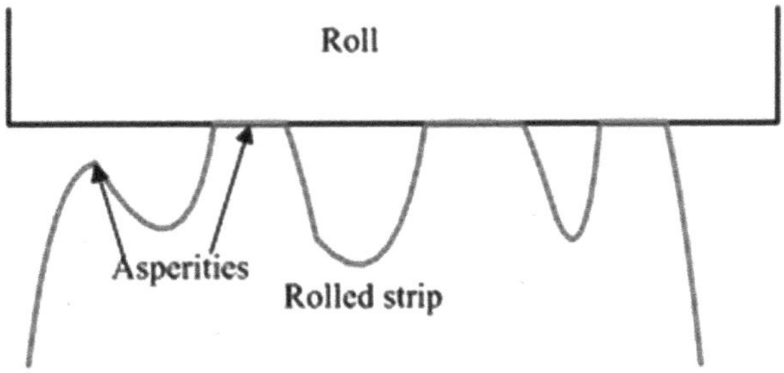

FIGURE 9.14 Schematic deformation models of asperities.

increasing pressure, the plastic deformation fields around individual asperities merged. Due to surface roughness it occurs at discrete points, and the force between the bodies is the sum of the individual forces on contacting asperities which constitute the pressure distribution. Fig. 9.14 shows the schematic deformation model of asperities. It can be seen that the asperities flattening occurs near the contact interface between the roll and the rolled strip. When surfaces are loaded together under dry conditions or with boundary lubrication, most of the load is carried by the asperities in close contact. In these cases the friction between the roll and strip will be controlled by the growth of the true areas of contact.

Two factors complicate the picture of contact growth in rolling processes. First, at the high pressures involved, there is interaction between contacting asperities and second, the deformation of the rolled materials will affect the asperity deformation. Without any bulk deformation, the growth of contact area will be limited as the pressure increases. The effect of bulk plasticity was highlighted which showed how asperities were flattened when deforming short cylinders with bulk deformation directly below the asperities. The area of contact ratio was increased in model asperities with bulk deformation of the substrate, and the rate of asperity flattening with bulk straining was related to the spacing of the asperities and to the difference in pressure between loaded and unloaded parts.

9.3.2.1 Asymmetrical Rolling Analysis

Asymmetrical rolling is an advanced rolling technology to produce ultrathin strip due to existing cross-shear deformation zone. Fig. 9.15 shows the schematic illustrations of asymmetrical rolling where the speed of the upper roll is higher than that of the lower roll. The nonuniform distribution of longitudinal speed is significant across the strip thickness. Because of higher reduction of the upper strip layer resulted from higher speed, the upper neutral point is shifted along the rolling direction to near the deformation exit. Correspondingly the critical angle and forward slip are greater than that in symmetrical rolling. An increase of the roll speed ratio will result in an

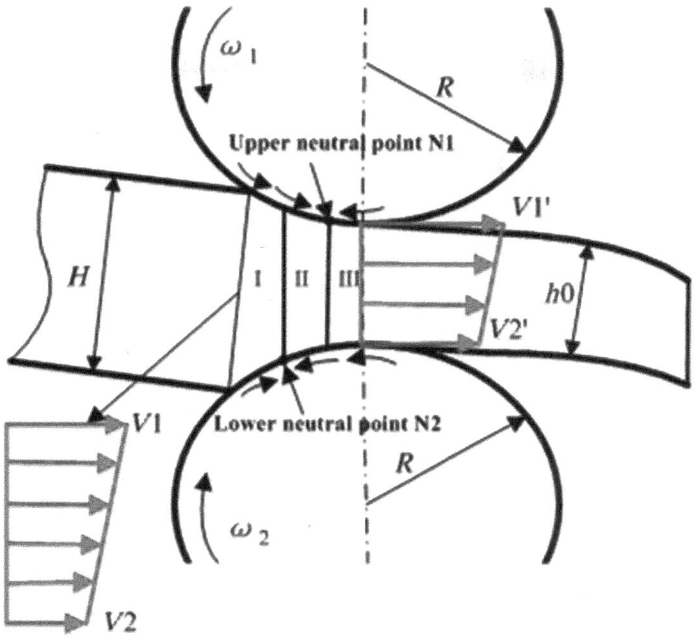

FIGURE 9.15 Schematic illustrations of asymmetrical rolling.

increase in the length of the shear zone. At the same time, due to unchanged energy balance for asymmetrical rolling in the vertical plane, the lower neutral point will be moved back to near the rolling entry. Variation in these parameters increases the longitudinal speed of the strip near the upper roll and reduces the longitudinal speed of the strip near the lower roll in the exit. In this case, the critical angle, forward slip, and longitudinal speed of the strip near at the lower roll in the exit plane are reduced. This process improves productivity of the rolling operation in reduction of the rolling force, pressure, torque, and improving properties of the strip surface. The neutral points are at different positions resulting in a cross-shear zone between them.

The analysis of asymmetrical rolling in the thin strip has been carried out within the present study with the aim of establishing the effect of the speed asymmetry factor, $\forall_A$, on rolling force, strip bending, and surface roughness during the rolling process.

As shown in Fig. 9.15, the whole plastic deformation region at the roll gap is divided into three distinct regions according to the directions of the friction force exerted on the strip from the upper and lower rolls: zone I for the entry region, zone II for the cross-shear region (CSR), and III for the exit region. From the volume constancy of the material, the positions of the upper and lower neutral points N_1 and N_2, have the following relationship as

$$x_{N1} = \sqrt{\forall_A x_{N2}^2 + (\forall_A - 1)\frac{h_0}{R}} \quad (9.6)$$

where $\forall_A = \frac{\omega_2}{\omega_1}$ is the speed ratio of two rolls; R the roll radius, mm; h_0 the thickness after rolling, mm; x_{N1} and x_{N2} are the distances of upper and lower neutral points to the vertical symmetry plane of two rolls, respectively.

9.3.2.2 Effect of Roll Materials

From the simulation, the rolling force is reduced during asymmetrical rolling. Fig. 9.16 illustrates the V-M effective stress in micro rolling. The harder roll has a higher value than that of soft one.

Fig. 9.17 shows rolling force distribution for four cases where the roll material and speed ratio are different. It can be seen that higher speed rate can reduce rolling force due to existing cross-shear deformation zone in asymmetrical rolling process.

In order to investigate the effect of elastic deformation of rolls on the foil deformation, two roll materials with different Young's modulus are used in this simulation: SKD 11 and cemented carbide. Fig. 9.18 shows the effect of roll Young modulus on the rolling force. It can be seen that more rolling force is required to obtain the same reduction in thickness for SKD 11 than that for cemented carbide. The higher the Young's modulus of the roll material, the lower is the roll flattening. Therefore, it is possible to obtain more reduction in thickness for the rolled strip.

9.3.2.3 Real Contact Area Evolution

The real area of contact can be defined as a function of the surface topography, material properties, and interfacial loading conditions as shown below.

$$A_{\mathrm{RCA}} = f(\text{Surface topography, material properties, loading conditions}) \quad (9.7)$$

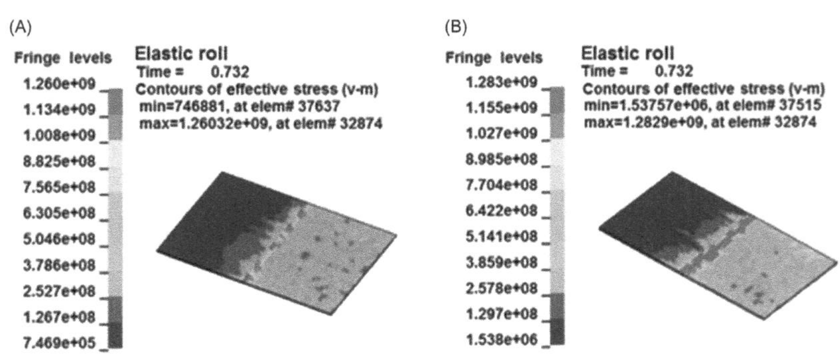

FIGURE 9.16 Simulation in micro rolling: (A) soft roll and (B) hard roll.

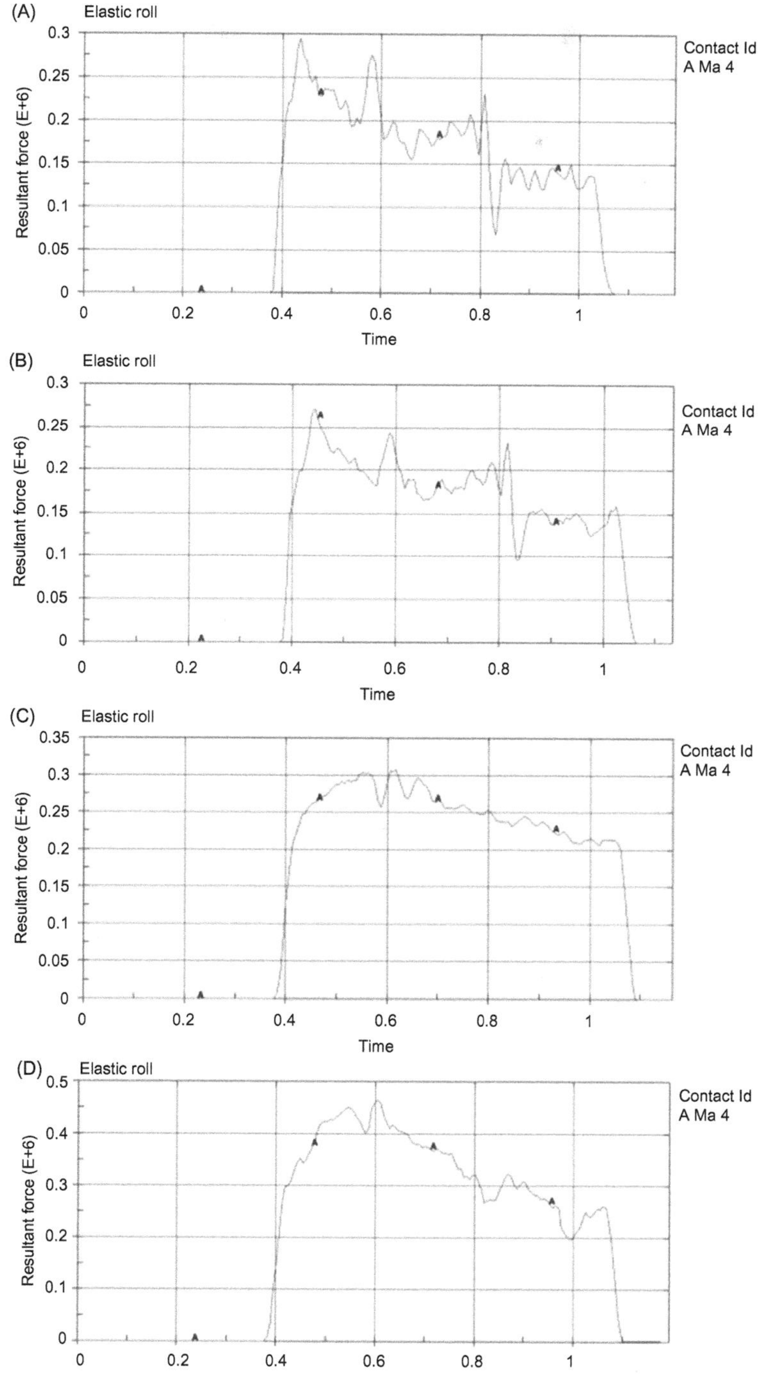

FIGURE 9.17 Rolling force in simulation for different rolling conditions: (A) cemented carbide, speed ratio of 1.5, (B) cemented carbide, speed ratio of 1.0, (C) SKD 11, speed ratio of 1.5, and (D) SKD 11, speed ratio of 1.0.

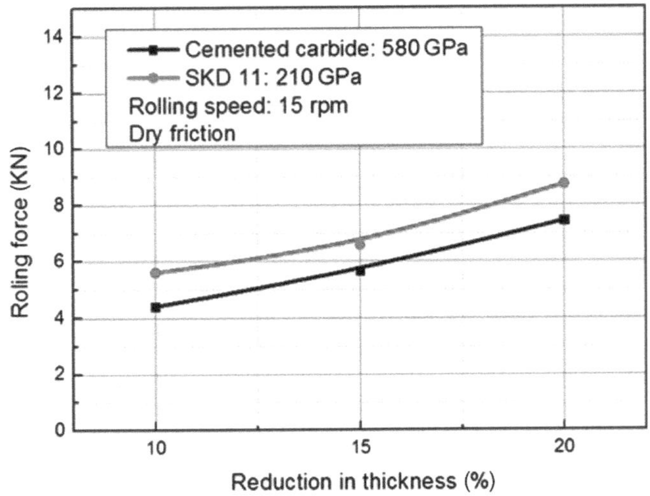

FIGURE 9.18 Rolling force versus roll materials: SKD 11 and cemented carbide.

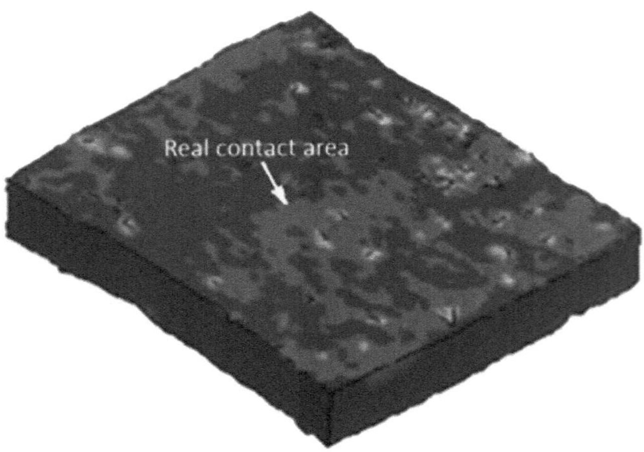

FIGURE 9.19 Real contact area under rolling pressure [11].

When two surfaces move relative to each other, the adhesion of these asperities and other sources of surface interactions contribute to friction force. A real contact region consists of contact spots, the total area of which is a small fraction of the nominal contact area which is the minimal connected region enclosing all the contact spots. Fig. 9.19 shows the real contact area under asymmetrical rolling. The total contacting length of the profile, as well as the equivalent stress at each node contacting the roll is recorded. The former is used to calculate the contact area ratio, while the data of the equivalent stress are collected for further analysis. The randomness comes from the random nature of surface roughness, whereas the different sizes of spots occur due to the multiple scales of surface roughness.

For a given load, the size of spots depends on the surface roughness and the mechanical properties of the contacting materials. The existing spots will increase in size with an increase of rolling pressure, new spots will appear, and two or more spots may coalesce to form a larger spot. The calculation of real contact area ratio is listed below,

$$\alpha = \frac{A_r}{A_a} \tag{9.8}$$

where A_r and A_a are real contact area and apparent area, respectively.

Fig. 9.20 compares the evolution of area of contact ratio with normal pressure obtained by the FEM models and Keife's model [12]. To obtain a general understanding, mean normal pressures applied are all normalized by over the yield stress in plain strain. It is clear that the contact ratio increases as a function of normalized pressure for the two models. The curve of present model is relatively close to that of Keife's model with a maximum difference of 10.5% at p/2k of 0.7. The difference between the present model and the Keife model apparently becomes larger before the normal pressure reaches 1.2. Both curves have similar trends and the differences between two models might be explained as different asperity slopes and different methods being used in the simulations.

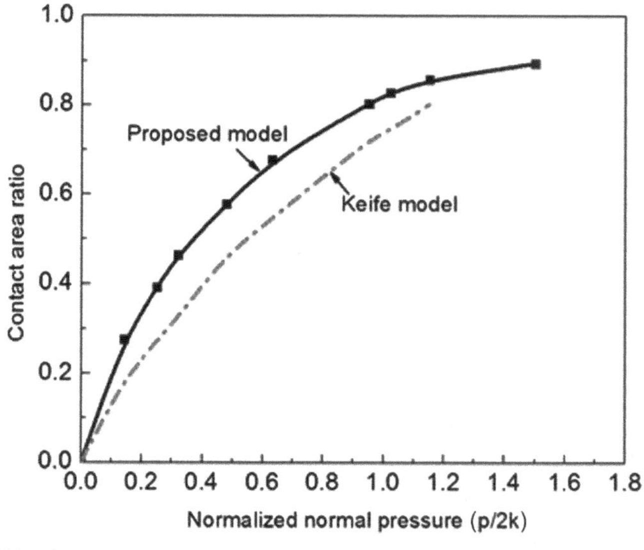

FIGURE 9.20 Contact area ratio versus the normalized normal pressure [11].

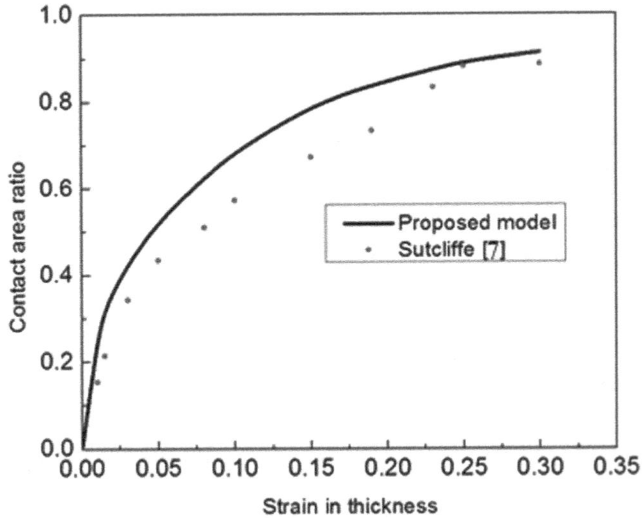

FIGURE 9.21 Contact area ratio dependence on strain in thickness [11].

Real contact area fraction as a function of strip strain shown in Fig. 9.21 and Sutcliffe [7] experimental results is also plotted. The asperities on the surfaces to interact is in a nonuniform way, exerting forces on each other and leading to frictional forces and wear. The key to predicting the impact of surface topography is the ability for the asperities to deform in response to rolling pressure and for the program to solve those asperity deformations with as little solution error as possible. With an increase of strain in thickness, the contact area ratio increases accordingly. However, this trend is much more flattening when the contact area ratio reaches to near 0.9.

9.3.2.4 *Surface Roughness Evaluation During Rolling*

Fig. 9.22 shows the change of the surface roughness ratio with normalized normal pressure. R_a' is the surface roughness after flattening while R_a is the initial surface roughness. It is clear that the roughness decreases with an increase of normal pressure because the roll flattens the asperities and peaks much more. High pressure results in more real contact area and the more asperity flattening will occur.

The surface roughness ratio of the upper and the bottom sides with strip thickness and speed ratio is shown in Fig. 9.23. Results show that the surface roughness ratio increases with strip thickness and speed ratio. Nevertheless, both sides of the surface have nearly conformed to the roll surface after the rolling. Since the surface roughness on both the incoming and outgoing strip is close to that of the rolls, any effect of surface roughness is overshadowed by scatter in the measurements. Nevertheless, there appears to be a slight

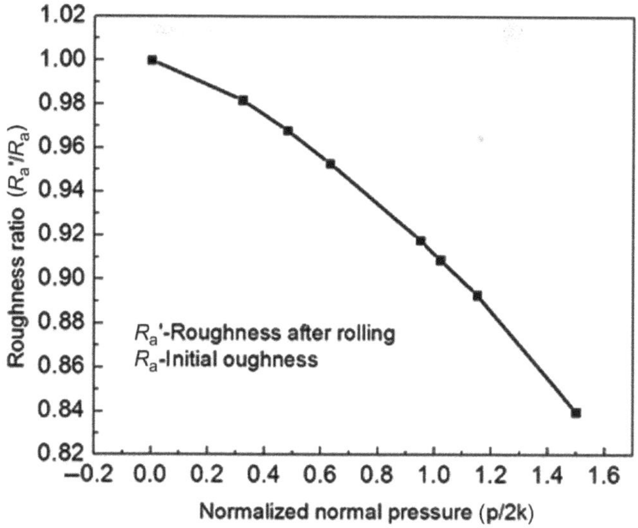

FIGURE 9.22 Surface roughness changes with normal pressure [11].

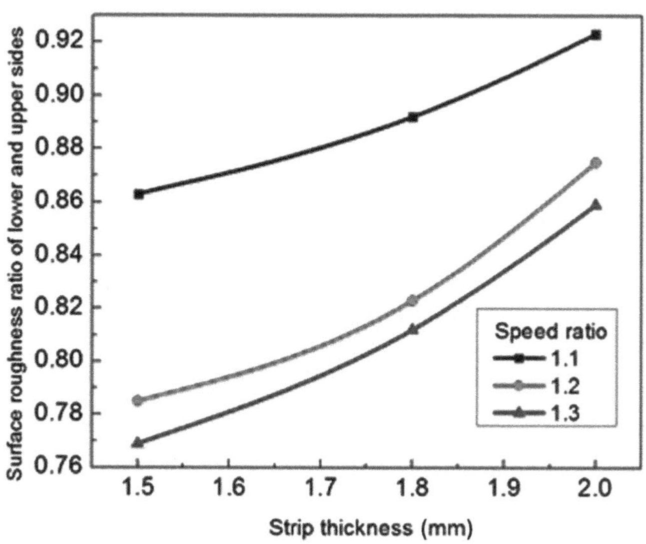

FIGURE 9.23 Roughness amplitude of the rolled strip [11].

increase in the surface roughness at very small, which may be associated with a breakdown of lubrication.

Fig. 9.24 shows the surface roughness versus pass reduction under different initial thicknesses. It can be seen that the surface roughness decreases with an increase of pass reduction and the decrease of initial strip thickness.

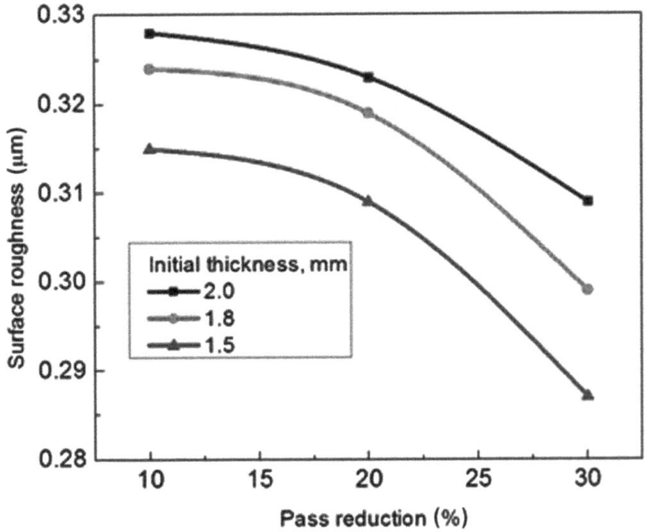

FIGURE 9.24 Surface roughness versus pass reduction under different initial thickness [11].

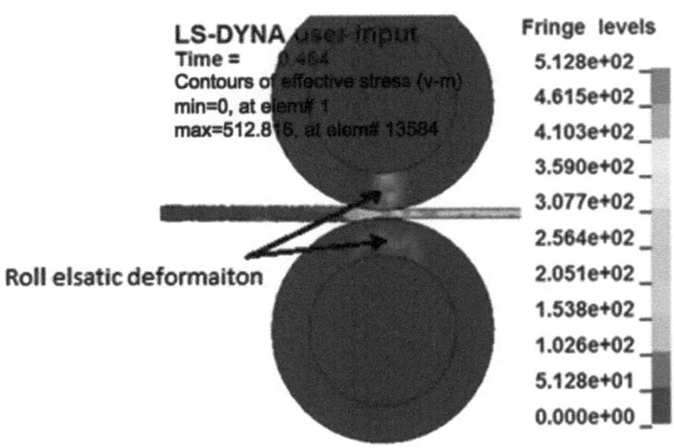

FIGURE 9.25 Asymmetrical rolling deformation [11].

With the increasing rolling pass reduction, more plastic deformation occurs. The reprinting between the roll and the strip will be more significant which results in the decrease of the surface roughness of rolled strip.

9.3.2.5 *Rolling Characterization of Asymmetrical Rolling*

The effect of rolling speed ratio is investigated in this section.

Fig. 9.25 shows the rolling state in asymmetrical deformation, the roll elastic deformation is also shown in this figure. It can be seen that the maximum effective stress is around 512 MPa at time of 0.464 s.

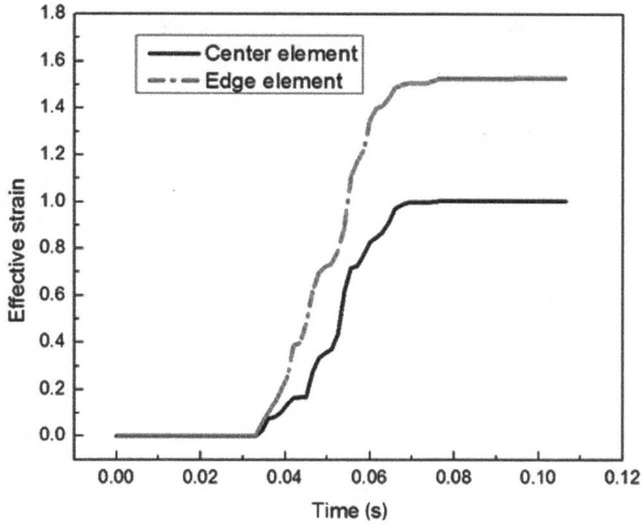

FIGURE 9.26 Effective strain comparison between the element in the central and the strip edge.

Fig. 9.26 shows the effective strain comparison between the elements in the central and the strip edge. It can be seen that the effective strain is higher in the strip edge (1.5) than that in the central (1.0). This is because of the roll elastic deformation which results in an inhomogeneous rolling deformation across the strip width. More deformation exists in the strip edge which will be a potential factor for edge defect occurrence.

Fig. 9.27 shows the rolling pressure under different roll speed ratio across the contact length. It can be seen that the maximum rolling pressure is in the position with contact length of 4 mm. With an increase of roll speed ratio, the rolling pressure decreases. When the roll speed ratio is 1, the neutral points become a single one.

Strip curvature in asymmetrical rolling is an important evaluation item for strip shape quality. If the radius of outgoing curvature due to differences in the shear and axial strains are defined as r_1 and r_2, respectively, the resultant radius of curvature at exit expresses as [13]

$$\frac{1}{R_e} = \frac{1}{r_1} + \frac{1}{r_2} \tag{9.9}$$

To distinguish between the upward and downward curvatures of the outgoing strip, the curvature index is defined as

$$K = \frac{R_e - \frac{h_0}{2}}{R_e + \frac{h_0}{2}} \tag{9.10}$$

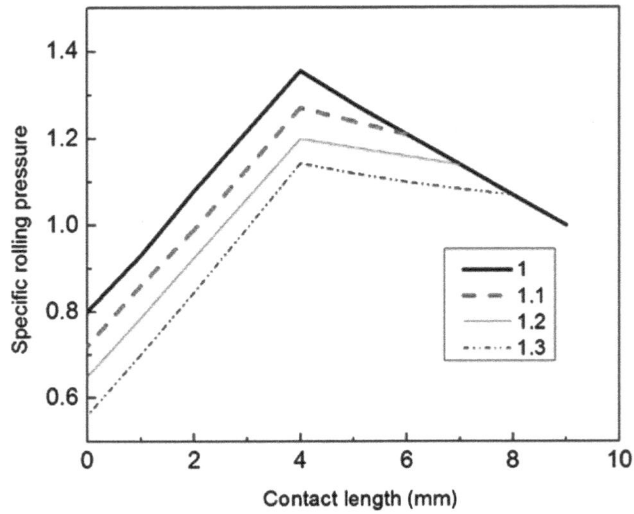

FIGURE 9.27 Specific rolling pressure for various roller speed ratio.

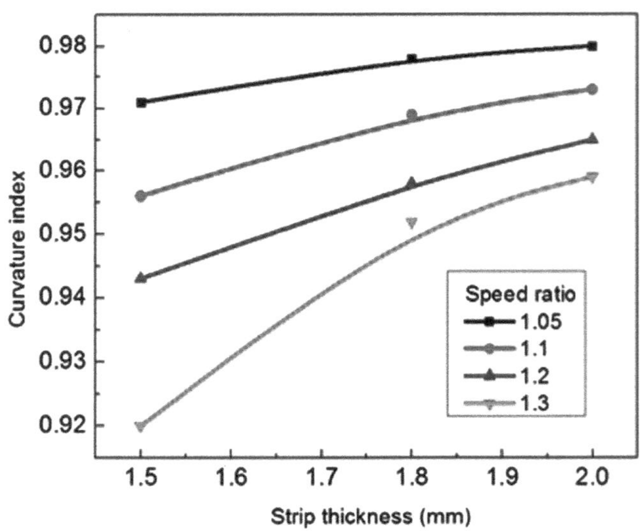

FIGURE 9.28 Relation between the curvature index and initial thickness for different speed ratios.

In Fig. 9.28, variations of the curvature index for different input thicknesses at three specified speed ratios are depicted. It can be seen, in a small amount of speed mismatch, change in the initial thickness has no vigorous effect on the outgoing curvature. However, at high speed mismatch values, the curvature index effectively changes by increasing the initial thickness of the strip. It is again seen that the radius of curvature at exit increases by increasing the initial thickness of the strip. It should be noted that although there is the same amount of 6% velocity mismatch between the work rolls for cases with speed ratios of 1.05 and 1.0, downward curvatures but with different radii are produced for both cases. The analysis shows that the strip bends most often toward the lower-speed roll. Therefore, the forward tension and backward tension are necessary in asymmetrical rolling to maintain better strip shape profiles.

In summary, a 2D elastoplastic finite model is developed to simulate foil rolling process. Roll contact profile is approximate to be circular arc with 0.2 mm entry thickness. With the decrease of entry foil thickness, the contact arc profile changes to be noncircular and the proportion of neutral zone for the whole roll bite increases. Entry plastic reduction occurs in the shorter zone whereas the exit plastic reduction occurs in relative longer zone. Contact arc length reduces and contact pressure peak moves to entry side when the COF increases. Vertical displacement reduces obviously whereas the neutral zone increases when the COF increases. As the reduction rate increases, the contact pressure distribution hill becomes sharper, and the contact arc length and neutral zone become longer.

Cross-shear deformation zone in asymmetrical rolling can reduce the rolling force and it benefits getting thinner rolled strip. The area of contact ratio increases as a function of normal pressure. The analysis shows that the strip bends most often toward the lower-speed roll. Increasing the value of the speed asymmetry factor causes an increase in the strip curvature, while the force parameters decrease with an increase of the asymmetry factor. The length of CSR in the roll bite increases and the neural points at the upper and lower work rolls vary significantly when the roll speed ratio increases. Surface roughness decreases with an increase of the pass reduction.

REFERENCES

[1] N.A. Fleck, K.L. Johnson, Towards a new theory of cold rolling thin foil, Int. J. Mech. Sci. 29 (1987) 507–524.

[2] W.Y.D. Yuen, A. Dixon, D.N. Nguyen, The modeling of the mechanics of deformation in flat rolling, J. Mater. Process. Technol. 60 (1996) 87–94.

[3] L.C. Zhang, On the mechanism of cold rolling thin foil, Int. J. Mach. Tools Manufact. 35 (1995) 363–372.

[4] Z.Y. Jiang, H.T. Zhu, A.K. Tieu, Mechanics of roll edge contact in cold rolling of thin strip, Int. J. Mech. Sci. 48 (2006) 697–706.

[5] T.A.M. Langlands, D.L.S. McElwain, A modified Hertzian foil model: approximations based on perturbation methods, Int. J. Mech. Sci. 44 (2002) 1715–1730.

[6] M.P.F. Sutcliffe, Flattening of random rough surfaces in metalforming processes, J. Tribol. 121 (1999) 433–440.

[7] M.P.F. Sutcliffe, Surface asperity deformation in metal forming processes, Int. J. Mech. Sci. 30 (11) (1988) 847–868.

[8] H.R. Le, M.P.F. Sutcliffe, Analysis of surface roughness of cold-rolled aluminium foil, Wear 244 (2000) 71–78.

[9] Y.M. Hwang, G.Y. Tzou, Analytical and experimental study on asymmetrical sheet rolling, Int. J. Mech. Sci. 39 (1997) 289–303.

[10] L. Hao, Z.Y. Jiang, D.B. Wei, X.W. Chen, Finite element analysis of roll bit behaviors in cold foil rolling process, AIP Conf. Proc. 1532 (2013) 478.

[11] H.B. Xie, K. Manabe, Z.Y. Jiang, A novel approach to investigate surface roughness evolution in asymmetric rolling based on three dimensional real surface, Finite Elem. Anal. Des. 74 (2013) 1–8.

[12] H. Keife, Jingyu Shi, D.L.S. McElwain, T.A.M. Langlands, Foil rolling: a new interpretation of experimental results, J. Mater. Process. Technol. 121 (2002) 50–59.

[13] M. Qwamizadeh, M. Kadkhodaei, M. Salimi, Asymmetrical sheet rolling analysis and evaluation of developed curvature, Int. J. Adv. Manufact. Technol. 61 (2012) 227–235.

Chapter 10

Simulation of Micro Deep Drawing

10.1 INTRODUCTION

Driven by the urgent requirement on the miniaturized products that can bear harsh working environments, such as small or tiny devices, micro metal products have been required [1,2]. Several obstacles relating to tiny sizes need to be overcome to fulfill the superiority of microforming processes [3]. Once scaled down to a microscale, influences of different impact factors in microforming change from that in normal forming processes, and some ignorable factors in macro forming become obvious and significant [1,4–7]. The decrease of flow stress with small sample sizes has been observed and corresponding theories, such as the surface layer model and the modified surface layer model [8,9], have been developed as explanations. Modifications to the traditional material models, such as the Swift model and the Hall–Petch model [10,11], have been conducted to fit the strain–stress relationship in the microscale. However, those developed or modified models can partly explain size effects, and none of them can present deformation with size effects on the microscale. A large ratio of grain size to sample size means that deformation behavior is determined by a few grains and therefore, randomness is increased resulting in significant data scatter [12].

To reveal the size effects further, the Voronoi models were developed [8–10]. The sample was divided into small subareas according to the Voronoi structures and the small subareas were assigned with different mechanical properties. Although the distribution and sizes of the Voronoi cells can be adjusted by a density function, the cells distribution can hardly match the real microstructures of the samples. As there are no clear classification criteria, all the Voronoi cells are organized randomly. Grain boundaries are generally ignored in the Voronoi simulation models. Therefore, the Voronoi models are not accurate enough compared to the real micro samples. Thus, a similar microscale simulation model based on the real microstructure of the sample is required, where there are exact shapes and sizes employed to the crystals on the sample. The grains and grain boundaries are considered separately.

Microforming Technology. DOI: http://dx.doi.org/10.1016/B978-0-12-811212-0.00010-8

Surface roughness, a nonintrinsic property, though being very slight, can be relatively large compared to the minute sizes of micro products. Surface roughness can be considered as a part of the geometry and affects the microforming process and then the products, whereas the research on surface roughness has mainly focused on the friction fields [13–15]. Lubricant pockets model and the normal pressure-based dynamic coefficient of friction (COF) model [13,16] have been developed. Consequently, the influence of surface roughness is primarily expressed by the COF in the normal microforming FE models. A real surface topography or a simplified rough surface was introduced into FE models, and these models could exhibit detailed surface roughness information [17]. Nonetheless, the meshes based on this complex surface topography were of tetrahedron or triangular meshes. As a result, forming simulation could barely be conducted due to the large quantity and poor quality of meshes in these models. Therefore, an FE model that considers influences of the surface roughness from a geometry viewpoint and counts friction in the microscale is highly needed.

In this chapter, micro deep drawing (MDD) of circular cups was chosen as the research objective as the cup-like micro products have been widely applied. The real microstructure and the Voronoi MDD simulation models were developed with consideration of both grains and grain boundaries based on image-based modeling methods. Furthermore, the surface roughness of the blank was considered and represented by thickness distribution in this Voronoi model. Springback was also simulated after MDD calculation. Consequently, effects of surface roughness on the MDD process and the drawn cups were studied. The results from the proposed models were compared with the results from a normal deep drawing model and experimental data. Influences of grains and grain boundaries on drawn cups were discussed.

10.2 SIMULATION PROCEDURE

10.2.1 Basic Models

A normal deep drawing FE model was firstly developed as a fundamental model in LS-DYNA. Fig. 10.1 displays (A) the fundamental deep drawing model and (B) its geometry sketch. There were a punch, a die, a blank holder, and a blank that was the only deformable part in this model. One quarter of the machine was modeled to accelerate computational speed. Accordingly, symmetrical boundaries were applied on the symmetrical lines of the blank. Moreover, all the parts were modeled on shell elements, and the blank was meshed with high-quality quadrilateral meshes through carefully allocating mesh seeds on each line of the blank. These high-quality quadrilateral meshes are beneficial for forming simulation with large deformation. Additionally, the blank applied a fully integration elemental formula

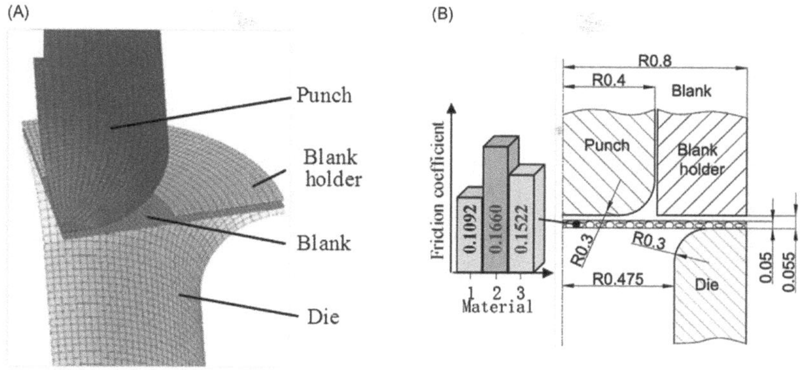

FIGURE 10.1 (A) Basic deep drawing model and (B) its geometry illustration [18].

while others were assigned a default elemental formula. Seven integration points through thickness direction were employed to accurately capture thickness direction stress–strain distribution and provide accurate and adequate information for subsequent springback simulation. Regarding the friction, a surface-to-surface forming contact algorithm specialized for metal forming simulation was applied. Further, a constant COF was utilized and its value was tested by a Universal Tribometer from Rtec Instruments. A 3-parameter-Barlat material model was employed for the blank with a thickness of 50 μm [19], while all the other parts were of a rigid material model. The parameters of each blank material model were tested 10 times using repeated tensile tests. According to the plane stress assumption of the material model, a plane anisotropy index (the *R*-value), defined as the instantaneous strain ratio of width to thickness (Eq. (10.1)), was utilized to update thickness. Moreover, a yield stress criterion, defined as Eq. (10.2), was employed for the blank.

$$R = \frac{\frac{dW}{d\varepsilon}/W}{\frac{dT}{d\varepsilon}/T} \tag{10.1}$$

$$\Phi = a|K_1 + K_2|^m + a|K_1 - K_2|^m + c|2K_2|^m = 2\sigma_Y^m \tag{10.2}$$

$$K_1 = \frac{\sigma_x + h\sigma_y}{2} \tag{10.3}$$

$$K_2 = \sqrt{\left(\frac{\sigma_x - h\sigma_y}{2}\right)^2 + p^2\tau_{xy}^2} \tag{10.4}$$

where σ_Y is the yield stress, σ_x and σ_y are stress along x and y directions, respectively, τ_{xy} is the shear stress, a, c, h, and p are the material parameters which can be obtained by Eqs. (10.5)–(10.8). Moreover, p can be searched by iterations of Eq. (10.8) at the condition of $\phi = 45$ degrees. For isotropic material, $a = c = h = p = 1$. Further, $m = 8$ for face centered cubic material.

$$a = 2 - 2\sqrt{\frac{R_{00}}{1 + R_{00}} \frac{R_{90}}{1 + R_{90}}} \tag{10.5}$$

$$c = 2 - a \tag{10.6}$$

$$h = \sqrt{\frac{R_{00}}{1 + R_{00}} \frac{1 + R_{90}}{R_{90}}} \tag{10.7}$$

$$g(p) = \frac{2m\sigma_Y^m}{\left(\frac{\partial \Phi}{\partial \sigma_x} + \frac{\partial \Phi}{\partial \sigma_y}\right)\sigma_\phi} - 1 - R_{45} = 0 \tag{10.8}$$

where R_ϕ is the R-value at angle of ϕ.

This model employed normal material model for the blank and therefore cannot represent any size effects. Furthermore, the blank had a constant thickness, thus surface roughness and its influences cannot be revealed.

10.2.2 Voronoi Models

As the basic model cannot consider any size effects, a Voronoi model was then developed. The Voronoi tessellation is a special spatial partition where each Voronoi cell has its own space and does not overlap the other cells. This partition is comparable to crystal microstructure of metals. Therefore, the basic idea of this Voronoi model is dividing the blank into small regions according to the Voronoi tessellation where each region has similar shape and size to grains on the blank and assigning individual material properties to each region [20]. Hence, the Voronoi model can represent the blank on a grain scale and include size effects through adjusting distribution of small regions with different material properties. To generate the Voronoi FE model, group of Voronoi generating seeds with the same number of grains on the blank were randomly distributed on the blank area. Due to the relationship between the random generating seeds and the Voronoi structure, the initial Voronoi tessellations were irregular and far from the blank microstructure. As the metal was annealed and its grains were equiaxed, a constant density function was introduced to generate a centroid Voronoi where Voronoi cells are of similar shape and size. Furthermore, a Laplacian optimization algorithm was adopted to further optimize the geometry of this centroid Voronoi structure. Therefore, short lines and small faces were eliminated. Finally, the optimized centroid Voronoi was mapped onto the meshed blank and each shell element on the blank was reassigned its part ID index. Consequently, the meshes belonging to one Voronoi cell were identified, and then individual mechanical properties were attached to these elements. Although the same material model, the 3-parameter-Barlat model, was used, meshes in different Voronoi cell parts had different material parameters. The whole blank model thus had different properties at different areas. Furthermore, the material parameters covered the whole range of material properties from the repeated tensile tests and individual grain's properties randomly lay in this material properties range. Moreover, for simplification, all the Voronoi cells were randomly classified into five groups and each group, defined as one independent part in LS-DYNA, had similar number of Voronoi cells. Each Voronoi cell part had different material properties and a unique response to a given deformation resulting separate strain. Consequently, strain and stress distributions were affected by each grain part on the blank. Moreover, as each Voronoi cell was similar to the grains on the blank, this Voronoi model can represent the blank at a grain scale reflecting the material size effects. Fig. 10.2 displays the Voronoi structures and their corresponding FE models with average grain sizes of 10, 20, and 40 μm, respectively.

The Voronoi structure was generated and its geometry information was stored in MATLAB. Subsequently, the blank meshes' information was read into MATLAB from a "k" document which is the LS-DYNA executable document. By comparing elements and Voronoi cells geometrical information, each element's ID information was reassigned and exported to the same "k" document by a MATLAB program. Therefore, the Voronoi model of the blank was built and ready for calculation. This Voronoi model considered the material size effects but not the surface roughness. Therefore, the surface roughness introduced phenomena were not well presented.

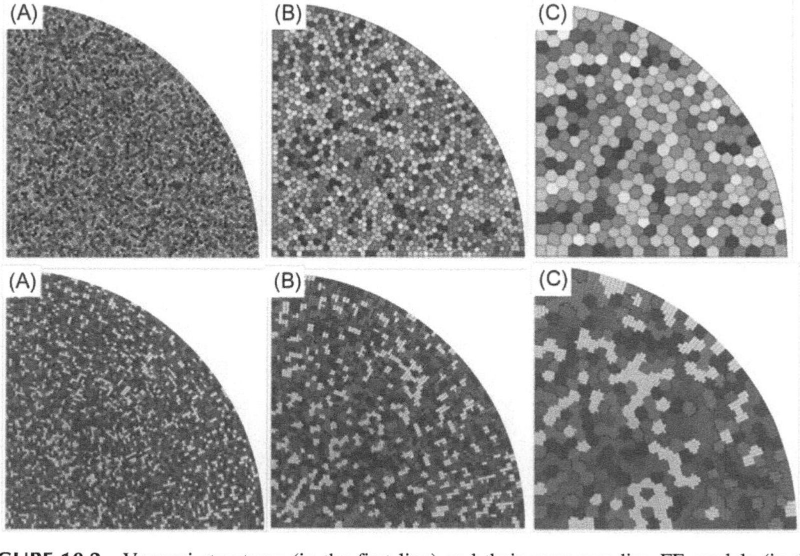

FIGURE 10.2 Voronoi structures (in the first line) and their corresponding FE models (in the second line): average grain sizes of (A) 10, (B) 20, and (C) 40 μm [18].

10.2.3 Voronoi Models With Consideration of Surface Roughness

As the surface roughness asperity heights are relatively large compared to the thickness of the blank in microscale, these asperities can be a part of the thickness. Each Voronoi tessellation owning a distinct thickness can then represent a rough surface on a grain scale. Moreover, as the Voronoi cells were randomly arranged, five separate thickness values for the Voronoi cell groups were able to characterize the surface roughness from a statistic viewpoint. Furthermore, the FE model could satisfy detailed surface roughness parameters, e.g., the arithmetic average height (R_a), due to the interlaced distribution of cells in each group.

Four surface roughness parameters, the arithmetic average height (R_a), the root mean square roughness (R_q), the skewness (R_{sk}), and the kurtosis (R_{ku}) [21], were firstly utilized to characterize surface roughness and build equations. Additionally, one asperity value was set as the weighted mean value of the other four asperity values (Eq. (10.9)). Therefore, five equations were established and all five asperity values in this model can be solved. As the selected surface roughness parameters are based on lines, they need to be extended to areas. First, the R_a, as shown in Eq. (10.10), was extended through the width direction, as shown in Eq. (10.11). Consequently, R_a based on area (Eq. (10.12)) can be obtained and its discretization form, Eq. (10.13), was utilized for numerical calculation. All the other surface roughness parameters were modified as that of R_a. Eqs. (10.14)–(10.16) show the numerical forms of R_q, R_{sk}, and R_{ku}, respectively.

$$Z_5 = \frac{\sum s_i Z_i}{\sum s_i} \tag{10.9}$$

$$R_a = \frac{1}{l}\int_0^l \left|Z(x)\right| dx \tag{10.10}$$

$$R_a = \frac{1}{l}\frac{1}{w}\int_0^w \int_0^l \left|Z(x,y)\right| dxdy \tag{10.11}$$

$$R_a = \frac{1}{S}\oint \left|Z(s)\right| ds \tag{10.12}$$

$$R_a = \frac{\sum s_i \left| Z_i - \frac{\sum s_i Z_i}{\sum s_i} \right|}{\sum s_i} \tag{10.13}$$

$$R_q = \sqrt{\frac{\sum s_i \left(Z_i - \frac{\sum s_i Z_i}{\sum s_i} \right)^2}{\sum s_i}} \tag{10.14}$$

$$R_{\mathrm{sk}} = \frac{1}{R_q^3}\left[\frac{\sum s_i\left(Z_i - \frac{\sum s_i Z_i}{\sum s_i}\right)^3}{\sum s_i}\right] \tag{10.15}$$

$$R_{\mathrm{sk}} = \frac{1}{R_q^4}\left[\frac{\sum s_i\left(Z_i - \frac{\sum s_i Z_i}{\sum s_i}\right)^4}{\sum s_i}\right] \tag{10.16}$$

where l is the sampling length, w is the sampling width, Z is the height of peaks or valleys, Z_i is the average asperity height of an individual Voronoi cell s_i, $S = \sum s_i$ is the total sampling area, and s_i is the area of an individual cell.

Actual surface roughness was measured by a laser-assisted high-resolution digital microscope (VK-X1000/X2000 series from KEYENCE) and listed in Table 10.1. As surface roughness parameters were the mean value for three repeated measurements, it is possible that the surface roughness equations are unsolvable for the exact solution. Therefore, they were solved in MATLAB iteratively and a group of approximate solutions within tolerable errors could be found. Moreover, R_p and R_v, the highest peak and lowest valley heights of a surface, were used to limit the searching range of the numerical solution. As the normalized blank thickness was 50 μm, the actual thicknesses (T_i) in the simulation model were therefore the sum of the normalized blank thickness and surface roughness asperity height (Eq. (10.17)). Thickness values of each group of Voronoi cells were shown in Fig. 10.3D. Due to the errors in iterations being quite small, these thickness values can be used in the simulation embodying the real surface roughness. Fig. 10.3 exhibits the initial thickness distribution of each blank model.

$$T_i = 50 + Z_i \tag{10.17}$$

Shell thickness is important for the surface-to-surface forming contact algorithm specialized in LS-DYNA as it affects contact force and friction.

TABLE 10.1 Surface Roughness of Each Blank Material

Material	R_a (μm)	R_q (μm)	R_{sk}	R_{ku}	R_p (μm)	R_v (μm)
1	0.160	0.220	0.729	6.081	1.807	4.517
2	0.231	0.304	0.157	5.692	2.237	6.341
3	0.217	0.293	−0.318	5.628	1.784	4.478

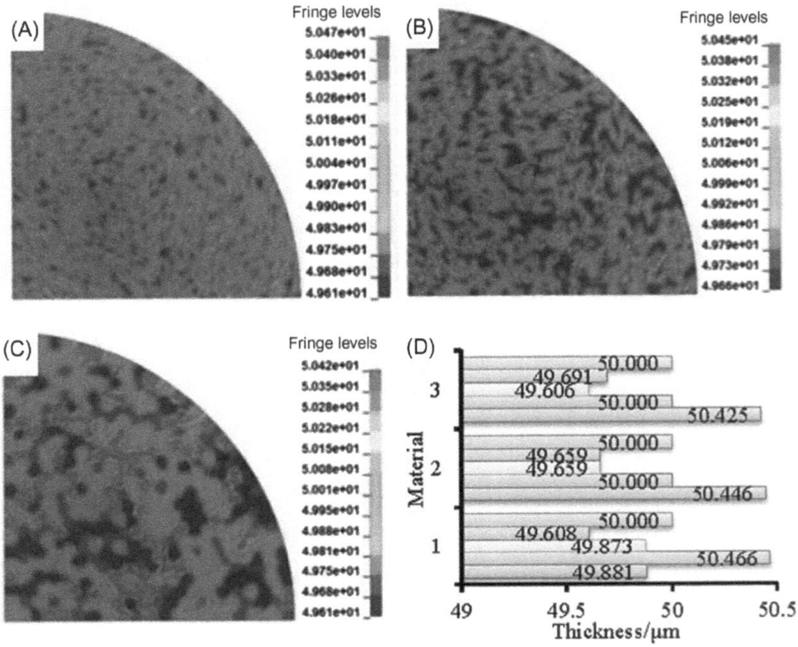

FIGURE 10.3 Initial thickness distribution of each blank model: (A) material 1, (B) material 2, (C) material 3, and (D) thickness values of each material group [18].

A penalty contact method was applied to load contact force and the contact force (the penalty force) was calculated from the penetration distance between two contact faces (Eq. (10.18)). Moreover, the penetration distance is relative to the blank thickness, shown in Eq. (10.19). Due to the surface roughness (the shell thickness in the models), the penetration distance at a local area was different, and therefore the contact force was various for the nodes in a local area as illustrated in Fig. 10.4. Consequently, the friction at a different node was different. While for models with constant blank thickness, the penetration distance was the same for all nodes in a local area due to a constant shell thickness, and therefore the contact forces were the same for these nodes. Although the constant COF was used, friction force at different nodes at a local area can be unequal due to the uneven contact force. Therefore, friction conditions on the microscale can be presented in this new model.

$$\boldsymbol{f}_s = -l \cdot k_i \cdot \boldsymbol{n}_i \tag{10.18}$$

$$l = \boldsymbol{n}_i(\xi_c, \eta_c) \cdot [\boldsymbol{t} - \boldsymbol{r}(\xi_c, \eta_c)] < 0 \tag{10.19}$$

$$k_i = \frac{f_{si} K_i A_i}{\max(\text{shell diagonal})} \tag{10.20}$$

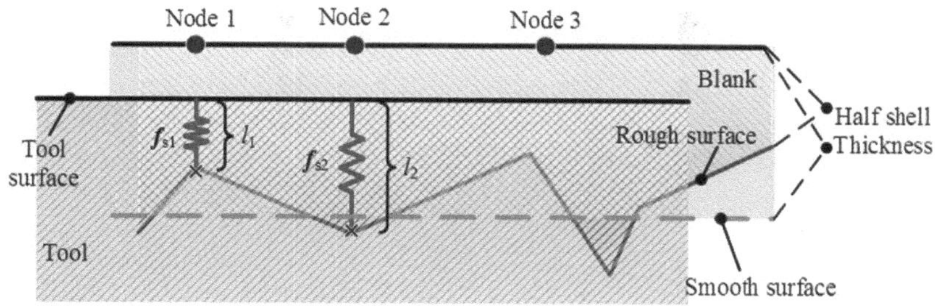

FIGURE 10.4 Penalty contact in a local area [18].

where l is the penetration distance, $\boldsymbol{n}_i$ is the normal vector at the ith node, $\boldsymbol{t}$ is the half thickness projected to the normal of tool surface, $\boldsymbol{r}$ is the vertical distance between a node on the blank to the tool surface, $\boldsymbol{f}_s$ is the penalty force (contact force), k_i is the stiffness factor of master face, f_{si} is the contact stiffness factor, K_i is the contact element modulus, and A_i is the elemental area.

The Voronoi model with variable thickness considered the material size effects and presented precise friction conditions for nodes in different Voronoi cells. Furthermore, roughness influences on drawing process and drawn cups were included as surface asperities were considered as one part of geometry. Therefore, this model considers more details in MDD and can offer precise simulation results.

10.2.4 Microstructural Images–Based Modeling

The modeling method uses the microstructural images with clear grain boundaries from various instruments, such as metallurgic microscope and electron backscatter diffraction (EBSD) [18]. The grains were detected and then all the grain boundaries were identified through an image processing program in MATLAB. Fig. 10.5A shows a typical microstructure image from the EBSD. The microstructural image was first read into MATLAB and then transferred into the binary image where there are only black grain boundaries and white grains. During the transformation, noise and small holes were eliminated. Therefore, nongrain information on the image was omitted. Subsequently, the binary image was treated with the Moore-Neighbor tracing algorithm modified by Jacob's stopping criteria. All the grains, the individual closed subareas, were then detected and sorted, as shown in Fig. 10.5B, and their information were stored in MATLAB. During the detection, each grain's area, geometrical center and geometrical orientation can be calculated. The blue ports in Fig. 10.5B are the grain's geometrical centers.

Simultaneously, a normal finite element model was generated where all the elements were in a single part with the same mechanical properties. The elemental and nodal information was then imported into MATLAB. The

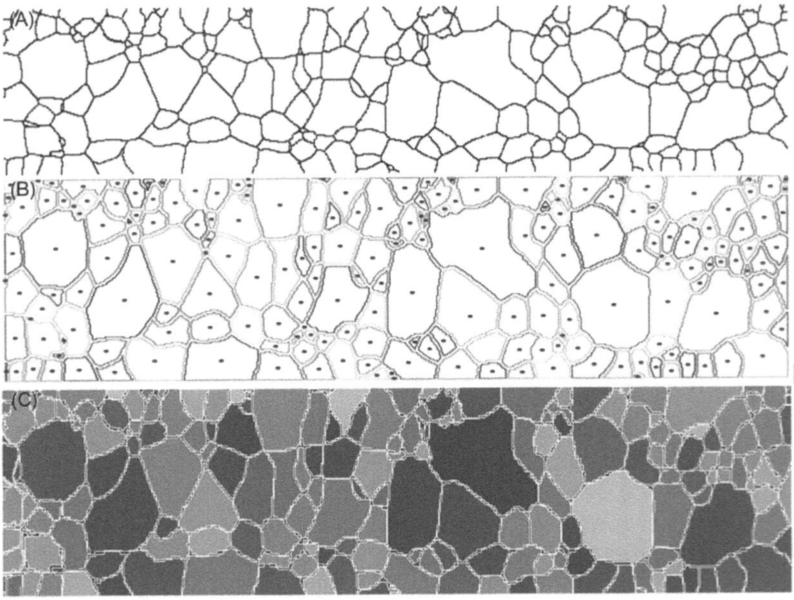

FIGURE 10.5 (A) Microstructure of a sample from EBSD, (B) its corresponding geometry detected by MATLAB, and (C) corresponding simulation model.

center of each element was used to present the elemental positon. Then, the elements were mapped onto the grains detected by MATLAB. By comparing all the grains and the elements positions, the elements inside one grain were assigned the corresponding grain's ID number. Accordingly, the elements in one grain were set as one individual part with separate mechanical properties. By contrast, the elements that belong to none of the gains were defined as the grain boundaries. Finally, the updated relating elemental information, part and material properties indexes, was written into the pre-built normal finite element model. Fig. 10.5C displays the final finite element model based on the microstructure as shown in Fig. 10.5A.

As the simulation model is on the microscale, there are a great quantity of grains and their accordant mechanical properties in the simulation model. Thus, the simulation model can be very large and over the executable size of the commercial simulation software. Therefore, all the grains should be categorized as several groups, and each group has one type of mechanical properties. The overall simulation model size can then be greatly reduced due to the limited information about material properties. In the aforementioned Voronoi models, the grains were randomly classified or groups based on the grains' statistic information. By contrast, the grains areas and their geometrical orientations can be utilized for reasonable arrangement in the new model. Generally, samples with relatively large grains have relatively low flow resistance and low tensile strength. Accordingly, large grains were set low mechanical properties, while small grains were assigned high mechanical properties. Furthermore, the grain boundaries were allocated mean mechanical properties of all grain groups. Fig. 10.6A shows the area distribution of the grains on the sample, and Fig. 10.6C displays the corresponding finite element simulation model. Five colors indicate five area levels of grains and five grain groups. Regarding grain orientation, although the orientation is calculated based on grain's geometry, this geometry indicates the former processing of the material and thus the preferred deformation direction of the grain. The geometrical orientation is the angle (in degrees ranging from 0 to 180 degrees) between the x-axis (horizontal axis) and the major axis of the ellipse that has the same second-moments as the detected grain. Thus, all the grains orientations were analyzed and all the grains were classified into several groups accordingly. In addition, the anisotropic material model can be utilized and relating parameters indicating anisotropy were set according to the orientation angle. For example, the in-plane anisotropic value, the R-value defined as Eq. (10.1), is generally used for presentation of material anisotropy. At any angle $\varnothing$, $R_{\varnothing}$ can be calculated from Eq. (10.2). According to the 3-parameter Barlat yielding criteria, Eq. (10.3), anisotropic properties of the grains with different orientation angles can be considered.

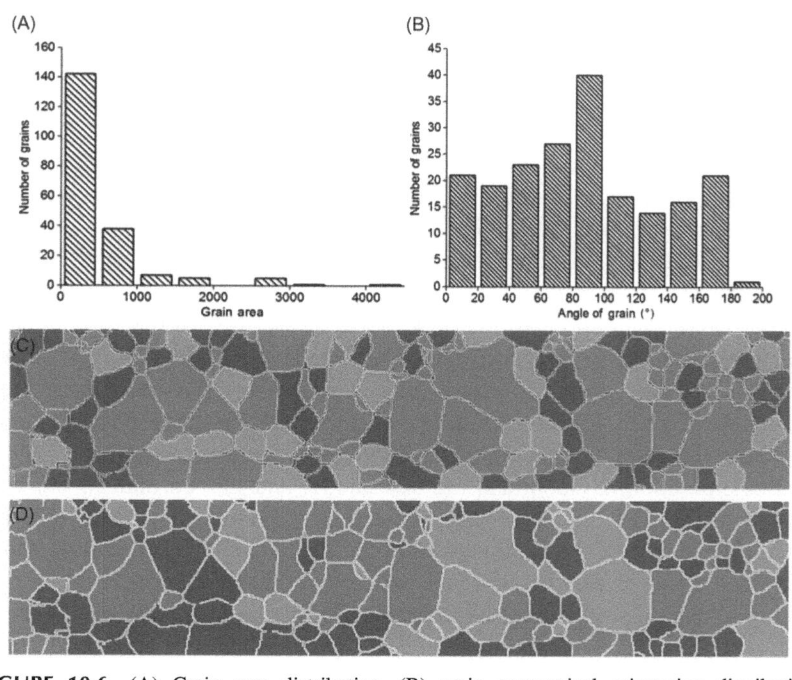

FIGURE 10.6 (A) Grain area distribution, (B) grain geometrical orientation distribution, (C) simulation model based on grain areas, and (D) simulation model based on grain orientations.

Fig. 10.6B presents the orientation distribution of the grains and Fig. 10.6D exhibits the corresponding simulation model.

10.2.5 Springback Models

After MDD simulation, only the drawn cup was employed for a following springback simulation. The material model, the elemental information, and the parts arrangements were the same as that in the deep drawing models while the solver was changed from explicit to implicit solver in LS-DYNA. The thickness and strain–stress information at the end of the drawing simulation was also included in this springback model. As seven integration points were employed, the thickness and strain–stress information were accurate and adequate for springback simulation. Furthermore, two points on the symmetrical edges were fixed according to the symmetrical boundary conditions to limit the rigid movement of the cup. Consequently, the residual stress can be released gradually in a few calculation steps. Dynamic step adjustment and artificial stability were added to improve the calculation. As Voronoi models were still used in the springback simulation, size effects on springback behavior can be presented. Furthermore, effects of surface roughness on the springback could be observed as well.

10.3 SIMULATION RESULTS

10.3.1 Drawn Cup Thickness Comparison

The blanks with thickness of 50 μm were annealed at 1100°C and drawn into micro cups. The microstructure of the blank was of equiaxed crystals and the average grain size was 40 μm. Fig. 10.7A displays the mouth of a drawn cup and Fig. 10.7B–D exhibit maximum thickness distribution of the drawn cups from the developed new model, a Voronoi model without consideration of grain boundaries and a normal model without account of material inhomogeneity, respectively. The maximum wall thickness, at the cup mouth area, was selected for comparison between the simulation and the experimental results. Without consideration of the localized deformation the maximum thickness was averaged resulting in the lowest peak thickness values among all the simulation. By contrast, the new model and the Voronoi model considered microscopic heterogeneity and higher maximum thickness than that in the normal model was obtained from these two models. Without buffer of the grain boundaries, the Voronoi model presented the largest thickness among all simulation. Fig. 10.8 compares the maximum thickness between both simulation and average experimental results. The developed model presented closer results to the experimental results than the other simulation models.

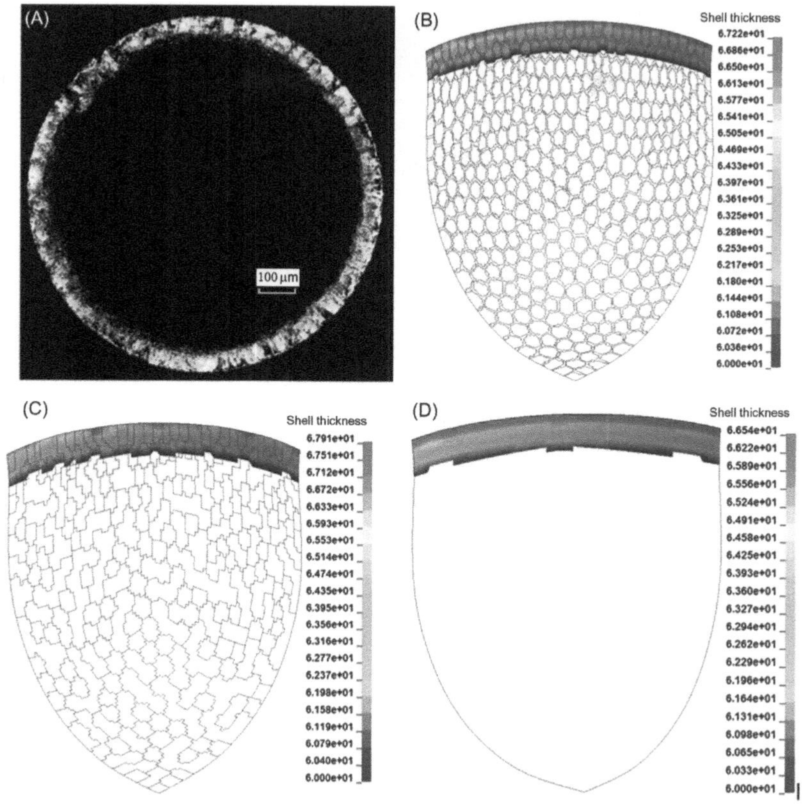

FIGURE 10.7 (A) Drawn cup with 1100°C annealed blank and maximum thickness distribution from the simulation with (B) the developed new model, (C) the Voronoi model, and (D) the normal model.

10.3.2 Springback and Shape Quality Comparison

For convenience, the basic model, the Voronoi model with a constant blank thickness and the Voronoi model considering surface roughness were named as models I, II, and III, respectively. As the SUS304 sheets were heated at three different temperatures (975°C, 1050°C, and 1100°C), the sheets were classified as materials 1, 2, and 3 correspondingly. Therefore, a given simulation model with a specified material can be expressed by a Roman number and an Arabic number, e.g., "II 3" meaning Voronoi model with a constant blank thickness (model II) using 1100°C heated SUS304 sheets (material 3).

Due to the difficulty in measuring springback, a springback judgment, an average power of displacement before and after springback of each node (Eq. (10.21)), was defined to approximate the total springback behavior quantitatively.

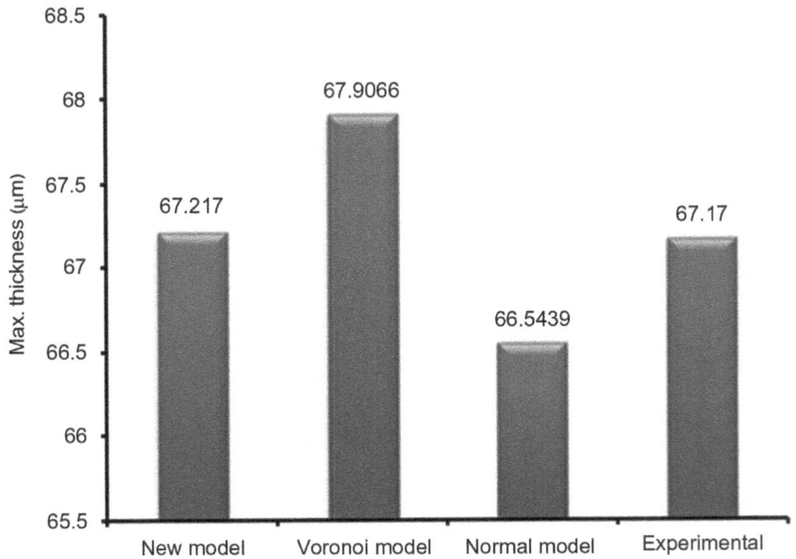

FIGURE 10.8 Comparison of the maximum thickness among different simulation models and average experimental results.

$$J_s = \frac{\sum(d_{ix}^2 + d_{iy}^2 + d_{iz}^2)}{n} \tag{10.21}$$

where J_s is the springback judgment, d_i is the displacement of node i, and n is the total number of the nodes on the cup.

Fig. 10.9 displays the springback judgment for each simulation model and material. The overall springback in the Voronoi models was higher than that of the common models without considering size effects or surface roughness. Due to the introduction of the Voronoi model, the blank had different properties at different areas. The weak and thin areas on the blank undertaken the same stress had great strain due to their low flow resistance. Therefore, the stored strain energy was high for the Voronoi models and the largest for the Voronoi models considering surface roughness among all the simulation models. Moreover, the unevenness of material properties and thickness distribution on the blank intensified the unevenness of strain–stress distribution on the blank. After springback, all the residual stress–strain stored in the blank was released and therefore, the Voronoi models accounting the surface roughness underwent the greatest springback. The Voronoi models with a constant thickness owning less unevenness had smaller springback value than that of the models considering surface roughness, while the normal models had the smallest springback value.

The diameter of the drawn cup mouth after springback was chosen to evaluate the accuracy of the simulation. First, the nodes' coordinates of the

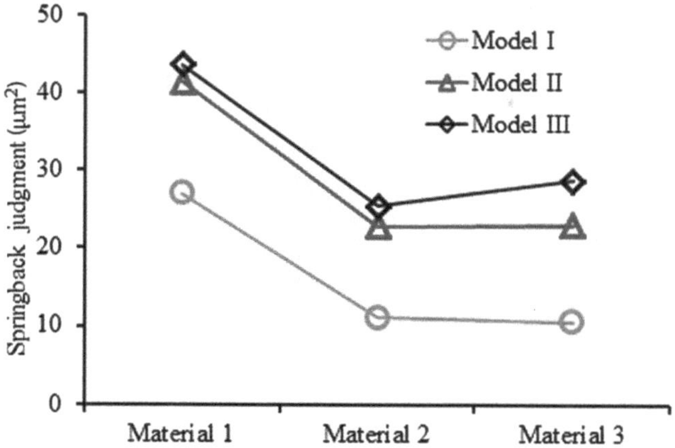

FIGURE 10.9 Springback judgment [18].

cup were exported from springback simulation results and the nodes at the top were identified according to their *Z*-coordinates. Second, these selected nodes were fitted by a circle in MATLAB using a second-order polynomial equation to obtain the diameter of the fitted circle. Finally, the drawn cups in experiments were observed under the same microscope to that for surface roughness measurement. The diameter was determined as the mean value of the maximum and minimum diameters of the top layer on the cup. Fig. 10.9 displays the drawn cups colored according to cup height in experiments as it is easy to identify the mouth of the cup. Furthermore, the mean value of three repeated measurements was utilized as the final experimental result. Fig. 10.9D shows the diameters of each simulation and experimental conditions. Voronoi models considering surface roughness had closer diameter values to the experimental results than the other two kinds of models. Furthermore, diameters of the three simulation models had a decreasing trend for a given material. Because of the weak areas on the blank with low flow resistance, they coordinated the overall deformation and strain–stress development. Therefore, the compression strain on the top of the cup, where the compression is the largest on the blank, was easy to transfer to the adjacent area. Therefore, the springback on the top of cup, the extending deformation of the cup mouth, was small due to the reduced compression strain.

The surface roughness represented by the thickness differences further reduced the diameter due to the introduced thin areas with low flow stress. Consequently, the diameter of cup mouth in the Voronoi model considering surface roughness was the smallest of all simulation models. Furthermore, with low flow stress, materials 2 and 3 had smaller cup mouth diameters than that of material 1 for each simulation model. This can also be explained by the compression transferring. Low flow stress material has low flow resistance and is easy to transfer strain. Therefore, the great compress at the top of cup was transferred to the neighboring areas on the blank resulting in a tiny amount of outwards springback and consequently small cup mouth diameter. Although all the simulation results were higher than the experimental results, the differences were small and acceptable. These differences come from the underestimation of the inner diameter of the cup mouth in experiments. One face of the blank at the out edge area of the blank was contacted with the die for all drawing process, while the other side was contacted with the blank holder for only the initial time of the drawing process and then was free for the later drawing period. The different friction condition on each side of the blank at the blank edge area caused an unclear inner edge of the cup mouth which introduced the error in measuring the inner diameter of the cup mouth.

As minimum thickness determines the success of a drawing, the minimum thickness of each simulation case was compared in Fig. 10.10. The Voronoi models had decreased minimum thickness for each material group and the uneven thickness distribution further reduced the minimum thickness. As the thin and weak areas on the blank underwent further thickness direction strain due to the low flow resistance, the minimum thickness in the model III was the smallest, while that of the model I was the largest among all simulation models. Thus the drawability was overestimated if the surface roughness and material size effects were ignored. With low flow stress, strain and stress are easy to transfer from one element to subsequent element. Therefore, strain distribution in the models II and III for the material 3 was smoother than that of the other two material groups, and thus it was more difficult for strain localization. Accordingly, models II and III in material group 3 had higher minimum thickness on the drawn cups than that in the

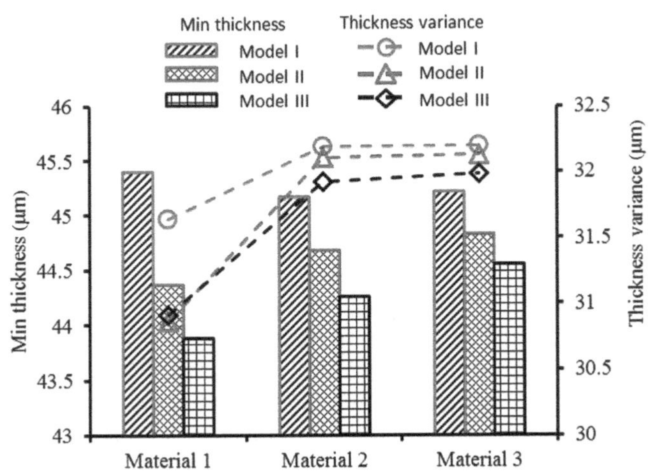

FIGURE 10.10 Minimum thickness and thickness variance of the drawn cup [18].

other two material groups. By contrast, the normal model, ignoring size effects and surface roughness, had an inverse trend to that of models II and III. With high flow stress, the strain distribution in model I was smoother than that of the other two models, as it overlooked the strain localization introduced by the size effects and surface roughness. Consequently, normal models with high flow stress material presented by the large minimum thickness.

To further judge the quality of the cups, variance of the thickness distribution that describes the deviation of the shell element thickness to their mean thickness (Eq. (10.22)) was selected. Large variance means great thickness unevenness and consequently poor cup quality. Fig. 10.10 indicates that Voronoi models decreased the unevenness and rough surface further diminished the thickness unevenness. The weak and thin areas on the blank acted as deformation buffering and coordinated the deformation on the blank. Therefore, the thickness distribution tended to be smooth. The contact force also influenced thickness evenness as the thin elements bore smaller contact forces while the thick elements undertaken high contact forces at a local area. These contact forces flattened the thickness distribution and led to a smaller thickness variance of model III than that of model II. Additionally, with high flow stress and long elongation, models with material 1 had smaller thickness variance than the models using the other two materials. That is because that good material properties mean high ability to flat the stress-strain during the drawing process leading to an even thickness distribution.

$$D = \frac{\sum (t-\bar{t})^2}{n} \tag{10.22}$$

where D is the variance of thickness of the drawn cup, n is the total number of elements, t is the thickness of an element, and $\bar{t}$ is the average thickness of all elements on the cup.

Shape accuracy is another important factor for the micro products quality. The drawn cup's nodes information from springback simulation results was therefore exported for shape accuracy analysis. A distortion between cup's surface and a reference surface was defined as the shape accuracy judgment. A high distortion value means poor shape accuracy. Accordingly, the distortion between the inner cup surface and the punch surface was defined as an inner distortion of the drawn cups, while the distortion between the outer cup surface and a virtual surface that was a surface offset 50 μm outwards the punch surface was defined as the outer distortion of the cups. As the initial blank thickness was 50 μm, the virtual surface was offset by 50 μm based on an idea that the ideal drawn cup should have the same thickness to the initial blank thickness. Based on the cup's geometry, all the cup's nodes were classified into three segments and each segment had its particular calculation method of shape distortion judgment. Fig. 10.11A illustrated the classification of the nodes and the basic sizes of the reference surfaces.

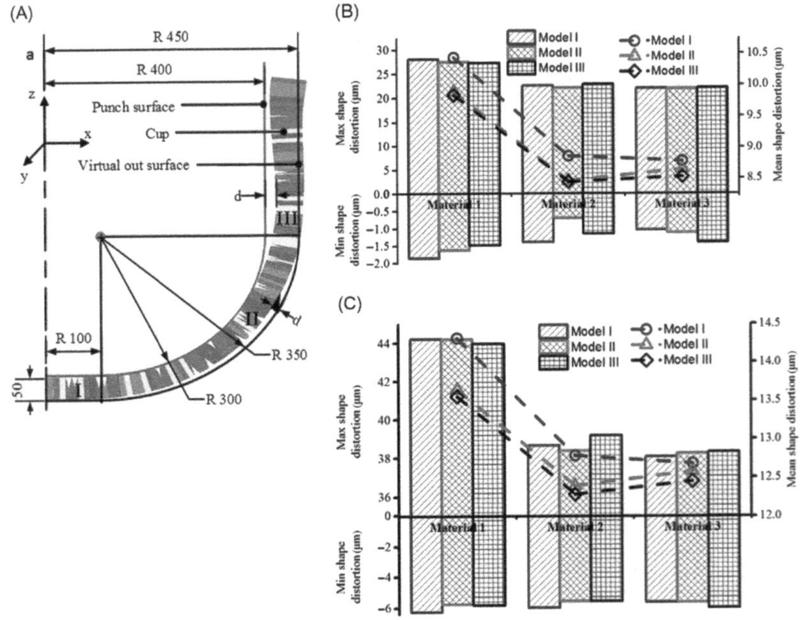

FIGURE 10.11 (A) Illustration of shape distortion, (B) shape distortions compared to the punch surface, and (C) shape distortion from the virtual surface [18].

In the part I area, the reference surfaces were horizontal and therefore height difference was selected as the judgment. In parts II and III, the radial differences in the spherical and cylindrical coordinates were chosen as the judgments, respectively.

Given the thickness unevenness, the thickness at each node was included for shape accuracy calculation. The nodes' coordinates and thickness information was firstly exported to Excel and then all the nodes were sorted based on the *Z*-coordinate and subsequently the cylindrical radial. Therefore, all nodes belonging to the three parts can be divided. Eqs. (10.23)–(10.25) list the calculation method for inner shape distortion, whereas Eqs. (10.26)–(10.28) list the equations for outer shape distortion. Moreover, the mean value of all nodes' absolute distortion (Eq. (10.32)) was utilized as the judgment of the shape accuracy.

$$d_{\mathrm{I}}^{\mathrm{in}} = z_b - z_p - \frac{t_i}{2} \tag{10.23}$$

$$d_{\mathrm{II}}^{\mathrm{in}} = R_b - R_p - \frac{t_i}{2} \tag{10.24}$$

$$d_{\mathrm{III}}^{\mathrm{in}} = r_b - r_p - \frac{t_i}{2} \tag{10.25}$$

$$d_{\mathrm{I}}^{\mathrm{out}} = z_b - z_v + \frac{t_i}{2} \tag{10.26}$$

$$d_{\mathrm{II}}^{\mathrm{out}} = R_b - R_v + \frac{t_i}{2} \tag{10.27}$$

$$d_{\mathrm{III}}^{\mathrm{out}} = r_b - r_v + \frac{t_i}{2} \tag{10.28}$$

$$R_b = \sqrt{(r_b - r_c)^2 + (z_n - z_c)^2} \tag{10.29}$$

$$r_b = \sqrt{x_n^2 + y_n^2} \tag{10.30}$$

$$r_c = 100 \tag{10.31}$$

$$s = \frac{\sum |d|}{n} \tag{10.32}$$

where d_a^b is the judgment in the part a (a = I, II, or III) compared to the reference surface b (b = in or out) of the cup, x, y, and z are the node's coordinates, t is the node's thickness, R_b is the spherical radius, r_b is cylindrical radius, r_c is the radius of horizontal part of the punch, n is the total number of nodes, and s is the shape accuracy judgment.

For all simulation conditions, the average inner shape distortion was lower than that of the outer shape distortion. Further, the maximum and minimum inner shape distortion values were smaller than that of the outer shape distortion. During the MDD simulation, the inner cup surface that contacts with the punch surface bore stricter restriction than that of the outer cup surface due to that the gap between the die and the punch is larger than the blank thickness. Furthermore, at the bottom part of the cup the outer surface was almost free of limitation whereas at the top part of the cup, both inner and outer cup surface undertook similar constraint due to the thickening of the blank causing the blank contacting with both the punch and the die surfaces. Therefore, the outer cup surface had greater shape distortion than that of the inner cup surface, in terms of the average, the maximum and the minimum shape distortion values.

Further, models employing materials 2 and 3 presented low maximum and mean shape distortion values, due to that the materials had low flow stress and were relatively soft. With small deformation resistance, these materials are easy to follow the shape of the punch and their thickness distributions are smooth. This led to low average and maximum shape distortions when using the two materials.

Generally, the Voronoi models had smaller mean shape distortion values than that of the normal simulation models. The weak areas on the blank coordinated the deformation and made the blank grasping the shape of the punch. Moreover, with a low thickness variance at the end of drawing simulation, model II had smaller average shape distortion than the normal simulation models. The thin areas in the model III further adjusted the deformation and thickness distribution. Therefore, it resulted in the smallest mean shape distortion among all simulation models. Furthermore, the thin elements had small contact forces while the thick elements had high contact forces in a local area. These forces adjusted the shell elements to fit their contacting surfaces (tool surfaces) and led to smaller mean shape distortion in model III than that in model II with constant thickness.

As the maximum drawing force combined with the ultimate tensile strength can be used to predict the drawability, the maximum drawing force was compared in Fig. 10.12.

As the material parameters in Voronoi models were from the tensile tests while that used in the normal models was the mean value of those tensile tests results, the drawing forces for each simulation model in each material group were close. Although different areas on the blank had different material properties in the Voronoi models, their overall effects on drawing force were balanced resulting a close value to that in the normal models. Furthermore, the thickness differences impacted on the contact force and then the friction force which is an important part of drawing force. However, there were both thin and thick areas on the blank and their opposite effects on the friction force were neutralized. Additionally, the maximum drawing force in the models using material 1 with high flow stress was higher than the models using the other two materials. That is because that the tensile strength of material 1 was higher than that of materials 2 and 3. Moreover, the slight differences between the experimental and simulation results were acceptable. This may come from the tolerance in tensile tests results and the COF employed in the simulation models to the real situations.

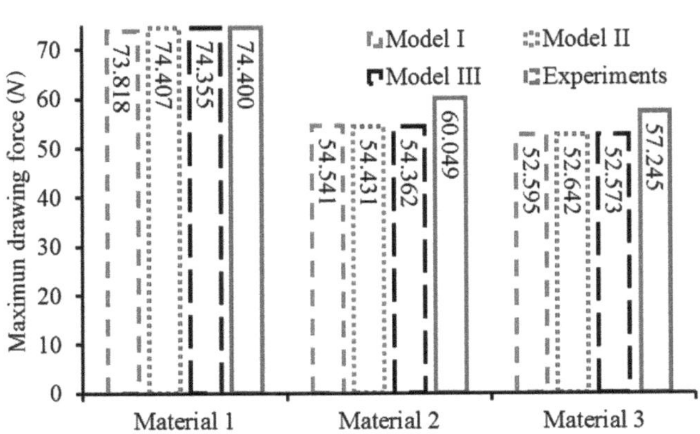

FIGURE 10.12 Maximum drawing forces [18].

10.3.3 Fracture and Earing in Micro Deep Drawing

Fracture is one major failure of deep drawing and it is generally occurred on the drawn cup near punch nose area due to the maximum tensile stress without compression stress and limited positive friction at this area. Accordingly, tensile stress determined the thickness development at the critical area. Eight percent of thickness reduction was set as a critical value below which the fracture possibility greatly increases. Fig. 10.13 displays the thickness distribution of the critical area on the drawn cups with thickness less than 46 μm ($=50\times(1-8\%)$). Fig. 10.14 displays proportion of the critical area with thickness under 46 μm to the whole cup. Uneven thickness distribution can be observed from all simulation results. Big grains in each real microstructural model presented the minimum thickness at the critical area while random grains in the Voronoi models had the minimum thickness. Further, the minimum thickness values and the area where thickness was less than 46 μm was increased with the growth of grains. Big grains were assigned low mechanical properties and thus had weak resistance to thinning, especially for those surrounded by small grains with high material properties. General

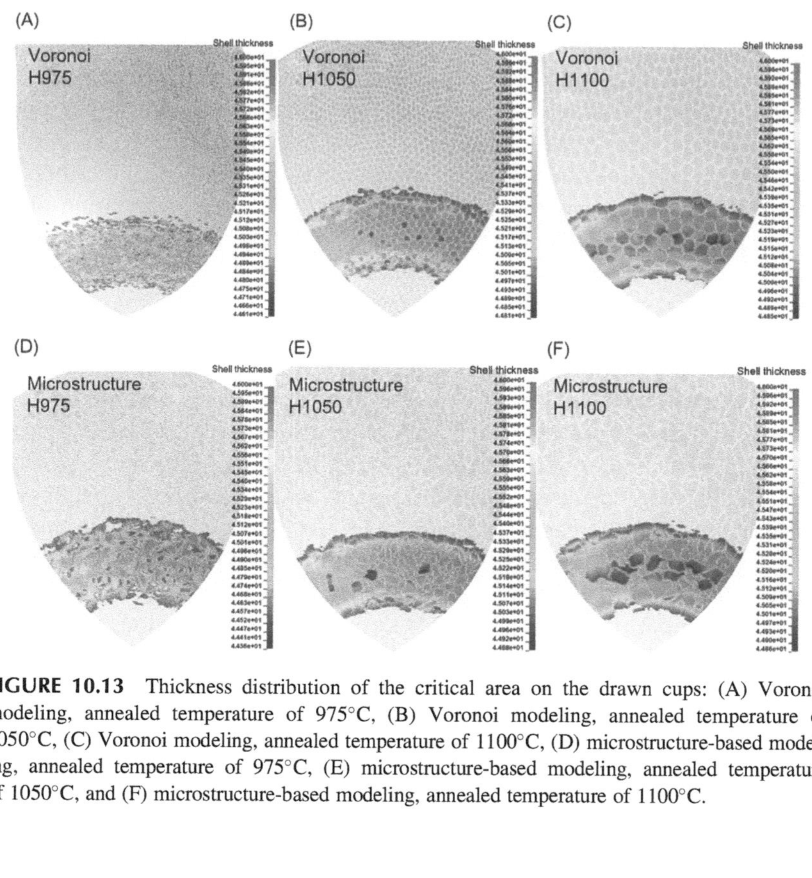

FIGURE 10.13 Thickness distribution of the critical area on the drawn cups: (A) Voronoi modeling, annealed temperature of 975°C, (B) Voronoi modeling, annealed temperature of 1050°C, (C) Voronoi modeling, annealed temperature of 1100°C, (D) microstructure-based modeling, annealed temperature of 975°C, (E) microstructure-based modeling, annealed temperature of 1050°C, and (F) microstructure-based modeling, annealed temperature of 1100°C.

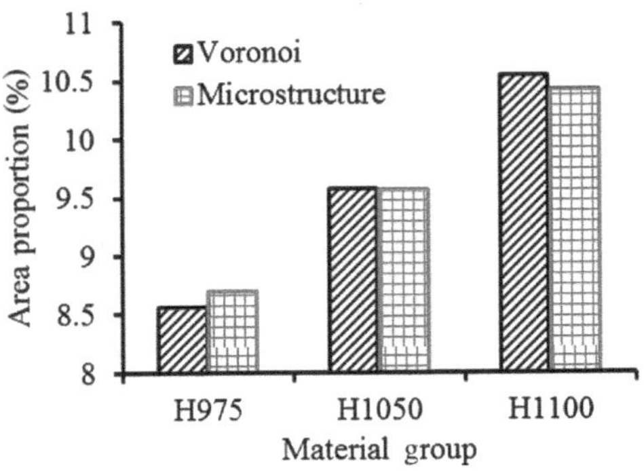

FIGURE 10.14 Proportion of the area with thickness under 46 μm to the whole cup.

increase of the area with thinner thickness than 46 μm among all three kinds of blanks revealed that statistic material properties were reflected in the models. Therefore, the general thickness distribution trend was similar while the local thickness was different between the Voronoi and real microstructural models.

The thickness distribution was resulted from the strain–stress distribution, and it further influenced springback of the drawn cup and consequently the final quality of the drawn cups. Fig. 10.15 displays displacement of the drawn cups after springback with overall displacement over one micrometer as one micro's displacement was ignorable. Unevenly distributed displacement reflected significant material anisotropy. Considerably different springback values then caused obvious shape distortion of the final drawn cups. Although localized springback was observed, general trend, e.g., the decreasing maximum displacement among all material groups was the same between the two kinds of simulation models. Refined-grain blanks resulted in high strain energy stored in the drawn cup, and thus with releasing this energy during the springback, the cup deformed further. Additionally, the strain energy can be reflected by the critical thickness area, as shown in Fig. 10.15.

The drawn cups' quality, chosen as the earing phenomenon in this study, were then analyzed and compared with experimental results. Nodal position information of the cup's mouth was utilized for the earing calculation. An average difference between the nodal heights and their average height was defined as earing parameter, as shown in Eq. (10.6). Regarding experimental results, a low pass filtering and noise reduction were conducted before earing calculation. Fig. 10.16 shows the earing values of all simulation and experimental conditions. The earing values between the two simulations and experimental results were close, and their developing trend was the same. However, the real microstructural simulation models had higher accuracy than that of the Voronoi models when compared with experimental data. The 1050°C heated material group had the most significant earing value compared to the other groups. Low temperature annealing blanks had small grain sizes. This is beneficial for material isotropy, and therefore restriction of

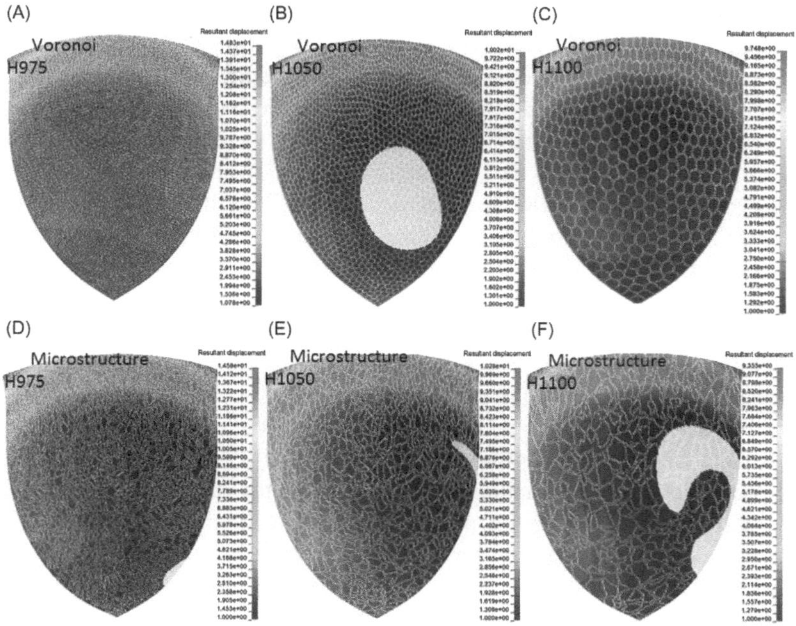

FIGURE 10.15 Displacement of the drawn cups after springback: (A) Voronoi modeling, annealed temperature of 975°C, (B) Voronoi modeling, annealed temperature of 1050°C, (C) Voronoi modeling, annealed temperature of 1100°C, (D) microstructure-based modeling, annealed temperature of 975°C, (E) microstructure-based modeling, annealed temperature of 1050°C, and (F) microstructure-based modeling, annealed temperature of 1100°C.

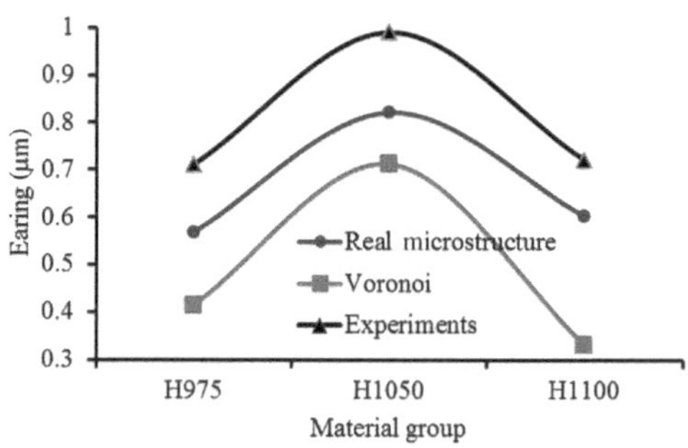

FIGURE 10.16 Earing values of all simulation and experiments.

earing. By contrast, high temperature heating made the grain sizes grow, and lowered the material properties which reduced the springback and the earing value as a consequence.

$$\text{Earing} = \frac{\sum_{i=1}^{n} |z_i - \bar{z}|}{n} \tag{10.33}$$

where Earing is the earing parameter, z_i is the height of the ith node, n is the number of nodes, and $\bar{z}$ is the average height of all nodes.

In summary, the new FE model that includes surface roughness information and considers size effects was successfully developed combined with image-based modeling methods. The effects of surface roughness and the influences of size effects on the MDD were studied based on SUS304 sheets. Simulation results of the new models were close to the experimental results in terms of the diameters of cup mouths and the maximum drawing forces. However, variance of material properties on the blank in the Voronoi models increased the overall springback value and the surface roughness additionally enlarged the springback behavior. By contrast, thickness evenness was increased in the Voronoi models and further improved in the Voronoi models considering surface roughness. Moreover, models accounting both size effects and surface roughness presented the best shape accuracy, while the shape accuracy in the normal models was the worst.

Products' quality has been evaluated in terms of thickness evenness, shape accuracy, and the earing of drawn cups. Large grains own lower material properties and deformation resistance. Thus, the large grains surrounded by small grains are potential places for serious thinning and fracture. The grain boundaries act as a deformation buffer which balances the deformation between the large and small grains. Consequently, although with differences in material properties, the cups can be drawn without early fracture or significant earing.

REFERENCES

[1] Z. Xu, L. Peng, X. Lai, M. Fu, Geometry and grain size effects on the forming limit of sheet metals in micro scaled plastic deformation, Mater. Sci. Eng.: A 611 (2014) 345–353.

[2] L. Peng, P. Yi, P. Hu, X. Lai, J. Ni, Analysis of micro/mesoscale sheet forming process by strain gradient plasticity and its characterization of tool feature size effects, J. Micro Nano-Manufact. 3 (2015) 11006.

[3] Q. Yi, A. Brockett, Y. Ma, A. Razali, J. Zhao, C. Harrison, et al., Micro-manufacturing: research, technology outcomes and development issues, Int. J. Adv. Manufact. Technol. 47 (2010) 821–837.

[4] Z. Fang, Z.Y. Jiang, X. Wang, C. Zhou, D.B. Wei, X.H. Liu, Grain size effect of thickness/average grain size on mechanical behaviour, fracture mechanism and constitutive model for phosphor bronze foil, Int. J. Adv. Manufact. Technol. (2015) 1–10.

[5] A. Molotnikov, R. Lapovok, C.F. Gu, C.H.J. Davies, Y. Estrin, Size effects in micro cup drawing, Mater. Sci. Eng.: A 550 (2012) 312–319.

[6] F. Vollertsen, Size effects in micro forming, Key Eng. Mater. 473 (2011) 3–12.

[7] F. Vollertsen, H. Schulze Niehoff, Z. Hu, State of the art in micro forming, Int. J. Mach. Tools Manufact. 46 (2006) 1172–1179.

[8] N. Ma, Study on Size Effects and Simulation Techniques of Micro-forming, Huazhong (Central China) University of Science and Technology, Wuhan, People's Republic of China, 2005.

[9] N. Ma, X.H. Dong, The size effects on process design of micro deep drawing, J. Shanghai Jiaotong Univ. (Sci.) 12 (2007) 323–327.

[10] G.C. Wang, W. Zheng, T. Wu, H. Jiang, G.Q. Zhao, D.B. Wei, et al., A multi-region model for numerical simulation of micro bulk forming, J. Mater. Process. Technol. 212 (2012) 678–684.

[11] F.H. Yeh, C.L. Li, Y.H. Lu, Study of thickness and grain size effects on material behavior in micro-forming, J. Mater. Process. Technol. 201 (2008) 237–241.

[12] M. Koç, Tr Özel, Micro-manufacturing: Design and Manufacturing of Micro-products, John Wiley & Sons, Hoboken, NJ, 2011.

[13] Z. Hu, Realisation and application of size dependent FEM-simulation for deep drawing of rectangular work pieces, CIRP J. Manufact. Sci. Technol. 4 (2011) 90–95.

[14] J.H. Deng, M.W. Fu, W.L. Chan, Size effect on material surface deformation behavior in micro-forming process, Mater. Sci. Eng.: A 528 (2011) 4799–4806.

[15] F. Vollertsen, Effects on the deep drawing diagram in micro forming, Prod. Eng. Res. Dev. 6 (2012) 11–18.

[16] F. Vollertsen, Z. Hu, Tribological size effects in sheet metal forming measured by a strip drawing test, CIRP J. Manufact. Sci. Technol. 55 (2006) 291–294.

[17] J. Tang, A.K. Tieu, Z. Jiang, A simulation of surface roughness in hot strip rolling, Materials Processing and Design: Modeling, Simulation and Applications-NUMIFORM 2004-Proceedings of the 8th International Conference on Numerical Methods in Industrial Forming Processes, AIP Publishing, 2004, pp. 630–635.

[18] L. Luo, Z.Y. Jiang, D.B. Wei, K. Manabe, X. Zhang, D. Wu, et al., Effects of surface roughness on micro deep drawing of circular cups with consideration of size effects, Finite Elem. Anal. Des. 111 (2016) 46–55.

[19] F. Barlat, K. Lian, Plastic behavior and stretchability of sheet metals. Part I: a yield function for orthotropic sheets under plane stress conditions, Int. J. Plast. 5 (1989) 51–66.

[20] W.L. Chan, M.W. Fu, J. Lu, J.G. Liu, Modeling of grain size effect on micro deformation behavior in micro-forming of pure copper, Mater. Sci. Eng.: A 527 (2010) 6638–6648.

[21] E. Gadelmawla, M. Koura, T. Maksoud, I. Elewa, H. Soliman, Roughness parameters, J. Mater. Process. Technol. 123 (2002) 133–145.

Chapter 11

Simulation of Micro Hydromechanical Deep Drawing

11.1 INTRODUCTION

To improve the drawability, fluid pressure has been used in the macro scale [1]. In particular, hydromechanical deep drawing (HDD) has been widely used as the forming technique instead of female die. This can fabricate the complex shape component without a female die which is difficult to fabricate in the micro scale. The drawability can also be improved because of the effects of fluid pressure such as the friction holding effect and hydrodynamic lubrication [2]. The radial pressure applied to the blank edge can also increase the drawability [3]. Due to these advantages, it is expected that the micro sheet hydroforming effectively solves the above-mentioned problems in micro sheet forming. Previously, conventional HDD technology was applied to the micro scale and micro cups with a diameter of 0.8 mm have been successfully fabricated [4,5]. The size effects on deformation behavior in MHDD have not been explained until now. Extensive investigation of the size effects in MHDD is necessary to further improve the drawability.

The assistance of hydraulic pressure is generally considered as a positive factor to normal scale hydro deep drawing process, especially for the drawability and friction in the process. However, once scaled down to a micro scale, the effects of hydraulic pressure change due to the size effects on material properties and friction conditions. Compared with micro deep drawing (MDD) process, micro hydro deep drawing is a promising micro metal forming technology with great advantages, such as the ability to fabricate complex 3D structures within one stroke and the high quality of surface finishing and shape accuracy of the micro products. However, the size effects, such as the decrease of flow stress and increase of data scatter of experimental results with tiny samples sizes, become significant and greatly influence the micro forming processes [6]. Moreover, influences of hydraulic pressure change from that in the normal scale hydro deep drawing

Microforming Technology. DOI: http://dx.doi.org/10.1016/B978-0-12-811212-0.00011-X

process due to the reduced sizes and the material size effects. With limited sizes, the fluid development is restricted as a laminar flow determined by the small Reynolds number. Moreover, it is proportional to the characteristic channel diameter, while the characteristic channel diameter is minute in the micro hydro deep drawing system. Furthermore, the frictional condition changes timely during the micro hydro deep drawing process [4,7]. Regarding the materials bore the hydraulic pressure, resistance of bending and wrinkling changes due to the material size effects. Therefore, investigation on the impacts of hydraulic pressure on the micro hydro deep drawing process and its products is significant. Fluid behavior can induce the hydrodynamic lubrication and reduce the friction force significantly. The characteristics of fluid behavior in the micro scale, however, are still not clear. Not only hydrodynamic lubrication but also the boundary and mixed lubrications exist during the MHDD process. To analyze these lubrication conditions in micro forming, the surface topography, especially at the open lubricant pockets (OPLs) and closed lubricant pockets, must be considered [4].

In this chapter, finite element method (FEM) is adopted and modified to study the influence of processing parameters on micro drawing process and obtained products. A normal hydro deep drawing model was developed, and then a Voronoi blank model was generated to consider the size effects of materials. In addition, a surface layer model was created based on the Voronoi blank model. Subsequently, different processing parameters, such as hydraulic pressure and material inhomogeneity, were investigated. Simulation results indicate that a proper hydraulic pressure improves drawability and shape accuracy of the drawn cups. Moreover, material inhomogeneity affects the quality of drawn cups. In addition, the simulation methods of tribological size effects in MDD and MHDD are proposed, and the developed model has confirmed the tribological size effects accordingly.

11.2 SIMULATION PROCEDURE

11.2.1 Geometric Model

A hydro deep drawing model with scaled down sizes was firstly developed. Fig. 11.1 illustrates the sketch and dimensions of the MHDD system. The parameters in this model were the same as that in experiments, and all the parts in this FEM model were of shell elements and only the blank was deformable compared to other rigid parts.

Due to the axisymmetrical geometry, one-quarter of the drawing system was modeled to reduce the computational time. There were the punch, die, blank holder, and the blank in the simulation model. Moreover, the blank was deformable governed by the 3-parameter Barlat formula. In the simulation, the die and the blank holder were fixed with a constant

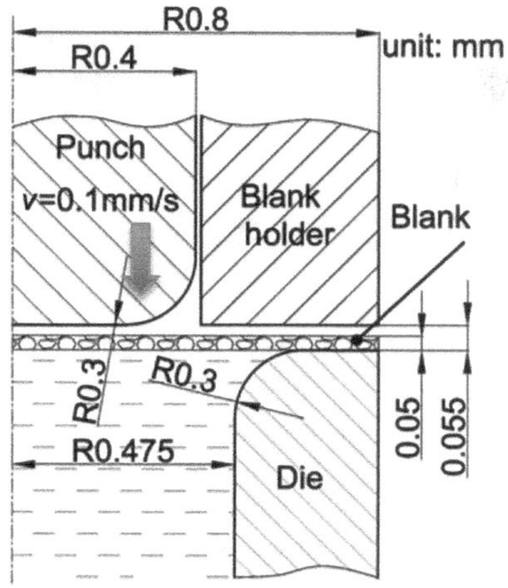

FIGURE 11.1 Geometry illustration of micro hydro deep drawing.

gap of 0.055 mm, while the punch was set with a drawing speed of 0.1 mm/s. Although shell elements were used for the blank, seven integration points through thickness direction were assigned to accurately capture thickness strain and stress. Furthermore, a constant hydraulic pressure load was applied on the blank based on the assumption that the leakage during the micro hydro deep drawing process is small and the pressure loss is limited. Although the deformation of the blank causes fluid volumetric compression which increases the hydraulic pressure, the relatively large initial volume under the blank and in the pipeline reduces the increase of pressure to an ignorable level. Thus, the constant pressure is suitable for this simulation. Moreover, as the punch moved very slowly, the deformation of the blank did not cause obvious pressure change. Therefore, a uniform hydraulic pressure on the blank can be used in this situation. Although the hydraulic pressure was applied on thickness direction of the blank, the in-plane stress assumption was still suitable as the hydraulic pressure was much smaller than the metal's initial yielding stress.

11.2.2 Surface Layer Modeling

Given the size effects, a Voronoi blank model was generated as a centroid Voronoi structure with properly adjusted tessellation size distribution can be comparable to the microstructures of annealed metals. Stainless steel SUS304 sheets with a thickness of 50 μm were annealed at 975°C, 1050°C, and 1100°C in an argon gas protection atmosphere for 2 min, respectively. Therefore, grains on the sheets were of equiaxed crystals and their average

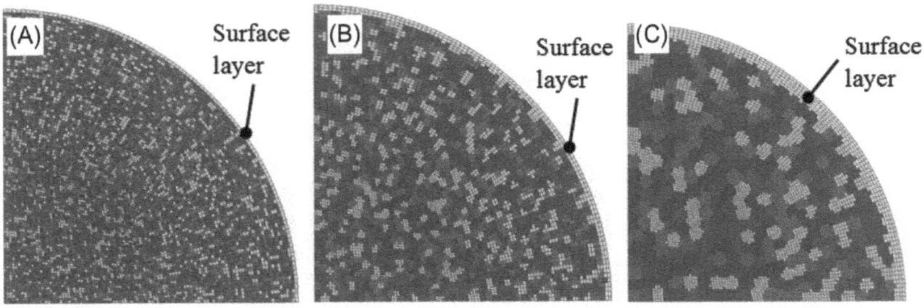

FIGURE 11.2 Voronoi FEM models with a surface layer for sheets experienced different annealing temperatures: (A) 975°C, (B) 1050°C, and (C) 1100°C [7].

sizes were 10, 20, and 40 μm corresponding to the increasing annealing temperatures, respectively. Consequently, three Voronoi models with the average tessellation sizes of 10, 20, and 40 μm, respectively, for those three kinds of heated SUS304 sheets were generated as shown in Fig. 11.2. Different material properties were then assigned to individual tessellation parts. For simplification, all the tessellations were divided into five groups, and each group had its own material parameters. All the material parameters were obtained from tensile tests, and each color in Fig. 11.2 indicates one tessellation group. As the centroid Voronoi tessellations had similar shape and size to the grains on the sheets, while different tessellation parts had different mechanical properties, this Voronoi model can present the blank on a grain scale. Thus, the material size effects can be included into the Voronoi FEM models.

A surface layer with only one grain in the radial direction of the blank was additionally built based on the Voronoi FEM models. Due to the small limitation of dislocation on the surface layer, the flow stress of the surface layer with the decrease of sample sizes was reduced 10% from the average tensile tests results. The surface layer on the edge of the blank was set as the open lubricant area with relatively high coefficient of friction (COF), while the other part of the blank was treated as closed lubricant field with relatively low COF. As this model was based on the developed Voronoi model, the material size effects were remained while special friction behavior was considered.

The drawn cups were then conducted a springback simulation as tiny springback was considerable compared to the minute cup sizes. The residual stress and strain energy were released gradually due to small deformation of the drawn cups. According to the request on the springback simulation, two nodes at symmetrical lines of the drawn cup were fixed, and for each simulation case the same two nodes were set. As the Voronoi models were remained in this springback model, material inhomogeneity was kept and presented in the springback process.

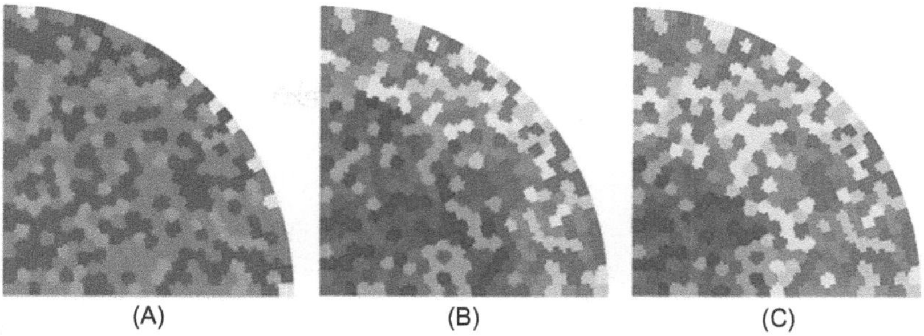

FIGURE 11.3 Voronoi models for 1100°C annealed SUS304 foils with open and closed lubricant areas under different hydraulic pressures: (A) 5–10 MPa, (B) 15–20 MPa, and (C) 25–30 MPa [7].

11.2.3 Open and Closed Lubrication Modeling

The open and closed lubricant theory was adopted in the Voronoi model. The grains located outer of a critical radius on the blank were defined as the open lubricant area with a large COF, while grains at the inner part of the blank were classified as the closed lubricant area with a small COF. Moreover, with an increase of the hydraulic pressure, the open lubricant area was enlarged, whereas the closed lubricant area was reduced. Fig. 11.3 displays the Voronoi model where the bright area is the open lubricant area, and the other is the closed lubricant area based on different hydraulic pressures [8].

11.3 SIMULATION RESULTS

All the simulation was run on the same computer with the same LS-DYNA solver and their results were compared.

11.3.1 Friction Size Effects

To model the tribological size effects in MHDD, the inner and outer pockets which have different COF are considered on the blank surface in FE model. Also, the low COF at the outer pockets is assumed in MHDD to consider the lubrication effect of fluid medium [9]. Table 11.1 lists the COF between the blank and blank holders and die at each lubrication condition.

Fig. 11.4 shows the comparison of normalized punch force–stroke curve between experiment and analysis. The punch force–stroke curves obtained by experiment in MDD and MHDD are in good agreement with the analytical results of MDD with dry friction and MHDD with no leakage. However, the punch force–stroke curves between the MDD and MHDD are almost the same. The difference of effective punch force shows the difference of friction force. It means there is a high friction force in the experiment of

TABLE 11.1 The COF Between the Blank and Blank Holders and Die at Each Lubrication Condition [9] (μ_s: Static COF; μ_d: Kinetic COF)

Lubrication Conditions	Inner Pockets	Outer Pockets
MDD with dry friction	μ_s: 0.35, μ_d : 0.30	μ_s: 0.35, μ_d: 0.30
MDD with lubrication	μ_s: 0.05, μ_d : 0.03	μ_s: 0.35, μ_d: 0.30
MHDD with no leakage (blank-side wall of die)	μ_s: 0.05, μ_d : 0.03	μ_s: 0.05, μ_d: 0.03
MHDD with no leakage (blank-BH, flange of die)	μ_s: 0.35, μ_d : 0.3	μ_s: 0.35, μ_d: 0.30
MHDD with leakage (blank-die)	μ_s: 0.05, μ_d : 0.03	μ_s: 0.05, μ_d: 0.03
MHDD with leakage (blank-BH)	μ_s: 0.35, μ_d : 0.30	μ_s: 0.05, μ_d: 0.03
MHDD with radial pressure	μ_s: 0.35, μ_d : 0.30	μ_s: 0.05, μ_d: 0.03
MHDD with perfect lubrication	μ_s: 0.00, μ_d : 0.00	μ_s: 0.00, μ_d: 0.00

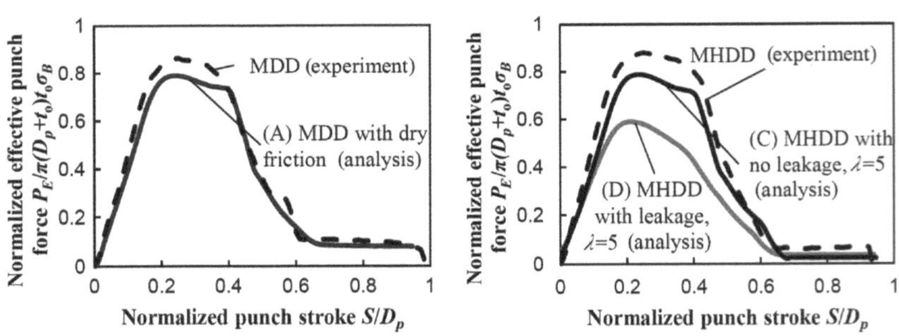

FIGURE 11.4 Comparison of normalized punch force–stroke curves of the experimental and analysis [8].

MHDD due to no leakage of the fluid medium between the blank and die. Generally, the leakage makes the friction force low at die shoulder and flange area. However, the contact pressure between the blank and die becomes high at the die shoulder area due to small D_p/t in MHDD [10]. This prevented the leakage and caused the high friction force in MHDD.

Fig. 11.5 illustrates the effects of scale factor on the normalized punch force–stroke curves at MDD with lubrication and MHDD with radial pressure. The shape of punch force–stroke curves in $\lambda = 1, 2$ is as similar as that with $\lambda = 50$ at MDD and MHDD. In these conditions, only the inner or outer pockets exist in the flange area. Therefore, the COF in the flange area is almost uniform. On the other hand, in $\lambda = 5$, the inner and outer pockets are mixed in the flange area. In the initial process, the inner pockets mainly

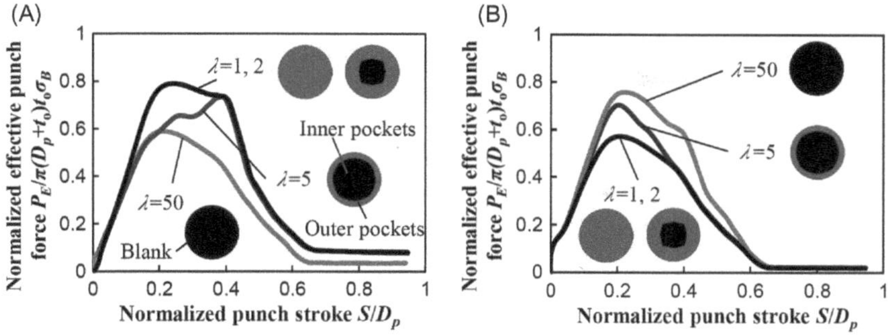

FIGURE 11.5 Effects of scale factor on normalized punch force–stroke curve at different lubrication conditions [8].

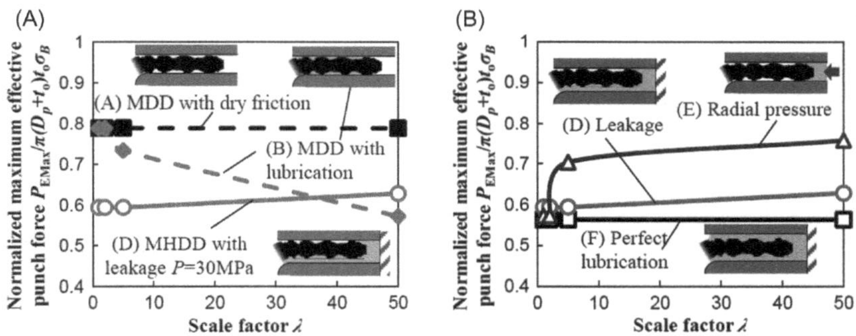

FIGURE 11.6 Tribological size effects at different lubrication conditions in MDD and MHDD [8].

exist at die shoulder and flange area and affect the tribological behavior significantly. Therefore, the punch force–stroke curves are as similar with that in macro scale. However, in the middle process, the ratio of outer pockets increases. As a result, the tribological behavior shifts to that in micro scale. This behavior appears at both MDD and MHDD. This causes the maximum punch force shifts as shown in Fig. 11.5A. These results indicate the ratio of the outer pockets to the flange area during the forming process influences the tribological behavior of the MHDD.

Fig. 11.6A shows the tribological size effects in MDD and MHDD. With the decrease in the size, the friction force increases in case of MDD with lubrication because the ratio of outer pockets increase. When $\lambda = 1$, 2, the maximum effective punch forces in MDD with the dry friction and lubrication become the same because only the outer pockets exist at flange area. On the other hand, with the decrease in the size, the friction force in MHDD decreases. It can be seen the tribological size effects in MHDD has an opposite behavior with MDD. In MHDD, the fluid medium is provided to the outer pockets whose ratio is high in micro scale. This caused the decrease in friction force in MHDD. This effect has been theoretically confirmed [10].

Fig. 11.6B shows the effect of the lubrication type on friction force in MHDD. The decrease of friction force in radial pressure condition is much larger than that in leakage condition. Also in radial pressure condition, the friction force significant decreases from $\lambda = 5-1$. It is because the contact pressure between the blank and die at die shoulder is higher than that between the blank and blank holder in the small D_p/t. Therefore, it can be seen that the decrease in COF at die shoulder is especially important to decrease the friction force in MHDD. According to the above results, the friction force can decrease with the decrease in size in MHDD, while it only increases in MDD. The friction force can be reduced by filling the fluid medium in the outer pockets in MHDD.

11.3.2 Effect of Hydraulic Pressure

The minimum thickness of the drawn cups is an important factor in cup's quality, and it is generally occurred on the blank corresponding to the punch nose area. Due to the hydraulic pressure, the minimum thickness was affected. Fig. 11.7 shows the minimum thickness of the sheets undergone different heat treatments under different hydraulic pressures. The minimum thickness reached a peak value under 10 MPa hydraulic pressure condition for the sheets heated to 975°C. The COF reduced due to the hydraulic oil lubrication and, therefore, the friction force between the blank and the die decreased. Moreover, the hydraulic pressure forced the blank to grasp the punch and increased the friction between the blank and the punch. Hence, the tensile stress on the blank reduced. Furthermore, the higher the hydraulic pressure was, the smaller the tensile stress was. The overall effect of the change in friction was the increase of the minimum thickness with

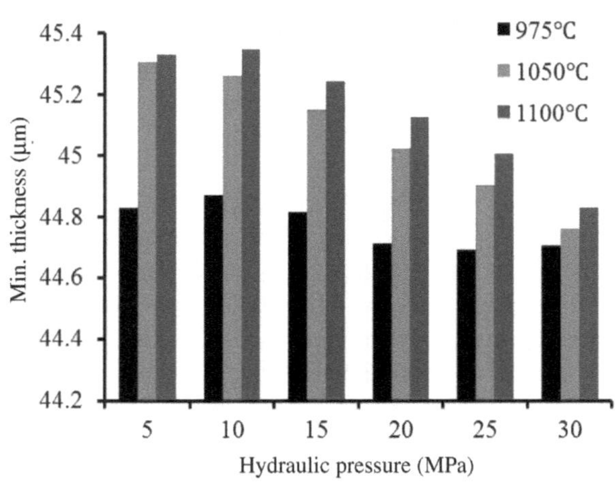

FIGURE 11.7 The minimum thickness of sheets undergone various annealing temperatures.

hydraulic pressure. However, once the hydraulic pressure increased further, the minimum thickness decreased. The high pressure pushed the blank to deform toward the punch to a certain degree and changed the strain path of the blank near the punch nose. Consequently, this change led to change in thickness development and accordingly decreasing the minimum thickness. For the sheets with annealing temperature of 1100°C, there was a similar trend in the minimum thickness to that of the sheets experienced annealing temperature at 975°C. However, a monotonous decrease of the minimum thickness in the annealing temperature of 1050°C indicated that the randomly distributed grain parts with different material properties affected the strain–stress development and finally impacted the minimum thickness. If the material on the blank near the punch nose was soft and had low deformation resistance, the blank at this area underwent significant thinning during the initial drawing process and presented great thinning at the end of the drawing process. Moreover, the high pressure increased this initial thinning on the blank near the punch nose, and consequently, it worsened the thickness decrease during the entire drawing process. Therefore, the minimum thickness of treated sheets at the temperature of 1050°C trended to decrease with the increase of hydraulic pressure.

The strain path change can be proved by the drawing force. Fig. 11.8A shows the drawing force of sheets undergone annealing temperature of 975°C under different hydraulic pressures. The general drawing force developing tendency for simulation under each hydraulic pressure was slightly different at the first half drawing period, whereas that in the second half period was similar. For simulation with hydraulic pressures of 5 and 10 MPa, the drawing forces had a concave shape at the first half drawing process, whereas drawing forces of high hydraulic pressure conditions had a convex shape. The lower or higher the hydraulic pressure was, the more obvious the concave or convex shapes obtained. The concave drawing force curve showed that the blank was not strained to a high level after prebugling.

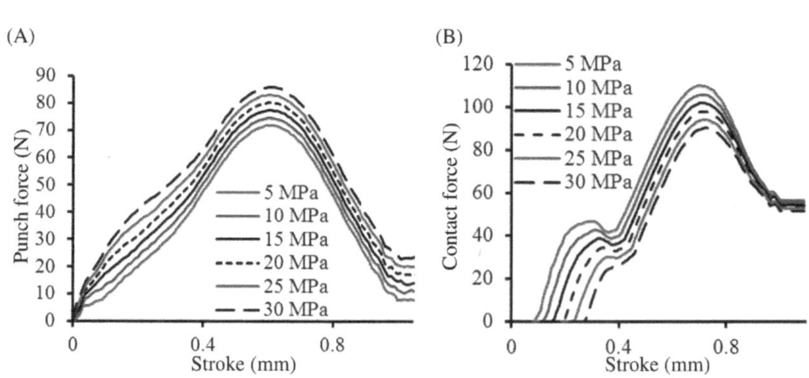

FIGURE 11.8 (A) Drawing forces and (B) contact forces between the blank and the die.

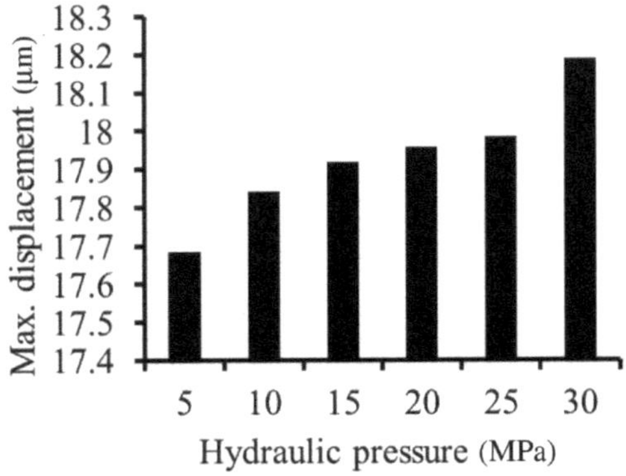

FIGURE 11.9 The maximum nodal displacement of drawn cups with annealing temperature of 975°C after springback.

Therefore, the strain hardening did not cause significant flow stress growth. By contrast, the convex curve indicated that the blank bore high strain and had significant strain hardening. The thickness on the blank close to the punch nose suffered relatively significant thinning, and this initial thinning led to the decreased minimum thickness. In addition, high pressure reduces the effective radii of the blank covering the die shoulder and the punch nose. Thus, the decreased fillet radii aggravated thinning of the blank at the fillets areas. The later contact between the blank and the die under high hydraulic pressure, as shown in Fig. 11.8B, proved this decrease of effective fillet radii near the die shoulder and the punch nose. Furthermore, high hydraulic pressure resulted in increasing the high maximum drawing force and residual force which impacts the springback behavior of the drawn cups directly.

The maximum displacement of a node on the drawn cup after springback was chosen as a simple judgment of springback behavior. Fig. 11.9 shows the maximum nodal springback displacement of drawn cups with annealing temperature of 975°C under different hydraulic pressures. The maximum springback displacement increased simultaneously with the increase of hydraulic pressure. High hydraulic pressure meant great elastic deformation energy stored in the drawn cups. During the springback process, the residual stress and the deformation energy were released via small nodes displacements. Therefore, the produced cups under high hydraulic pressure had great nodal peak springback displacement.

The shape accuracy of the drawn cups under higher hydraulic pressure after springback, however, was higher than that of the drawn cups with lower hydraulic pressure, regardless of the large peak nodal springback displacement. Fig. 11.10 shows the profile of the drawn cups with annealing temperature of 975°C under different hydraulic pressures. Due to the hydraulic pressure, the shape of the drawn cups after springback was close

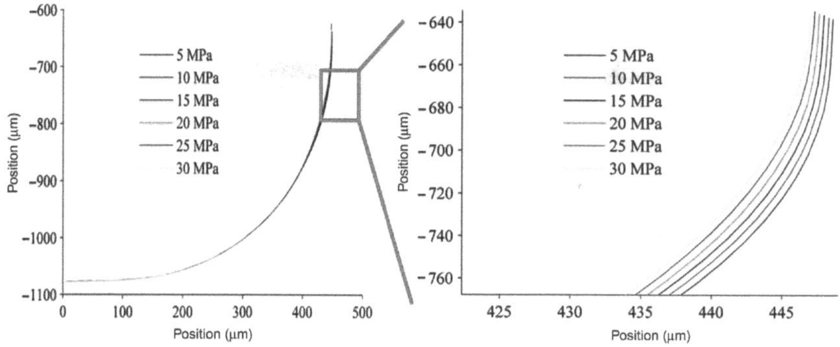

FIGURE 11.10 Shape of the drawn cup with annealing temperature of 975°C after springback.

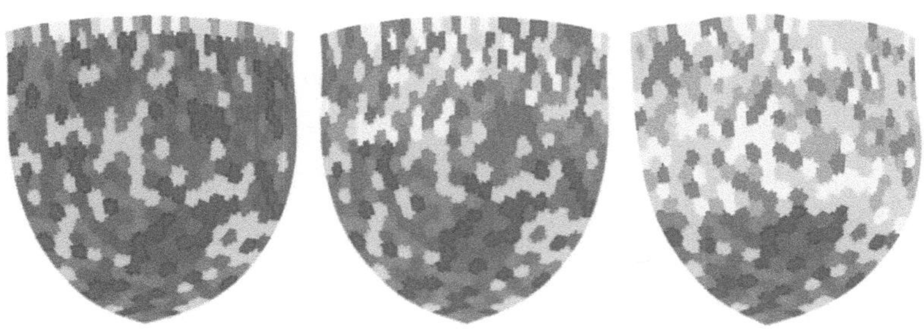

FIGURE 11.11 Drawn cups with 1100°C annealed SUS304 foils under different hydraulic pressures: (A) 5 MPa, (B) 15 MPa, and (C) 30 MPa.

to the punch. Higher pressure resulted in closer cup shape to the punch compared to the drawn cups with lower hydraulic pressure. Therefore, in spite of the relatively large peak nodal springback displacement, the overall effect on the final shape accuracy of the drawn cups was improved with high hydraulic pressure.

11.3.3 Wrinkling and Earing

The coordinates of the nodes on the cup's mouth in simulation were exported. Therefore, the wrinkling and earing can be presented by the coordinates. Fig. 11.11 displays the drawn cups with 1100°C annealed SUS304 foils under different hydraulic pressures, and Fig. 11.12 shows the drawn cups with different foils under 30 MPa hydraulic pressures.

The *x*- and *y*-coordinates of the nodes on the drawn cup's mouth were used for wrinkling measurement. All the nodes were firstly projected to a fitted circle with its center on the origin point (0, 0). Consequently, the difference between the radii of each node and the fitted circle was defined

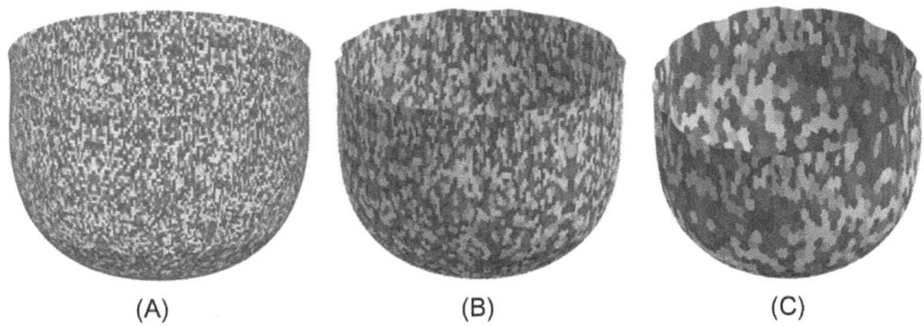

FIGURE 11.12 Drawn cups with different foils drawn under 30 MPa hydraulic pressures: (A) H975, (B) H1050, and (C) H1100.

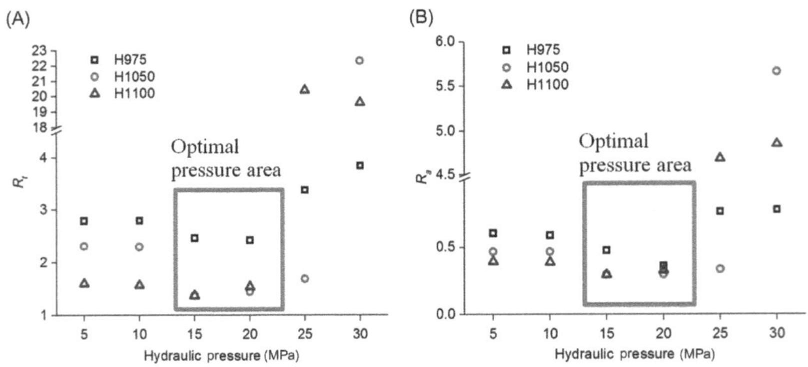

FIGURE 11.13 R_t and R_a values of wrinkling in different simulation cases: (A) R_t and (B) R_a.

as the wrinkles. Finally, the wrinkling phenomenon can be expressed by these wrinkles similar to the roughness parameters. Accordingly, the maximum height of the profile (R_t) was defined as the difference between the largest and the smallest radii of the nodes on the drawn cup's mouth, as shown in Eq. (11.1). Moreover, the arithmetic average height (R_a) was expressed as Eq. (11.2). Therefore, high values of R_t and R_a indicate significant wrinkling phenomenon. Fig. 11.13 displays the R_t and R_a values of each simulation model under different hydraulic pressure conditions.

$$R_t = R_p + R_v \tag{11.1}$$

$$R_a = \frac{1}{n}\sum_{i=1}^{n} |r_i| \tag{11.2}$$

where R_t, R_p, R_v, and R_a are the maximum height, the height of peaks, the height of valleys, and the arithmetic average height, respectively. r_i is the radius of the ith node and n is the total number of the nodes.

With an increase of hydraulic pressure in the simulation, the wrinkling phenomenon of each SUS304 foil group was firstly decreased and then increased once the hydraulic pressure reached a critical value. Moreover, the critical value decreases with the increase of annealing temperature. Therefore, the foils annealed at high temperature experienced wrinkling at low hydraulic pressures. Once the wrinkling occurs, both the R_t and R_a values increased sharply. Furthermore, the foils annealed at high temperatures with low initial yielding stress and ultimate tensile strength presented slight wrinkling at hydraulic pressures below the critical pressure compared to the foils with low temperature annealing.

The hydraulic pressure offers additional blank holding force and improves friction condition. Therefore, the wrinkling phenomenon becomes weak with increasing hydraulic pressure. However, once the hydraulic pressure reached a critical value, disturbance on the blank's compression stability introduced by the hydraulic pressure compensates the benefits of the hydraulic pressure. Thus, wrinkling occurs and becomes more significant with further increasing pressure. Below the critical hydraulic pressure, the foils heated at high temperatures have low deformation resistance and consequently can bear large strain without buckling. Furthermore, due to the low initial yielding stress and ultimate tensile strength, disturbance of the hydraulic pressure on the drawing process becomes more significant and is presented as wrinkling at low hydraulic pressures. Therefore, the optimal hydraulic pressure for micro hydro deep drawing process should be designed based on the foils' mechanical properties.

Similar to the analysis of wrinkling, the earing was analyzed based on the z-coordinates of each node. Moreover, the R_t and R_a values were utilized to index the earing phenomenon. Fig. 11.14 displays the R_t and R_a values of earing in different simulation cases.

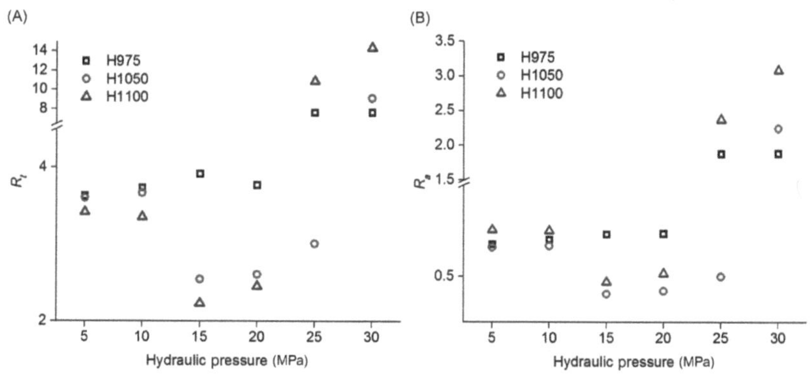

FIGURE 11.14 R_t and R_a values of earing in different simulation cases: (A) R_t and (B) R_a.

Similar to the wrinkling trend, the earing phenomenon weakened under low hydraulic pressure conditions and increased when the hydraulic pressure is over the critical value. Moreover, the earing was accompanied by the wrinkling and significant earing appeared under the same hydraulic pressure conditions as that of the obvious wrinkling phenomenon. The foils with high temperature annealing had weak earing, while the earing occurred under low hydraulic pressure situation in terms of the R_t value. Furthermore, once significant earing occurred, the earing at high temperature annealed foils was more significant than that of the heated foils under low temperature.

The Voronoi model introduces material anisotropy and leads to different strain distributions on the cup's mouth. Therefore, the R_t and R_a values are not zero when the hydraulic pressure is lower than the critical value. Below the critical pressure, the benefits of hydraulic pressure overcome its disadvantages, and thus, the earing phenomenon weakens with an increase of hydraulic pressure. However, once reached the critical value, significant wrinkling occurs. Consequently, the strain development and the geometry of the cup change resulting in significant earing. The foils with strong resistance to the wrinkling also obviously restrict the earing process. Therefore, the foils heated at 975°C under 30 MPa pressure presented weaker earing than that of the other foil groups.

In summary, the Voronoi FEM model with the one-grain surface layer developed in this chapter can present the micro hydro deep drawing process successfully. The developed simulation models can simulate the micro hydro deep drawing process with a consideration of material size effects and frictional size effects. There is a critical hydraulic pressure value, and below the value no obvious wrinkling and earing occur. Additionally, the hydraulic pressure is beneficial to the drawing process at low pressures. High hydraulic pressure increases the springback in terms of the maximum nodal displacement, while the final shape of the drawn cups under high pressure is closer to the punch than that of the drawn cups with low hydraulic pressure. The analytical results of punch force–stroke curves are in good agreement with the experimental results. With the decrease in the size, the friction force can be reduced in MHDD because of the existence of fluid medium in outer pockets.

REFERENCES

[1] L.H. Lang, Z.R. Wang, D.C. Kang, S.J. Yuan, S.H. Zhang, J. Danckert, et al., Hydroforming highlights: sheet hydroforming and tube hydroforming, J. Mater. Process. Technol. 151 (2004) 165–177.

[2] K. Nakamura, T. Nakagawa, Fracture mechanism and fracture control in deep drawing with hydraulic counter pressure—studies on hydraulic counter pressure forming I, J. JSTP (in Japanese) 25 (1984) 831–838.

[3] K. Nakamura, T. Nakagawa, Radial pressure assisted hydraulic counter pressure deep drawing—studies on hydraulic counter pressure forming II, Jpn. Soc. Technol. Plast. (in Japanese) 26 (1985) 73–80.

[4] K. Manabe, H. Sato, T. Furushima, D.B. Wei, N. Mathew, Z.Y. Jiang, Deformation behavior in micro sheet hydroforming process, Steel Res. Int., Special edition (2012) 651–654.

[5] H. Sato, K. Manabe, K. Ito, D.B. Wei, Z.Y. Jiang, Development of servo-type micro-hydromechanical deep-drawing apparatus and micro deep-drawing experiments of circular cups, J. Mater. Process. Technol. 224 (2015) 233–239.

[6] I. Irthiea, G. Green, S. Hashim, A. Kriama, Experimental and numerical investigation on micro deep drawing process of stainless steel 304 foil using flexible tools, Int. J. Mach. Tools Manuf. 76 (2014) 21–33.

[7] L. Luo, Z.Y. Jiang, D.B. Wei, K. Manabe, Influences of hydraulic pressure on wrinkling and earing in micro hydro deep drawing with consideration of size effect, International Conference on Materials Processing Technology 2016, March 28–29, 2016, Tokyo, Japan.

[8] H. Sato, K. Manabe, D.B. Wei, Z.Y. Jiang, Numerical modeling of size effect in micro hydromechanical deep drawing, AIP Conf. Proc. 1567 (2013) 926.

[9] H. Sato, K. Manabe, D.B. Wei, Z.Y. Jiang, Effect of fluid pressure in micro hydromechanical deep drawing, Proceedings of the 6th International Conference on Tube Hydroforming (2013) 313–317.

[10] H. Sato, M. Manabe, D.B. Wei, Z.Y. Jiang, Analysis of axisymmetric cup forming of metal foil and micro hydroforming process, ASME International Mechanical Engineering Congress and Exposition (IMECE), ASME, United States, 2013.

Chapter 12

Simulation of Micro Bending

12.1 INTRODUCTION

When the scale of specimens is downsized to micro level, the parameters which may not be significant in conventional bending begin to play a vital role in influencing the accuracy of bending, due to size effects [1,2]. Since size effects are significantly important in the micro world, grain heterogeneity of specimen has to be considered regardless of conducting physical experiment or modeling. Grain heterogeneity can be described by the scatter of grain behavior which can be attributed to different grain sizes, shapes, and orientations. When building models, the scatter can be embedded into single virtual grain, and different grains will affect separately and mutually during simulation process [3,4]. In recent years, Voronoi diagram has been introduced to generate polycrystalline aggregate. This method subdivides a space tessellation into Voronoi polyhedral, which is similar to morphogenetic process of nucleation and growth from random seeds. Voronoi tessellation has an extensive application in scientific research, and specifically in materials science, polycrystalline microstructures are commonly represented by Voronoi diagram because of its geometric features [5,6]. However, limited computational research has illustrated the complete 3D Voronoi tessellation realization when finite element (FE) model is carried out, since it has been used as a functional approach in polycrystalline materials research.

The main objective of this chapter is to introduce an FE model for simulating micro V-bending with considerations of polycrystalline aggregate and grain heterogeneity. Polycrystalline aggregate is implemented in commercial software ABAQUS/CAE in terms of Voronoi tessellations, and grain heterogeneity is reflected by the scatter of mechanical properties of grains. To obtain related information on grain size and grain's mechanical properties which are necessary for building up bending FE model, corresponding heat treatment and micro tensile tests are conducted. The efficiency of the developed model is verified by comparing the simulation results with the experimental ones. By applying digital Voronoi diagram program described in Chapter 6 and combining it with experimental data, it is promising to explore an appropriate FE simulation method to provide a better understanding of micro bending process of metals.

Microforming Technology. DOI: http://dx.doi.org/10.1016/B978-0-12-811212-0.00012-1

12.2 SIMULATION PROCEDURE

12.2.1 Experiments

12.2.1.1 Heat Treatment

Phosphor bronze C5191 also known as copper tin foil C5191 with thickness of 70 μm is chosen for micro V-bending experiment. The bending specimens measuring $2 \times 1 \times 0.07\ mm^3$ are prepared by wire-cut electrical discharge machining (EDM) from as-received phosphor bronze foil, and then annealed at 650°C for 4 h followed by furnace cooling. Heat treatment is carried out under argon air protection in order to avoid oxidation. The microstructure of specimen along length direction after annealing treatment is shown in Fig. 12.1, which is obtained through etching with $FeCl_3$ (5 g) + HCl (15 ml) + H_2O (85 ml) solution for 2 s. It can be seen that the average grain size is around 162.8 μm and the grain size in the direction of foil thickness is restricted, which means only one partial grain exists in thickness direction. When simulating such microstructure in FE model, it is easy to observe each grain's influence on the whole deformation behavior because the grain size is big enough to distinguish.

12.2.1.2 Micro Tensile Tests

The main purpose of micro tensile tests is to gain the stress–strain relationship of the annealed C5191 foil, and then to extend the scatter of the material's mechanical properties to the scatter of mechanical properties of grains. The dimensions of tensile specimens are shown in Fig. 12.2. The specimens are also annealed under the same heat treatment process described above. Micro tensile tests are conducted at room temperature on INSTRON Micro Tester 5848 with the crosshead velocity of 0.005 mm/s, and a video extensometer is utilized to record and measure the strain precisely.

Based on the research of Lu et al. [3], the statistical distribution of grain's properties can be divided into seven classes. Among these classes of grain's properties, the upper and lower bound properties and the mean properties can be calculated [4]. In order to obtain the grains' stress–strain

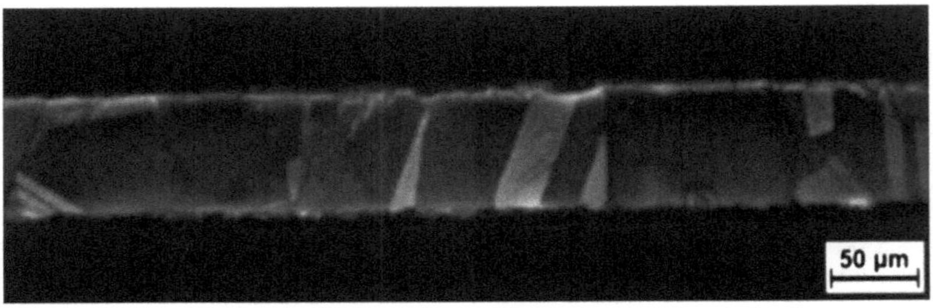

FIGURE 12.1 Microstructure of annealed phosphor bronze foil.

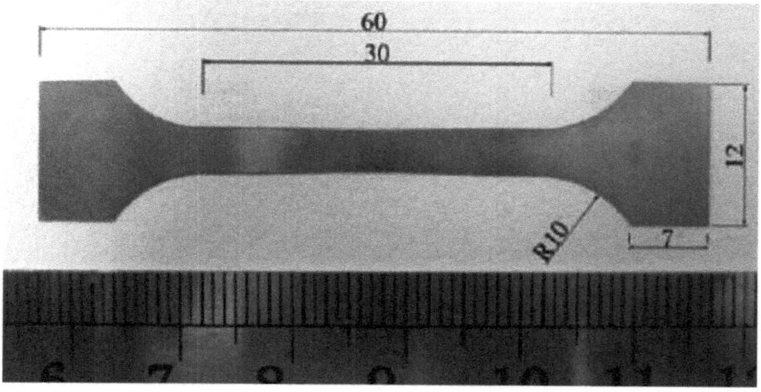

FIGURE 12.2 Specimen for micro tensile tests.

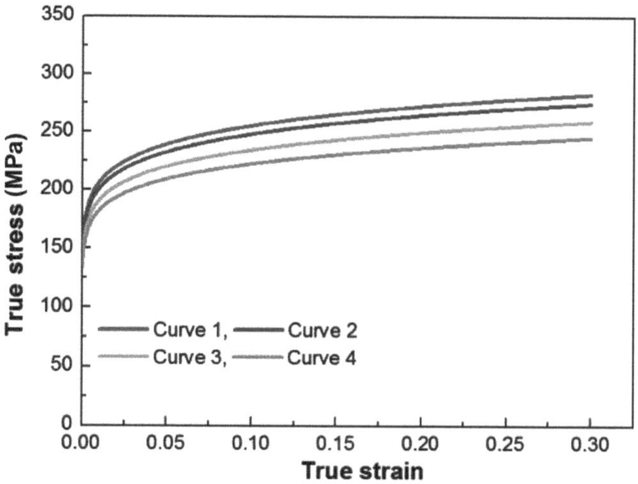

FIGURE 12.3 Material mechanical properties from tensile tests.

relationship, four groups of true stress–strain curves are obtained after tensile tests, which are marked as curves "1," "2," "3," and "4," respectively, as shown in Fig. 12.3. It can be seen that the tensile curves are not repeatable due to size effects. In microforming, the minimum unit is no longer a whole piece of specimen but each grain contained in the specimen because the specimen is not regarded as homogeneous material. The scatter range is determined by the properties of grains which can be influenced by grain size, orientation, shape, and forming process. Therefore, the effects could be attributed to different mechanical properties of grains. In other words, if a certain tensile test result goes to curve "1," then it is believed that there must be some grains possess mechanical properties which are higher than curve "1." Similarly, there also must be some grains possess

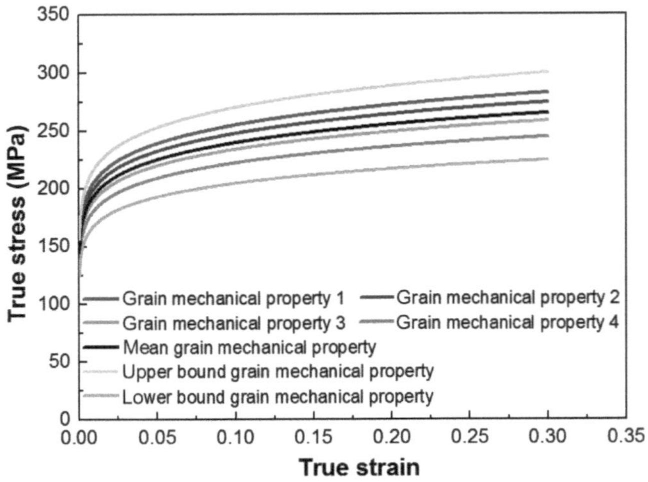

FIGURE 12.4 Evaluating grain plastic properties from tensile test.

mechanical properties which are lower than and equal to curve "1." All these microscopic mechanical properties of grains work together, producing the macroscopic curve "1."

Based on these four testing results, the mean mechanical properties, the upper and the lower bounds of mechanical properties of grains can be calculated through Eqs. (6.37) and (6.38). Finally, seven mechanical properties of grains are obtained, and the scatter of mechanical properties is displayed in Fig. 12.4.

12.2.2 Voronoi Tessellation Implement in ABAQUS

The detailed methodology for how to implement Voronoi diagram into ABAQUS/CAE has been described in Section 6.4.3 of Chapter 6. Here, the corresponding application is shown.

Fig. 12.5 displays the 3D Voronoi diagram implementation in software MATLAB and ABAQUS with 30 grains (grain size 290.9 μm) and 300 grains (grain size 96.1 μm) in a $2 \times 1 \times 0.07\ \text{mm}^3$ specimen, respectively. The grain density can be calculated by the following equation:

$$\rho_G = \frac{3}{4} \cdot \frac{l \times w \times h}{\pi (D_{\text{ave}}/2)^3} \tag{12.1}$$

where l, w, and h are the specimen's dimensions of length, width, and height, respectively, and D_{ave} is the average gran size.

When there is only one grain in the thickness direction which is the same condition with all specimens, there is no need to take thickness value

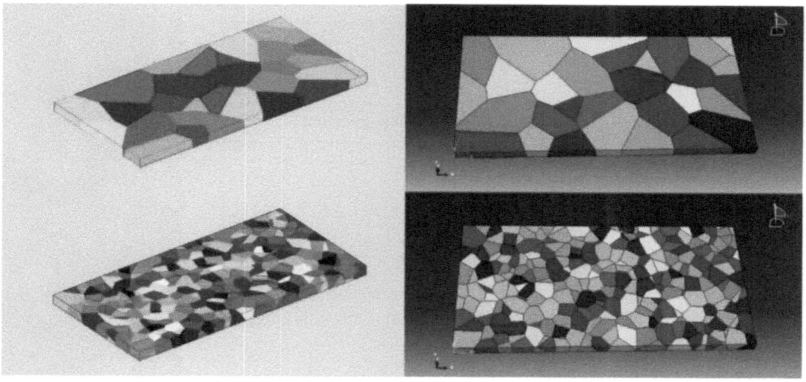

FIGURE 12.5 3D Voronoi tessellation implementation in MATLAB (left) and ABAQUS/CAE (right) with 30 and 300 grains, respectively.

into account. Regarding 2D specimen, the grain density equation can be expressed as follows:

$$\rho_G = \frac{l \times w}{\pi (D_{\text{ave}}/2)^2} \tag{12.2}$$

It can be seen that each Voronoi tessellation can be successfully generated and visualized as "grains" in ABAQUS/CAE.

12.2.3 Grain Heterogeneity

The realization of grain heterogeneity in Voronoi diagram by Python script file has been detailed in Section 6.4.3 of Chapter 6. In this simulation, micro V-bending specimen sizing $2 \times 1 \times 0.07\ \text{mm}^3$ is designed with 75 grains according to the physical average grain size of 162.8 μm and Eq. (12.2). The quantities of grains which possess different mechanical properties are calculated via multiplying each $f(\sigma, \varepsilon)$ from Eq. (12.2) by the total number of 75 grains. Except grains of the mean grain's mechanical properties, other grains' quantities can be obtained by the as-above products. The mean mechanical properties can be calculated via subtracting the grain quantity of the other six mechanical properties from total 75 grains. After calculation, the quantities of grains possessing the upper and lower bound mechanical properties are both 6, and both 13 grains are equipped with mechanical properties "1" and "2." Moreover, mechanical properties "3" and "4" are occupied by both 10 grains individually and the rest of 17 grains possess the mean mechanical properties. In this way, all 75 grains which consist of a bending specimen are provided with certain mechanical properties.

For the 75 grains, green (17 grains) represents the mean plastic property, gray (13 grains) and white (13 grains) symbolize mechanical properties "2"

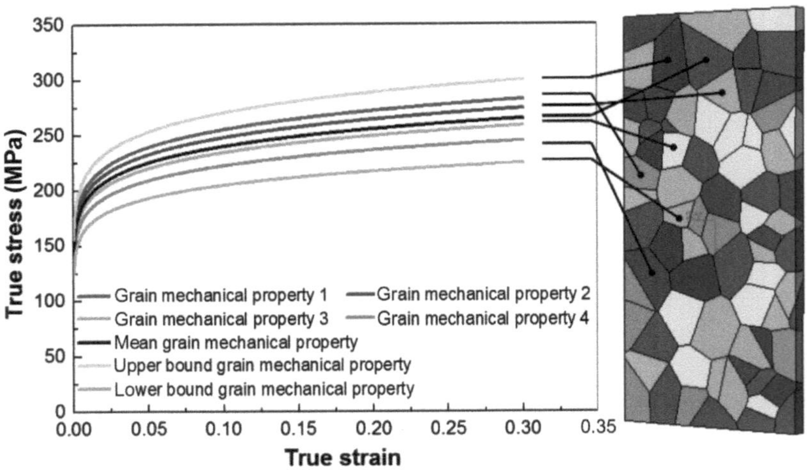

FIGURE 12.6 Grain properties randomly assigned on Voronoized bending sample.

and "3," and yellow (10 grains) and red (10 grains) stand for mechanical properties "1" and "4," respectively. The upper and lower bound grain plastic properties are represented by dark blue (6 grains) and light blue (6 grains), respectively, as shown in Fig. 12.6. In order to investigate the influence of grain heterogeneity on springback, as it can be seen that the FE model is not set up as a traditional asymmetrical one. Instead, the model is close to real physical experimental condition as the right and the left sides of the sample are not equal in terms of grain size and the scatter of mechanical properties of grains.

12.2.4 Numerical Simulation Procedure

Micro V-bending process is simulated with an implicit FEM package: ABAQUS/Standard. The processing parameters in the simulation are the same as those in physical experiments, and the value of coefficient of friction (COF) is set to be 0.02 [7]. The FE model of micro V-bending with Voronoized specimen is shown in Fig. 12.7A. Fig. 12.7B illustrates the grain heterogeneity in Voronoized specimen, among which different colors represent different mechanical properties of grains.

The meshing of bending specimen is displayed in Fig. 12.8. Due to tiny edges in Voronoi tessellations, some regions in the specimen are very complicated for structured and sweep meshing. Therefore, the modified tetrahedral element C3D10M which is robust to solve large deformation and contact problems is adopted in this simulation, and two analytic rigid surfaces are used to simulate the punch and die since they are stiff components.

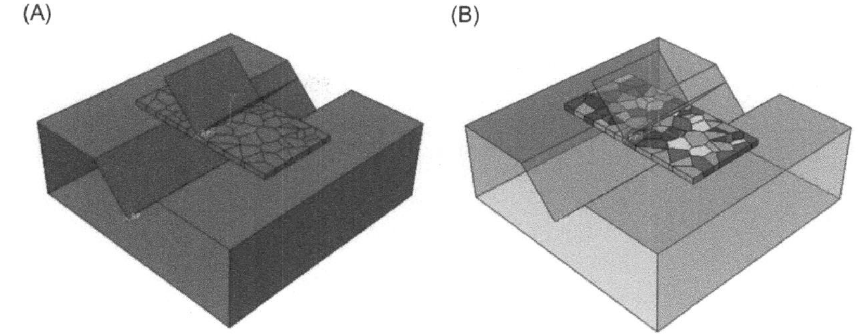

FIGURE 12.7 FEM simulation of micro V-bending with (A) Voronoi tessellations and (B) grain heterogeneity.

FIGURE 12.8 Meshing of Voronoized bending specimen.

12.3 SIMULATION RESULTS

12.3.1 Results Analysis

Fig. 12.9 illustrates the final result of micro V-bending simulation. It is noticeable that inhomogeneous deformation occurs during bending process. The middle deformation zone presents different colors, which means that different grains in this area have experienced different deformation behaviors due to grain heterogeneity. This can be explained that in the whole bending process, some grains firstly reach their yield stress points, which lead to plastic deformation occurs prior to other grains. By this analogy, some grains have low yield stress points, and this means they may still be under elastic stress condition even the specimen has started the plastic deformation. So, this kind of grain heterogeneous deformation can play a significant role in affecting springback result and needs to be considered when conducting numerical simulation of microforming process. The angle is measured by computer-aided design (CAD) software, and it is observed

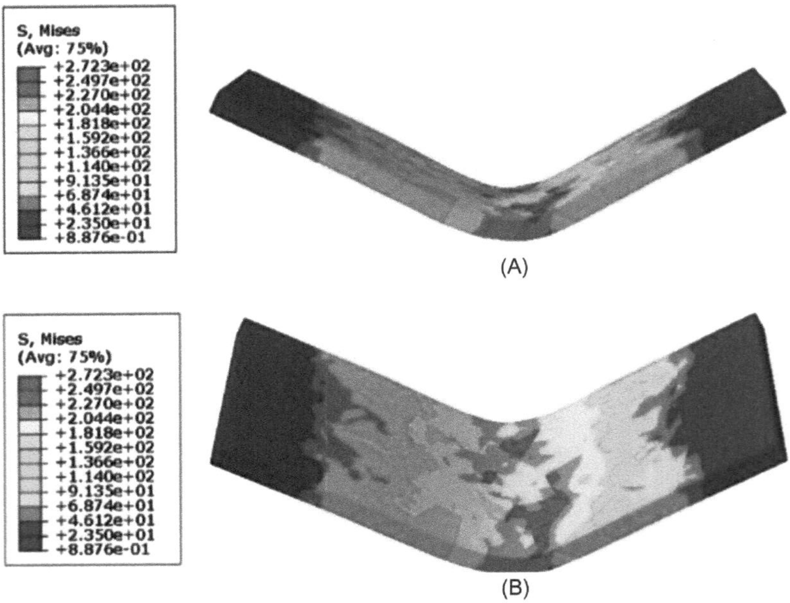

FIGURE 12.9 (A) Final angle after springback and (B) von Mises stress distribution.

that the right and the left sides of specimen experience different levels of bending and springback, which means the two halves of bending specimen deform differently due to grain heterogeneity.

During conventional bending test, central main area's deformation behavior is most decisive to the deformation process and the final angle. In micro V-bending, random distribution of grain's plastic properties in specimen has an important influence on simulation results. Grain heterogeneity can be quite dominant even the difference is small. This kind of phenomenon could also be caused by distinctive meshing in each grain in the FE model due to distinctive grain shape and size. With the consideration of grain heterogeneity in micro V-bending, the mechanical properties of grains in the main deformation area can straight forwardly affect the final angle and springback results. However, Python script cannot fix the distribution of the plastic properties of grains during the process of assigning plastic properties to grains, which means plastic properties are fixed in terms of their grain numbers but irregular in terms of distribution. Take upper bound mechanical properties of grains as an example, it is known that it is assigned to six grains but unassured that where these 6 grains will distribute in the specimen. So, in order to minimize the uncertainty of distribution of mechanical properties of grains in the main deformation area, it is assumed that each grain's mechanical property has one-seventh opportunity to be assigned to grains located in central deformation zone. This hypothesis is based on that

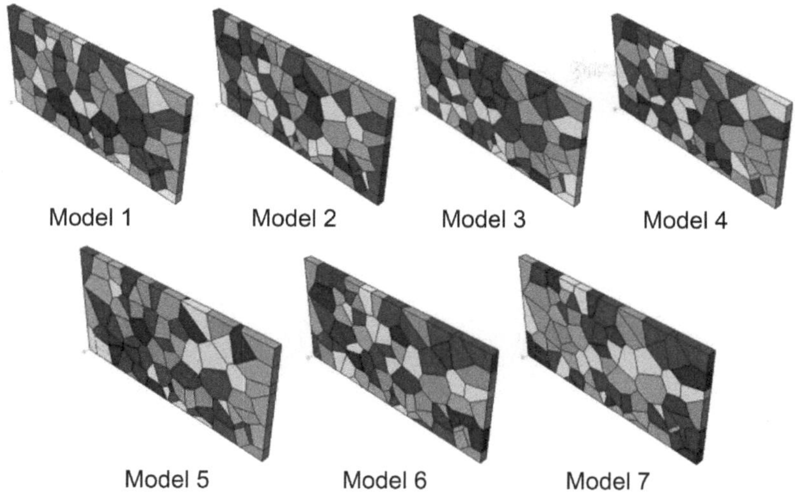

FIGURE 12.10 Bending specimens with different grain heterogeneity distributions.

TABLE 12.1 Springback Angles from FE Simulation (Degree)

Distribution Group	1	2	3	4	5	6	7	Average
Springback angle	30.52	27.37	32.09	31.78	33.44	28.65	34.12	31.14

seven plastic properties are obtained by experiment and calculation, and they are randomly distributed in bending specimens. Specimens with different random grain heterogeneity distributions are exhibited in Fig. 12.10, which are called models "1," "2," "3," "4," "5," "6," and "7," respectively.

Seven groups of micro V-bending FE simulations have been conducted with above seven specimens individually. Springback angles of seven simulations and an average value are measured and calculated, as shown in Table 12.1.

12.3.2 Simulation Verification

12.3.2.1 Micro V-Bending Experimental Results

In order to verify the simulated results, micro V-bending experiments are conducted on the desk-top servo press machine DT-3AW. This equipment has position accuracy within $\pm 2\,\mu m$ and a set of self-designed bending models are employed, as shown in Fig. 12.11. The machine can provide 30 KN

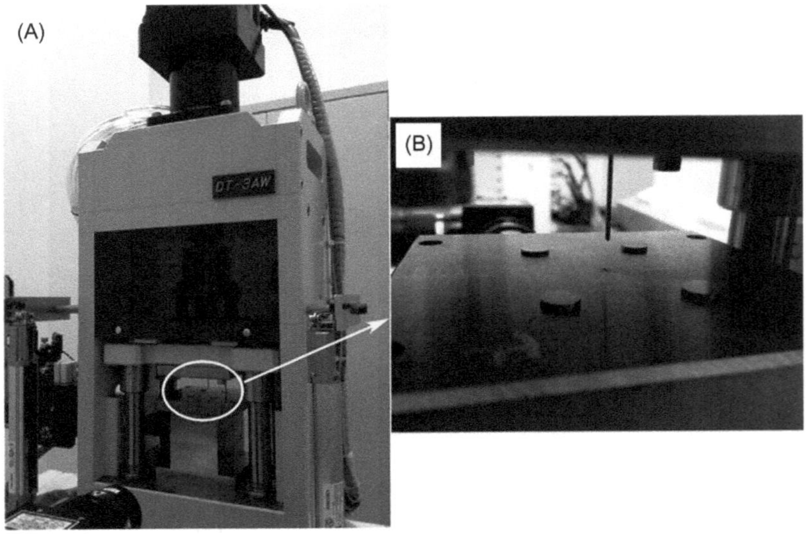

FIGURE 12.11 (A) Desk-top servo press machine DT-3AW and (B) micro V-bending set.

maximum force and ±0.002 mm guidepost position accuracy. In addition, micro V-bending tooling set (punch and female die) is high-precisely manufactured from D2 tool steel with very good dimensional stability (±2 μm) to conquer the difficulty of positioning the bending set.

The delicate punch and die are manufactured with 90 degrees convex and concave, respectively, and these two elements are separately installed into the upper and lower molds. The micro punch is fabricated with high dimensional accuracy of ±1 μm by micro grinding, and the other pertinent geometric parameters are: die gap = 0.8 mm, die depth = 0.5 mm, punch corner radius = 0.01 mm, and punch stroke = 0.43 mm. Before conducting experiment, the surfaces between the punch and the die are well lubricated by machine oil. Micro V-bending tests are performed at room temperature and annealed specimens are placed in the middle of die and punch with punch speed of 0.5 mm/s and punch stroke of 0.43 mm. Fig. 12.12 illustrates the micro V-bending process from loading to bending, and then unloading. Tight shots of these three processes are taken and key components of punch, specimen, and die are enlarged in ellipses. In order to obtain a high accuracy of the springback effect, bending tests are repeated three times. After bending tests, each bended specimen is observed and its angle is measured under VHX-1000 KEYENCE microscope for three times then averaged.

Fig. 12.13 shows one bended specimen. It can be observed that springback has occurred after unloading, from initial bending angle of 90 degrees to a final value greater than 90 degrees. In order to minimize the possibility

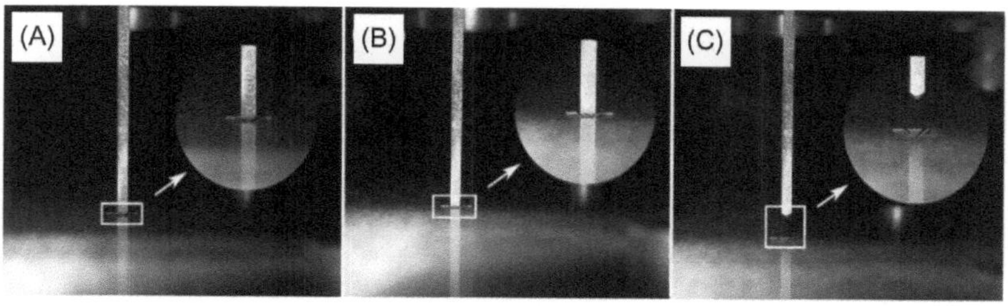

FIGURE 12.12 Micro V-bending process: (A) loading, (B) bending, and (C) unloading.

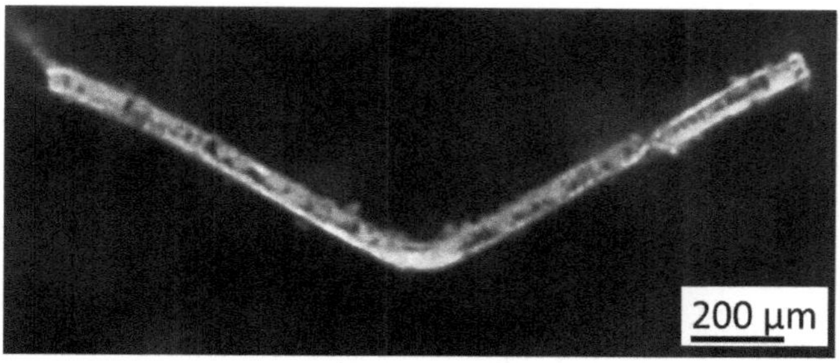

FIGURE 12.13 Bended angle after springback measured in VHX-1000 KEYENCE.

TABLE 12.2 Measured and Averaged Springback Angles for Bended Specimens (Degree)

Times of Measurement	1st	2nd	3rd	Average
Specimen no. 1	31.60	31.95	32.31	31.95
Specimen no. 2	32.56	31.85	31.63	32.01
Specimen no. 3	30.94	30.74	31.09	30.92

of measuring error, each bended specimen is measured three times. The measured springback angles from micro V-bending experiments are displayed in Table 12.2.

12.3.2.2 Comparison between Experimental and Simulation Results

Seven simulated springback angles are averaged to compare with three experimental values, as illustrated in Fig. 12.14. The results indicate that the springback angles from FE simulations are in good agreement with experimental results. This can be attributed to the grain heterogeneity which reflects bending specimen's mechanical properties at microscopic level.

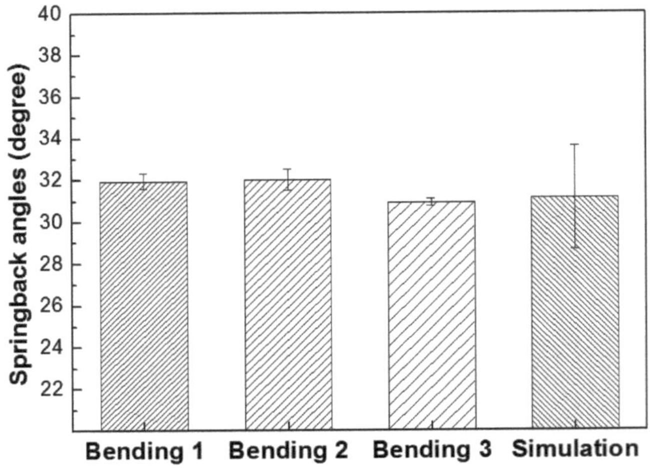

FIGURE 12.14 Comparison between experimental and simulated springback angles.

In other words, the adoption of the scatter of grain's mechanical properties in the model can compensate the errors from tensile tests. This is because more or less errors can occur during tensile tests, and the application of these results in FE models will lead to further inaccuracy of simulation results. The scatter of mechanical properties of grains provides a range of mechanical properties, covering the possible errors from tensile tests consequently.

The findings can also be explained by the grain heterogeneity. As mentioned above, Python scripts cannot control the distribution of mechanical properties of grains in the specimen. So, in simulation, if many upper or lower bound grain properties are distributed in central deformation area, the bending (springback) results could differentiate from experimental ones. Conversely, if enough grains assigned with mean mechanical properties are distributed in central deformation area, the numerical simulation results should have a good agreement with practical ones. Also, it is found that the average springback angle and experimental results have a good agreement. The scatter of springback angles can be successfully reduced by averaging the testing values. This is because the influence of random distribution of grain heterogeneity is lessened. The grain heterogeneity, in essence, is mechanical properties of grains irregularly spread in the whole specimen. To a single specimen in FE simulation, due to the uncertainty of the distributions of grain's mechanical properties, the corresponding springback angle may be larger, lower or close to counterpart experimental values. However, with more numbers of distributions are taken into FE simulation, the scatter can be minimized by averaging, and the average springback angles should be one of the closest results. From this perspective, grain heterogeneity is the main reason for the scatter of springback angles.

REFERENCES

[1] J. Liu, M. Fu, J. Lu, W. Chan, Influence of size effect on the springback of sheet metal foils in micro-bending, Comput. Mater. Sci. 50 (2011) 2604–2614.

[2] J.-T. Gau, C. Principe, M. Yu, Springback behavior of brass in micro sheet forming, J. Mater. Process. Technol. 191 (2007) 7–10.

[3] H. Lu, D. Wei, Z. Jiang, X. Liu, K. Manabe, Modelling of size effects in microforming process with consideration of grained heterogeneity, Comput. Mater. Sci. 77 (2013) 44–52.

[4] W. Chan, M. Fu, J. Lu, J. Liu, Modeling of grain size effect on micro deformation behavior in micro-forming of pure copper, Mater. Sci. Eng. A 527 (2010) 6638–6648.

[5] Z. Fan, Y. Wu, X. Zhao, Y. Lu, Simulation of polycrystalline structure with Voronoi diagram in Laguerre geometry based on random closed packing of spheres, Comput. Mater. Sci. 29 (2004) 301–308.

[6] I. Simonovski, L. Cizelj, Automatic parallel generation of finite element meshes for complex spatial structures, Comput. Mater. Sci. 50 (2011) 1606–1618.

[7] K. Li, W. Carden, R. Wagoner, Simulation of springback, Int. J. Mech. Sci. 44 (2002) 103–122.

Chapter 13

Simulation of Micro Compression

13.1 INTRODUCTION

Many explorations on the issue of size effects have been conducted using both experimental and numerical methods, and the corresponding constitutive models have been developed as well. Hall [1] and Petch [2] studied the effects of grain size on the yield stress and found a linear relationship between the yield stress and the reciprocal of the square root of grain size. Miyazaki et al. [3] investigated the effects of specimen size on the flow stress of polycrystalline Cu-Al alloy and an affecting zone model was proposed to explain why the flow stress decreased with the dimensional reduction of specimen. Leu [4] established a flow stress model as a function of the ratio of the sheet thickness to grain size based on the affecting zone model, the pile-up theory of dislocation and Hall–Petch relationship. Geiger et al. [5] proposed the surface layer model, which divided the specimen into inner and surface layer portions, to explain the reduction of flow stress in metal microforming process. A systematic influence of the sample size was proposed by Kals et al. [6] by defining a quantitative value for the share of surface grains. Lai et al. [7] developed a mixed model by combining the modified Hall–Petch relation and surface layer model. Based on the composite and surface layer models, Liu et al. [8] developed a constitutive model considering grain boundary strain strengthening.

Grain heterogeneity was taken into account by Chan et al. [9] to investigate the decrease of flow stress and scatter of grain behavior due to different grain sizes, shapes, and orientations. Grain heterogeneity was identified using nanoindentation tests to simulate the micro cross wedge rolling process and an inhomogeneity coefficient was introduced [10]. Voronoi tessellation was introduced to generate polycrystalline aggregate using probability theory. This method subdivides a space tessellation into Voronoi polyhedral, which is similar to morphogenetic process of nucleation and growth from random seeds [11,12]. The concept of Voronoi polygon has been used extensively in material science, especially for modeling of random microstructure of grain aggregates [13,14]. Zhang and Tong [15] presented 3D numerical modeling as a general investigation tool that featured an aggregate

Microforming Technology. DOI: http://dx.doi.org/10.1016/B978-0-12-811212-0.00013-3

computation including element type and geometrical representation of grain boundary. Based on Voronoi tessellation, lognormal and gamma distribution were used to describe the distribution of grain size in polycrystalline by Zhu et al. [16] and Dalla et al. [17].

The main objective of this chapter is to introduce a constitutive material model with consideration of intragranular heterogeneities that involve the difference between the grain interior and boundaries and the scatter of deformation behavior among grains in micro compression process. Temperature is considered to model the warm deformation behavior. 3D aggregate of polycrystalline is represented by a Voronoi tessellation. Applying the novel material model and 3D full-scale polycrystalline structure, micro compression tests of pure copper are numerically simulated by finite element method (FEM). In order to verify the accuracy of the developed material model, physical experiments are conducted to compare the simulation results with experimental ones. Scatter of grain heterogeneity is analyzed in terms of flow stress and surface asperity of the deformed parts.

13.2 SIMULATION PROCEDURE

13.2.1 Material Constitutive Model

13.2.1.1 Composite Model

The methodology for studying grain size effects on flow stress is based on the composite model developed by Kocks [18] and Meyers and Ashworth [19]. This model has been proven to be a valid and effective approach to describe the relationship between material yield stress and grain size. Composite model supposes that the deformation polycrystalline is composed of a hard grain boundary region and a soft grain interior region, as shown in Fig. 13.1.

Based on composite model, the flow stress of polycrystalline aggregate can be calculated as follows:

$$\sigma_s = (1-f)\sigma_i + f\sigma_b \tag{13.1}$$

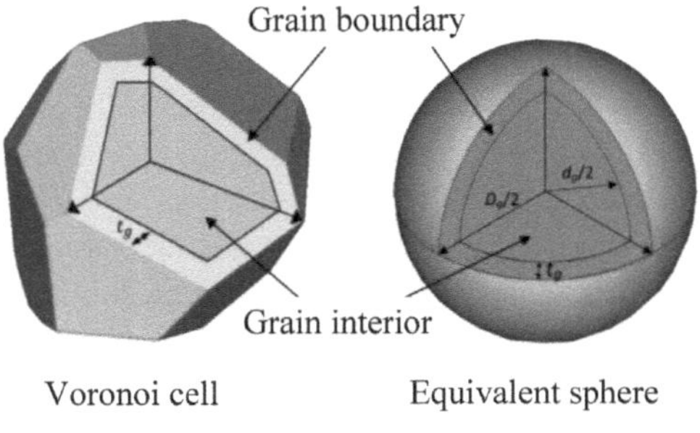

FIGURE 13.1 Grain interior and grain boundary model.

where σ_s is the flow stress of polycrystalline aggregate, f is the volume fraction of grain boundary, and σ_i and σ_b are the flow stresses of grain interior and boundary, respectively.

Assuming that the grain size is D_g^n and the thickness of grain boundary layer is t_g, then the relationship between the thickness of grain boundary and the grain size can be described by the following equation [20]:

$$t_g = kD_g^n (0 < n < 1) \tag{13.2}$$

where k and n are regarded as constants for a specific material. According to the previous research [21], k and n are determined to be 0.133 and 0.7, respectively, when the material is copper.

In this chapter, an equivalent sphere (Fig. 13.1) is proposed to take the place of Voronoi cell when grain size is calculated. The volume of sphere V_{es} is equal to that of the Voronoi cell V_{vc} and the grain size is equal to the diameter of the equivalent sphere. Therefore,

$$V_{es} = V_{vc} = \frac{4}{3}\pi \left(\frac{D_g}{2}\right)^3 \tag{13.3}$$

3D simplex integration method is used to calculate the volume of each Voronoi cell. Assuming there are m grains involved in the inner portion of specimen, the volume fraction of grain i ($i = 1, 2, \ldots, m$) in the inner portion of specimen is given in the following equation:

$$V_c(i) = \frac{4\left(\frac{D_g(i)}{2}\right)^3}{3(r_s - D_{gm})^2 (h_s - 2D_{gm})} \tag{13.4}$$

where D_{gm} is the average grain size, $D_g(i) = \left(6v_{vc}(i)/\pi\right)^{1/3}$, and r_s and h_s are the radius and height of the specimen, respectively, as shown in Fig. 13.2.

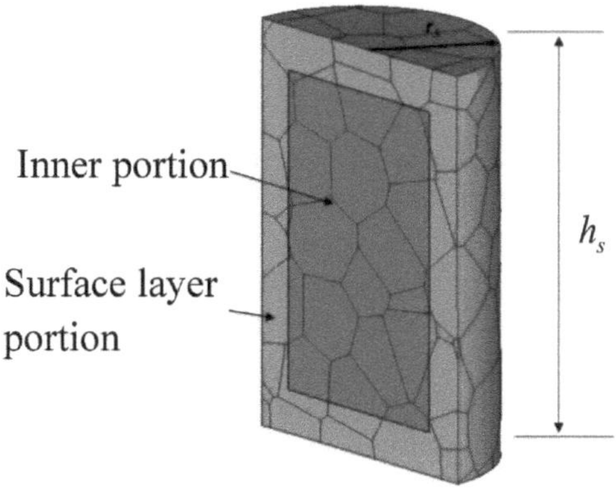

FIGURE 13.2 Surface layer and inner portions of a cylindrical specimen.

Substituting Eq. (13.3) into Eq. (13.4), the volume fraction of the boundary of grain i can be expressed as:

$$f_c(i) = 1 - \frac{V_{\text{in}}(i)}{V_{\text{vc}}(i)} = \left(\frac{2t_g(i)}{D_g(i)} - 1\right)^3 + 1 = \left(2kD_g^{n-1}(i) - 1\right)^3 + 1 \tag{13.5}$$

where $f_c(i)$ is the volume fraction of the boundary of grain and $V_{\text{in}}(i)$ is the volume of grain interior. The total volume fraction of the grain boundaries is obtained by:

$$f = \sum_{i=1}^{m} v_c(i) \times f_c(i) = \sum_{i=1}^{m} \frac{4\left(\frac{D_g(i)}{2}\right)^3 \left(2kD_g^{n-1}(i)^3 + 1\right)}{3\left(r_s - D_{\text{gm}}\right)^2 \left(h_s - 2D_{\text{gm}}\right)} \tag{13.6}$$

13.2.1.2 Constitutive Model Considering Temperature

The plastic deformation of metals can be regarded as the motion and accumulation process of dislocations, when these dislocations have to overcome both short- and long-range obstacles. In order to discern this nonuniform deformation, the theory of dislocation was proposed by Ashby [22]. It refers to geometrically necessary dislocation (GND) and statistically stored dislocation (SSD). Both the evolutions of GNDs and SSDs contribute to the short-range effect which accordingly results in material hardening. Based on the investigations on face-centered cubic (FCC) metals [23,24], the evolution of SSDs during crystallographic slip will increase the influence of short-range obstacles that can be overcome by thermal activation. Meanwhile, GNDs contribute to this short-range effect as well. In contrast, it is regarded that GNDs are responsible for long-range obstacles that are essentially independent on temperature (namely, athermal). According to the temperature effect on the evolution of dislocations, the universal flow stress σ_u that is defined by the resistance to dislocation motion can be expressed as:

$$\sigma_u = \sigma_{\text{ath}} + \sigma_{\text{th}} \tag{13.7}$$

where σ_{ath} is the athermal component of flow stress that reflects mainly the long-range obstacles and σ_{th} is the thermal component of flow stress that represents mainly the short-range barriers. The short-range barriers include forest dislocations, point defects, alloy elements, solution atoms, impurities and deposits, and the long-range obstacles may include grain boundaries, far-field dislocation forests, and other microstructural elements with far-field influence.

The Hall–Petch relationship between the reciprocal of activation volume and grain size has been employed to describe the size effects. Based on the intrinsic characteristics of dislocation and microstructure, Zerilli and Armstrong [25] made the athermal stress to Hall–Petch relationship for FCC metal. Jiang et al. [26] developed the Hall–Petch relationship associated with the dislocation pile-up theory and flow stress-grain size relationship as:

$$\sigma = \sigma_0^2(\varepsilon) + k^2(\varepsilon, d)d^{-\frac{1}{2}} = \sigma_0 + 1.4N\left[\frac{c_2\varepsilon^{m-n/2}}{\boldsymbol{b}(c_1)^{\frac{1}{2}}}\right]d^{-\frac{1}{2}} + N(C_1\varepsilon^n)^{\frac{1}{2}} \tag{13.8}$$

where σ_0 is the stress due to initial defects, k is the material parameter influenced by grain size d and stain ε, N, c_1, c_2, m, and n are material constants with consideration of dislocation density, and $\boldsymbol{b}$ is Burgers vector. In Eq. (13.8), the term $N(C_1\varepsilon^n)^{1/2}$ is the component reflecting short-range effect, so it should be ignored in order to describe the athermal flow stress:

$$\sigma_{\text{ath}} = \sigma_0 + 1.4N\left[\frac{c_2\varepsilon^{m-n/2}}{\boldsymbol{b}(c_1)^{\frac{1}{2}}}\right]d^{-\frac{1}{2}} \tag{13.9}$$

Klepaczko and Chiem [27] proposed a general balance equation to describe the strain-induced evolution of material microstructure based on the accumulation and recovery of dislocations. Gao and Zhang [23] further developed this general law in association with the effect of extreme strain rate and proposed a unified constitutive model for thermal stress as follows:

$$\sigma_{\text{th}} = c_3\sqrt{1-\exp(-k_0\varepsilon)}\left\{1-\left[-c_4T\ln\left(\frac{\dot{\varepsilon}}{\dot{\varepsilon_0}}\right)\right]^{\frac{1}{q}}\right\}^{\frac{1}{p}} \tag{13.10}$$

where c_3 is the reference thermal stress, c_4, q, and p are parameters relevant to material thermal property, k_0 is the annihilation factor at $T=0$, and $\dot{\varepsilon_0}$ is the reference strain rate.

Then, by combining Eqs. (13.9) and (13.10), the following constitutive equation for FCC metals can be obtained:

$$\sigma_u = \sigma_0 + 1.4N\left[\frac{c_2\varepsilon^{m-n/2}}{\boldsymbol{b}(c_1)^{\frac{1}{2}}}\right]d^{-\frac{1}{2}} + c_3\sqrt{1-\exp(-k_0\varepsilon)}\left\{1-\left[-c_4T\ln\left(\frac{\dot{\varepsilon}}{\dot{\varepsilon_0}}\right)\right]^{\frac{1}{q}}\right\}^{\frac{1}{p}} \tag{13.11}$$

During deformation, a dislocation density of GNDs accumulates and makes the grain boundary strengthening, but the grains located at the free surface of micro specimen have less hardening effect than that of inner grains [28]. Therefore, the portion of flow stress attributed by dislocation density of GNDs in surface layer can be ignored, and the flow stress of surface layer σ_{surf} can be expressed as:

$$\sigma_{\text{surf}} = \sigma_0 + c_3 \cdot \sqrt{1-\exp(-k_0\varepsilon)}\left\{1-\left[-c_4T\ln\left(\frac{\dot{\varepsilon}}{\dot{\varepsilon_0}}\right)\right]^{1/q}\right\}^{1/p} \tag{13.12}$$

With respect to the grains allocated at the inner region of specimen, the flow stress of inner portion σ_{inner} can be described by that of polycrystalline aggregate, i.e., $\sigma_{\text{inner}} = \sigma_u$. The grains allocated at the free surface have less hardening effect than that of inner grains. Therefore, the grain boundary strengthening effect in surface layer can be ignored. In micro deformation process, the surface layer model is proposed to account for the decrease of flow stress due to miniaturization. The specimen is subdivided into two portions: surface layer and inner part, as shown in Fig. 13.2. As a result, the geometrical size effects of specimen are taken into account in the composite model. The material athermal parameters (σ_0, N, c_1, c_2, m, and n) are taken the same values as those given in the corresponding references [23,24]. On the other hand, the material parameters of the thermal portion (c_3, c_4, k_0, p, q, and $\dot{\varepsilon}_0$) are selected with the optimal values given in the corresponding references [23,29] to ensure the data validity. The strain and strain rate are determined according to the physical micro compression tests. All constants required in the proposed constitutive relationship are listed in Table 13.1.

13.2.1.3 Surface Layer Model

Surface layer model described in Chapter 6, Simulation Models in Microforming, is utilized to explain the geometrical size effects on the decrease of flow stress during miniaturization. The specimen for micro compression is subdivided into two portions: surface layer and inner part, as shown in Fig. 13.2.

The flow stress of the material is composited by that of inner portion and surface layer of specimen as follows:

$$\sigma = \gamma\sigma_{\text{inner}} + (1 - \gamma)\sigma_{\text{surf}} \tag{13.13}$$

where σ is the flow stress of material, σ_{inner} and σ_{surf} are the flow stress of the grain located at inner and surface region, respectively, and γ is the fraction of inner portion which can be expressed as:

$$\gamma = \frac{\left(r_s - D_{\text{gm}}\right)^2 (h_s - 2D_{\text{gm}})}{r_s^2 h_s} \tag{13.14}$$

The flow stress of inner portion can be described by that of polycrystalline aggregate, and the grains located at the free surface have less hardening effect than that of inner grains. Therefore, the grain boundary strengthening effect in surface layer is ignored. The flow stress of surface layer portion is equal to that of grain interior [8]. As a result, the geometrical size effects are taken into account in the composite model. The flow stress in microforming can be expressed as:

$$\sigma = (1 - \gamma f)\sigma_i + \gamma f \sigma_b \tag{13.15}$$

TABLE 13.1 Constitutive Parameters for Pure Copper

Parameters	σ_0 (MPa)	N (MPa cm)	c_1 (cm^{-2})	c_2	c_3 (MPa)	c_4 (K^{-1})	b (cm)	M	N	k_0	p	q	$\dot{\varepsilon}_0$ (s^{-1})	$\dot{\varepsilon}$
Optimized results	95	7.4×10^{-4}	6.3×10^{11}	0.95	808	1.0×10^{-4}	2.56×10^{-8}	0.5	1.2	10	0.45	1.0	1.76×10^{8}	0.065

A parameter F is introduced to express the following relationship:

$$F = \gamma f \frac{4}{3r_s^2 h_s} \sum_{i=1}^{m} \left(\frac{D_g(i)}{2}\right)^3 \left[\left(2kD_g^{n-1}(i) - 1\right)^3 + 1\right] \tag{13.16}$$

Then, Eq. (13.15) can be rewritten as:

$$\sigma = (1 - F)\sigma_i + F\sigma_b \tag{13.17}$$

where F is defined as a size-effect factor and can be figured out when the dimension of micro part and its grain size are specified. Therefore, in the developed constitutive model, the flow stresses of grain interior and grain boundary are unknowns. No less than two stress–strain curves of deformed material with different grain sizes are required to determine σ_i and σ_b. The values of σ_i and σ_b at a certain strain can be obtained. The flow stress curves of grain interior and boundary are finally determined. When the dimension of cylindrical sample is Φ 0.8 × 1.2 mm^2, a series of flow stress curves with different grain sizes are obtained based on Eq. (13.17), as shown in Fig. 13.3.

13.2.1.4 Grain Heterogeneity

When the specimen size is scaled down to microscale, deformation behavior will be characterized by the deformation regions with only a few grains. The distributions of different grains with different orientations, sizes, and shapes have a significant influence on the size effects that cause inhomogeneous deformation behavior and scattering of material properties.

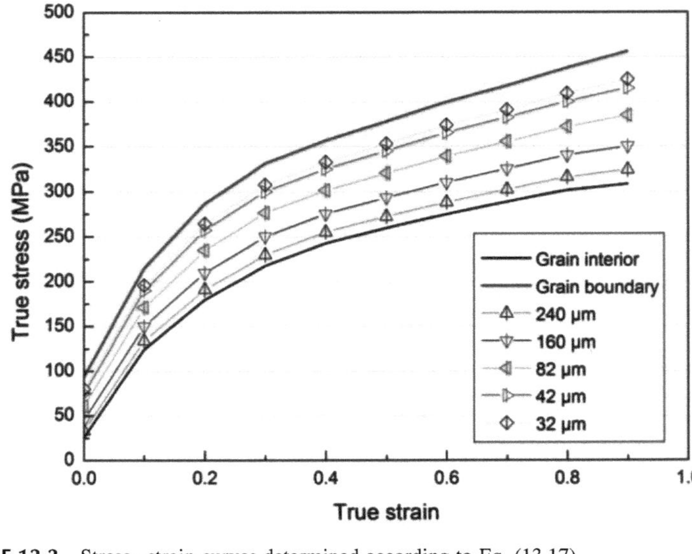

FIGURE 13.3 Stress–strain curves determined according to Eq. (13.17).

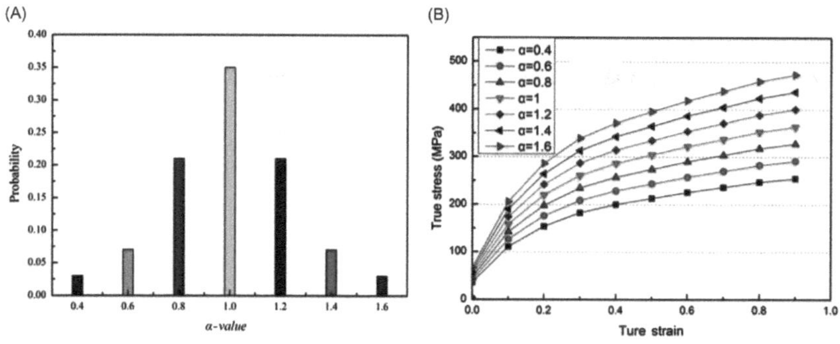

FIGURE 13.4 Stress–strain curves based on scatter of grain's deformation behavior: (A) statistical distribution of α-value and (B) flow stress of inhomogeneous material.

In this chapter, nanoindentation is adopted to identify the grain heterogeneity. Hardness is the ability of a material to resist plastic deformation via scratch or indentation. Hardness of grains obtained from nanoindentation test is an effective parameter to estimate the deformability of individual grain inside the micro part. According to the grain hardness, grain heterogeneity can be figured out. Statistical distribution of hardness is divided into seven classes (Fig. 13.4A). The ratio of the hardness at each class divided by the average hardness is defined as an inhomogeneity coefficient α to make the stress–strain curves vary following the statistical distribution. α is relevant to the grain size of workpiece and temperature of compression process, and the α values represent the different deformabilities of grains, that is the ability of grains to deform under a given level of applied stress. In order to have a deep understanding on the deformation behavior during micro compression, grains are meshed and 3D fully aggregates are computed with FEM. The material constitutive model illustrated in Fig. 13.4B is attributed to the specific region, i.e., individual grain, randomly.

13.2.2 Finite Element Model Setup

The deformed part is represented based on a space tessellation into 3D Voronoi diagram which can describe grain boundary and generation process of grain vividly. As demonstrated in Chapter 6, Simulation Models in Microforming, Voronoi diagram can be defined as n distinct regions V_i based on an open set Ω and n different seeds z_i $(i = 0, 1,\ldots, n-1)$ as:

$$V_i = \{w \in \Omega | d(w, z_i) < d(w, z_j), j = 0, 1, \ldots, n-1, j \neq i\} \tag{13.18}$$

where d is a function of distance. Voronoi polyhedral are the influence zones of these seeds that are their mass center.

A 3D domain is decomposed into a certain number of subdomains. Then, the seeds are spread into these subdomains. Assuming the center of a cube is

(x, y, z), the seed will lie $\psi\delta a/2$ away from the central point. Exactly, the coordinate of the seed position should be $(x + \psi\delta a/2, y + \psi\delta a/2, z + \psi\delta a/2)$. δ is the maximum distance between the seed and the cubic center and ψ is a random number between -1 and 1. Both δ and ψ are so-called shape factors which will determine the shapes of Voronoi cells. A critical distance d between different seeds is defined to prevent each seed from being excessively close to one another. After that, 3D Voronoi polyhedra are generated by taking these seeds as the generating points. The grain radius D_g is defined as the cubic root of the grain volume and the average grain radius D_{gm} is defined as the cubic root of average grain volume within the 3D grain.

During meshing process, space is discretized and described with a 3D digital image composed of voxels (pixels in 3D) that can be labeled with the number of grains that they belong to. According to the characteristics of Voronoi diagram, a distance function is introduced and a segmentation algorithm associates each voxel to the nearest nucleus. Fig. 13.5A gives an

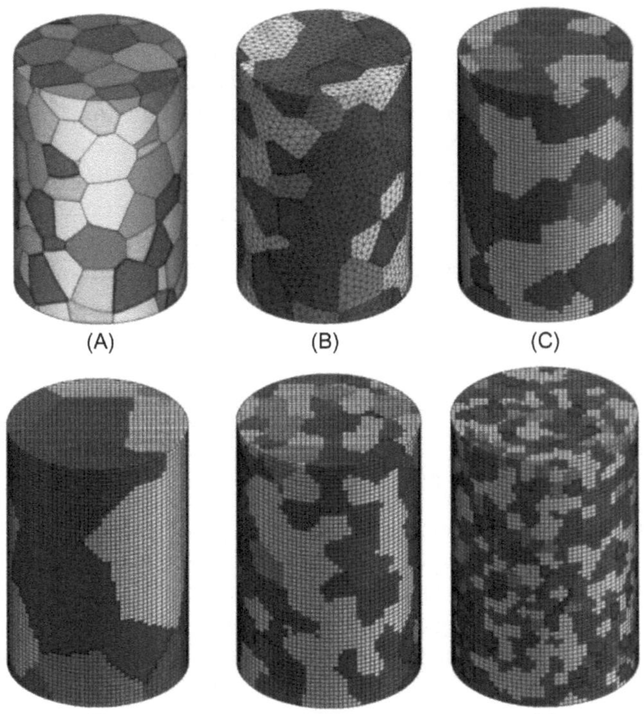

FIGURE 13.5 Aggregate of polycrystalline: (A) tessellation in 140 Voronoi polyhedral, (B) freely meshed model with 140 subdomains, and (C–F) microstructure mapping on a 1313 × 60 cubic element mesh and domain with subdomain numbers 140, 26, 385, and 2520, respectively.

example with an aggregate composed of 140 grains. This image can be discretized into FE using a regular mesh or free mesh as shown in Fig. 13.5B and C, respectively. Fig. 13.5B shows the FE model meshed freely with tetrahedral elements and Fig. 13.5C–F illustrates the aggregate of grains discretized into regular cubic elements. In Fig. 13.5C–F, the samples containing respectively 140, 26, 385, and 2520 grains are discretized into 1313×60 elements.

Mechanical fields can be quite sensitive to the element type and especially to the geometrical description of grain boundaries and mesh type. Although free meshing can describe accurately at the vicinity of the grain boundaries where large stress and strain gradient locate, it may generate abnormal elements and increase computational errors. In addition, tetrahedral elements are not competent to simulate plastic deformation because of their rigidity. In order to eliminate the effects of meshing, the specimen is meshed with regular cubic elements as an alternative. However, the regular meshing cannot describe grain boundary accurately even using large number of elements. Therefore, suitable meshing density should be specified to control the numerical simulation cost and geometrical errors of grain boundaries synchronously.

The tools are assured as a rigid body, which move at a velocity of 0.1 mm/s, and the coefficient of friction (COF) between tools and specimen is set to 0.11. When micro compression is performed at elevated temperatures, the temperature is assumed to be constant during deformation, and the COF is thought not to be affected by the temperature. Then, models are created with different grain sizes while the specimen size is kept a constant. The modeling of micro compression at elevated temperatures is developed on the basis of the following assumptions:

1. grains involved in the workpiece can be approximately described by Voronoi tessellation;
2. the micro workpiece is heated up to the expected temperature within a short time;
3. the heat generated by plastic deformation is ignored within the workpiece;
4. a constant temperature within the workpiece during compression;
5. a constant COF on the material–tool contact surface.

Fig. 13.6 illustrates the statistic distribution of inhomogeneity coefficient at different temperatures. Multiplication of the stress and the inhomogeneity coefficients can be used to obtain seven categories of stress–strain curves using the proposed constitutive model following this statistical distribution, which introduces grain heterogeneity into the constitutive relationship. These inhomogeneous material models are applied in FE simulation to study the deformation behavior at elevated temperatures.

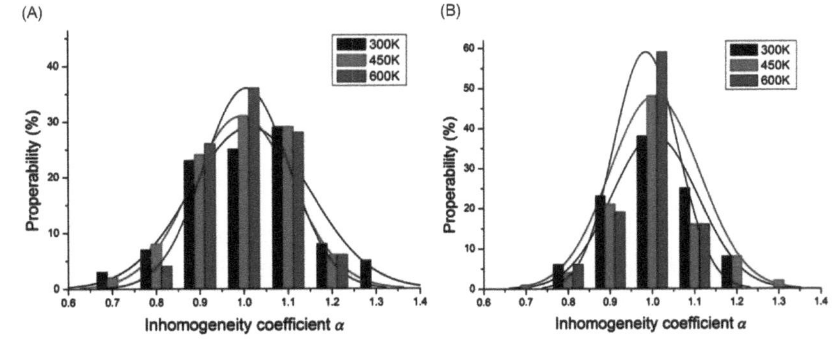

FIGURE 13.6 Statistic distribution of inhomogeneity coefficient influenced by grain size and temperature: (A) grain size of 200 μm and (B) grain size of 20 μm.

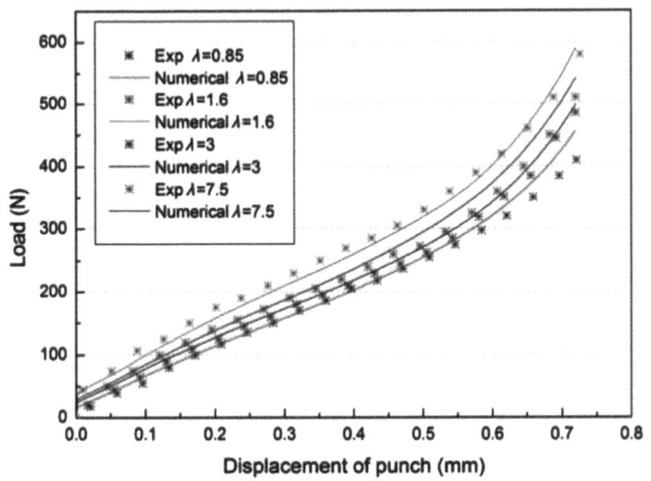

FIGURE 13.7 Punch load versus displacement.

13.3 SIMULATION RESULTS

13.3.1 Flow Stress

Fig. 13.7 compares the simulated and experimental results. In the figure, λ is the characteristic ratio of the specimen size to grain size, which is defined as:

$$\lambda = \frac{\text{Specimen radius}}{\text{Grain size}} \tag{13.19}$$

It can be seen from Fig. 13.7 that the simulation results agree well with the experimental ones. The punch load is found to increase gradually when λ increases from 0.85 to 7.5, which agrees with the feature that the increase of

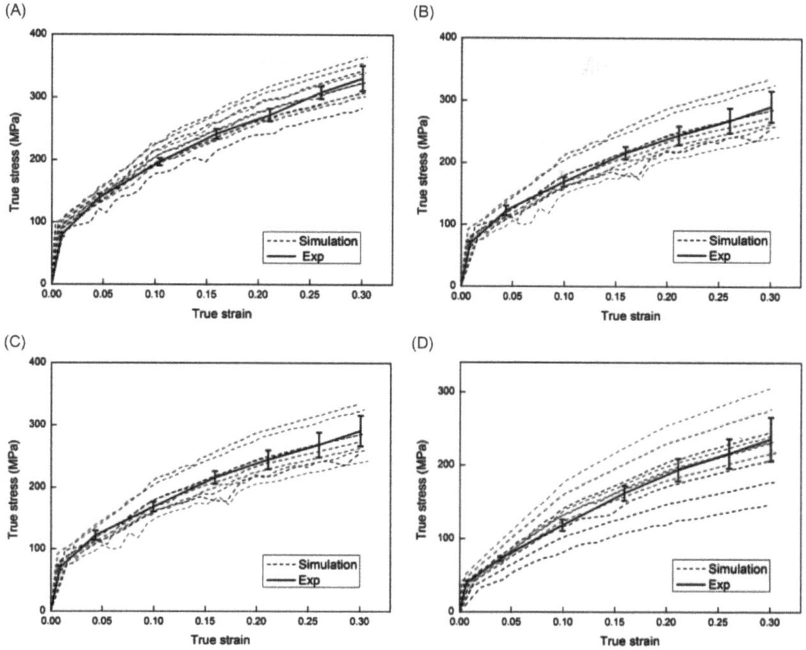

FIGURE 13.8 Comparison of stress–strain curves obtained from experiments and simulation: (A) $\lambda = 7.5$, (B) $\lambda = 3$, (C) $\lambda = 1.6$, and (D) $\lambda = 0.85$.

λ value will cause an increase of flow stress in a certain range of the ratio of the specimen size to grain size [30].

Fig. 13.8 shows a comparison of stress–strain curves obtained from experiments and simulation. It can be seen that both the experimental and simulated flow stresses decrease significantly as λ decreases. The four experimental stress values with corresponding strain of 0.3 are 330.9, 290.4, 265.7, and 235.7 MPa, for λ values of 7.5, 3, 1.6, and 0.85, respectively. The simulated flow stress curves obtained from the models distribute dispersedly due to grain heterogeneity and the scatter increases with an increase of grain size because an individual grain plays more significant role on material behavior when its amount decreases within the workpiece. According to Eq. (13.17), the effects of grain interior and boundary are integrated as a single body, so the volume fractions of grain boundary and surface layer grains vary while λ changes from 0.85 to 7.5. Consequently, grain boundary strengthening effect is enhanced and softening effect of surface layer grains is weakened due to the increase of λ, which accounts for the change of the mean flow stress obtained in the FE simulation and experimental results. Fig. 13.8 also shows that the error of flow stress becomes smaller with an increase of λ. The scatter of simulated flow stress is more significant than that of micro compression experiments and the scatter of the simulation

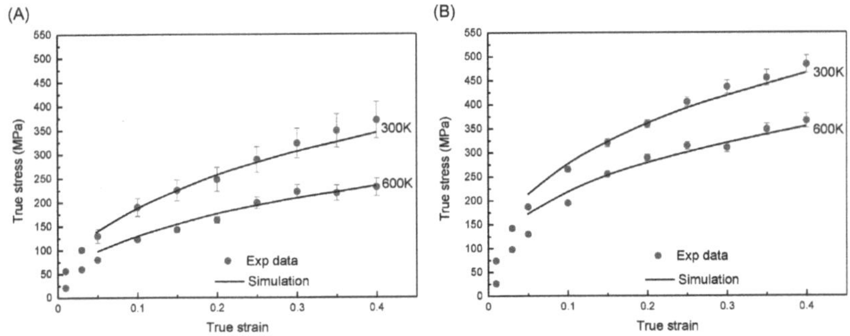

FIGURE 13.9 Comparison of stress–strain curves obtained from experiments and simulation at elevated temperatures: (A) grain size of 200 μm and (B) grain size of 20 μm.

increases with λ from 7.5 to 0.85. It can be seen that the scatter becomes less significant with an increase of λ, which means that the effect of individual grain on the deformation decreases with an increase of λ.

FE simulations are implemented at the temperatures of 300 and 600 K, respectively. The influences made by grain size and temperature on the flow behavior are found to be even more pronounced in micro compression, because the so-called size effects cause inhomogeneity of material flow behavior and the dimensional inaccuracy of the micro parts. Fig. 13.9 illustrates the simulated and experimental results for the pure copper specimens with different grain sizes of 200 and 20 μm at temperatures of 300 and 600 K, respectively. It can be seen that the simulated results have a good agreement with the experimental ones, indicating that the change in flow stress and work hardening behavior with variation of temperature and grain size can be well captured by using the proposed simulation method.

13.3.2 Surface Asperity

The scatter of an individual grain's plastic deformability affects the specimen's profile as well. Consequently, it supplies another method to penetrate size effects on the deformation behavior and the quality of micro products. Side surface asperity of specimen after compression has been measured in simulations. For comparison, relevant experimental work has been conducted.

When only a few grains are involved in deformation, deformation incompatibility between grains cannot be accommodated by enough strain gradients of neighboring grains. Such inhomogeneous deformation will influence the surface asperity of deformed micro parts. Fig. 13.10 shows the profiles of compressed specimens with homogenous model, and the models with λ ranging from 7.5 to 0.85. Based on the simulated contours of side surface, size effects on the increase of inhomogeneous deformation behavior with the decrease of λ can be obtained by simulation. It reveals that the side surface

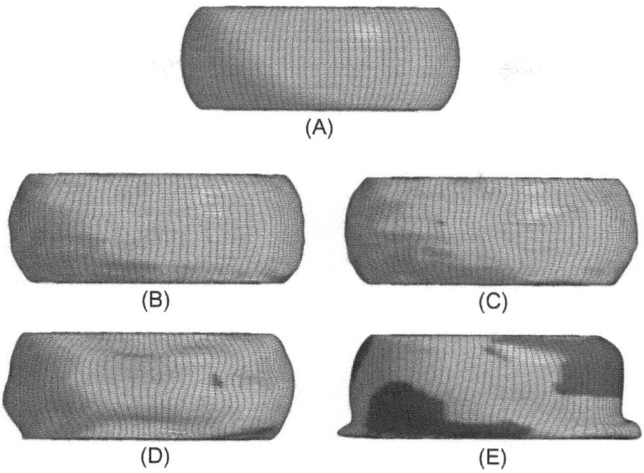

FIGURE 13.10 Profiles of specimens after compression: (A) homogenous model, (B) $\lambda = 7.5$, (C) $\lambda = 3$, (D) $\lambda = 1.6$, and (E) $\lambda = 0.85$.

becomes rougher with the decrease of λ when the specimen's size keeps constant. The half surface profile of the compressed samples in experiment is obtained by a 3D digital microscope, as shown in Fig. 13.11. It can be seen that the side surfaces become rougher when λ decreases, which has a good agreement with the simulation results.

In order to measure the surface asperity quantitatively, 12 generatrices uniformly spreading on the side surface of cylindrical specimen are determined, as illustrated in Fig. 13.12A. After deformation, surface becomes rougher comparing with the original sample surface. Assuming the deformation behavior of grains inside the sample is homogenous, side surface of the compressed sample is perfectly smooth. One generatrice of this assumed compressed sample is used as a reference. The absolute difference in radial direction between point P_1 the selected generatrice and point P_2 on the reference generatrice at the same height (Fig. 13.12B) is figured out. The mean value of the absolute differences obtained from all of the 12 generatrices is defined as H_{sa} and has been used to assess the surface roughening.

The experimental and simulated results of H_{sa} are compared, as shown in Fig. 13.13. It can be seen that the surface is rougher when λ value decreases, which means the effect of grain heterogeneity on surface roughness is significant when there are fewer grains inside specimen. When λ increases from 0.85 to 6.8, H_{sa} falls down from 0.025 to 0.004 dramatically. However, the variation of H_{sa} is not so sensitive to the change of λ when λ exceeds 6.8, which means the influence of individual grain on the surface asperity get into a decline if a certain grain size is reached.

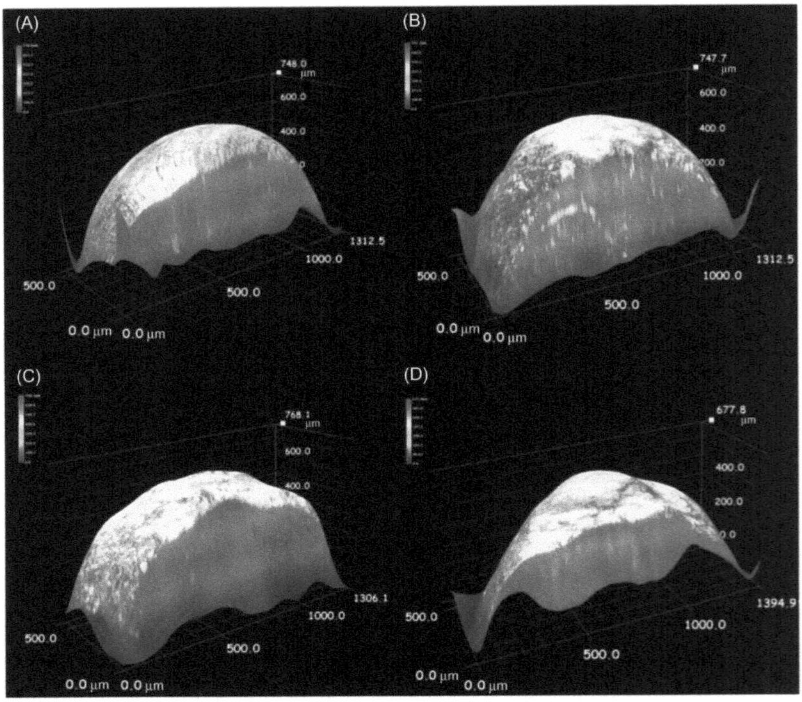

FIGURE 13.11 Profiles of specimens with various grain sizes in compression test: (A) $\lambda = 7.5$, (B) $\lambda = 3$, (C) $\lambda = 1.6$, and (D) $\lambda = 0.85$.

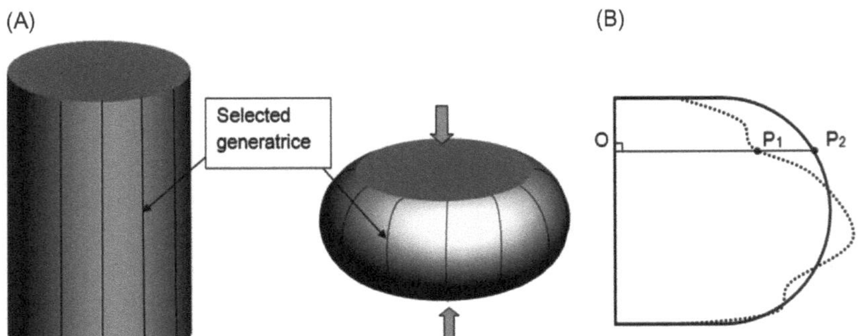

FIGURE 13.12 Illustration of the definition of H_{sa}: (A) generatrices for measuring and (B) radial difference between points on selected and reference generatrices.

It can be summarized that the developed FE model is effective for the simulation of micro compression processes at both room and elevated temperatures with high accuracy. The simulated results provide an important reference for the design of practical micro compression process, which is

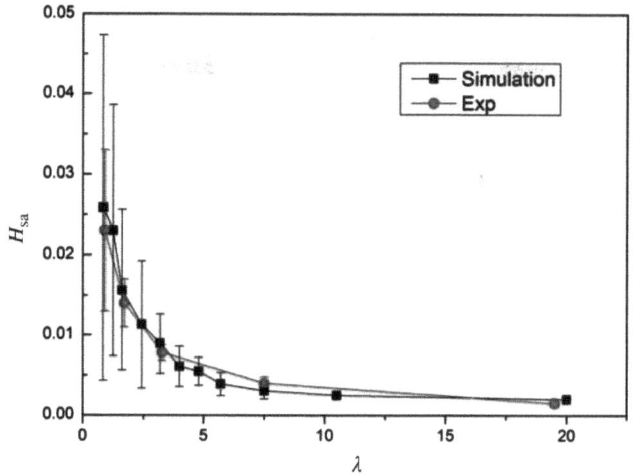

FIGURE 13.13 Influence of λ on H_{sa}.

essential and significant for the implementation of micro compression in practice with the purposes of reducing the inhomogeneous flow in forming and improving the quality of the formed parts through optimization of processing parameters.

REFERENCES

[1] E.O. Hall, The deformation and ageing of mild steel: III discussion of results, Proc. Phys. Soc. B 64 (1951) 747–753.

[2] N.J. Petch, The cleavage strength of polycrystals, Iron Steel Inst. 174 (1953) 25–28.

[3] S. Miyazaki, K. Shibata, H. Fujita, Effect of specimen thickness on mechanical properties of polycrystalline aggregates with various grain sizes, Acta Metall. 27 (1979) 855–862.

[4] D.K. Leu, Modeling of size effect on tensile flow stress of sheet metal in microforming, J. Manuf. Sci. Eng. 131 (2009) 011002-1–011002-8.

[5] M. Geiger, M. Kleiner, R. Eckstein, N. Tiesler, U. Engel, Microforming, CIRP Ann. Manuf. Technol. 50 (2001) 445–462.

[6] R. Kals, H.J. Pucher, F. Vollertsen, Effects of specimen size and geometry in metal forming, In: M.S.J. Hashmi (Ed.), Proceedings of the 2nd International Conference on Advances in Materials and Processing Technologies, Dublin, 1995, pp. 1288–1297.

[7] X. Lai, L. Peng, P. Hu, S. Lan, J. Ni, Material behavior modelling in micro/meso-scale forming process with considering size/scale effects, Comput. Mater. Sci. 43 (2008) 1003–1009.

[8] J.G. Liu, M.W. Fu, W.L. Chan, A constitutive model for modeling of the deformation behavior in microforming with a consideration of grain boundary strengthening, Comput. Mater. Sci. 55 (2012) 85–94.

[9] W.L. Chan, M.W. Fu, J. Lu, J.G. Liu, Modeling of grain size on micro deformation behaviour in micro-forming of pure copper, Mater. Sci. Eng. A 527 (2010) 6638–6648.

[10] H.N. Lu, D.B. Wei, Z.Y. Jiang, K. Manabe, T. Furushima, Study on size effect in micro cross wedge rolling, In: Steel Research International, Special Edition: 10th International Conference on Technology of Plasticity (ICTP 2011), 2011, pp. 962–967.

[11] G.F. Voronoi, Nouvelles applications des paramètres continus à la théorie de formes quadratiques, J. Reine Angew. Math. 134 (1908) 198–287.

[12] G.F. Voronoi, Nouvelles applications des paramètres continus à la théorie de formes quadratiques, J. Reine Angew. Math. 133 (1907) 97–178.

[13] O. Diard, S. Leclercq, G. Rousselier, G. Cailetaud, Evaluation of finite element based analysis of 3D multicrystalline aggregates plasticity: application to crystal plasticity model identification and the study of stress and strain fields near grain boundaries, Int. J. Plast. 21 (2005) 691–722.

[14] A.G. Crocker, P.E.J. Flewitt, G.E. Smith, Computational modelling of fracture in polycrystalline materials, Int. Mater. Rev. 50 (2005) 99–124.

[15] N. Zhang, W. Tong, An experimental study on grain deformation and interactions in an Al—0.5%Mg multicrystal, Int. J. Plast. 20 (2004) 523–542.

[16] B. Zhu, R.J. Asaro, P. Krysl, K. Zhang, J.R. Weertman, Effects of grain size distribution on the mechanical response of nanocrystalline metals: Part II, Acta Mater. 54 (2006) 3307–3320.

[17] F. Dalla Torre, H. Van Swygenhoven, M. Victoria, Nanocrystalline electrodeposited Ni: microstructure and tensile properties, Acta Mater. 50 (2002) 3957–3970.

[18] U.F. Kocks, The relation between polycrystal deformation and single-crystal deformation, Metall. Mater. Trans. 1 (1970) 1121–1143.

[19] M.A. Meyers, E. Ashworth, A model for the effect of grain size on the yield stress of metals, Philos. Mag. A 46 (1982) 737–759.

[20] H.-H. Fu, D.J. Benson, M.A. Meyers, Analytical and computational description of effect of grain size on yield stress of metals, Acta Mater. 49 (2001) 2567–2582.

[21] M. Geiger, S. Geißdörfer, U. Engel, Mesoscopic model: advanced simulation of microforming processes, Prod. Eng. 1 (2007) 79–84.

[22] M.F. Ashby, The deformation of plastically non-homogenous materials, Philos. Mag. 21 (1970) 339–424.

[23] C.Y. Gao, L.C. Zhang, Constitutive modelling of plasticity of FCC metals under extremely high strain rate, Int. J. Plast. 32–33 (2012) 121–133.

[24] N. Hansen, B. Ralph, The strain and grain size dependence of the flow stress of copper, Acta Metall. 30 (1982) 411–417.

[25] F.J. Zerilli, R.W. Armstrong, Dislocation-mechanics-based constitutive model for material dynamics calculation, J. Appl. Phys. 5 (1987) 1816–1825.

[26] Z.H. Jiang, J.S. Liang, B. Baudelet, A dislocation density approximation for the flow stress-grain relation of polycrystals, Acta Metall. Mater. 43 (1995) 3349–3360.

[27] J.R. Klepaczko, C.Y. Chiem, On rate sensitivity of F.C.C. metals instantaneous rate sensitivity and rate sensitivity of strain hardening, J. Mech. Phys. Solids 34 (1986) 29–54.

[28] A. Arsenlis, D.M. Parks, R. Becker, V.V. Bulatov, On the evolution of crystallographic dislocation density in non-homogeneously deforming crystals, J. Mech. Phys. Solids 52 (2004) 1213–1246.

[29] A.M. Lennon, K.T. Ramesh, The influence of crystal structure on dynamic behaviour of materials at high temperature, Int. J. Plast. 20 (2004) 269–290.

[30] S. Mahabunphachai, M. Koç, Investigation of size effects on material behaviour of thin sheet metals using hydraulic bulge testing at micro/meso-scales, Int. J. Mach. Tools Manuf. 48 (2008) 1014–1029.

Part IV

Practice of Microforming

Chapter 14

Practice of Micro Cross Wedge Rolling

14.1 EQUIPMENT AND TOOLS FOR MICRO CROSS WEDGE ROLLING

14.1.1 Design of MCWR Machine

The modern cross wedge rolling (CWR) machine in conventional domain was invented in the Czech Republic for automotive part production [1,2], and the CWR machine is compact and integrated with a material heating and loading system. Two rolls are enclosed in a weld mainframe, where one roll is fixed and the other can be adjusted perpendicular to the axis to control the relative position. The power out of an electric motor is transferred to two rolls by gears. The clutch/brake system in the powertrain allows the forming rolls to stop at any time during production, which is beneficial to tooling setup and workpiece try-out. In Russia and China, configurations of two-roll CWR machines are different from those in the Czech and Japan. Universal joints are installed in the CWR systems to connect the power output from the speed reducer instead of direct gear driving, and the mainframe is made of cast steel. It is more stable and can resist deformation induced in the rolling process.

In the micro-device manufacture, the development of sustainable manufacturing requires the disengagement of the link between the increased micro-device manufacture and the rise in energy consumption [3–5]. It requires the replacement of the old technology and manufacturing processes with new technology that is less energy intensive per unit of manufactured product. Therefore, it is promising to apply CWR in the area of microforming because of high product rate, minimized material consumption, and excellent mechanical properties when producing stepped shafts. There are three common CWR machine configurations used in industry as shown in Fig. 14.1 [1,6].

Flat wedge CWR machines were extensively investigated in Germany in the 1970s [4]. The University of Wollongong (UOW) series flat wedge CWR machines were widely accepted and exported to foreign countries [1]. Flat CWR machines, however, require less sophisticated tooling structures so that tool manufacturing is much easier compared to two-roll machines. On the

Microforming Technology. DOI: http://dx.doi.org/10.1016/B978-0-12-811212-0.00014-5

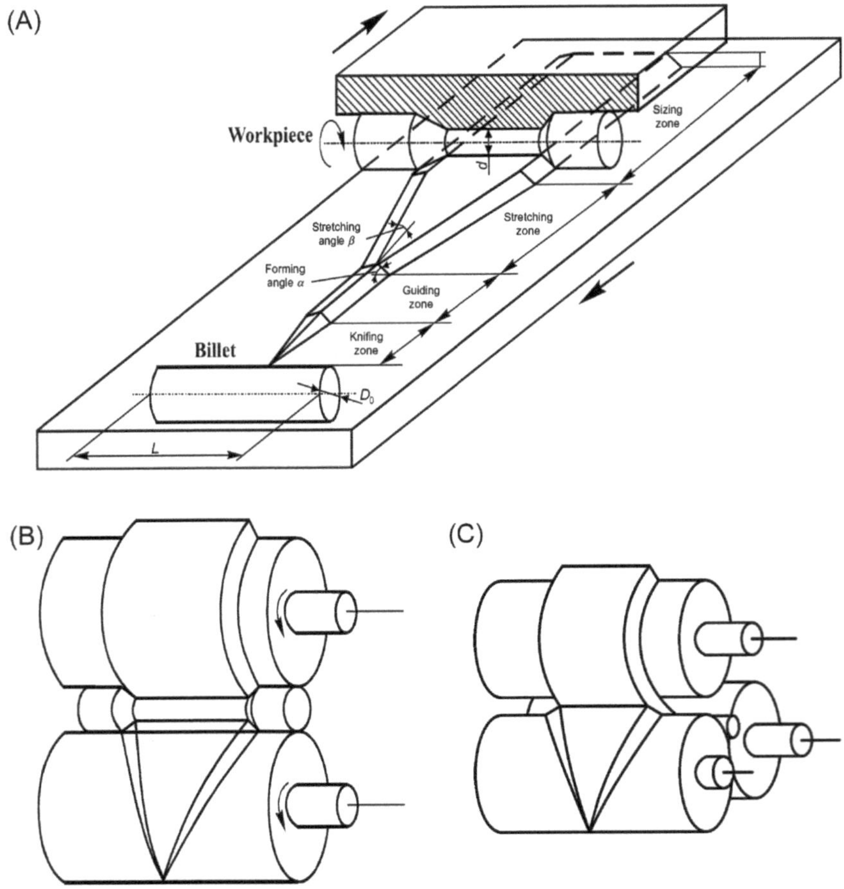

FIGURE 14.1 Three common CWR machine configurations used in industry: (A) Flat CWR, (B) two-roll CWR, and (C) three-roll CWR [1,6].

basis of optimum results of processing parameters, the accuracy of MCWR equipment is the most essential issue when it was designed and manufactured. According to basic requirement of microforming machine, an MCWR machine was designed and manufactured. The forming machine is critical and essential for exploration of MCWR technique. The geometrical, controlling, and positioning accuracy are the main challenges in the microforming process. With respect to these problems, the more accurate components were selected, e.g., the ball screw was used to transmit the driving power. Moreover, the processing optimization and investigation were conducted.

The determination of complete geometrical dimensions is the first step of tooling design and manufacturing. Based on the product information, the suitable MCWR equipment can be selected based on the manufacturing cost, the potential for failure and the system accuracy. Once the preliminary machine design is ascertained, a simplified 3D CAD model is generated for finite element (FE) simulation and optimization. Subsequently, FE simulation is utilized to predict the internal stress and strain concentration and eliminate the surface asperity, which is necessary to establish a set of optimum processing parameters as shown in Chapter 7, Simulation of Micro Cross Wedge Rolling. After FE verification with ANSYS software, the detailed designs are carried out including full parameter calculation, segment design, and user-accessory design. 3D full-scale model of these designed components are created and assembled on a virtual MCWR machine as shown in Fig. 14.2A to simulate the production environment by commercial CAD software-Pro Engineer [1]. Fig. 14.2B shows the factual device of MCWR machine. A simple and effective automatic control system was designed for this equipment. It is composited of positioning, displacement, and velocity controls; meanwhile, the position limitation, and overload protection were also considered and involved in this design as shown in Fig. 14.2B. The pushing force on the vehicle and die-set dislocations were captured and exported by the load cell and displacement sensor, respectively.

A couple of stepper motors is utilized to provide driving force due to the unique characteristics of stepper motor. A stepper motor (Fig. 14.3) is a brushless, synchronous electric motor that converts digital pulses into mechanical shaft rotation. The stepper motor can only take one step at a time and each step is the same size. Since each pulse causes the motor to rotate a precise angle, typically 1.8 degrees, the motor's position can be controlled without any feedback mechanism. As the digital pulses increase in frequency, the step movement changes into continuous rotation, with the speed of rotation directly proportional to the frequency of the pulses.

14.1.2 MCWR Forming Process

The practical design of MCWR machine with flat wedge tools was outlined in order to simplify fabrication of tools and reduce influencing parameters. As shown in Fig. 14.2A, to produce a stepped shaft micro part, a cylindrical wire is placed between two moving and wedge-shaped dies, which move relative to each other. The wire suffers plastic deformation (essentially, localized compression) during its rotation between the moving dies. In flat wedge CWR, there are five stages of the workpiece deformation corresponding to the five zones of the wedge tool geometry as shown in Fig. 14.4: (I) knifing zone, L1; (II) guiding zone, L2; (III) stretching zone, L3; (IV) sizing zone, L4; and (V) unloading zone, L5. The five cross sections from initial cylindrical rod to finished part of workpiece during the forming process have also been schematically shown in the right side of Fig. 14.4.

The cross section of simple cross rolling where the rolling mills are replaced with tow plates is shown in Fig. 14.5. The calculation process of rotation criteria is similar to that of MCWR process with roll. Four forces ($2P$ and $2T$) are symmetrically subjected to the workpiece by two rolling

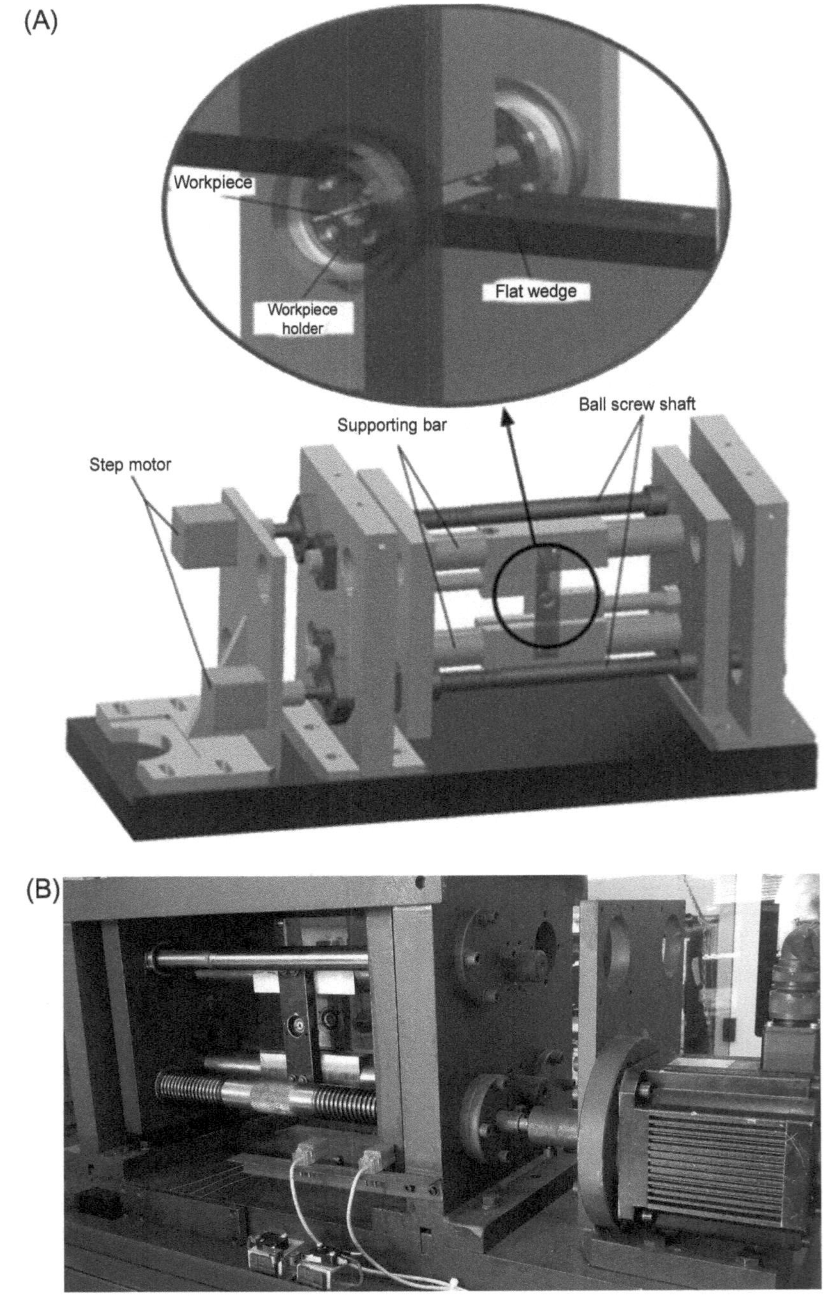

FIGURE 14.2 MCWR prototype machine: (A) three-dimensional model and (B) manufactured MCWR machine.

FIGURE 14.3 Step motor.

plates. P is the normal force, which is perpendicular to the interfaces of tools and workpiece. T is the friction force which is perpendicular to P and proportional to P,

$$T = \mu P \tag{14.1}$$

where μ is the coefficient of friction (COF).

In order to rotate the cylindrical workpiece, the frictional moment M_T (generated by the friction T between the roll and the workpiece) must overcome the normal M_P (generated by the contact force P), which is

$$M_T \geq M_P \tag{14.2}$$

or

$$Ta \geq Pb \tag{14.3}$$

where a and b are the perpendicular distances between two couples of forces T and P, respectively. Substitute for T and P in terms of a and b, we have

$$\mu \geq \frac{b}{a} \tag{14.4}$$

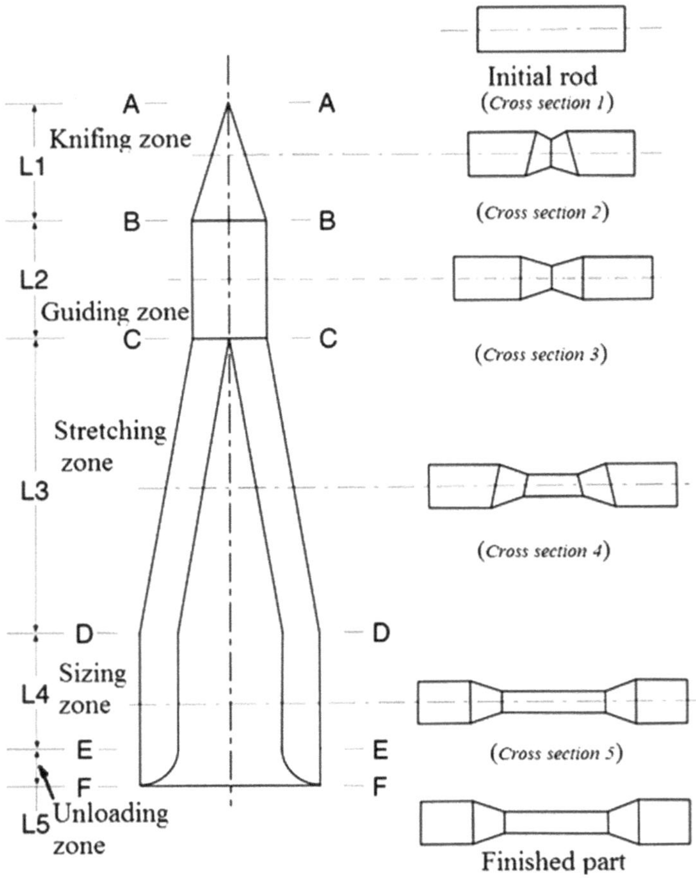

FIGURE 14.4 The schematic diagram of symmetrical CWR tool.

Plate wedge parameters:

$$a = 2r = d \tag{14.5}$$

$$b = (\Delta r^2 + \Delta r d)^{1/2} \tag{14.6}$$

and

$$\mu \geq \frac{b}{a} = \left(\left(\frac{\Delta r}{d} \right)^2 + \frac{\Delta r}{d} \right)^{1/2} \tag{14.7}$$

where Δr is an important parameter which can determine the critical COF. Eq. (14.7) is so-called rotation criteria, the excessive slip (Fig. 14.6) may occur if the processing parameters do not follow these criteria.

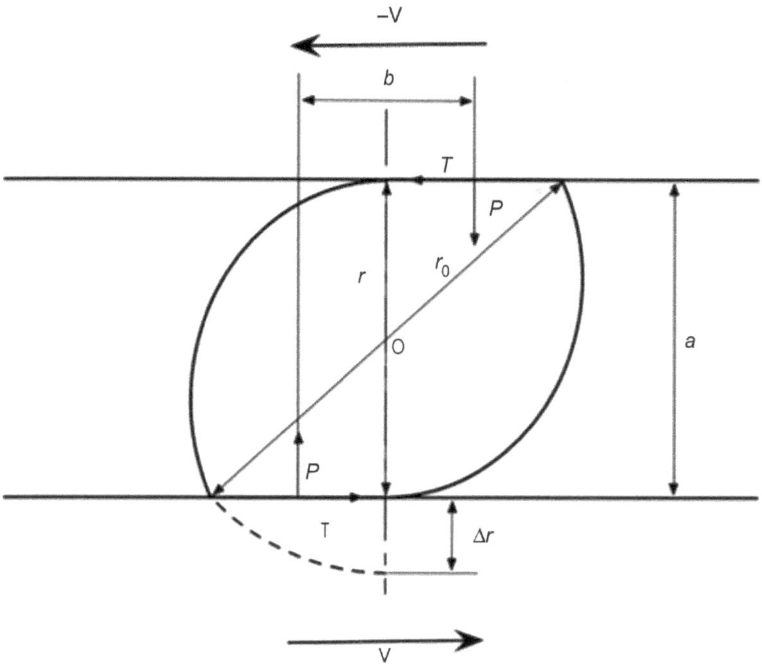

FIGURE 14.5 The schematic diagram of cross rolling process.

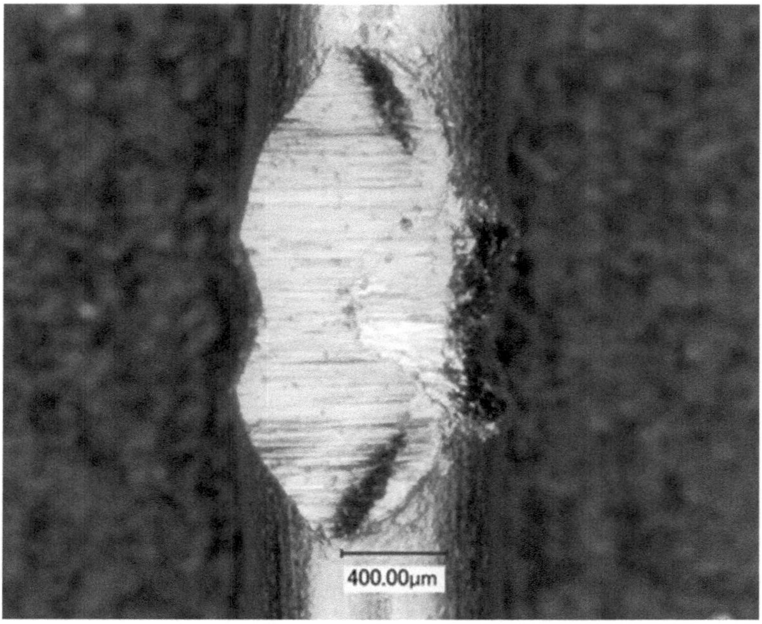

FIGURE 14.6 Excessive slip of MCWR workpiece.

A classical CWR is unfolded as shown in Fig. 14.7. When the workpiece rotates half a round, the spreading length of the rotating distance is L,

$$L = \frac{1}{2}\pi d_k \tag{14.8}$$

where the equivalent rotating diameter $d_k = d_1 + 2g\Delta r$ is with rotating coefficient $g = 0.5$ usually. Then the transient width extension S can be expressed as

$$S = \frac{1}{2}\pi d_k \tan\beta \tag{14.9}$$

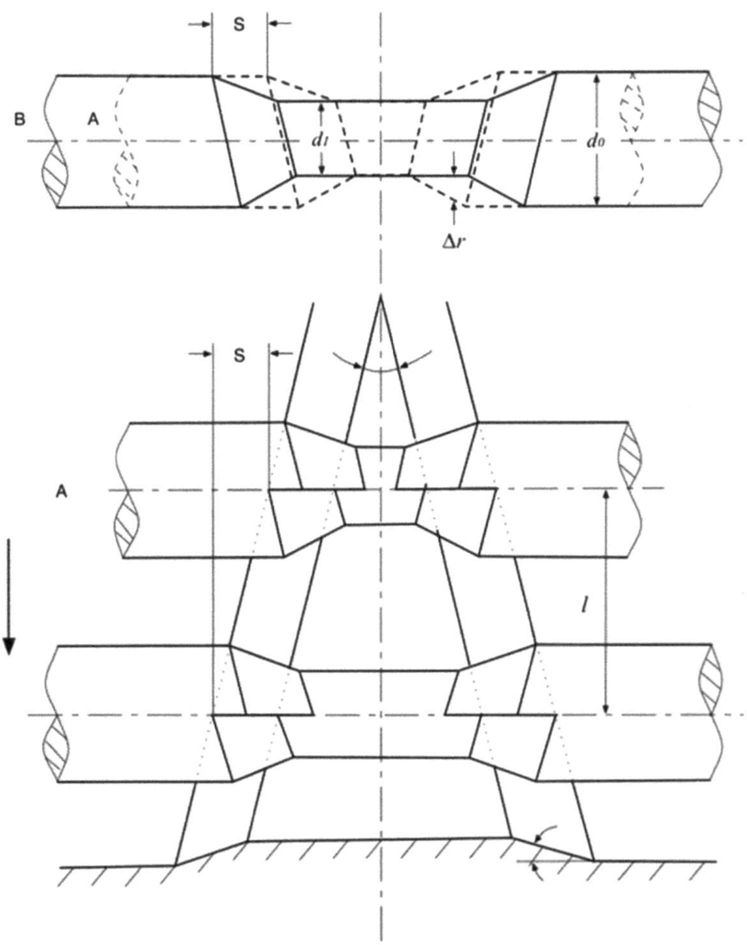

FIGURE 14.7 Width extension of CWR.

The largest compressive distance is

$$\Delta r = S \tan\alpha \tag{14.10}$$

Therefore,

$$\Delta r = \frac{1}{2}\pi d_k \tan\,\alpha\,\tan\,\beta \tag{14.11}$$

Substituting for Δr in Eq. (14.7), the following equations can be obtained

$$\tan\,\beta \geq \frac{d\left(\mu^2+1\right)^{1/2}-1}{\pi d_k \tan\,\alpha} \tag{14.12}$$

and

$$\beta \leq \tan^{-1}\left(\frac{2\mu^2}{\pi \tan\,\alpha(1+gr)}\right) \tag{14.13}$$

Eq. (14.13) indicates the friction at the interface of tools and workpiece is one of the critical factors that influence rotating conditions, and it determines the design of geometrical parameters such as forming angle and spreading angle. Fig. 14.8 illustrates the relationship between the COF μ and spreading angle β as described in Eq. (14.13).

In order to guarantee the rotation of workpiece or application of large spreading angle, the tool surface should be manufactured roughly in certain extent. Forming angle α also plays a significant role on the determination of

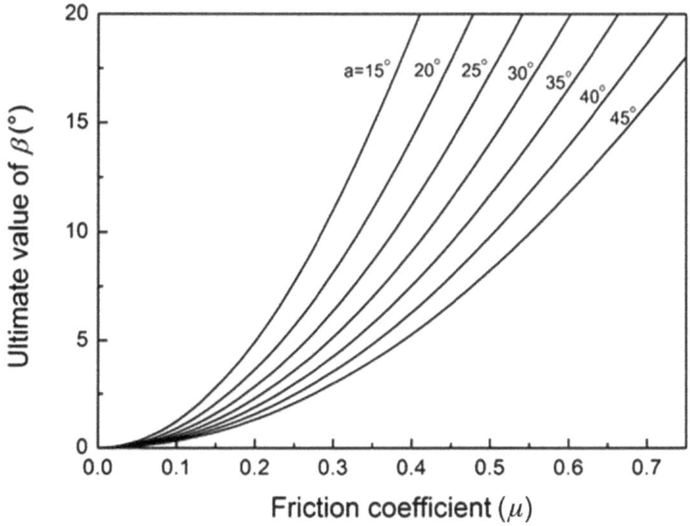

FIGURE 14.8 Influence of friction and forming angle on spreading angle.

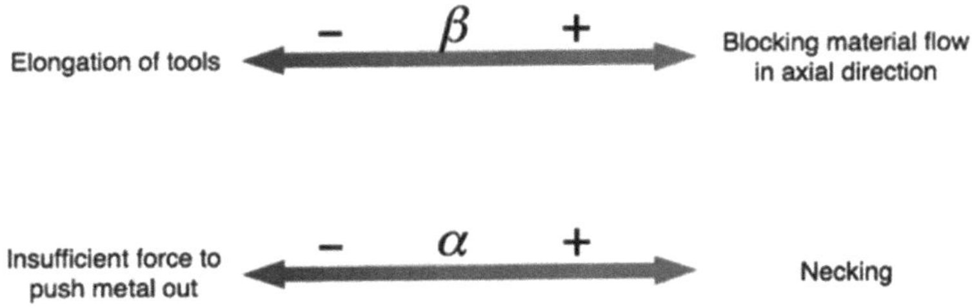

FIGURE 14.9 Influences of forming angle α and spreading angle β on MCWR process.

spreading angle as shown in Fig. 14.8. However, forming angle should range between 20 and 35 degrees to prevent products from central void or necking in practice, and the spreading angle is not sensitive to the variation of forming during this range. Eqs. (14.12) or (14.13) shows the influence of reduction on spreading angle and there is a negative relevance between them (see Fig. 14.9).

Generally, the cross-sectional area reduction ΔA is an elemental parameter in CWR and represents the amount of plastic deformation that occurs radially in workpiece. The cross-sectional area reduction of cylindrical workpiece is given by

$$\Delta A = 1 - \left(\frac{d}{D}\right)^2 \tag{14.14}$$

where D and d are the diameters of the workpiece before and after deformation, respectively.

In the conventional CWR, the area reduction is usually set less than 75%; otherwise, some problems such as necking, nonrotation may occur. However, when area reduction is smaller than 35% the rolling process cannot be accurately controlled. Moreover, the central void may appear due to an insufficient reduction, because the plastic deformation occurs only in the vicinity of workpiece surface, and there is little deformation in axial direction if the reduction is inadequate. Consequently, the central metal will undergo the impact of repeated tensile stress and shearing stress leading to internal defects. In macroscale, the optimum cross-sectional area reduction varies between 50% and 65%.

According to the deformation characteristics of MCWR, forming and stretching angles significantly influence the rotation criteria, internal void, necking, and rolling force. Therefore, these critical parameters should be selected based on above mechanical analysis.

Based on the influencing factors discussed above, the geometrical parameters of forming tool in the CWR process are listed in Table 14.1.

TABLE 14.1 Relationship Between Area Reduction and Forming and Stretching Angles

Area reduction (%)	80–70	70–60	60–50	50–40	<40
Forming angle α (°)	18–24	22–30	26–32	>28	>28
Stretching angle β (°)	4–8	5–9	7–9	5–9	<8

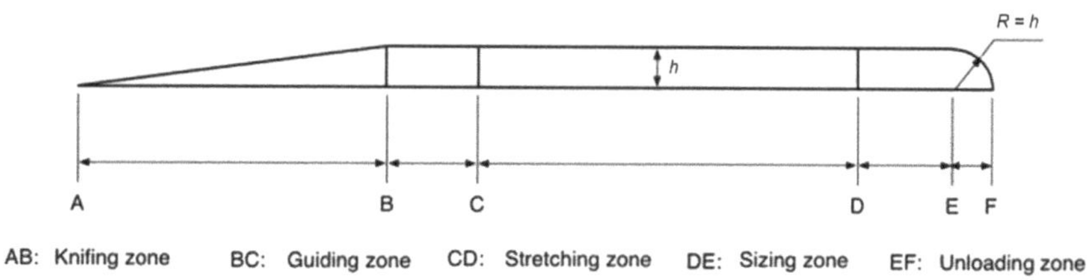

FIGURE 14.10 Unloading zone of MCWR tool.

14.1.3 Design of Forming Tool

In order to cause the workpiece to roll, the cross rolling should be analyzed at first, because the rotational conditions are developed based on that of simple cross rolling under the following assumptions: (i) the roles of leading plate and leading tube are neglected; (ii) the influences of weight and inertia are ignored; and (iii) the workpiece center is on the rolling line.

To simplify the research on the characteristics of MCWR, the wedge tools are designed symmetrically and the classical design is illustrated in Fig. 14.10 where five zones—knifing zone, guiding zone, stretching zone, sizing zone, and unloading zone in the flat wedge tool—have been revealed.

Knifing zone (A–B: L_1). The height of wedge is defined as h. During the knifing zone (A-B stage), the cylindrical workpiece is forced into the wedges and starts to rotate. A V-shaped groove will be generated gradually on the workpiece. The largest height is $\Delta r = r_0 - r_1$ and h must be bigger than Δr to ensure the machine safety. The length of knifing zone L_1 can be calculated as follows:

$$L_1 = h\cot\alpha\cot\beta \tag{14.15}$$

The rotation criteria should be mainly taken into account when α and β are determined.

Guiding zone (B–C: L_2). During the guiding zone, the wedge cross section and height h keep constant with $\beta = 0$. A V-shaped groove of Δh in height will be generated along the entire circumference of workpiece to improve the plastic deformation at the beginning of stretching zone.

The length of guiding zone is calculated as follows:

$$L_2 > \frac{\pi}{2} d_k \tag{14.16}$$

which can assure that the workpiece rotate at least half a circle during this stage.

Stretching zone (C–D: L_3). In stretching zone, the wedge height does not change while the widths of wedge top and bottom increases gradually. The stretching zone is the most significant stage of MCWR deformation. The compression of workpiece diameter and elongation of workpiece in axial direction are both achieved during this zone. The evolution of workpiece profile is illustrated as the cross section 4 in Fig. 14.4.

The length of stretching zone is calculated as follows:

$$L_2 = \frac{1}{2} l_1 \cot \beta \tag{14.17}$$

where l_1 is the length of workpiece with diameter d_1.

Sizing zone (D–E: L_4). The wedge height and width are constant, namely, $\beta = 0$. After this stage, the expected geometrical dimensions are obtained along the entire circumference of workpiece. The geometrical accuracy and surface roughness achieve the standard of final product (cross section 5 in Fig. 14.4). The length of sizing zone should be

$$L_4 > \frac{\pi}{2} d_k \tag{14.18}$$

which can assure that the workpiece rotate at least half a circle during this stage.

Unloading zone (E–F: L_5). The gap between two wedges varies during rolling process due to the elasticity of entire machine. The moment when the workpiece departs from tools, pressure disappears and two wedges get close to each other suddenly, which will leave marks on the workpiece surface. Therefore, the unloading zone is designed to prevent the occurrence of this defect. As shown in Fig. 14.5, the profile of unload zone is a quarter of circle with the radius of h.

The actual designed plate tools of MCWR are shown in Fig. 14.11. The tools are a pair of wedge-shaped plates in order to simplify the machine design and guarantee accuracy.

14.2 MICRO CROSS WEDGE ROLLING PRACTICE

14.2.1 Surface Roughness Analysis

The used workpieces are copper, stainless steel, and aluminum cylinders with diameter of ∅ 0.8 mm. The tool steel surfaces were prepared using electron-spark treatment and three kinds of surface roughness (R_a) 3.2, 1.6,

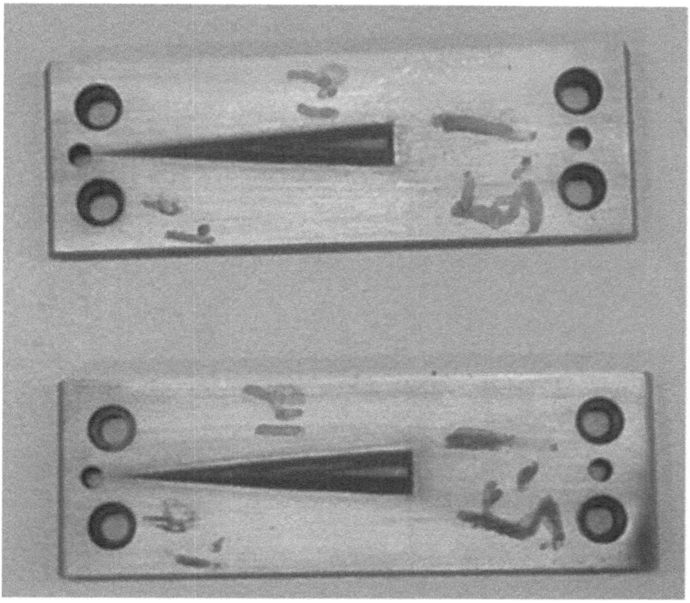

FIGURE 14.11 Plate tools of CWR.

and 0.8 μm are generated, respectively. Prior to the MCWR test, the surface morphologies of tools are observed and measured by 3D digital microscope, and the tool surface was flushed by high pressure sprayer to remove adhesive on workpiece. MCWR tests were conducted with each kind of tool for 10 times. After tests with cross-sectional area reduction of 52.73%, the tool surface was compared with that before tests as shown in Figs. 14.12–14.14, which indicate the surface wear looks obvious due to deformation process. In order to account for the phenomena of material transfer quantitatively during MCWR process, further measurements were conducted repeatedly and the data obtained from the measurements were calculated to identify surface roughness R_a.

Surface roughness is the arithmetic average of values at randomly selected spots on the surface of an object. The arithmetical average roughness (R_a) is selected to describe the surface roughness of tool and workpiece. A portion stretching based on the line coordinates over a reference length in the direction in which the average line extends is cut out from the roughness curve (Fig. 14.15). This portion is presented in a new graph with the x-axis extending in the same direction as the average line and the y-axis representing the magnitude. When the roughness curve is represented by $y = f(x)$, R_a is the value in microns (μm) found from the formula shown in Fig. 14.15.

A net composited of 20 lines as illustrated in Fig. 14.16A was generated on the surface topology image and utilized to assess R_a of tools. The position

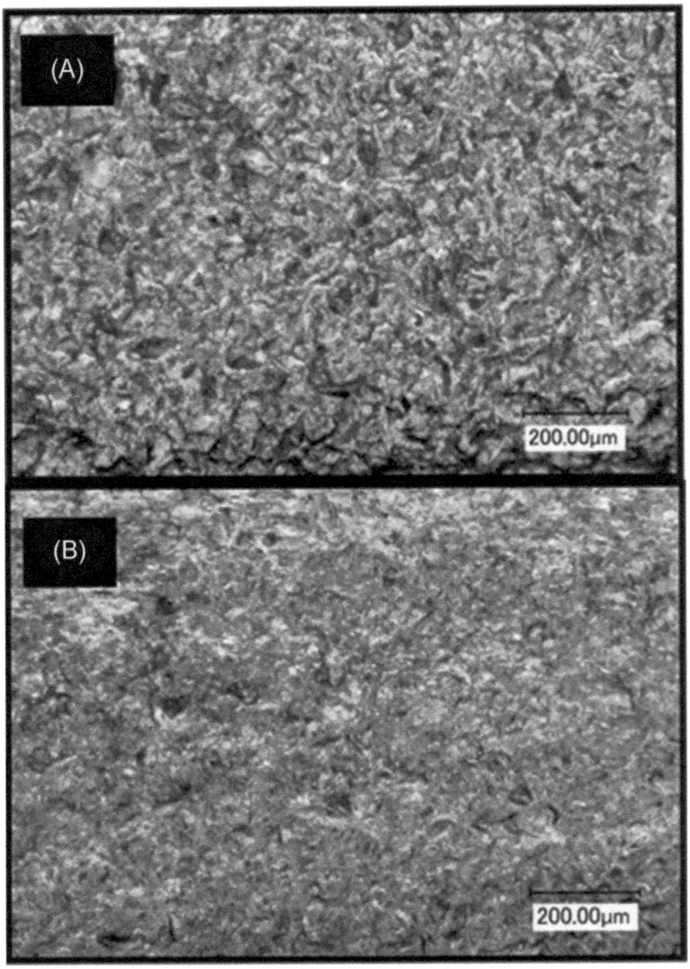

FIGURE 14.12 Surface morphology of workpiece after material transfer during sliding contact, (A) copper and (B) aluminum.

coordinates of these 20 lines distributing evenly were captured by 3D digital microscope (Fig. 14.16B). Then, the tool surface roughness was the meaning value of R_a obtained from these 20 lines with assistance of formula in Fig. 14.15.

When comparing three groups of pre- and post-contour maps of tool surface topology, significant scratch can be seen on the post-contour maps (Fig. 14.17). Sliding contact clearly produced visible wear tracks on all three kinds of tool surfaces with different surface roughness. Table 14.2 shows the tool roughness parameters of the pre- and posttests and the scatter in the

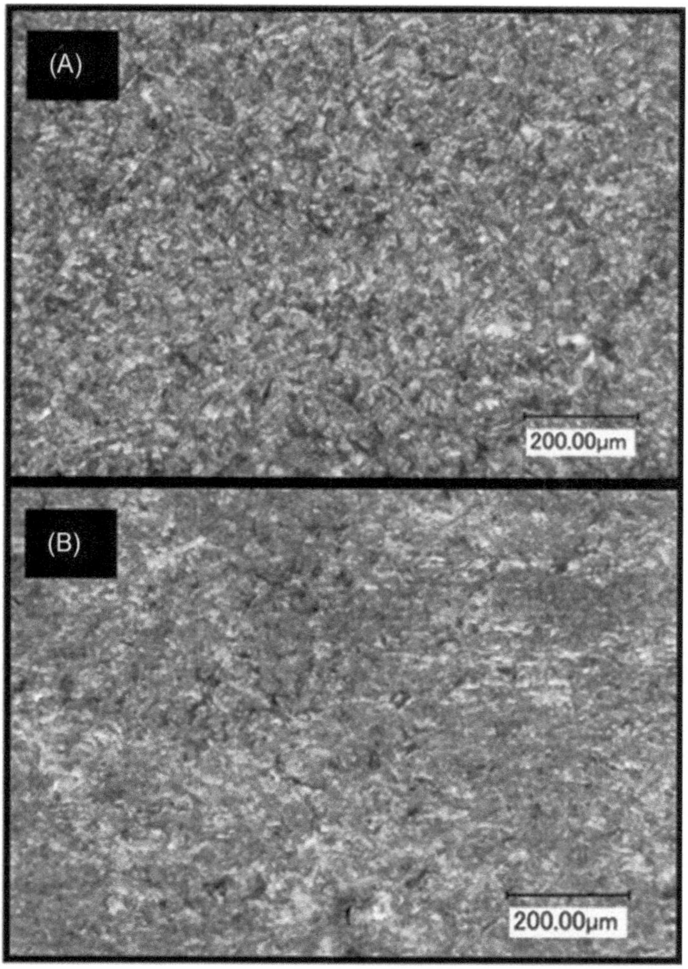

FIGURE 14.13 Comparison of tool surface morphology with original $R_a = 3.2$ μm: (A) pretest tool surface and (B) posttests tool surface.

measurements was typically less than 10%. As expected, the surface roughness values increase after tests due to material transfer.

The result indicates the material transfer and adhesion tend to be alleviated when the tool surface roughness (R_a) increases. This is due to the fact that material transfer is controlled by the abrasion action of the sharp protruding scratch ridges scraping against the softer workpiece material surface. In the case of rougher tool surface, the material is mainly to the larger scratches where the workpiece material mainly adheres in front of the sharping ridges and builds up large transfer lumps resulting in a rougher tool surface.

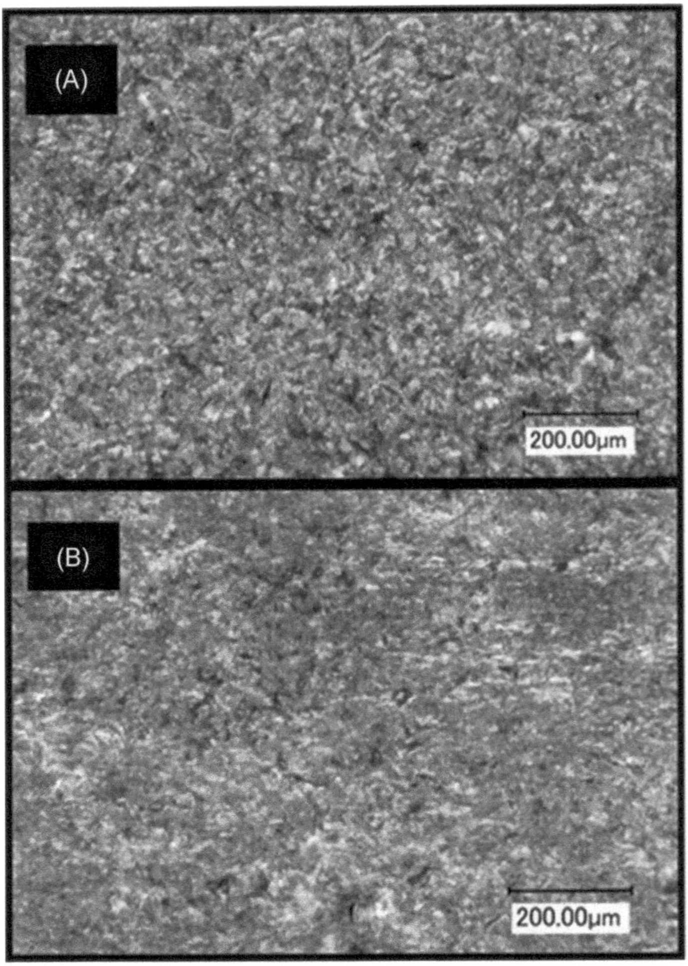

FIGURE 14.14 Comparison of tool surface morphology with original $R_a = 1.6$ μm (A) pretest tool surface and (B) posttests tool surface.

As illustrated in Fig. 14.18, the tool surface profiles of posttests undulate dramatically in contrast to that of pretests. When the initial roughness value is 3.2 μm, the maximum height of the topological peek even approximately reaches 8 μm after rolling tests (Fig. 14.18A and B). The effects of material transfer and adhesion on roughening tool surface become not so significant when the initial tool surface roughness reduces to 0.8 μm, where the frequency of profile waves increases, but the amplitude remained the same.

The results demonstrate that the transfer occurs on the tool/workpiece interface depending on the tool surface topography. This kind of transfer to rough tool surface occurs in connection to surface scratches, where these act

$$Ra=\frac{1}{L}\int_{x=0}^{x=L}|y|\,dx$$

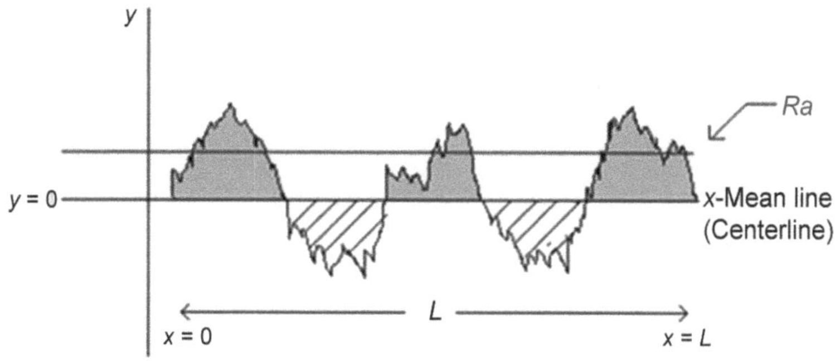

FIGURE 14.15 Calculation of arithmetical average roughness (R_a).

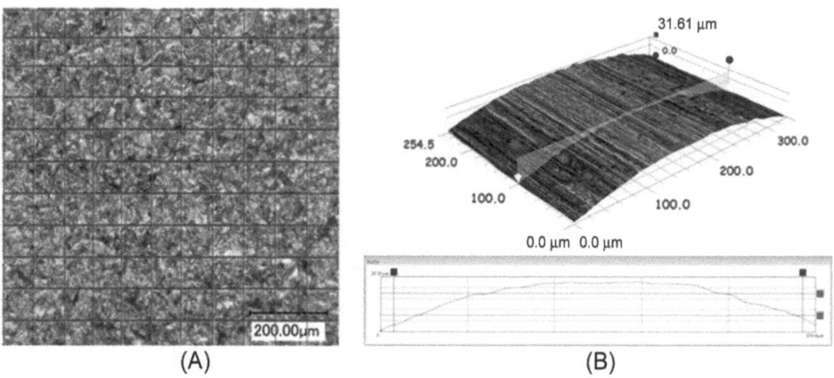

FIGURE 14.16 Measurement of surface roughness: (A) 20 lines distributing on the surface morphology image and (B) capturing position coordinates of the selected lines.

as mechanically scrape off softer material and result in lumps off on the tool surface. When the tool surface becomes finer, there is a tendency to pick up small amounts of workpiece. However, there is slight excessive slip at the interface when the stainless steel is rolled with an area reduction less than 50% for the tool surface roughness R_a of 0.8 μm, which means that the tool surface cannot be refined any more. Consequently, the harder tool surface should be as smooth as possible to reduce the mechanical interaction with the mating workpiece. However, the tool surface roughness should be optimized to find a compromise between the rotation criteria and the material transfer.

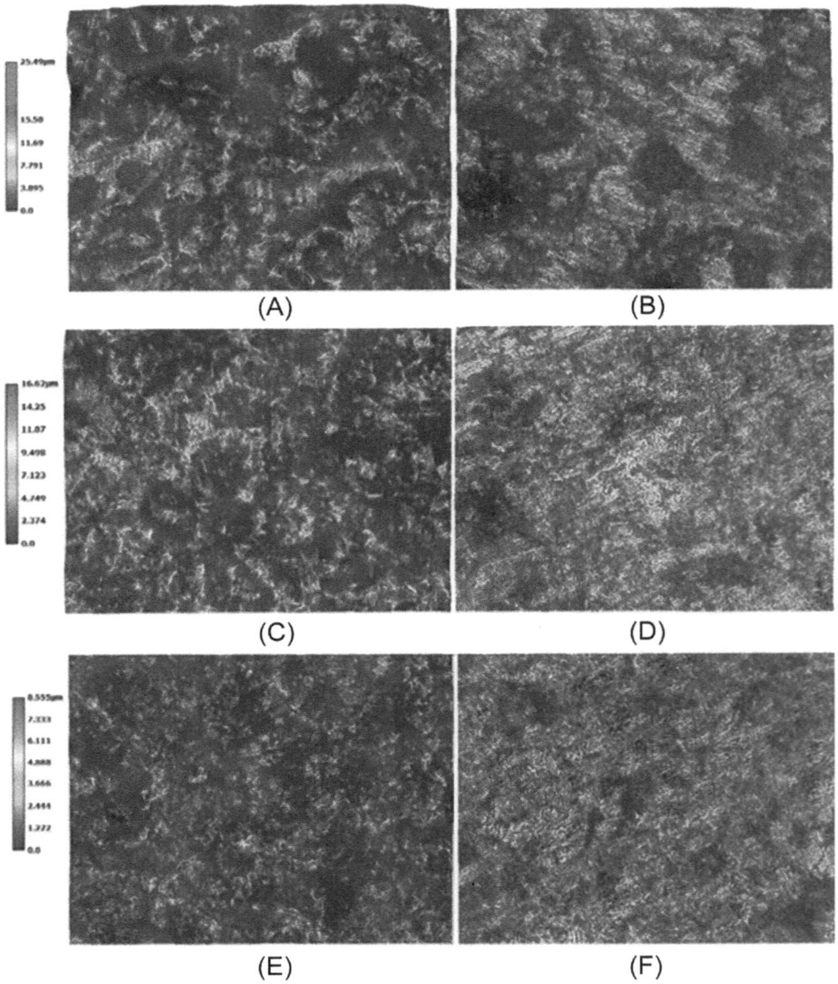

FIGURE 14.17 Workpiece surface morphology before and after MCWR of aluminum with different initial roughness value: (A, B) $R_a = 3.2$ μm, (C, D) $R_a = 1.6$ μm, (E, F) $R_a = 0.8$ μm, and (A, C, and E) are pretest surface topographies and (B, D, and F) are post ones.

TABLE 14.2 Tool Surface Roughness R_a of Pre- and Post-MCWR Tests

Presurface Roughness (R_a, μm)	Postsurface Roughness (R_a, μm)		
	Copper	Aluminum	Stainless Steel
3.2	3.353	3.460	3.508
1.6	1.694	1.725	1.731
0.8	0.835	0.866	0.864

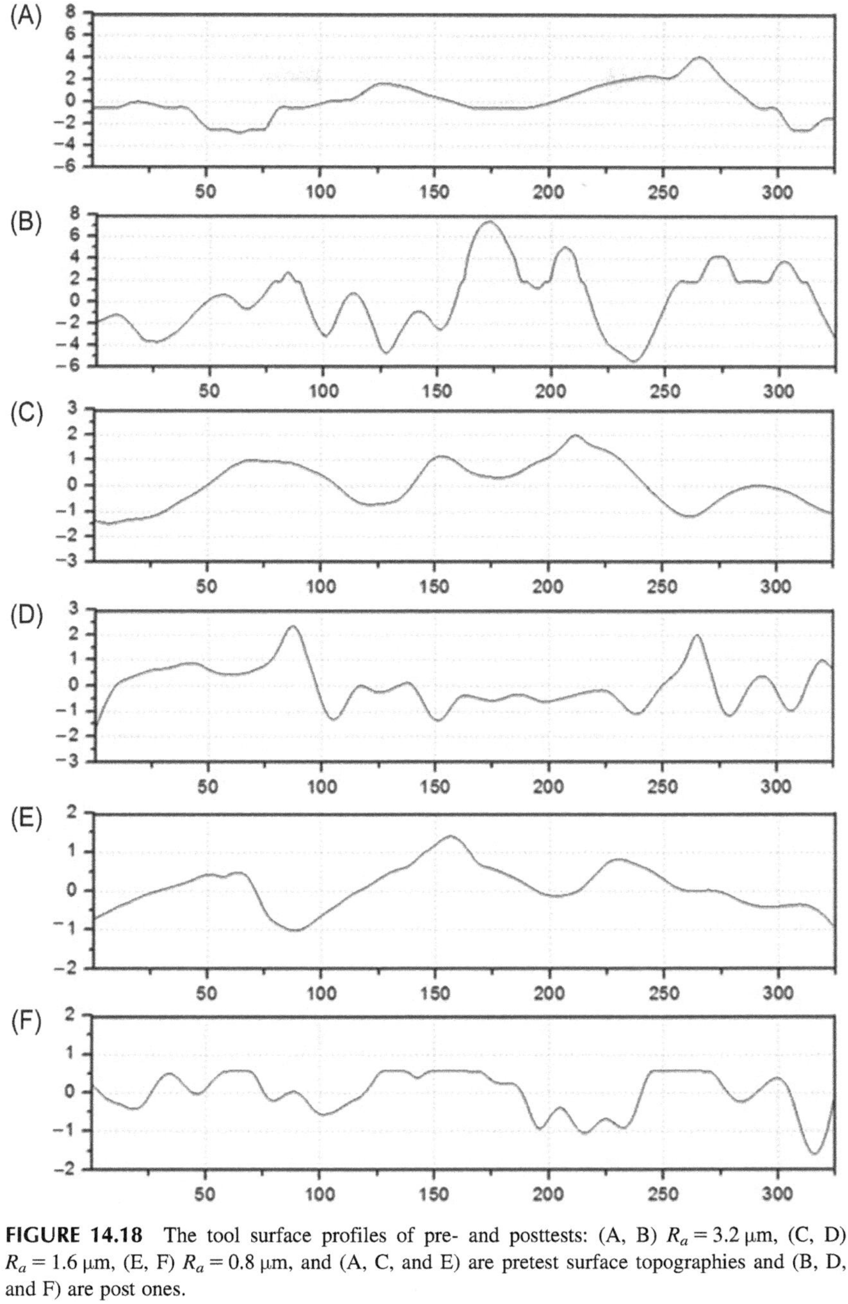

FIGURE 14.18 The tool surface profiles of pre- and posttests: (A, B) $R_a = 3.2\ \mu m$, (C, D) $R_a = 1.6\ \mu m$, (E, F) $R_a = 0.8\ \mu m$, and (A, C, and E) are pretest surface topographies and (B, D, and F) are post ones.

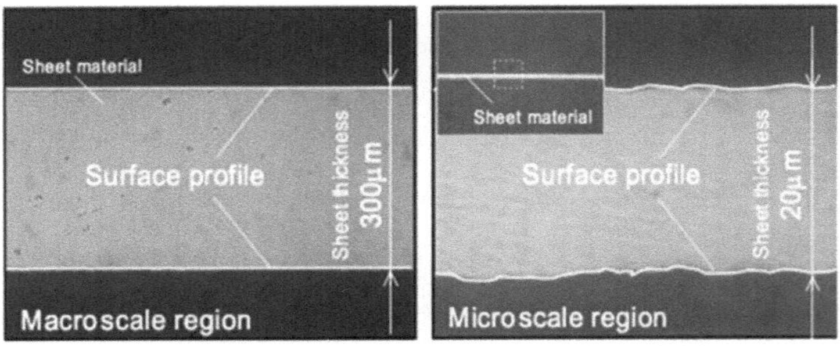

(A) Macroscale region (Thickness = 300 mm) (B) Microscale region (Thickness = 20 mm)

FIGURE 14.19 Comparison of cross sections of macro- and micro-sheet materials [8].

In the miniaturization of dimensions for metal forming, the relative ratio of the surface asperities of tools and blanks to the outside dimensions becomes larger than that in the case of the conventional macroscale process. This means that the surface asperities may affect frictional behavior during micro fabrication, so that it would more significantly affect processing characteristics and accuracy of micro products [7]. Shimizu et al. [8] conducted micro deep drawing tests to investigate the influencing factors of the product surface morphology. Fig. 14.19 illustrates the comparison of the relative ratio of the surface asperities to sheet thickness between the macro- and microscale sheet materials. In spite of the same absolute values of surface roughness in both scales, there is a big difference in the relative ratios to the sheet thickness.

In order to achieve micro products of high quality, technological advancements in surface treatment and microstructure fabrication require a fundamental understanding of rough surface contact, adhesion, and friction on the meso- and microscale. The types of forces that influence microscale devices are different from those that influence devices with conventional scale. This is because size effects in microforming bear a significant influence on the physical phenomena that dictate the dynamic behavior of that system. For example, microscale systems are influenced by inertia effects to a much greater extent than microscale systems, while smaller systems are more influenced by surface effects. Therefore, strong adhesion, friction, and wear induced by surface effect are major problems limiting both the fabrication yield and operation lifetime of many micro-electromechanical systems devices [9]. Because the complicated interactive behavior at interface of tools and workpiece is classical dry contact friction in MCWR, the interaction between tools and workpiece will significantly affect the workpiece surface asperity. According to the geometrical range of MCWR, the factors in terms of surface, asperity, component, and device should be taken into account to investigate the contact behavior.

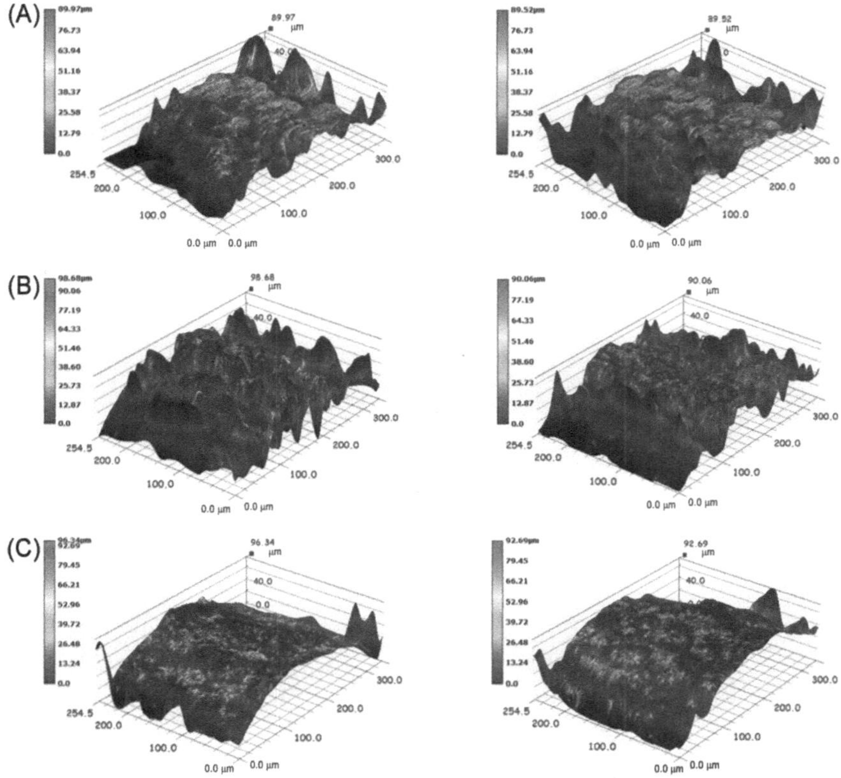

FIGURE 14.20 Workpiece surface morphology after MCWR with tool roughness $R_a = 3.2$ μm: (A) steel, (B) copper, and (C) aluminum.

As illustrated in Figs. 14.20–14.22, when the initial surface morphology of workpiece varies, there is no significant variation of the surface morphology after deformation. However, the final surface of stainless steel, copper, and aluminum workpieces trends to be glossier with a decrease of surface roughness of wedge tools. It is noted that the initial surface of the harder one in the contact pair influences the micro product surface more significantly than that of the softer one. When the surface roughness of tools R_a is 3.2 μm, it shows that the aluminum workpiece surface is much smoother than that of the other two and its R_a is approximately 1.5 μm.

The surface roughness values of steel and copper workpiece are both close to 2.5 μm as shown in Fig. 14.23A. Fig. 14.23B shows the characteristics of workpiece surface roughness when the microstructure of workpiece varies in term of grain size. As can be seen, the surface roughness of workpiece shows a sensitive relationship to the grain size when the tool surface is rough, where the workpiece surface becomes rougher significantly with a

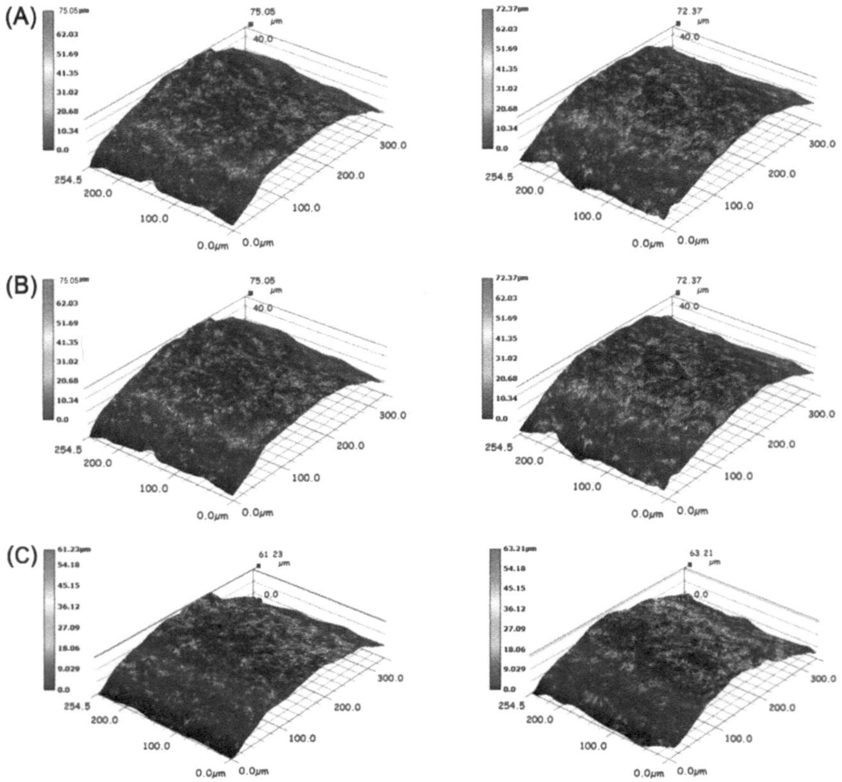

FIGURE 14.21 Workpiece surface morphology after MCWR with tool roughness $R_a = 1.6\ \mu m$: (A) steel, (B) copper, and (C) aluminum.

decrease of grain size. In contrast, the value of surface roughness R_a levels off with the variation of grain size when the tool surface is smooth ($R_a = 0.8$ or $1.6\ \mu m$). The scatters of R_a increase slightly due to the enlargement of grain inside copper workpiece. The product surface roughness, when the tool surface R_a is $3.2\ \mu m$, displays a slightly stable R_a scatters.

The cross-sectional area reduction represents the amount of deformation during rolling process. According to the characteristics of MCWR, the deformation amount determines the number of times that workpiece counters and contacts with tools to complete the whole rolling process. Thus, the reduction has an inevitable impact on the work surface morphology to some extent as shown in Fig. 14.23C. Fig. 14.23C demonstrates the influence of area reduction on the final surface topology of products quantitatively. When the tool surface roughness is $3.2\ \mu m$, there is significant increase of workpiece surface roughness with the area reduction and the R_a scatters increase as well. However, when the tool surface roughness is 1.6 or $0.8\ \mu m$, the fluctuation of workpiece surface roughness is limited.

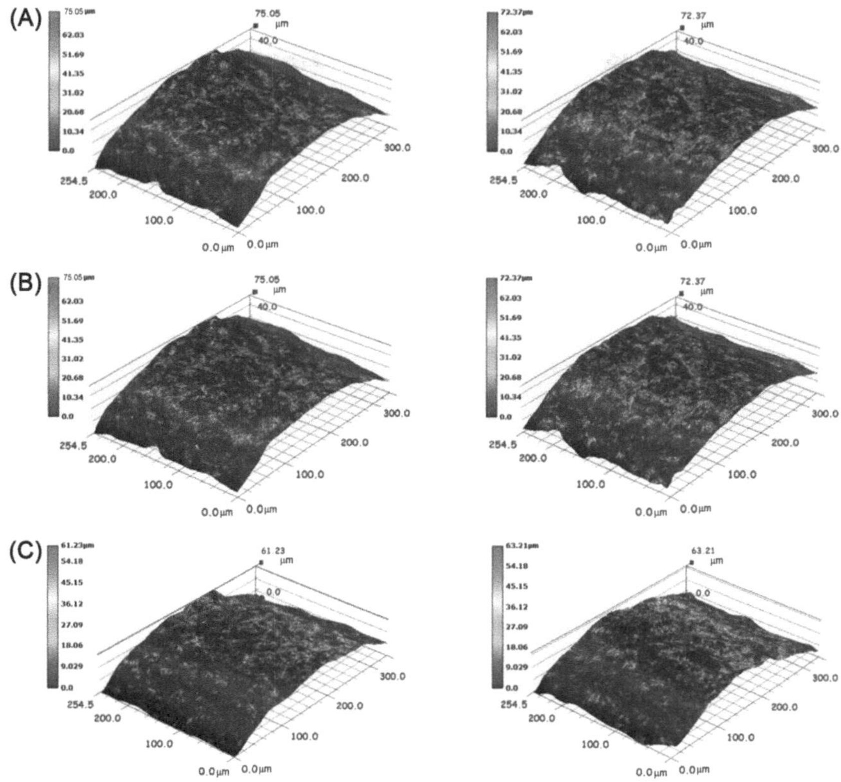

FIGURE 14.22 Workpiece surface morphology after MCWR with tool roughness $R_a = 0.8$ μm: (A) steel, (B) copper, and (C) aluminum.

14.2.2 Springback of Micro-Metal Workpiece in MCWR

Springback phenomenon basically exists in whatever kind of metals forming process and it is a significant characteristic of every process that involves material plastic deformation. The recoverable deformation of deformed part occurs during unloading can cause dimension deviation and error [10,11]. In industrial practice, deformations due to springback are compensated manually by doing extensive measurements on prototype parts and altering the tool geometry. Efficient springback compensation requires an accurate springback prediction. In virtual product design, this compensation will be based on the springback prediction provided by theoretical analysis.

In macroscale, the coverable strain is very small comparing with entire strain. However, springback is particularly critical in the metal microforming process because it causes size inaccuracy of micro products [12]. Therefore, the capability of predicting the final geometry of a micro-metal part as well as the amount of springback is significant in microforming. However, the

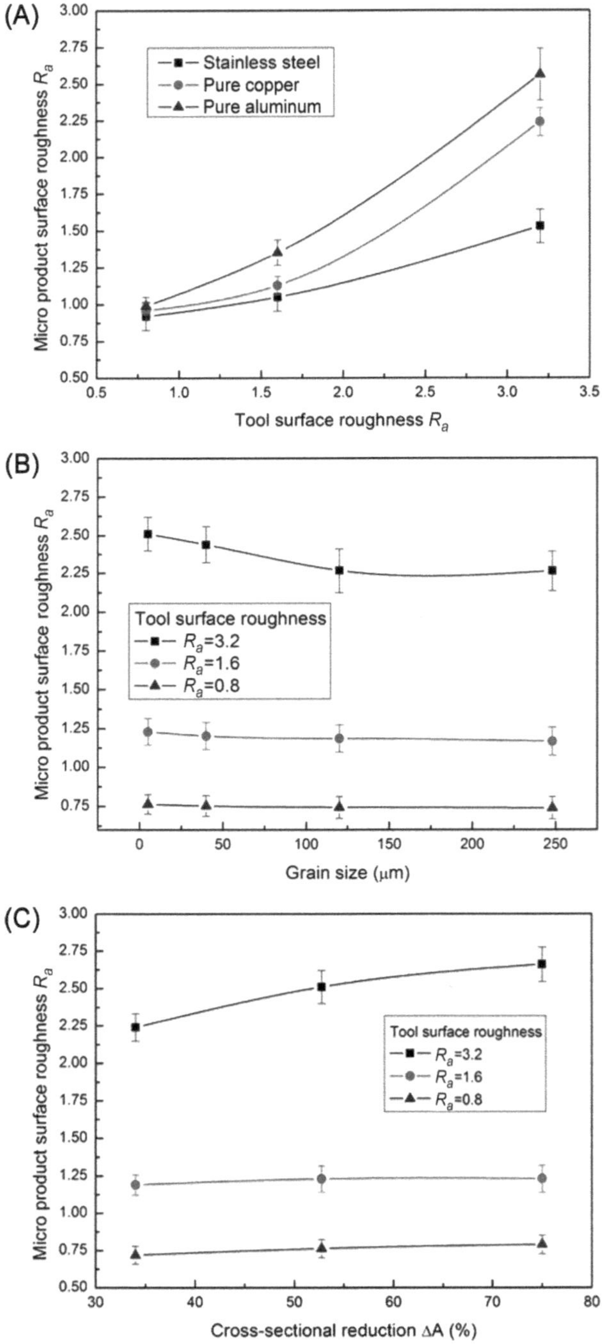

FIGURE 14.23 Influence of (A) tool surface roughness on product surface roughness, (B) grain size on product surface roughness, and (C) cross-sectional area on product surface roughness.

detailed grain structures and size effects of the materials are not considered in the modeling of springback [13]. The knowledge obtained based on such an understanding is inadequate to explain springback occurring in microforming, which is heavily related to the material microstructures at the micro-level, including the grain size and grain orientations of the blank materials. Therefore, it is necessary to develop a new plastic theory to explain size effects on the springback behavior. Literature studies have revealed that the design concepts regarding to the springback of conventional metal forming cannot be applied directly to the forming of micro products [14]. In this chapter, the influence of springback on geometrical inaccuracy is focused on in MCWR with different blank materials and processing. The springback behavior is analyzed with assistance of nonlinearity and strain gradient theories. Based on the experimental and numerical results, the material microstructure, tool geometry, and processing parameters are optimized to predict and reduce the springback in MCWR, which can benefit the product accuracy.

In metal MCWR, the deformation and material flow are more complicated than that in metal-sheet bending. Hence, springback of the micro cylindrical workpiece (Fig. 14.24A) cannot be quantitatively eliminated just by angle or deflect in one direction.

The differences between the actual and desired deformation were measured in both radius and axial directions. Thus, the springback can be described as follows:

$$S_a = l_d - l_2 \tag{14.19}$$

$$S_r = \frac{(d_2 - d_d)}{2} \tag{14.20}$$

where S_a and S_r are defined as indicators of springback in radius and axial directions, respectively. d_2, l_2, d_d, and l_d are illustrated in Fig. 14.24. The digital microscope of VHX-1000X was used to measure the geometrical profile of micro product from MCWR process (Fig. 14.24B). In MCWR experiment, the forming and spreading angles of plate wedge tool are 38 and 5 degrees, respectively. There are three kinds of wedge tools with surface roughness (R_a) of 0.8, 1.6, and 3.2 μm, and the surface topology of these wedge tools is presented in Fig. 14.24.

The cylindrical workpiece of pure copper was utilized as experimental material and its original diameter is 0.8 mm. In order to prevent excessive slip at interface of workpiece and tools, the workpiece surface was rubbed with ultra-fine sandpaper to remove the oxide coating. The workpiece was deformed into stepped shaft as illustrated in Fig. 14.25B. The cylindrical workpieces are rolled into stepped shafts with three different final diameters (d), viz., 0.4, 0.55, and 0.6 mm. The cross-sectional area reduction is defined as Eq. (14.14), and these cylindrical workpieces were compressed by the area reductions of 34%, 52.73%, and 75%, respectively.

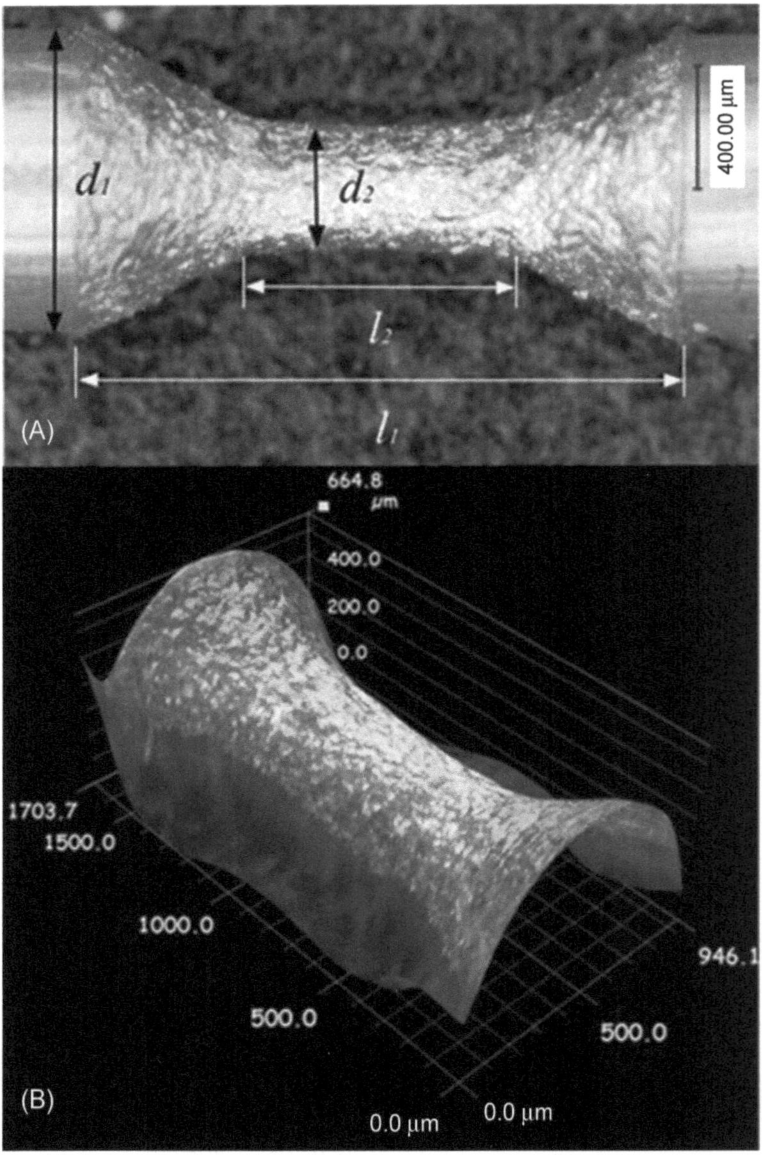

FIGURE 14.24 Dimensions and morphology of micro products obtained from 3D digital microscope.

As illustrated in Fig. 14.25B, when the wedge tools loads are removed after forming step, a portion of the deformation recovers causing a change in micro product shape. In other words, springback happens, which should be measured and controlled especially in microforming process. In order to

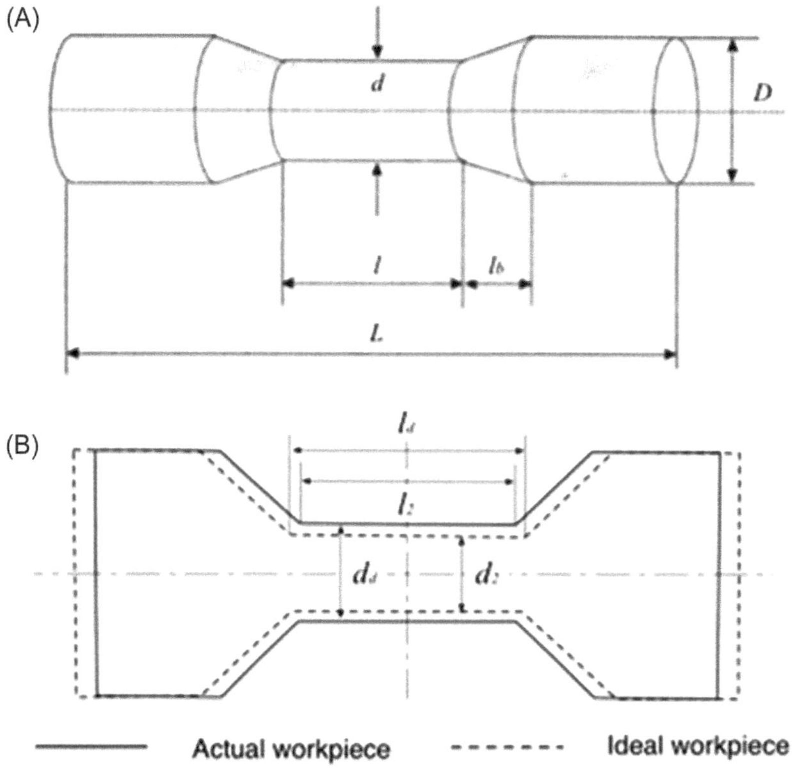

FIGURE 14.25 Geometrical dimensions of MCWR product: (A) geometry of MCWR product and (B) geometrical deflection from ideal profile to actual profile due to springback.

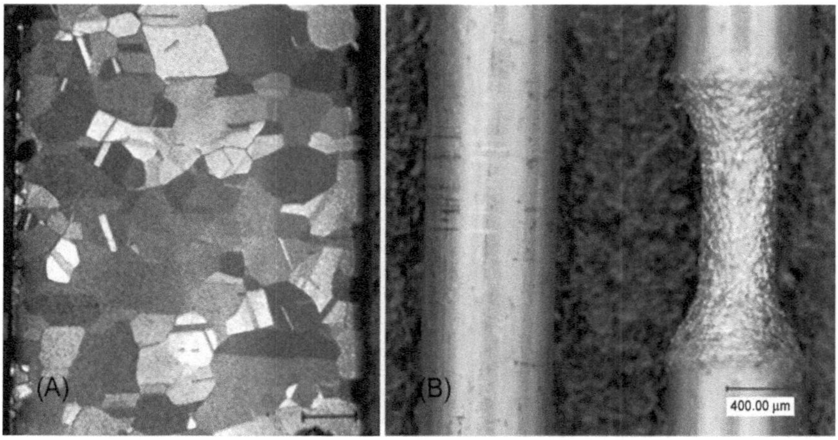

FIGURE 14.26 Blank material and product of MCWR: (A) microstructure of workpiece and (B) final profile of MCWR product [15].

investigate the influence of microstructure on springback in microscale, the blank materials in annealed under different conditions were used. These specimens underwent heat treatment at 900 K for different holding time to enlarge meaning grain size ranging between 6 and 248 μm [15]. Fig. 14.26A shows the microstructure of workpiece with grain size of 120 μm. These pure copper wires were carefully subdivided by wire cutting.

After deformed by MCWR machine, the morphology and the geometrical dimensions of deformed workpieces were obtained from the 3D digital microscope. As shown in Fig. 14.27, the actual profile of MCWR product was compared to the expected profile. The absolute difference of geometrical dimensions between these two profiles is considered as a parameter to assess the springback quantitatively.

According to classical metal forming theory, the main factors determining springback error are the material intrinsic properties and the ratio of the deformation to the specimen size. In micro domain, the strain gradient has a more dominate effect on the elastic stiffness, when the specimen size approaches to the intrinsic material length. The size effects in microforming can be discussed from the view of dislocation dynamics and strain gradient plasticity [16].

In order to discern this nonuniform deformation, the theory of dislocation mechanics was proposed [17]. It refers to geometrically necessary dislocation (GND) and statistically stored dislocation (SSD) (Fig. 14.28). Both the evolutions of SSDs and GNDs were proved to contribute to the short-range effect, which accordingly results in material hardening. The former was

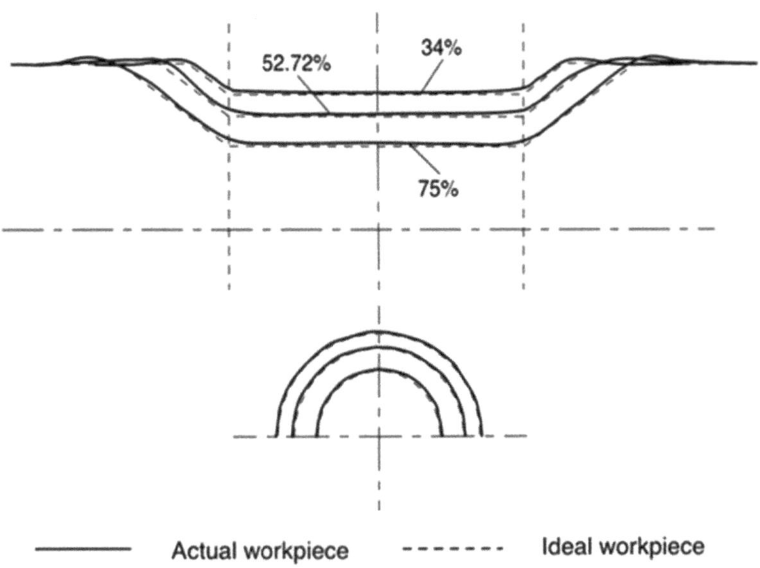

FIGURE 14.27 Comparison between actual and expected profiles of MCW products.

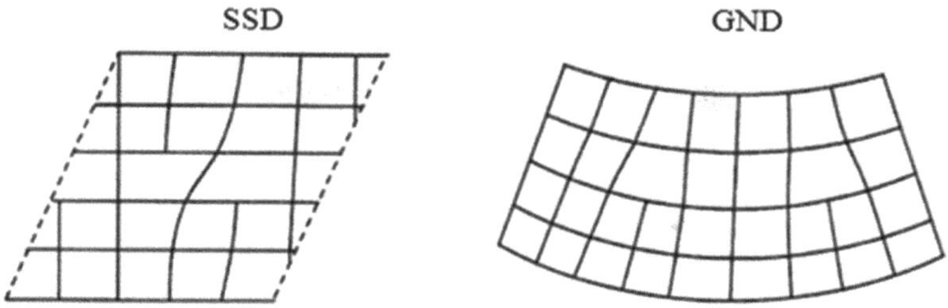

FIGURE 14.28 Illustration of statistically stored dislocations during shear and geometrically necessary dislocations during bending.

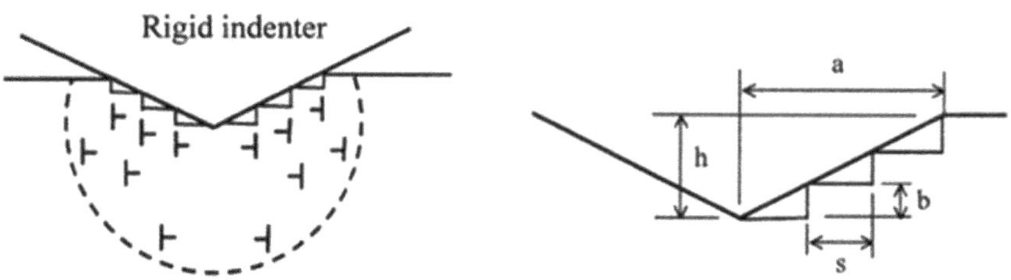

FIGURE 14.29 Geometrically necessary dislocations during indentation.

proved to distribute evenly within the whole grain and the latter to concentrate on the vicinity of grain boundary region.

Fig. 14.29 shows the GNDs during indentation. Nix and Gao [18] reported that for geometrical (strain compatibility) considerations, the density of the GNDs is

$$\rho_G = \frac{3}{2b}\nabla\varepsilon \tag{14.21}$$

where ρ_G is proportional to strain gradient (scale dependent) whereas the density of SSD, and ρ_S is dependent upon the average strain in the indentation.

In Taylor model of plasticity, dislocations are emitted from Frank–Read sources [19]. The dislocations were stuck in what is called the Taylor network because of interaction with one another, but when externally applied stress reaches the order of Peierls stress for the dislocations, they are activated and the plastic yield is initiated. The magnitude of the Peierls stress τ_p is proportional to the dislocation's Burgers vector b divided by a distance between dislocation lines s [20,21],

$$\tau_p = \frac{Gb}{2\pi s} \tag{14.22}$$

where G is the elastic shear modulus. An approximate relation of the shear yield strength τ to the dislocation's density at a moment when yield is initiated is given by

$$\tau = \frac{cGb}{s} = \frac{cGb}{\sqrt{\rho_T}} \tag{14.23}$$

where c is an empirical material constant ranging from 0.1 to 0.5 and ρ_T is the total dislocation density which is a function of strain and strain gradient.

$$\rho_T = \rho_S(\varepsilon) + \rho_G(\nabla\varepsilon) \tag{14.24}$$

where ρ_S and ρ_G are total dislocation density of SSDs density and GNDs density, respectively. Then, Taylor's relationship between the yield shear stress and the dislocation density is

$$\tau = cGb\sqrt{\rho_S + \rho_G} = \left(\frac{cGb}{\sqrt{\rho_S}}\right)\sqrt{1 + \frac{\rho_G}{\rho_S}} \tag{14.25}$$

$$\sigma = M\tau \tag{14.26}$$

where M is the Taylor coefficient reflecting the ratio of the tensile yield stress to shear yield stress.

During unloading, the atomic bond contraction (linear elastic) and microplastic recovery (nonlinear) simultaneously occurs. As a result, the total strain recovery during unloading can be written as

$$\varepsilon_{\text{springback}} = \varepsilon_{\text{elastic}} + \varepsilon_{\text{microplastic}} = \frac{\sigma}{E} + \varepsilon_{\text{mp}} \tag{14.27}$$

where σ is the stress at the beginning of unloading, E is the Young's modulus, and ε_{mp} is the recoverable component of plastic strain [21]. The nonlinear strain portion during unloading has been termed as micro plasticity [22]. In order to better understand the springback nonlinearity, chord modulus was defined as the slope of the line connecting two stress points before and after unloading, viz., E_U in Fig. 14.30.

Since inelastic strain comes from the motion of mobile dislocations and GNDs play a dominate role on the microplastic deformation, it can be proposed that the inelastic portion of metal springback has a directly relation with GND density. The relationship between the amount of springback and the mean grain size was graphically illustrated in Fig. 14.31. It reveals that both the axial and radial springbacks of MCWR product increase with the grain size of the bullet associated with the cross-sectional reductions of 34%, 52.73%, and 75%. It is apparent that there is an increasing tendency of springback scatter when the grain size increases from 6 to 248 μm. It also can be observed that there is no significant influence of cross-sectional area reduction on springback. Comparing Fig. 14.31A with Fig. 14.31B, the radial springback is smaller than that in the axial direction. In addition, the numerical results are in good agreement with the experimental data.

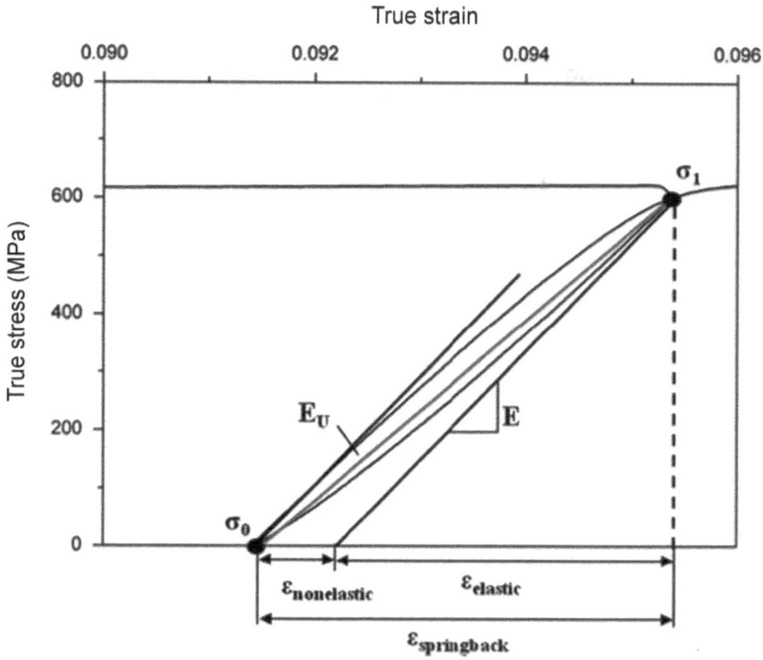

FIGURE 14.30 A schematic illustration of the unloading test.

When the microstructures within the micro workpieces are the same, the GND density is similar. However, the workpieces with the same microstructure undergone different extents of plastic deformation exhibit different springback behaviors. With the increase of deformation, strain penetrates more deeply into the grains resulting in activation of more dislocations and intertwinement of repelling dislocations which will release more inelastic springback in the unloading phase. Consequently, the MCWR workpieces springback is more significantly when the area reduction increases.

In summary, the systematically design and analysis of MCWR have been successfully completed. High-quality micro-stepped shaft parts have been smoothly obtained. The following demonstrates the main contents of this chapter: (i) the design strategy has been introduced in terms of driving part, power transamination part, and forming tools; (ii) the geometrical and processing parameters including cross-sectional reduction, forming, and stretching angles of wedge tools were optimized with assistance of FE simulation; (iii) the surface roughness evolution has been identified and tool surface wear was investigated in order to guarantee tool working life and avoid to causing workpiece defects; and (iv) springback of micro-metal workpiece in MCWR has been identified and quantified successfully.

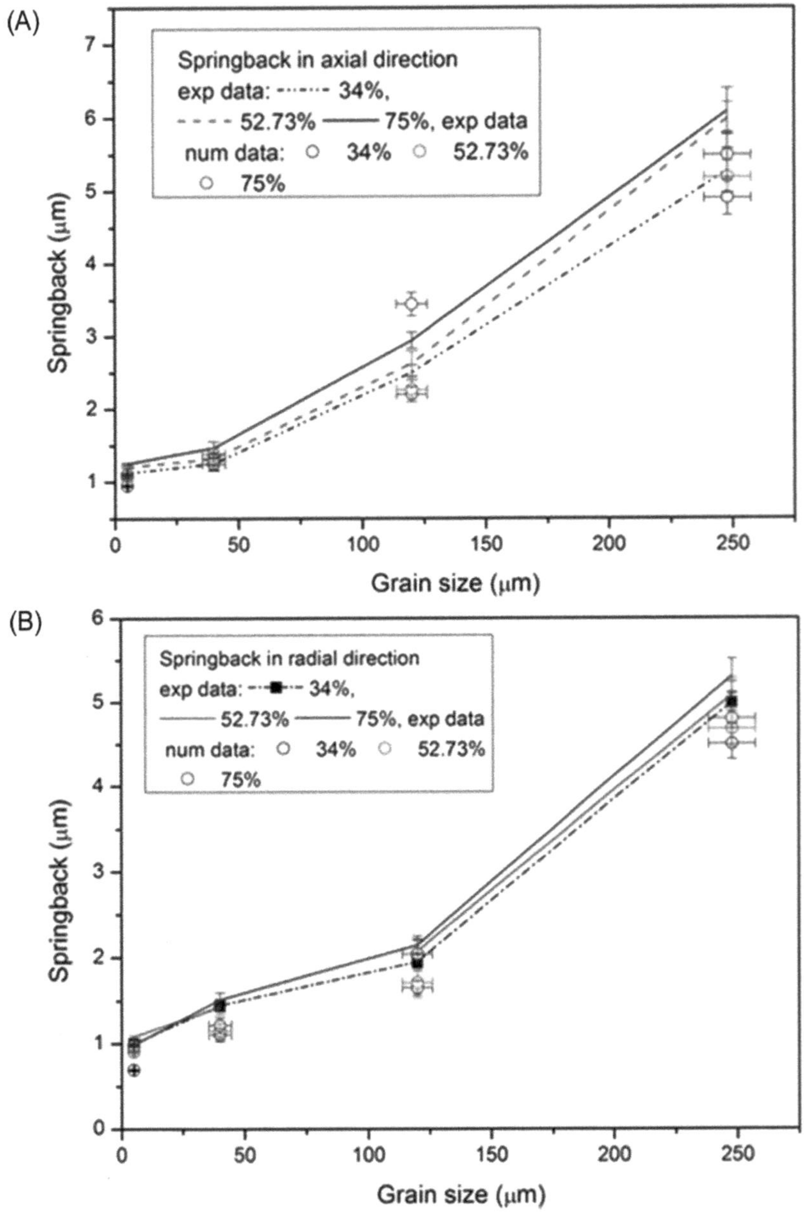

FIGURE 14.31 Influence of cross-sectional area reduction on springback in: (A) axial direction and (B) radial direction.

REFERENCES

[1] H.N. Lu, A study on the micro cross wedge rolling of metals, Doctoral Dissertation, University of Wollongong, Australia, 2013.

[2] A.F. Baun, Forgings of Cross Wedge Rolling, Mashgiz, Moscow, 1959.

[3] Z. Pater, A study of cross wedge rolling process, J. Mater. Process. Technol. 80–81 (1998) 370–375.

[4] Z. Pater, Tools optimisation in cross wedge rolling, J. Mater. Process. Technol. 138 (2003) 176–182.

[5] H.N. Lu, D.B. Wei, Z.Y. Jiang, K. Manabe, T. Furushima, Study on size effect in micro cross wedge rolling, Steel Res. Int. 82 (2011) 962–967.

[6] Q. Li, M. Lovell, Cross wedge rolling failure mechanisms and industrial application, Int. J. Adv. Manuf. Technol. 37 (2008) 265–278.

[7] U. Engel, Tribology in microforming, Wear 260 (2006) 265–273.

[8] T. Shimizu, Y. Murashige, K. Ito, K. Manabe, Influence of surface topographical, interaction between tool and material in micro-deep drawing, J. Solid Mech. Mater. Eng. 3 (2009) 397–408.

[9] M. Ramesh, S. Kailas, K. Simha, Micro and macro contact mechanics for interacting asperities, Sadhana 33 (2008) 329–338.

[10] F. Vollertsen, Categories of size effects., Prod. Eng. Res. Dev. 2 (2008) 377–383.

[11] M.A. Osman, M. Shazly, A. ElMokaddem, A.S. Wifi, Springback prediction in V-die bending: modelling and experimentation, J. Achiev. Mater. Manuf. Eng. 38 (2010) 179–186.

[12] M. Geiger, M. Kleiner, R. Eckstein, N. Tiesler, U. Engel, Microforming, CIRP Ann.—Manuf. Technol. 50 (2001) 445–462.

[13] H. Li, X. Dong, Y. Shen, A. Diehl, H. Hagenah, U. Engel, et al., Size effect on springback behavior due to plastic strain gradient hardening in microbending process of pure aluminum foils, Mater. Sci. Eng.: A 527 (2010) 4497–4504.

[14] J.T. Gau, C. Principe, M. Yu, Springback behavior of brass in micro sheet forming, J. Mater. Process. Technol. 191 (2007) 7–10.

[15] H.N. Lu, D.B. Wei, Z.Y. Jiang, D. Wu, X.M. Zhao, Study on the influence of temperature on the surface asperity in micro cross wedge rolling, The 11th International Conference on Numerical Methods in Industrial Forming Processes: NUMIFORM 2013, May 2013, AIP Publishing, Shenyang, China, 2013, pp. 1032–1037.

[16] B. Bhuashan, M. Nosonovsky, Scale effects in friction using strain gradient plasticity and dislocation-assisted sliding (microslip), Acta Mater. 51 (2003) 4331–4345.

[17] M.F. Ashby, The deformation of plastically non-homogenous materials, Philos. Mag. 21 (1970) 339–424.

[18] W.D. Nix, H. Gao, Indentation size effects in crystalline materials—a law for strain gradient plasticity, J. Mech. Phys. Solid (1997) 411–425.

[19] J. Friedel, Dislocations, Pergamon Press, Oxford, New York, 1964.

[20] J. Weertman, J.R. Weertman, Elementary Dislocations Theory, The MacMillan Co., New York, 1966.

[21] R.M. Cleveland, A.K. Ghosh, Inelastic effects on springback in metals, Int. J. Plast. 18 (2002) 769–785.

[22] S. Chatti, N. Hermi, The effect of non-linear recovery on springback prediction, J. Comput. Struct. 89 (2011) 1367–1377.

Chapter 15

Practice of Micro Flexible Rolling

15.1 EQUIPMENT AND TOOLS FOR MICRO DEEP DRAWING

Fig. 15.1 shows the micro flexible rolling (MFR) mill systems, including rolling set system, driving system, and operation panel. The mill stand consists of a mill house containing two cylindrical work rolls, which perform the reduction. The rolling force can be measured by a load cell.

Fig. 15.2 shows the driving device. This universal joint has increased joint twist angle, improved drive dynamic performance, increased drive service life, noiseless operation, and high performance in transferring the required torque.

Fig. 15.3 illustrates roll screw down system. Roll position is controlled by a closed loop position regulator that automatically adjusts the flow according to a hydraulic servo valve. There are regulators on both sides of operator and drive, respectively. Both sides are coupled together by software control. The control maintains the level of two sides when they move together. The hydraulic cylinders set the pass line and change gap opening while rolling to maintain strip thickness. The position regulator calculates the error between the reference and feedback position. The error is used by a speed balance control to keep either the operator or drive side from lagging behind during long movements of both sides. The control acts by reducing the servo valve flow command for the side with the smaller error and increasing it for the side with the larger error. Hydraulic cylinder control also includes a total force regulator with a differential position loop to maintain the stand level.

Fig. 15.4 shows control panel in the flexible rolling systems. This is a user-friendly color operation panel. Operator can input or revise rolling parameters via the touch screen panel.

Table 15.1 lists the basic parameters of MFR mill, where the rolling speed is 1–2 m/min and the rolling force limit is 15 KN.

Generally speaking, flexible rolling can be performed by changing the roll gap during rolling [1]. The difficulty is the precise control of size and shape of longitudinal section in each variable thickness region, and the control is usually accomplished using absolute automatic gage control. In

Microforming Technology. DOI: http://dx.doi.org/10.1016/B978-0-12-811212-0.00015-7

FIGURE 15.1 Micro flexible rolling systems.

FIGURE 15.2 Driving device.

FIGURE 15.3 Roll screw down system.

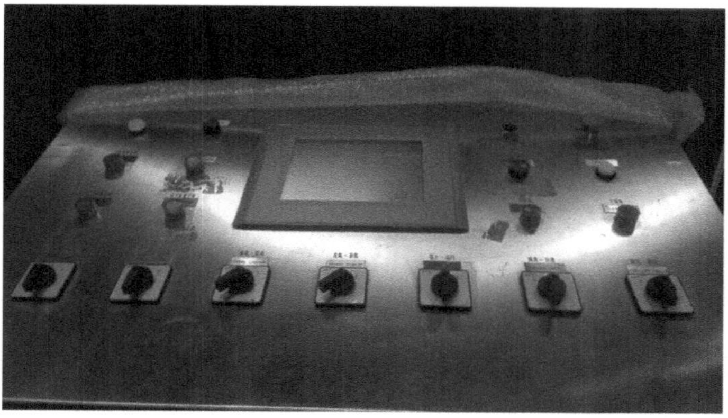

FIGURE 15.4 Control panel.

TABLE 15.1 Parameters of Micro Flexible Rolling Mill

Material Width (mm)	Material Thickness (mm)	Rolling Force Limit (KN)	Rolling Speed (m/min)	Roll Diameter (mm)
10–20	0.1–0.5	15	1–2	25

order to obtain the products with various thicknesses, rolling mill and its control system need to have the following characteristics and capabilities: rolling mill with high stiffness; stand and roll system with good manufacturing accuracy; main motor with sufficient power margin, good speed response characteristics, and regulating performance; hydraulic system with fast response, especially the servo valve and cylinder with high frequency response characteristics; the absolute Automatic Gauge Control (AGC) system and high-precision mathematical models; good detectors for rolling parameters, such as workpiece and rolling speed, displacement, force and energy, and data acquisition and processing system. It is better to have the self-learning function in the process control system.

15.2 MICRO FLEXIBLE ROLLING PRACTICE

15.2.1 Flexible Rolling

In MFR, the rolled material can be made to different shapes in order to obtain the best results to save metals, as shown in Fig. 15.5. These flexible rolled products would be superior to tailor welded blank (TWB) on

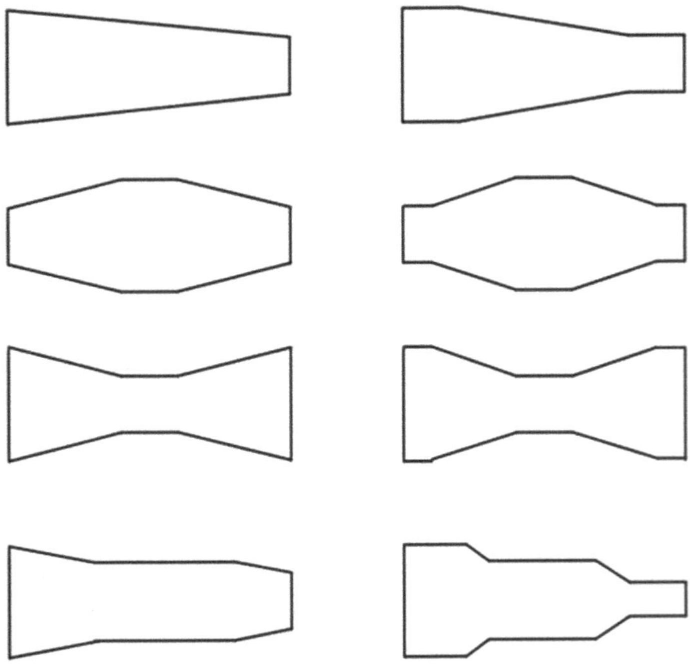

FIGURE 15.5 Products in flexible rolling.

connection strength, surface quality, and flexibility of thickness combination. Compared with conventional rolling, flexible rolling can be realized through equipped with advanced computer control systems and high-precision mathematical models as well as improved hydraulic system and automatic control technology [2].

Dynamic adjustment of roll gap is required to product different thickness in rolling direction when the flexible rolling is used. Therefore, a new rolling forming theory has to be set up to solve the new problems in flexible rolling. This will lead to new problems, and as a result, a new forming theory is needed. The formulas for deformation parameters, such as forward slip, spread, and rolling force, should be derived based on new forming conditions. As shown in Fig. 15.6, MFR is divided into three stages: common rolling (CR), upward rolling (UR), and downward rolling (DR). Thick region and thin region are rolled by CR, whereas the transition region (variable thickness) is rolled by UR or DR.

15.2.2 Materials and Rolling Procedure

The rolled thin strip is 6061 aluminum alloy with thickness of 0.792 mm, length of 150 mm, and width of 15 mm. The details about chemical

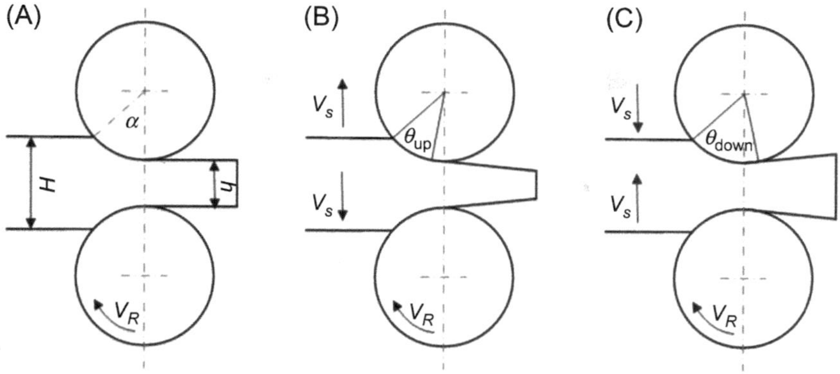

FIGURE 15.6 Three stages of flexible rolling: (A) common rolling, (B) upward rolling, and (C) downward rolling.

TABLE 15.2 Chemical Composition (wt%) of the Investigated Alloy

Si	Mg	Fe	Ti	Cu	Mn	Cr	Zn	Al
0.6	0.85	0.7	0.15	0.22	0.15	0.17	0.25	Balance

compositions of the rolled material are given in Table 15.2. Before flexible rolling, all the strips were suffered temper treatment. The specimens were divided into two groups based on the different rolling parameters with the purpose of studying the influences of thickness ratios (the ratio of thickness at thicker zone to that of thinner zone) on the microstructure and mechanical properties of flexibly rolled products in the transition zone. The rolling schedules of the groups 1 and 2 are as presented in Tables 15.3 and 15.4, respectively. Rolling tests were conducted on a flexible rolling mill at room temperature. The diameter and roll barrel length of the work roll are 25 and 40 mm, respectively.

Fig. 15.7 shows a microscope of the cross section of the flexibly rolled thin strip, from which thicker zone, transition zone, and thinner zone can be clearly distinguished. The thickness in the transition zone changes continuously with the progress of reduction. As a result, a continuous and smooth thickness distribution without stress peaks in the transition zone will lead to improved formability. It is quite different from TWB which produces discontinuous welding defects such as porosity, cracks, and voids that will affect the formability of the product.

Specimens for metallographic observations were cut from the rolled strips along both the rolling and normal directions, then mechanically ground and polished by 0.25 μm alumina (<40 nm). Subsequently, specimens were

TABLE 15.3 Rolling Schedule of the Group 1

Group	Specimens	Initial Thickness (mm)	Thickness at Thicker Zone (mm)	Reduction at Thicker Zone (%)	Thickness at Thinner Zone (mm)	Reduction at Thinner Zone (%)	Thickness Ratio	Rolling Speed (cm/min)
G1	S1	0.792	0.781	1.39	0.229	71.09	3.41	40
	S2	0.792	0.781	1.39	0.403	49.12	1.94	40
	S3	0.792	0.781	1.39	0.587	25.88	1.33	40
	S4	0.792	0.781	1.39	0.677	14.52	1.15	40

TABLE 15.4 Rolling Schedule of the Group 2

Group	Specimens	Initial Thickness (mm)	Thickness at Thicker Zone (mm)	Reduction at Thicker Zone (%)	Thickness at Thinner Zone (mm)	Reduction at Thinner Zone (%)	Thickness Ratio	Rolling Speed (cm/min)
G2	S5	0.792	0.779	1.64	0.348	56.06	2.24	20
	S6	0.792	0.779	1.64	0.348	56.06	2.24	40
	S7	0.792	0.779	1.64	0.348	56.06	2.24	80
	S8	0.792	0.779	1.64	0.348	56.06	2.24	160

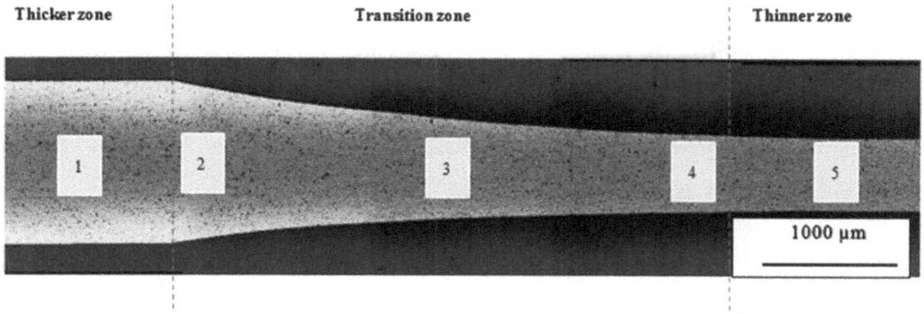

FIGURE 15.7 Assembled image of TRB strip by the 3D-laser scanning microscopy.

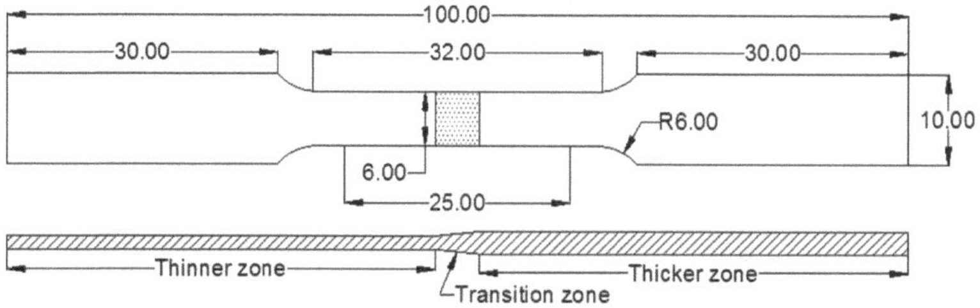

FIGURE 15.8 Geometry and dimensions of TRB tensile test specimens with a transition zone.

electrolytic etched with Barker's etchant (5 ml HBF_4 in 200 ml distilled water) at 24 V for 1.5 min by using stainless steel cathode according to ASTM E407 for grain structure observation under polarized light. The microstructure was examined by a KEYENCE VK-X100 3D-laser scanning microscope.

15.2.3 Experimental Methodology

15.2.3.1 Tensile Test

Tensile tests were conducted on an INSTRON 4206-006 universal testing machine with a strain rate of $8 \times 10^{-4}\ s^{-1}$ at ambient temperature to evaluate tensile properties of the rolled aluminum thin strips produced under a variety of rolling schedules. The tensile specimens were machined along the rolling direction with the rolled transition zone oriented in the middle of the gage section that is perpendicular to the tensile axis, as shown in Fig. 15.8. The dimensions of the specimens were determined according to ASTM standards (E8M), and this method has been widely used to evaluate the tensile properties of metal strips with varying thickness [3,4]. For each rolled thin strip, three specimens were prepared for the tensile tests.

A parameter β was well used to evaluate the quality and strength of flexibly rolled thin strips by reflecting the propensity of the specimens to experience strain localization in the transition zone and fail during forming [5]:

$$\beta = \frac{\varepsilon_T}{\varepsilon_O} \tag{15.1}$$

where ε_T is the strain occurred in the transition zone and ε_O is the overall tensile strain obtained according to the gage length.

15.2.3.2 Hardness Test

A Vickers hardness testing machine was employed to determine the hardness differences of all the rolled strip specimens in the thicker, thinner, and transition zones by applying a load of 1 N with a dwelling time of 12 s. Hardness measurements were performed on the cross-section surfaces of the rolled strips along the rolling direction. Five measured positions with three reduplicate tests were made for each specimen, as indicated in Fig. 15.7.

15.2.3.3 Fractography

The fracture surfaces of the specimens after tensile tests were observed on a QUANTA 600 scanning electron microscope (SEM) operated at 20 kV to determine the fracture features and failure modes.

15.2.4 Results and Discussion

15.2.4.1 Microstructural Evolution

The initial microstructure of the AA 6061-T6 alloy strip is shown in Fig. 15.9, from which the typical rolled structure with grains elongated can be clearly observed.

After flexible rolling, all the microstructure evolution of G1 specimens observed showed significantly distinct influences ranging from transition zone to thinner zone (Figs. 15.10–15.12). As compared with the initial microstructure, no apparent change in microstructure in the thicker zone due to the small thickness reduction, as shown in Fig. 15.10. To examine the features of microstructure evolution in the transition zones, each specimen was measured at position 2 as shown in Fig. 15.7. By contrast, the lamellar spacing of the elongated bands decreases gradually in the transition zones of S1, S2, S3, and S4 because the different rolling reductions acted on these sections (as shown in Fig. 15.11). As indicated in Fig. 15.11A, it is noted that S1 with the biggest thickness ratio of 3.41 induces the highest trend line slope of deformed grains in the surface layer. Also, the slope of G2 shows a downward trend with the decrease of thickness ratio, as shown in Fig. 15.12B–D. Furthermore, the microstructure evolution of the thinner zone (Fig. 15.12) performs a trend that the aspect

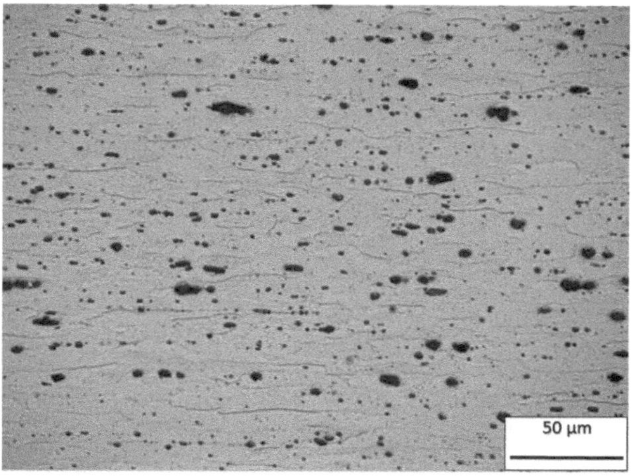

FIGURE 15.9 Initial microstructure of the AA 6061-T6 alloy strip.

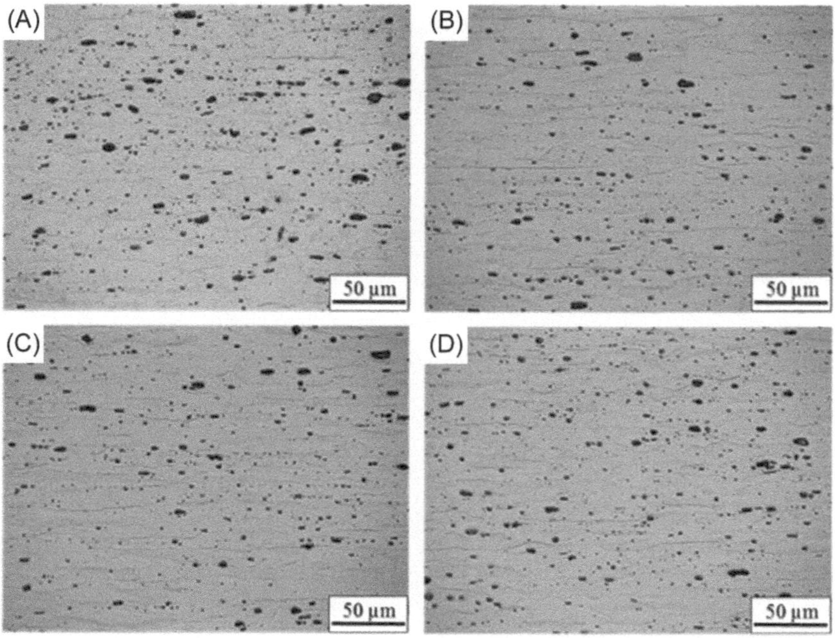

FIGURE 15.10 Microstructures of (A) S1, (B) S2, (C) S3, and (D) S4 in the thicker zones.

ratio of the flexibly rolled specimens was clearly aggravated from S4 to S1 gradually. This is also in accordance with the corresponding rolling schedule (Table 15.3) that a higher reduction always leads to a heavier severe deformation.

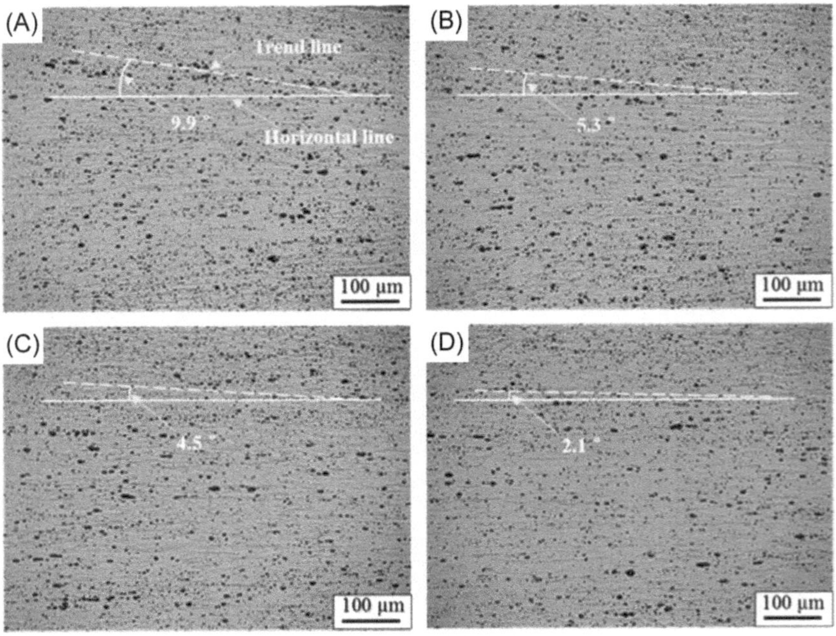

FIGURE 15.11 Microstructures of (A) S1, (B) S2, (C) S3, and (D) S4 in the transition zones.

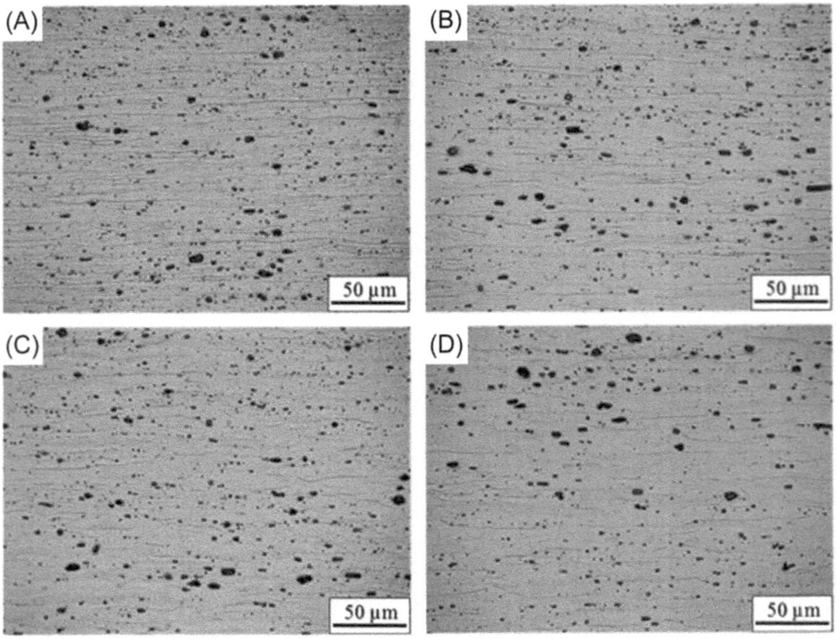

FIGURE 15.12 Microstructures of (A) S1, (B) S2, (C) S3, and (D) S4 in the thinner zones.

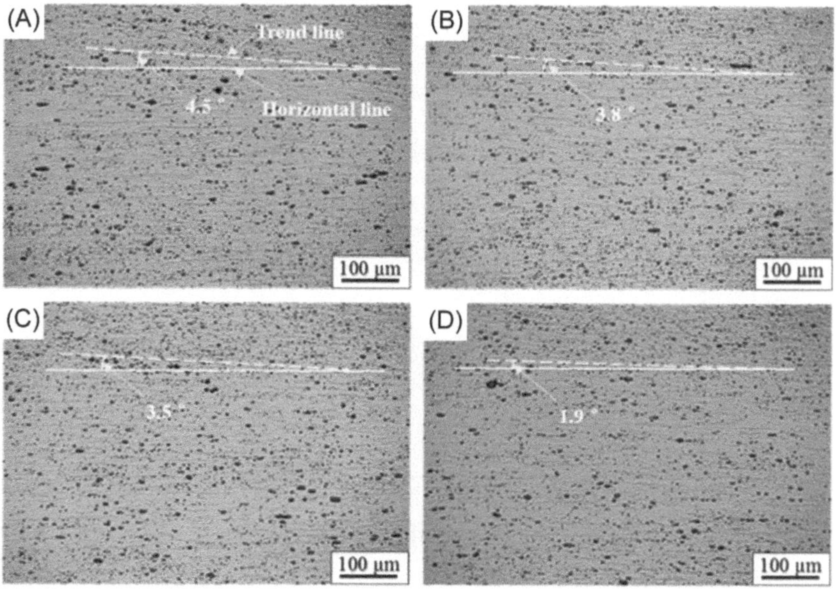

FIGURE 15.13 Microstructures of (A) S5, (B) S6, (C) S7, and (D) S8 in the transition zones.

For G2, the transition zone lengths have quite similar effects on the microstructure compared to that of G1 as shown in Fig. 15.12. Higher lamellar spacing density also occurred in the right side of each microstructure photo in Fig. 15.13, and the angle between trend line and horizontal line reduces from 4.5 degrees for S5 to nearly 1.9 degrees for S8. This can be understood the mechanisms in Figs. 15.16 and 15.17 and analyzed in the next part, which a longer transition zone length will lead to a more gently aspect ratio evolution of the deformed grains. All the behavior of deformed microstructures mentioned above can also be found the corresponding relationships to the mechanical properties of tensile tests, hardness tests, and fractographs discussed in the following sections.

15.2.4.2 Tensile Properties

Figs. 15.14 and 15.15 show the engineering stress–strain curves of G1 and G2, respectively. The tensile properties including yield strength (YS), ultimate tensile strength (UTS), and elongation are collected and presented in Table 15.5. The variations of tensile strength and elongation in the G1 group are related to the total reductions applied during flexible rolling. It can be seen that both the YS and UTS of the S1 specimen are the highest in the group of G1, and they decrease gradually with an order of S1, S2, S3, and S4. Differently, elongation shows an opposite trend as compared to the tensile strength. This means the strain hardening (cold working hardening)

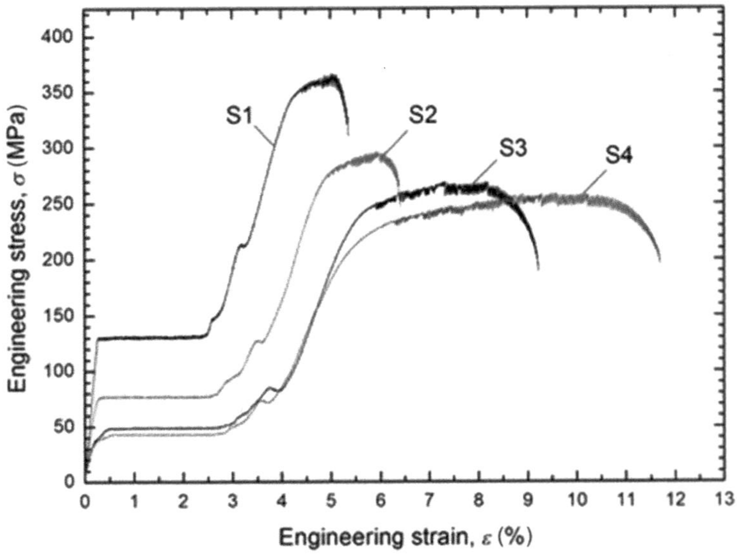

FIGURE 15.14 Engineering stress–strain curves of S1, S2, S3, and S4.

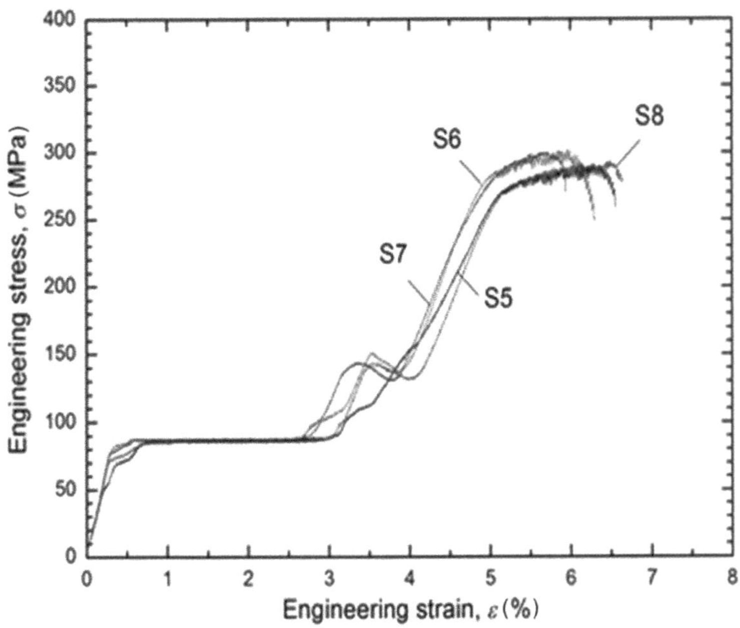

FIGURE 15.15 Engineering stress–strain curves of S5, S6, S7, and S8.

TABLE 15.5 Tensile Properties of G1 and G2 Specimens

Group	Specimen	Yield Strength (MPa)	UTS (MPa)	Elongation (%)
G1	S1	130.52	365.22	5.37
	S2	76.77	296.85	6.43
	S3	48.75	268.79	9.22
	S4	42.96	258.92	11.7
G2	S5	87.26	289.97	6.61
	S6	87.01	302.33	6.32
	S7	86.56	300.15	5.94
	S8	85.60	293.60	6.69

UTS: Ultimate Tensile Strength.

phenomenon appears since slips of deformed microstructure, dislocation tangle [6] which elongates the grains into high aspect ratio (Fig. 15.12) or crush during flexible cold rolling with plastic deformation under/below the recrystallization temperature [7]. In this case, strength and hardness of the S1 specimen will be enhanced in some extent while the ductility presents a relatively downward trend. In addition, for G2 as shown in Fig. 15.15 and Table 15.5, all of the tensile properties fluctuated around a specific value except somewhat drop in YS from 87.26 to 85.60 MPa. However, it is reasonable since the transition zone lengths only behave a little variation among the G2 specimens because of distinct rolling speeds.

It is worth mentioning that all the tensile curves exhibit discontinuous yielding behavior. This phenomenon can occur when the stress almost maintains a stable condition accompanied with an increase in strain. The dislocation interactions including low mobile dislocation density prior to deformation, a dramatically increase of dislocation quantity during deformation and dislocation proportion on an applied force play a dominated role in this phenomenon [8].

Figs. 15.16 and 15.17 illustrate the mechanisms under different tensile testing conditions. S1–S8 represent their corresponding deformation behavior with different rolling schedules. Once a tensile force is acted on both sides of a specimen, such as S1 (as shown in Fig. 15.16), it is a conventional deformation procedure that the slope of S1 (T1) will gradually experience the trend lines from T2 to T4, which fortunately agrees with the evolution of S2–S4. Similarly in Fig. 15.17, the slope of S5 will bring the same process from S6 to S8 in an inherent sequence. It means S1 and S5 which possess the biggest thickness ratio in G1 or the shortest transition zone length in G2, respectively, need to overcome the highest tensile force and deformation resistance to realize the same metal flow during this tensile process hence the higher YS and lower ductility. This is highly identical to the results of mechanical behavior. A similar phenomenon was found and reported by Kopp et al. [9] that a longer transition zone will give a rise to less wrinkling. In addition, higher deformed grains with the larger angles (as shown in Figs. 15.11A and 15.13A) gathered in these two specimens which will be with more likelihood to generate a higher dislocation density and stress concentrations should be another reason for this phenomenon. Consequently, S1 and S5 got the highest YS and S4 and S8 obtained the lowest ones in their own groups (as shown in Table 15.5). However, it is evident that the thickness ratio is more important to affect mechanical properties of tailor rolled blank (TRB) compared with the transition zone length.

Although all the necking sites occurred in the thinner zones, transition zones still made a contribution to participate the tensile deformation. This can be certified by the variations of transition zone lengths before and after tensile tests (as shown in Table 15.6). Clearly, the transition zone length after flexible rolling decreased with the increase of the thickness ratio for G1, but lead to an upward trend with the growing rolling speed for G2. After tensile tests, all the transition zone lengths of each specimen displayed the same changing trend compared to the flexibly rolled ones. However, the strain in the transition zone ε_T leads to the opposite trend ranging from 1.00% to 2.36%, which is in accordance with the change of the overall strain ε_0. This can also be understood the deformations along the different thickness zones for the G1 are relatively stable. Although there is no significant difference for the β of G1, a slight decrease could be observed from 0.27 (S2) to 0.20 (S4). The reason can be explained that a smaller thickness difference between the thicker zone and the thinner zone, a stronger bonding force in the transition zone hence a fewer strain ratio happened. It is worth to be noted that S1 got the lowest β value which is mainly because of the work hardening mentioned before. For G2, both of the strain in the transition zone and β displayed a pronounced drop even though no obviously influence on the overall strains of G2. This means the thickness ratio has a more important effect on the overall strain than that of the transition zone length, while the transition zone length plays a more significant role in the strain of this zone.

Moreover, the results of two group tensile tests implied that a smaller obliquity happened between the transition zone bevel and the tensile force direction, the transition zone will bear a higher tensile stress which results in a better ductility and a fewer strain happened in this zone. This is strongly in agreement with the discussion and analysis of Figs. 10.16 and 10.17 hereinbefore.

TABLE 15.6 Variation of Transition Zone Length and Corresponding Strains

Group	Specimen	Transition Zone Length After Flexible Rolling (mm)	Transition Zone Length After Tensile Test (mm)	Overall Strain ε_O (%)	Strain in the Transition Zone ε_T (%)	β	Fracture Site
G1	S1	4.91	5.16	5.37	1.00	0.18	Thinner zone
	S2	4.70	5.13	6.43	1.72	0.27	Thinner zone
	S3	3.94	4.49	9.22	2.19	0.24	Thinner zone
	S4	3.61	4.20	11.7	2.36	0.20	Thinner zone
G2	S5	4.23	4.83	6.61	2.40	0.36	Thinner zone
	S6	4.58	5.01	6.32	1.72	0.27	Thinner zone
	S7	5.45	5.57	5.94	0.48	0.08	Thinner zone
	S8	6.70	6.77	6.69	0.28	0.04	Thinner zone

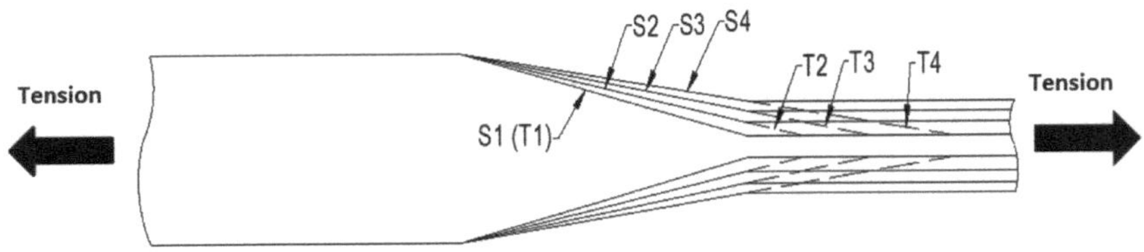

FIGURE 15.16 Deformation mechanisms of TRB with different thickness ratios in tensile tests.

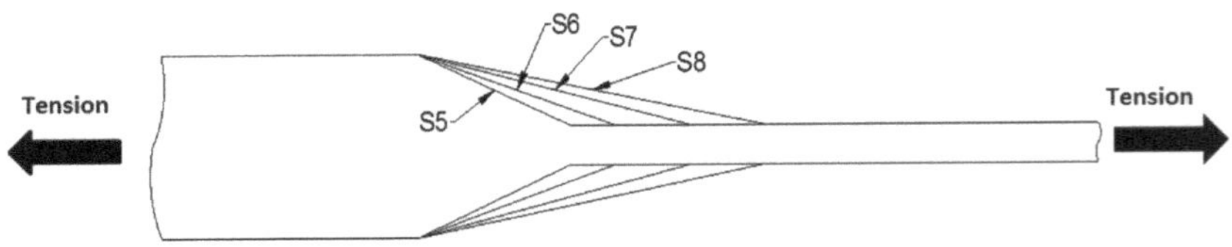

FIGURE 15.17 Deformation mechanisms of TRB with different transition zone length in tensile tests.

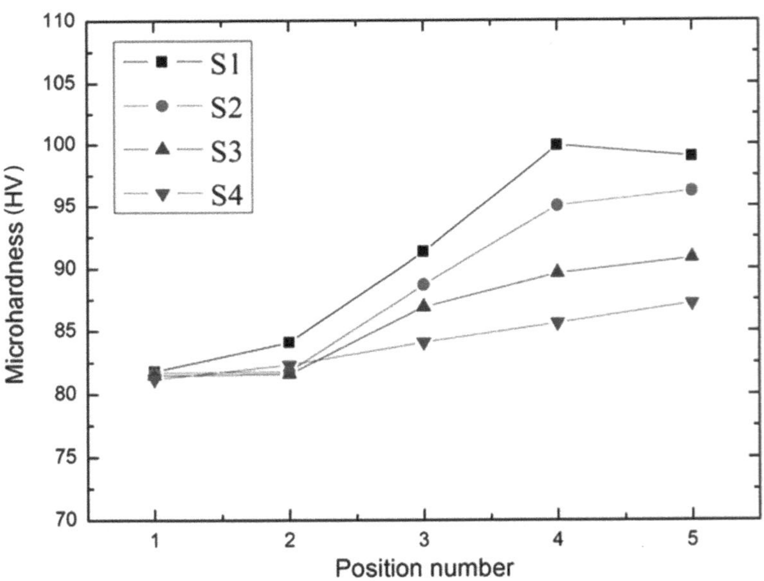

FIGURE 15.18 Changes in Vickers hardness of G1 specimens along the thickness direction in five measured positions.

15.2.4.3 Micro Hardness

The Vickers micro hardness test results of TRB in the two groups, which were carried out from thicker zone to the thinner zone and across the transition zone of each specimen cross-section as shown in Figs. 15.18 and 15.19. By contrast, all micro hardness profiles in the G1 specimens present the increasing

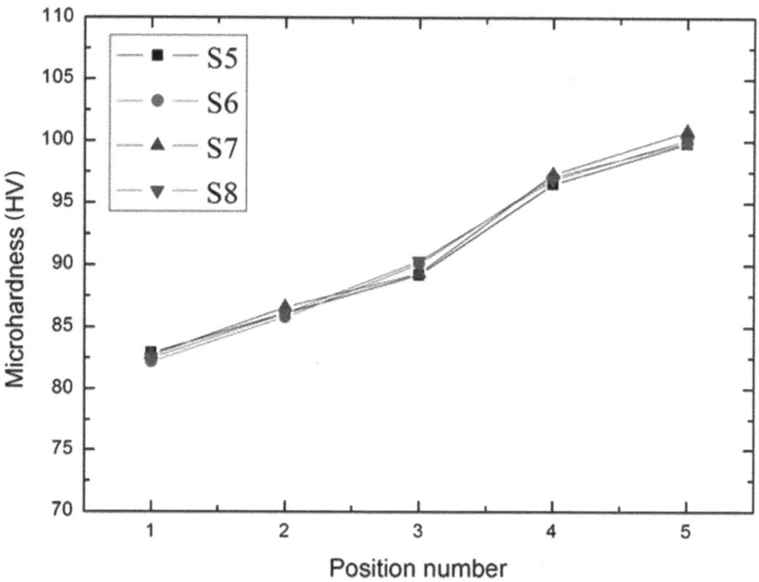

FIGURE 15.19 Changes in Vickers hardness of G2 specimens along the thickness direction in five measured positions.

trends from thicker zone (position 1) to thinner zone (position 5) except S1 which indicated a peak value in position 4. The reason should be there is a relatively stronger stress concentration induced by the biggest thickness difference area which results into a notch effect occurred in the discontinuous face [10]. This would also decrease the fatigue limit which can be eliminated by a corresponding heat treatment procedure such as annealing method. The degree of work hardening (discussed hereinbefore) occurred in thinner zone of S1 and S2 is relatively higher than that of S3 and S4 because of the higher thickness ratio. Hence, the micro hardness in the thinner zone with different thickness ratios of 3.41 (S1), 19.4 (S2), 1.33 (S3), and 1.15 (S4) display about 7–21% greater than that of thicker zone (as shown in Fig. 15.18). While a regular phenomenon, as depicted in Fig. 15.19, is that all the micro hardnesses exhibit the almost identical upward trend despite each specimen (S5–S8) varies the transition zone length. Meanwhile, there is hardly any change among these specimens since all the corresponding measured positions have the quite similar thicknesses.

15.2.4.4 Fractography

The fracture limits for all TRB specimens can be found in Figs. 15.20 and 15.21. Due to only quite limited differences for G2 specimens in tensile properties including YS, UTS, elongation, and toughness, thus the fractural photographs of G2 showed almost the same properties with heterogeneous

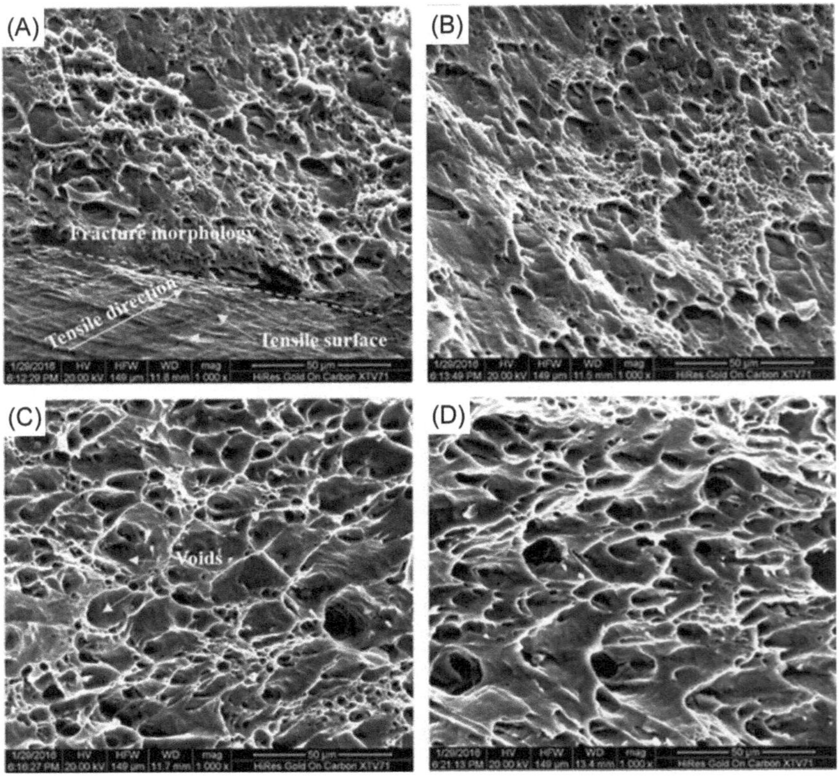

FIGURE 15.20 Fracture morphology comparisons of (A) S1, (B) S2, (C) S3, and (D) S4 by SEM.

sizes of voids. By contrast, fractographs of the tensile tests for all G1 specimens are shown in Fig. 15.20. It is apparent that the typical ductile fracture morphologies with round or elliptical ductile dimples displayed on the fracture surfaces in all cases. Note that the volume fraction of the fatal micro voids can directly affect the ductile fracture properties [11]. Generally, a ductile dimple with bigger diameter, larger depth, and higher uniform density voids contributes to a fully growth of the defects, which can exhibit a greater plasticity with a better fracture resistance to crack. Simultaneously, the second-phase particles or inclusions are the major factors for the forming of ductile dimples. They will always be found at the bottom of the ductile dimples [12] which are related to nucleation, grow, and coalescence of voids at hard particles. In fact, the voids will emerge by particle cracking consisting of radius, shape, and volume fraction [13], decohesion surrounding the constituent particles or pre-existing in the initial materials. For all fractographs in the present study, the fracture morphology of S4 showed much bigger, deeper, and more uniform ductile dimples followed by S3, S2, and

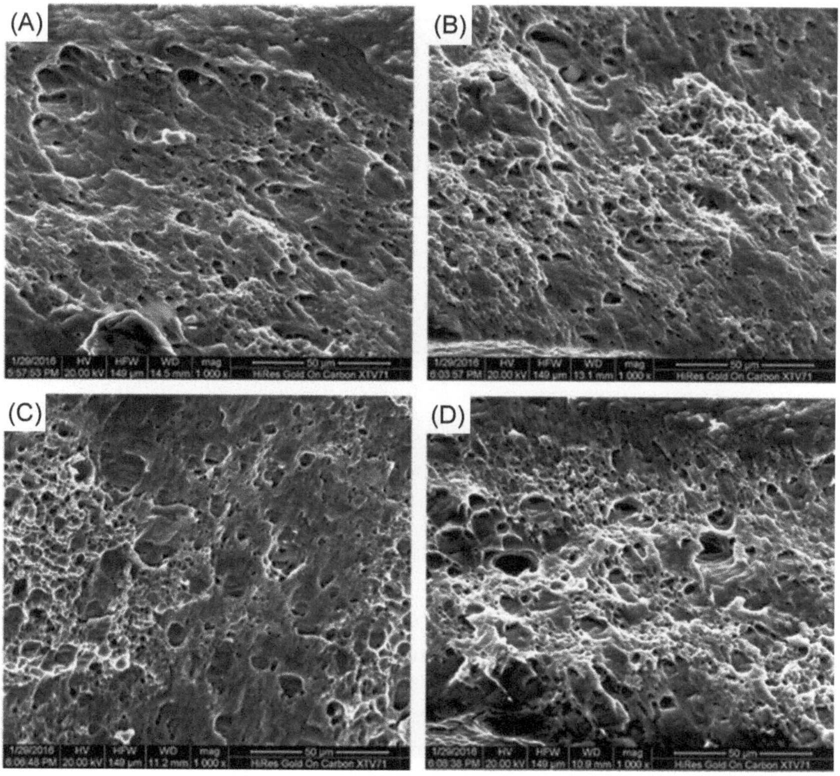

FIGURE 15.21 Fracture morphology comparisons of (A) S1, (B) S2, (C) S3, and (D) S4 by SEM.

S1 gradually, which predicates it has absorbed a stronger stored energy from the plastic deformation of flexible rolling due to the smaller grain aspect ratio, and then got a relatively better ductility. On the contrary, the fracture surface of S1 was full of heterogeneous size and depth voids hence it got the lowest elongation. This also verifies the results shown in Fig. 15.14 and Table 15.5 that the ductility (elongation to failure) increases gradually from 5.37% (S1) to 11.7% (S4). Therefore, plastic working is still a favorable and effective way to improve mechanical properties in discovering fracture mechanisms [11].

Based on the feasible amount of hardening degree, a pure shear deformation mode with the axisymmetrical stress–strain state could be prolonged or quite short [14]. Then, localized deformation would occur once the metal flow of the tensile specimen became unstable. For the same small necking area, the macro flow sequentially kept the axisymmetrical deformation while a micro localization could happen inside a thin planar layer composed with massive micro-shear bands. Generally, shear fracture angle (the angle of the

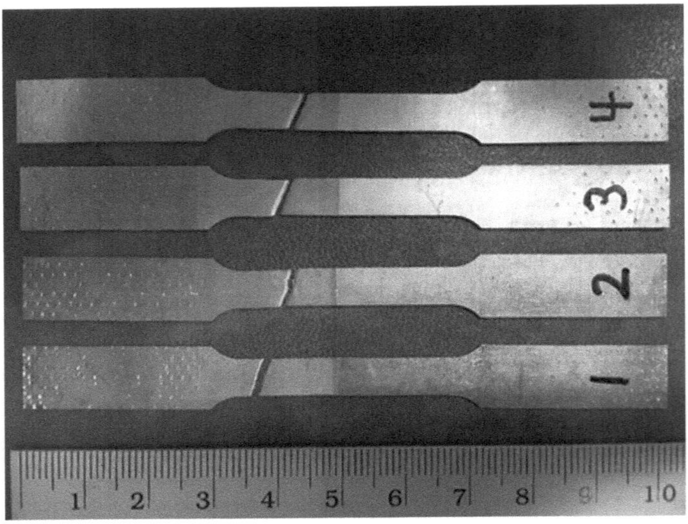

FIGURE 15.22 Shear fracture angle comparisons for G1 specimens.

fracture line with respect to the tensile direction) with 45 or 90 degrees should be most common in a typical uniaxial tensile test for a face-centered cubic metals and alloys. However, in this study as shown in Fig. 15.22, all the G1 specimens obtained the same fracture angle of approximate 65 degrees irrespective of the thickness ratios. This phenomenon is also occurred in the results of references [15,16] is suggested that the fracture mechanism of present materials can be shifted from necking fracture to shear fracture mode with various fracture angles. Thus, it is still needed to be discussed about the fracture angle on the fracture mechanisms in TRB.

In short, the microstructure of the continuous and smooth thickness distribution along the transition zones of thin aluminum strips are obtained after flexible rolling under different rolling schedules. The microstructural distribution in the transition area is quite sensitive to rolling reductions. The flexibly rolled thin strip with the biggest thickness ratio displays the highest YS and UTS along with lowest elongation due to strongest work hardening degree. A specimen with a bigger thickness ratio or shorter transition zone length should bear higher deformation energy to achieve the identical metal flow in the tensile testing process, hence resulting in a higher YS and a lower ductility.

The variation of β certifies that transition zones would always participate in the tensile deformation even though all the necking sites occurred in the thinner zones. This also means a better bonding condition with excellent mechanical behavior happened in the transition zones for TRB compared to some cracks in weld seams of TWB. Moreover, thickness ratio has a more significant influence on the change of ε_0, while transition zone length plays a more predominant role in the variation of ε_T. Hardness of both G1 and G2 presents an upward trend from thicker zone to the thinner zone. Besides, the hardness in the thinner zone is about 7–21% greater than that of thicker zone for G1 since different reductions they underwent lead to corresponding degrees of work hardening. S4 exhibited the biggest, deepest, and most homogeneous ones which bring out the greatest ductility followed by S3, S2, and S1 gradually. An unusual shear fracture angle with approximate 65 degrees occurred in all G1 specimens which imply mechanism shifted from necking fracture to shear fracture mode in TRB.

REFERENCES

[1] F.J. Qu, Z.Y. Jiang, H.N. Lu, Analysis of micro flexible rolling with consideration of material heterogeneity, Int. J. Mech. Sci. 105 (2016) 182–190.

[2] X.H. Liu, Prospects for variable gauge rolling: technology, theory and application, J. Iron Steel Res. Int. 18 (2011) 1–7.

[3] S.M. Chan, L.C. Chan, T.C. Lee, Tailor-welded blanks of different thickness ratios effects on forming limit diagrams, J. Mater. Process. Technol. 132 (2003) 95–101.

[4] S.K. Panda, D.R. Kumar, H. Kumar, A.K. Nath, Characterization of tensile properties of tailor welded IF steel sheets and their formability in stretch forming, J. Mater. Process. Technol. 183 (2007) 321–332.

[5] P.A. Friedman, G.T. Kridli, Microstructural and mechanical investigation of aluminum tailor-welded blanks, J. Mater. Eng. Perform. 9 (2000) 541–551.

[6] J. Banhart, Advanced Tomographic Methods in Materials Research and Engineering, Oxford University Publishing, Oxford, 2008.

[7] X. Wang, M. Guo, Y. Zhang, H. Xing, Y. Li, J. Luo, et al., The dependence of microstructure, texture evolution and mechanical properties of Al–Mg–Si–Cu alloy sheet on final cold rolling deformation, J. Alloys Compd. 657 (2016) 906–916.

[8] J.W. Zhao, Z.Y. Jiang, C.S. Lee, Effects of tungsten addition and heat treatment conditions on microstructure and mechanical properties of microalloyed forging steels, Mater. Sci. Eng., A 562 (2013) 144–151.

[9] R. Kopp, C. Wiedner, A. Meyer, Flexibly rolled sheet metal and its use in sheet metal forming, Sheet Metal 2005. 11th International Conference, SheMet'05, Erlangen, Germany (2005) 81–92.

[10] D. Anand, D.L. Chen, S.D. Bhole, P. Andreychuk, G. Boudreau, Fatigue behavior of tailor (laser)-welded blanks for automotive applications, Mater. Sci. Eng., A 420 (2006) 199–207.

[11] H. Toda, H. Oogo, K. Horikawa, K. Uesugi, A. Takeuchi, Y. Suzuki, et al., The true origin of ductile fracture in aluminum alloys, Metall. Mater. Trans. A: Phys. Metall. Mater. Sci. 45 (2014) 765–776.

[12] K.O. Pedersen, I. Westermann, T. Furu, T. Børvik, O.S. Hopperstad, Influence of microstructure on work-hardening and ductile fracture of aluminium alloys, Mater. Des. 70 (2015) 31–44.

[13] W.M. Garrison Jr, N.R. Moody, Ductile fracture, J. Phys. Chem. Solids 48 (1987) 1035–1074.

[14] V.M. Segal, Deformation mode and plastic flow in ultra fine grained metals, Mater. Sci. Eng., A 406 (2005) 205–216.

[15] F. Khakbaz, M. Kazeminezhad, Strain rate sensitivity and fracture behavior of severely deformed Al–Mn alloy sheets, Mater. Sci. Eng., A 532 (2012) 26–30.

[16] M. Wang, A. Shan, Effect of strain rate on the tensile behavior of ultra-fine grained pure aluminum, J. Alloys Compd. 455 (2008) L10–L14.

Chapter 16

Practice of Micro Ultrathin Strip Rolling

16.1 EQUIPMENT AND TOOLS FOR MICRO ULTRATHIN STRIP ROLLING

The research on multipass rolling deformation behavior of SUS 304 stainless steel foil has been carried out on the 2-high reversing micro rolling mill system which includes gear system, paralleled motor, gap adjustment device, micro rolls, shaft connection, and control system as shown in Fig. 16.1. The rolls are driven by two separated motors and this micro rolling mill is featured with adjustable and flexible rolling speed ratio for up and down rolls. The initial thicknesses of SUS 304 stainless steel foil are 50 and 100 μm with the foil width of 15 mm.

Multipass and reversing rolling is available in this rolling mill. This new micro rolling mill is scaled down and easy to move and install on the desk, and it is suitable for hard material rolling with low roll elastic deformation due to high rolling mill stiffness. The schematic rolling mill set is shown in Fig. 16.2 where the size of micro rolling mill is indicated. The length of rolling mill set is only 400 mm while the height is 130 mm. Various thickness reduction can be obtained through screwing the screw knob to adjust the roll gap.

Table 16.1 lists parameter conditions used in experimental and analytical models. The rolled foil is collected, measured, and compared after each pass. It is very important to use flat samples, especially relevant to the thin foil to ensure the measurement accuracy. The roll and the rolled foil are degreased with acetone before rolling in each pass. After rolling, the samples are carefully cut, trimmed, and stunk onto the smooth surface of one hard metal block for observation. It is noted that two materials of tool steel (SKD 11) and sintered hard alloy (cemented carbide) with high Young's modulus of 580 GPa are used as work roll materials in this rolling mill. The annealed austenitic SUS 304 stainless steel is prepared as experimental material and the chemical composition of SUS 304 stainless steel is listed in Table 16.2.

Microforming Technology. DOI: http://dx.doi.org/10.1016/B978-0-12-811212-0.00016-9

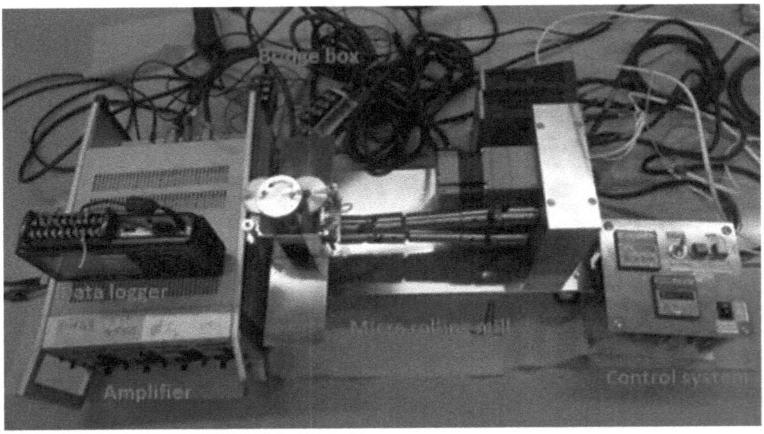

FIGURE 16.1 Desk-top micro rolling mills with flexible speed ratio.

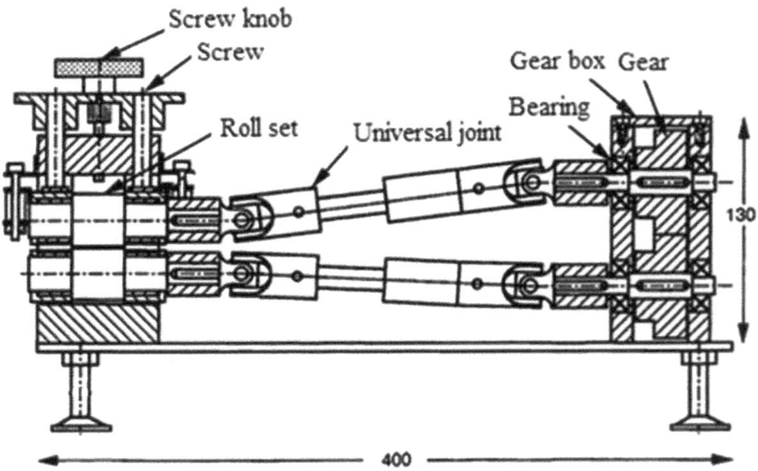

FIGURE 16.2 Schematic rolling mill, unit: mm.

TABLE 16.1 Parameter Conditions in Experimental and Analysis

Roll	Work Roll Diameter (mm)	Rolling Speed (rpm)	Roll Young's Modulus (GPa)	Passion Ratio
	27	0–100	580	0.30
Foil	Entry Foil Thickness (μm)	Reduction (%)	Young's Modulus (GPa)	Poisson Ratio
	50 and 100	0–50	210	0.30
Lubrication	Dry, O/W emulsion and oil lubricant (Niconic AWH) (viscosity: 44.13 mm^2/s in 40°C and 6.862 mm^2/s in 100°C)			

TABLE 16.2 Chemical Composition of SUS 304 Stainless Steel (wt%)

C	Si	Mn	P	S	Ni	Cr
0.05	0.44	1.18	0.033	0.001	8.22	18.23

16.2 MICRO ULTRATHIN STRIP ROLLING PRACTICE

16.2.1 Rolling Limit

The stainless steel foil is a hard material because of its high flow stress and low ductility. Such ultrathin foils are often obtained under low productivity with lots of rolling passes. Several attempts have been made to describe the thickness reduction limit in rolling caused by roll flattening. It has been shown that the rolling conditions such as roll diameter or Young's modulus of the work rolls influence the rolling limit [1,2]. There are several formula available for calculating the minimum thickness which can be rolled under a given roll diameter and a given yield stress in the rolled material, but in practice it is the maximum reduction, which can be obtained from the material of the given thickness and hardness [3].

$$\alpha = \frac{2\mu b}{h_1};\ g = \frac{\delta \pi E}{2(1-\nu^2)K_m \alpha b};\ \omega = \frac{p}{K_m};\ \beta = \frac{x}{b} \tag{16.1}$$

$$g = \int_0^1 \frac{\omega}{\alpha} \log \frac{1-\beta^2}{\beta^2} \mathrm{d}\beta \tag{16.2}$$

$$h \geq \frac{8(1-\nu^2)\mu K_m R}{\pi E} g_{min} \tag{16.3}$$

With front and back tensions, the minimum thickness can be formulated as below:

$$h_{min} = \frac{8(1-\nu^2)\mu (K_m - t_m) R}{\pi E} g_{min} \tag{16.4}$$

where b is the material width; h_{min} the minimum strip thickness after rolling, mm; α the coefficient related to the coefficient of friction (COF); R the radius of work rolls, mm; E the Young's modulus of the work roll, GPa; μ the COF; and K_m the deformation resistance, GPa.

The rolling conditions such as rolling force are set equally in each pass, and the foil thickness after each pass was measured and compared. Fig. 16.3 shows the rolling limit comparison of stainless steel foil under different rolling conditions via 2-high micro rolling mill. Generally it can be seen that the calculated minimum thickness limit from Eq. (16.4) is close to

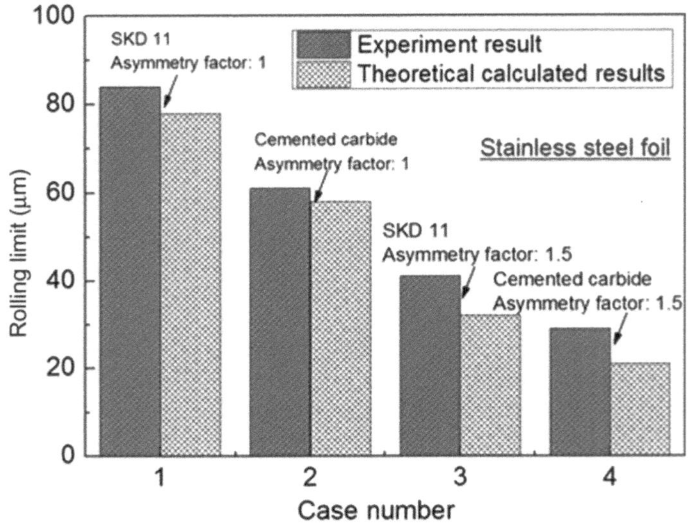

FIGURE 16.3 Rolling thickness limit under different experimental conditions: initial thickness: 100 μm; roll materials: SKD 11; and cemented carbide.

the experimental value. It is effective to improve the rolling limit of the micro rolling mill through increasing elastic modulus of the roll materials and using asymmetrical rolling because the elastic deformation of the work roll affects roll elastic deformation, then impacts the minimum thickness of the strip. Also cross-shear deformation area in asymmetrical rolling can reduce the rolling force and benefit getting thinner rolled foil. As for the work roll conditions, small diameter was likely to obtain high reduction in thickness from Eq. (16.4). A key feature of foil rolling is the fact that the elastic deformation of the work roll is strongly coupled to the plastic flow in the nip. Contact length is longer in foil rolling than that in thick strip rolling [4,5].

The foil cannot expand perpendicular to the direction of the highest pressure, i.e., toward the roll surfaces and expansion of the strip is in contradiction to the plastic flow rule. The relation between the relative surface speed of the foils and rolls must be consistent with the direction of the friction stress. When the thickness limit is reached, the curve of foil thickness becomes a horizon line, and however much force is applied, it is impossible to further reduce the thickness as shown in Fig. 16.4.

16.2.2 Rolled Material Profile Analysis

Sutcliffe and Rayner have conducted experimental measures of load and strip profile of cold rolling [6]. A comparison of the loads and rolled strip shapes has confirmed the general picture of thin strip rolling predicted by

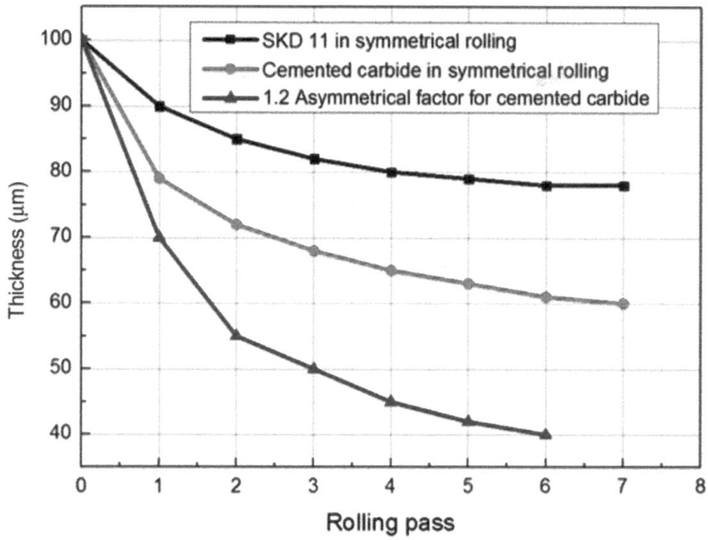

FIGURE 16.4 Rolling thickness limit for three different conditions.

Refs. [4,5]. No limiting gage is observed and the picture of a short region of reduction in the inlet is clearly observed. A relatively flat central region in strip profiles has been observed, and these results are the same with the Fleek's theory. Although the form of dependence of load on friction, reduction, and inlet strip thickness parameters is as predicted by theory, the measured loads are significantly lower than predicted for the thinnest strips [6]. In many cases, particularly at the higher loads, the length of the contact between rolls and strip is approximately equal to the roll radius (see Figs. 16.5 and 16.6). Clearly, the theoretical assumption that the roll can be modeled as an elastic half-space no longer takes effect. The sensitivity of the calculations to differences in the roll behavior was investigated by performing some calculations in which the undeformed roll shape was taken to be either circular or parabolic. Fig. 16.5 shows the rolling situation at a mediate strip thickness, while the thickness in Fig. 16.6 is thinner. Figs. 16.5A(ii) and 16.6A(ii) show substantial differences in the predicted strip reduction for these two roll shapes, despite the actual differences in profile being slight. As thickness decreases, the behavior of the contact is increasingly dominated by these details at the edge of the contact (Figs. 16.5A(i) and 16.6A(i)), in a way similar to that observed for elastohydrodynamic lubrication.

Figs. 16.7 and 16.8 illustrate the profile of rolled copper foil which was rolled from 0.12 mm down to 0.055 mm in thickness [7]. The foil strip was 22.5 mm wide. Two different roll forces, 225 and 740 kN, were applied. Two different holding times (0 and 15 s) were used for the case of a force of 740 kN. Fig. 16.7 shows the strip profile as the semithickness under

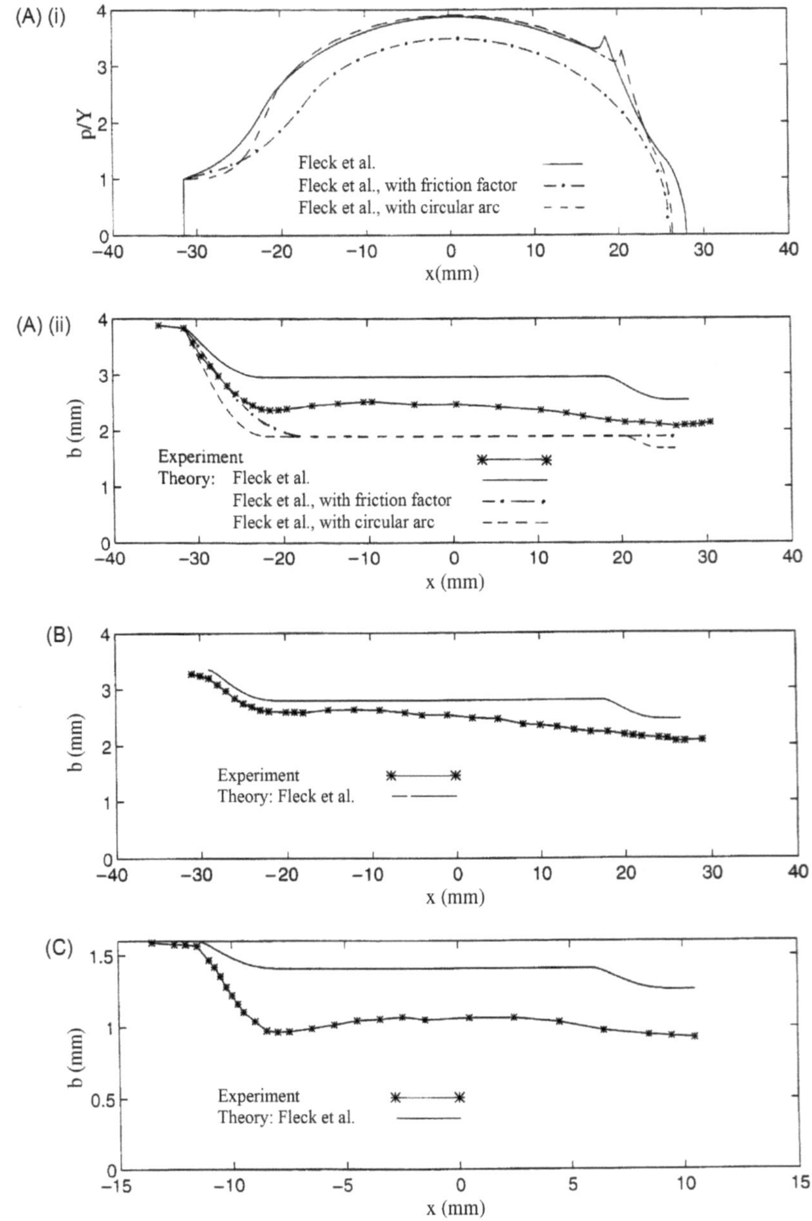

FIGURE 16.5 Mediate strip thickness results during rolling: (A) (i) theoretical pressure profile and (ii) strip profile, $B_0 = 3.0$, $W = 24$, $r = 0.47$, $U = 4.7$, $b_0 = 3.85$ mm, $R = 50$ mm, and $W' = 33$ kN/m. (B) Strip profile, $B_0 = 2.9$, $W = 31$, $r = 0.39$, $U = 4.9$, $b_0 = 3.35$ mm, $R = 50$ mm, and $W' = 37$ kN/m. (C) Strip profile, $B_0 = 3.1$, $W = 19$, $r = 0.39$, $U = 5.2$, $b_0 = 1.6$ mm, $R = 24$ mm, and $W' = 11$ kN/m (B_0: dimensionless inlet semithickness, W: strip width, r: overall reduction, U: dimensionless friction, b_0: semithickness of strip, R: roll radius, and W': line load) [6].

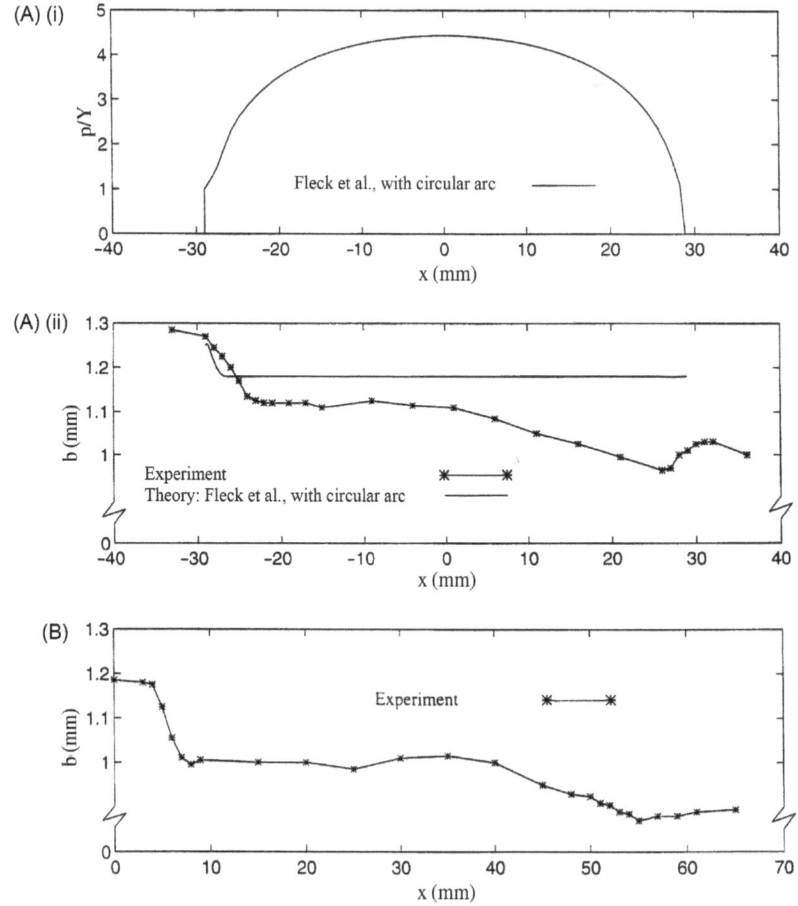

FIGURE 16.6 Thin strip rolling results: (A) (i) theoretical pressure profile and (ii) strip profile, $B_0 = 1.2$, $W = 32$, $r = 0.30$, $U = 4.7$, $b_0 = 1.25$ mm, $R = 50$ mm, and $W' = 32$ kN/m. (B) Strip profile, $B_0 = 1.4$, $W = 27$, $r = 0.14$, $U = 3.9$, $b_0 = 1.25$ mm, $R = 50$ mm, and $W' = 17$ kN/m [6].

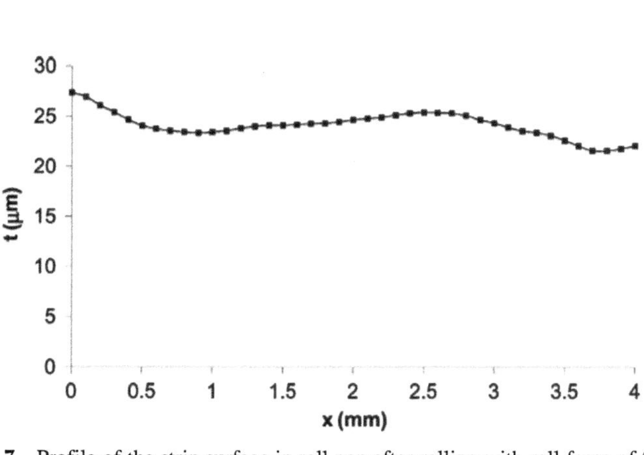

FIGURE 16.7 Profile of the strip surface in roll gap after rolling with roll force of 225 kN [7].

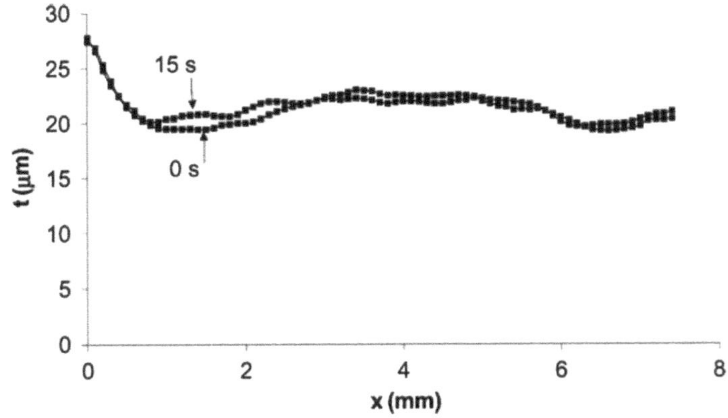

FIGURE 16.8 Profile of the strip surface in roll gap after rolling with roll force of 740 kN and two pressure releasing times: 0 and 15 s [7].

rolling force of 225 kN, and the pressure was released immediately when the roll rotation stopped. Fig. 16.8 illustrates the two profiles for the higher roll force of 740 kN: one for the case where the pressure was released immediately when the roll rotation stopped and the other for the case where the pressure was held for 15 s after the roll rotation stopped. The difference between these two cases can give some indication of the creep effect. It can be seen from Fig. 16.8 that the difference is small, so the creep effect in these experiments is small.

16.2.3 Lubrication Effect

16.2.3.1 Oil Lubrication in Metal Forming

In the rolling deformation region, both oil and water phases are mixed and entrained into the contact zone. The lubrication features depend not only on the surface roughness but also other factors such as relative sliding, surface waviness, lubricant viscosity, and thickness reduction. These factors influence the hydrodynamic state in rolling lubrication as well [8,9]. Fig. 16.9 illustrates schematically the contact areas between the roll and the rolled material and separated areas by an oil-film layer. Fig. 16.10 shows the micrograph of the rolled foil surface, in which there are many significant rolling traces along the rolling direction.

In the deformation regime, lubrication could be in a combined elasto-plasto-hydrodynamic lubrication: elastic deformation in the roll, plastic deformation in the strip, and hydrodynamical behavior in the lubrication oil when entrained by the motion of the strip and rolls. The pressure increases when the strip approaches the inlet zone before the contact zone by moving surfaces. The oil flow profile in the inlet zone is therefore a superposition of two flows: one is driven by the shearing force of the moving strip and roll

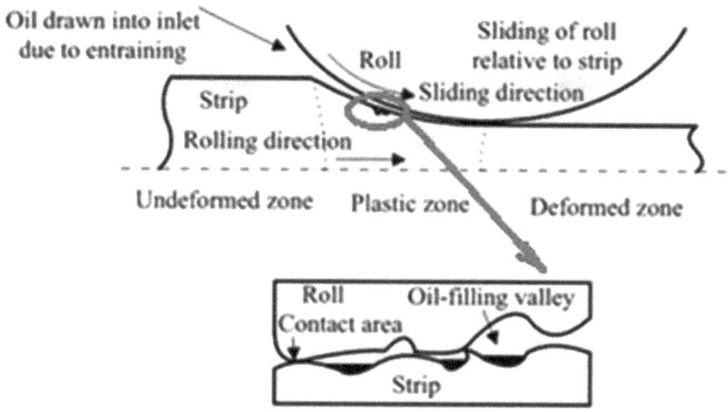

FIGURE 16.9 Schematic of lubrication mechanisms in cold rolling and details of contact.

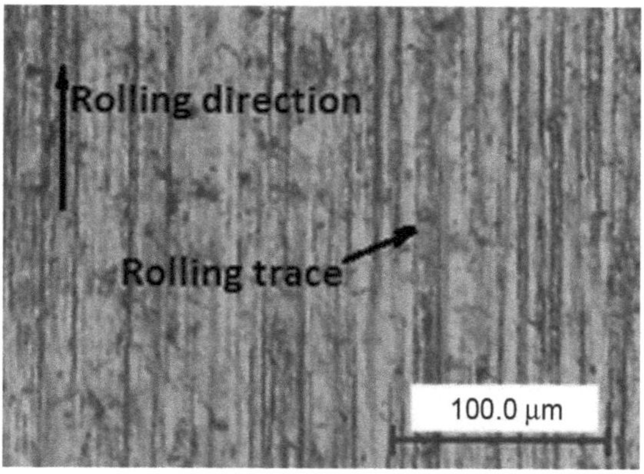

FIGURE 16.10 Micrograph of the rolled foil surface: (A) lubricant pressure and (B) contact area ratio in %.

surfaces and the other is driven by the pressure build-up in the inlet zone [10]. A full film exists in the valleys of the surface as shown in Fig. 16.9 where the rolling force and shear force are shared by asperities and the film. Rolling operations are normally run in mixed lubrication regime in which the load capacity is generated by the hydrodynamic pressure of the lubricant in between the valley of the roughness asperity and the pressure at the contacting asperity surfaces.

16.2.3.2 Mix Lubricant Analysis

In the mixed lubrication regime, the roll imprints its roughness onto the steel foil surface and the pressure increases with the decrease of oil-film thickness. The entry oil-film thickness is the oil-film thickness at a point where the pressure reaches the yield stress of the rolled foil. After entering the inlet zone, oil droplets undergo a competition between the lateral force that drives them toward the focus plane, and the force of the flow itself that at some point returns and flows back out of the entry zone. The existence of the lateral force leads to migration of oil droplets across streamline that eventually bends toward the plane of zero velocity passes through this plane and returns [10]. Since the lubricant film has lower shear yield stress than the steel foil and rolls, shearing will occur in the film instead of the rolled material. The lubrication film separates the surfaces and thereby prevents the mutual adhesion of the surfaces of the foil and rolls. Therefore, significant surface defects in lubrication rolling could be avoidable which caused by adhesion, such as material transfer between the foil and rolls.

For foil rolling, elastic deformation of the rolls needs to be included to predict the shape of the bite region. Because the roll bite is very long compared with the foil thickness, the rolling load or reduction is particularly sensitive to friction. Coupled solutions including roll deformation should be conducted in foil rolling. As the rolling speed falls, the area of contact ratio raises which could lead to a rise in friction and a very large increase in rolling load as well. Because of the considerable roll flattening during rolling, the predicted load is very sensitive to the details of both the friction and elastic roll deformation models, so that the reasonable agreement needs to confirm the accuracy of these models.

The mechanism of lubrication under asperity contacts has been termed micro plastomechanism of lubrication. Assumed that x-axis is the rolling direction, an averaged Reynolds equation could be derived to present the variation in the rolling direction of the reduced hydrodynamic pressure q_{v} of the lubricant as [11]:

$$\frac{\mathrm{d}(\gamma p_{\mathrm{v}})}{\mathrm{d}x} = 6\eta_0\gamma(v_{\mathrm{entry}} + v_{\mathrm{r}})\frac{h_{\mathrm{v}} - h^*}{h_{\mathrm{v}}^3(1 + 3{\psi_{\mathrm{c}}}^2/h_{\mathrm{v}}^2)} \tag{16.5}$$

where h_{v} is the mean film thickness; h^* the film thickness at the start of the roll bite, where the pressure gradient approaches zero; ${\psi_{\mathrm{c}}}^2$ the local combined roughness of the foil and rolls; η_0 dynamic viscosity; γ pressure-viscosity coefficient; v_{entry} roll surface speed; and v_{r} entry speed of the foil.

When the asperities are crushed in the bite, the film thickness falls which leads to an increase in valley oil pressure. Accordingly, this increase in valley pressure prevents the asperity crushing process. Fleek et al. [4,5] showed that there existed slipping and sticking zones between the roll and foil in foil rolling. The average shear stress τ is given by a weighted sum of the asperity and valley contributions and the shear stress in slipping regions can be estimated below [12,13]:

$$\tau = A\tau_{\mathrm{a}} + (1 - A)\tau_{\mathrm{b}} \tag{16.6}$$

$$\tau_{\mathrm{a}} = \delta\mu_{\mathrm{a}}p_{\mathrm{a}};\ \delta = \frac{\theta}{\sqrt{\theta^2 + {\theta_1}^2}};\ \tau_{\mathrm{b}} = \eta\dot{\gamma} \tag{16.7}$$

where A is the actual contact area ratio; μ_a the Coulomb COF; p_a the pressure on the asperity tops; θ the local roll slope; θ_1 a representative roll slope in a slipping region; τ_b is the pressure on the asperity valley regions, which can be expressed using the Newtonian viscosity; and the strain rate $\dot{\gamma}$ is estimated form the sliding speed and valley film thickness in the bite, and the viscosity is taken at the pressure in the roll bite.

$$p_a = \overline{p} + (1 - A)Y\Delta; \; p_b = \overline{p} - AY\Delta \tag{16.8}$$

where Y is the plane strain yield stress of the rolled foil, Δ the effective hardness of the asperity, and $\overline{p}$ the average pressure.

The contact area ratio, which is related to the average film thickness, is calculated on the basis of the asperity flattening theory by Chang and Wilson [14]:

$$\frac{dA}{dx} = -\frac{2x\delta}{M\left(2l(1-A) + Y\left(\frac{A}{(\overline{p} - p_v)f_1(A)} - \frac{f_2}{f_1}\right)\right)} \tag{16.9}$$

where δ is the thickness reduction, f_1 and f_2 are coefficients related to the contact area ratio. At a nondimensional position, there is some contact between the foil asperities and rolls, leading to a nonzero area of contact ratio and a mean pressure $\overline{p}$ above the valley pressure p_b. At the end of the inlet region when the mean pressure has reached the yield stress Y, bulk deformation begins. Further asperity flattening occurs in the transition zone which is driven by the difference between the asperity and valley pressures. Eqs. (16.6) and (16.9) were solved by using the integrating method of Runge Kutta. The primary value of contact area ratio A was assumed to be 0.5 at the beginning of the solution. The required pressure for the rolling was supposed to be the yield stress of the material. This trial and error is continued until a convergence criterion is achieved. New values of lubricant pressure and contact area are also obtained from Eqs. (16.5)–(16.9).

The lubricant pressure and contact area ratio under different pass reductions are illustrated in Fig. 16.11. It can be seen that both the lubricant pressure and contact area ratio increase with the increase of pass reduction. Compared with lubricant pressure, contact area ratio increases slowly with the increase of pass reduction, especially at low reduction, while the influence of pass reduction on the lubricant pressure is more significant. In calculation, the asperity slop is assumed to be constant and only asperity height crushing has been considered in rolling deformation.

16.2.3.3 Oil Film Thickness

In cold rolling with lubricant, the mechanism of film formation by emulsions at the entry is complicated. Friction and the surface quality of the rolled strip are closely related to the amount of oil drawn into the bite. In order to

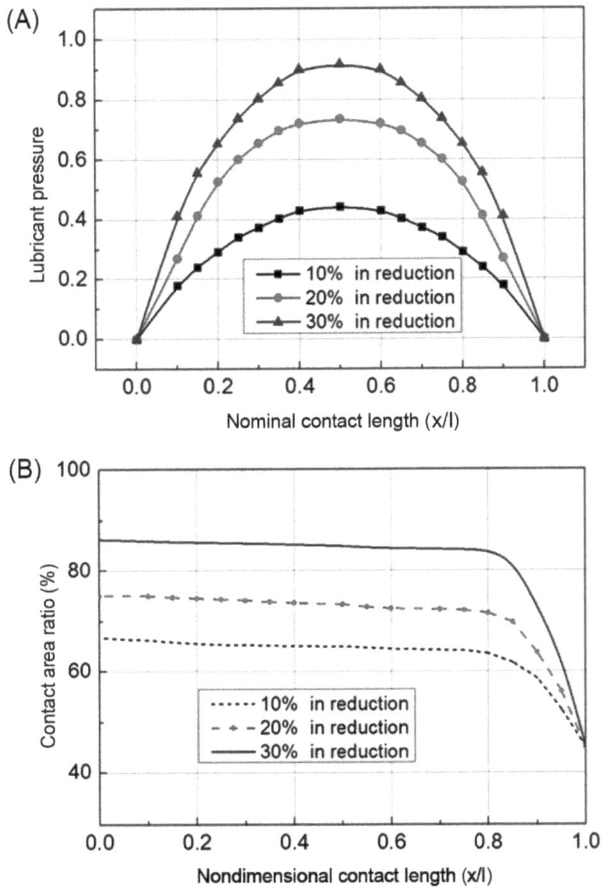

FIGURE 16.11 The lubricant pressure and contact area ratio distribution under different rolling reductions: (A) lubricant pressure and (B) contact area ratio [16].

investigate the oil-film thickness in lubricant rolling, the direct oil-dropping method could be applied: dropping a known quality of oil on the strip surface, rolling the strip in the mill, and taking a digital photograph of nearly ellipse-shaped mark left on the strip after rolling. The length and width of this emulsion mark are measured, and the final thickness of the oil film between the roll and the rolled material can be obtained through the drop volume (V_d) divided by the surface area of the mark (A_c). The measured oil thickness is an average value. Because the volume of drop shrinks during rolling, the thickness of the film at the end of the mark could be somewhat lower. However, this influence is small and the measured oil-film thickness could be obtained by:

$$h_m^* = (V_d/A_c)/(1-\delta) \tag{16.10}$$

In theory, the oil-film thickness could be determined by the lubricant rheological properties, the rolling speed and the roll geometry. Wilson and Walowit [14] calculated film thickness by applying an analytical solution of the Reynolds equation in the inlet zone and solved the continuity equation in the deformation regime. The simplified film thickness formula for smooth rolls and strip is shown below:

$$h^* = \frac{3\eta_0\gamma(v_{\text{entry}} + v_{\text{r}})R'}{L\{1 - \exp\left[-\gamma(\sigma_{\text{fm}} - \sigma_{\text{entry}})\right]\}} \tag{16.11}$$

$$R' = R\left(1 + \frac{16\omega}{\pi E(H - h)}\right) \tag{16.12}$$

A film thickness at the outlet of the contact zone can be calculated by

$$h_{\text{out}} = h^* \frac{v_{\text{entry}} + v_{\text{r}}}{v_{\text{out}} + v_{\text{r}}} \tag{16.13}$$

where h^* is the oil-film thickness in the inlet zone, σ_{fm} the average flow stress, σ_{entry} the tensile stress, ω the specific rolling force, h_{out} the oil-film thickness in outlet zone, and R' the deformed roll diameter in mm.

Fig. 16.12 shows the comparison between the oil-film thickness obtained from the Wilson model, analytical modeling, and the measured value. All oil-film thicknesses in experiment and calculation decrease with increasing pass thickness reduction. The results from the modeling are closer to the experimental values. It can be seen that the calculated value is always smaller than the experimental value, and this is because the film formation by the entraining action and surface roughness effect have not yet been considered in the calculation model.

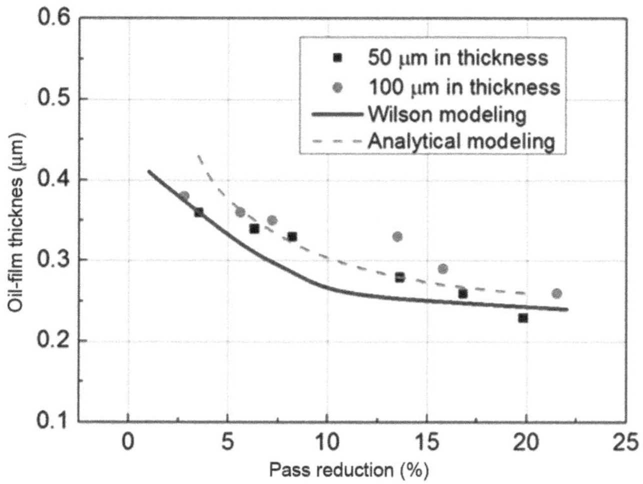

FIGURE 16.12 Measured film thickness for emulsion and film thickness based on the Wilson model [16].

Additionally, the oil-film thickness in 50 μm thick foil is lower than that in thickness of 100 μm. With increasing miniaturization, more open lubricant pockets appear and less lubricant is kept in the micro tribological contact zone for thinner foil. Therefore the oil-film thickness increases with increasing foil thickness.

The fraction of oil entrained into the contact zone significantly depends on the thickness reduction, it is reasonable that oil trapping between asperities of surface is subjected to viscous shearing, maintains its original viscosity and is easy to be entrained. A larger thickness reduction results in larger coalescence of the oil and the surface, so the oil thickness should be lower accordingly. There will be two reasons for this difference: the inlet zone and the roughness of the foil and rolls. The oil film becomes thicker with decreasing angle at the inlet of roll bite which had been identified in Ref. [12], but the effect is insignificant because of its small inlet angle in foil rolling.

This curve obtained from the model in Eq. (16.13) is roughly in agreement with the trend of the measured oil thickness as shown in Fig. 16.12. A matching parameter C depending on the surface roughness could be introduced to reduce the difference between the measured value h_m^* and the calculated value h_c. The relationship can be expressed approximately in linear:

$$h_m^* = Ch_c \tag{16.14}$$

The Wilson model to the oil-film thickness can be applied through this matching parameter in considering the surface roughness effect. An accurate agreement between the calculation value and the measured value is obtained by inserting this matching C, where C has a value of 1.05 for the smooth surfaces of 50 μm and 1.12 for the rough surfaces of 100 μm in this case.

Fig. 16.13 shows the measured oil-film thickness versus the rolling speed in two different lubricating cases: emulsion and neat oil. It can be seen that both

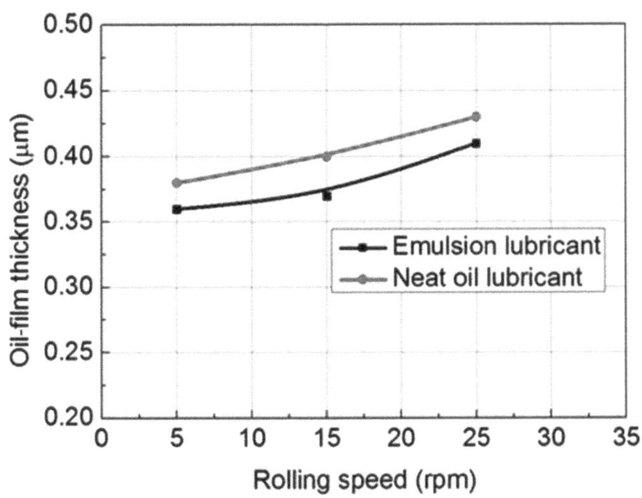

FIGURE 16.13 Oil-film thickness versus rolling speed in experiment ($H = 100$ μm; $r = 20\%$) [16].

oil-film thicknesses increase with increasing rolling speed slightly. The oil thickness in neat oil is higher than that in emulsion lubricant. Because the viscosity decreases in emulsion lubrication, the film thickness decreases due to the increase of asperity contact area. When the asperity contact area increases, the pressure in asperity contact area increases accordingly and the film pressure decreases. This is testified that the oil-film thickness would be slightly smaller than that in neat oil lubricant because of much lower viscosity in emulsions.

The rolling speed effect with wet lubricant indicates that the form of hydrodynamic lubrication is undergoing metal-to-metal contact. The rolling speed influences the viscosity of lubricant during rolling because the speed profile determines the oil drop movements. With increasing rolling speed, more lubricant is carried to the interface filling the surface valleys, separating the surface peaks and establishing the mixed lubrication. The oil that was already on the strip surface spreads thinner by the elongation of the strip and the replenished oil makes up for the difference.

16.2.3.4 Rolling Parameters for Various Rolling Conditions

The effect of rolling conditions on the rolling force has been investigated during rolling experiments. Fig. 16.14 illustrates the rolling force changes with rolling time. Two pressure sensors are installed under both ends of the roll neck, respectively [15]. The sensors are connected with the strain amplifier where the rolling force can be recorded by the data logger as shown in Fig. 16.1. It can be seen that the rolling force variation is large when biting the specimen in the rolling entry due to rolling vibration, and the average rolling force is around 6.73 kN for this experimental case. After that, the rolling force remains a constant during rolling process.

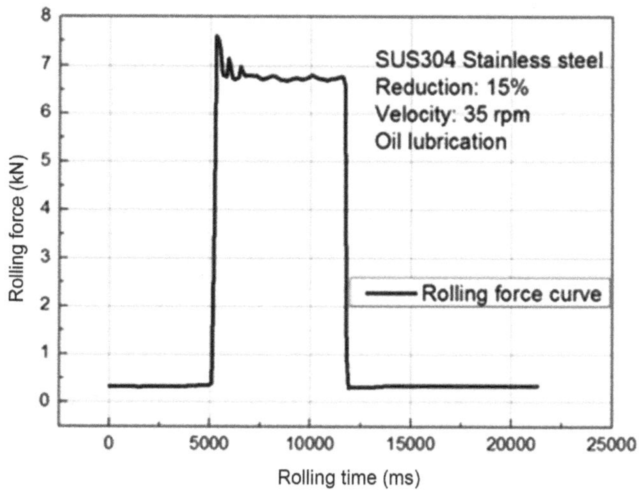

FIGURE 16.14 Rolling force variation with rolling time (rolling speed: 30 rpm; oil lubricant).

Fig. 16.15 shows the effect of lubricating condition on the rolling force at various rolling speeds. It can be seen that the lubrications are significantly effective in reducing the rolling force at low rolling speed from 10 to 30 rpm. However, the effect of the rolling lubricating conditions is insignificant at high rolling speed of over 30 rpm because the hydrodynamic lubrication prevails against the boundary lubrication. The rolling force distinctly decreases due to the reduced friction from the rolling experiments where the lubricating condition changes from the boundary condition to the hydrodynamic condition. This is verified that the rolling force decreases quickly in low rolling speed zones [16].

The effects of rolling speed and lubricating condition on limit pass reduction are shown in Fig. 16.16. It can be seen that the pass reduction is sensitively increased with increasing rolling speed under the constant rolling force without roll gap adjustment. The hydrodynamic lubrication becomes gradually predominant with the rolling speed and the friction is reduced accordingly. When using an oil or emulsion, the limit pass reduction is larger than that in dry friction rolling. Under high rolling speed, the limit pass reduction vibration is higher for dry friction while it is insensitive for emulsion lubrication. The low friction reduces the rolling force, which then reduces the elastic flattening of work rolls and enables a larger limit reduction.

16.2.3.5 Surface Topography Analysis After Micro Rolling

The surface roughness has a large effect on the film thickness and its effect is especially important. With low viscosity-liquids, the foil and the roll surface roughness are much larger than film thickness in the hydrodynamic emulsion, and surface roughness must have a substantial effect on lubrication

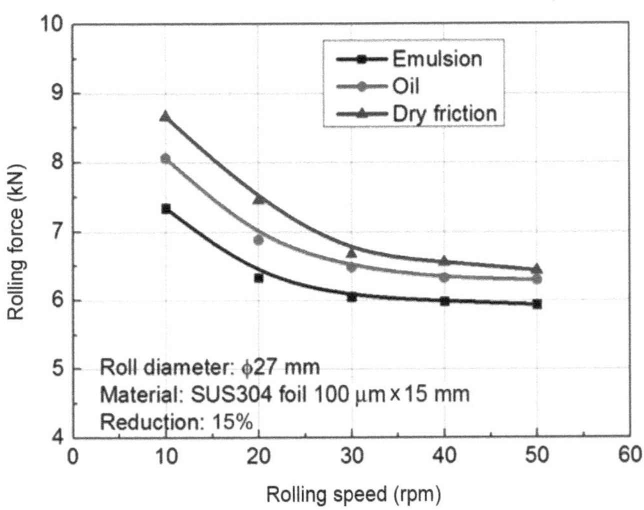

FIGURE 16.15 Effect of lubricating condition on rolling force under various rolling speeds [16].

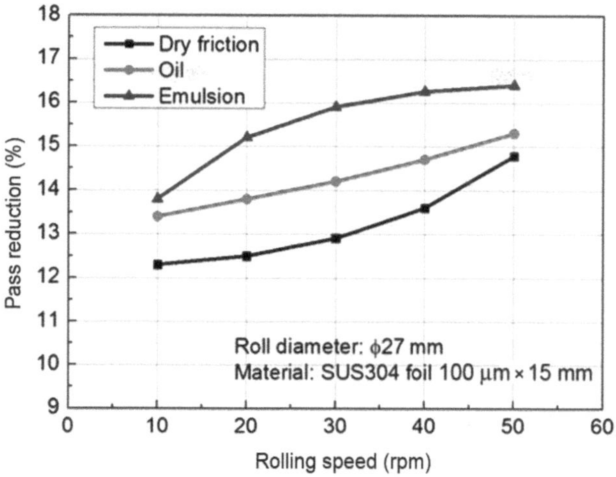

FIGURE 16.16 Effects of rolling speed and lubricating condition on limit pass reduction [16].

film thickness. Since the roll is much harder than the foil, the foil's asperities are expected to change shortly after roll biting. As a laboratory desk-top rolling mill, where the volume of the rolled metal is much less, the roll's surface roughness is not expected to change significantly, and the change can be neglected. Fig. 16.17 illustrates the rolled foil surface roughness variation with rolling lubricating conditions. Rolling trace and surface valley are significantly indicated in the surface topography. In the case without of lubricant, the roll surface is unevenly coated with foil, while in emulsion rolling there exists homogeneous coating with foil. The unevenly coated roll has rougher surface. In addition, the lubricating condition is remarkably affected by the roll surface because the forward slip in foil rolling is larger than that in strip rolling. It can be seen that the wavelength of surface roughness in friction rolling condition is larger than that in emulsion or oil rolling from Fig. 16.17. The structure consists of parallel grooves perpendicular to the rolling direction. By the theory of Wilson and Walowit [14], those grooves are formed by the differential deformation of the grains and should present a small depth, and consequently a small surface roughness.

Fig. 16.18 shows the Atomic-Force Microscopy (AFM) micrograph of rolled foil with and without lubricant during rolling. It can be seen that more flat areas are found on the surface of the rolled foil in lubricating rolling. Also the much rougher surface is located on the rolled foil surface in Fig. 16.18A of dry rolling.

Details of the surface roughness topography may influence the behavior of the foil during rolling. In addition, this topography plays important role in the tribological behavior in the roll bite. The effect of wavelength of roughness has been explored using measurements of the surface spectrum on strip rolled in the mixed lubrication regime on an experimental mill.

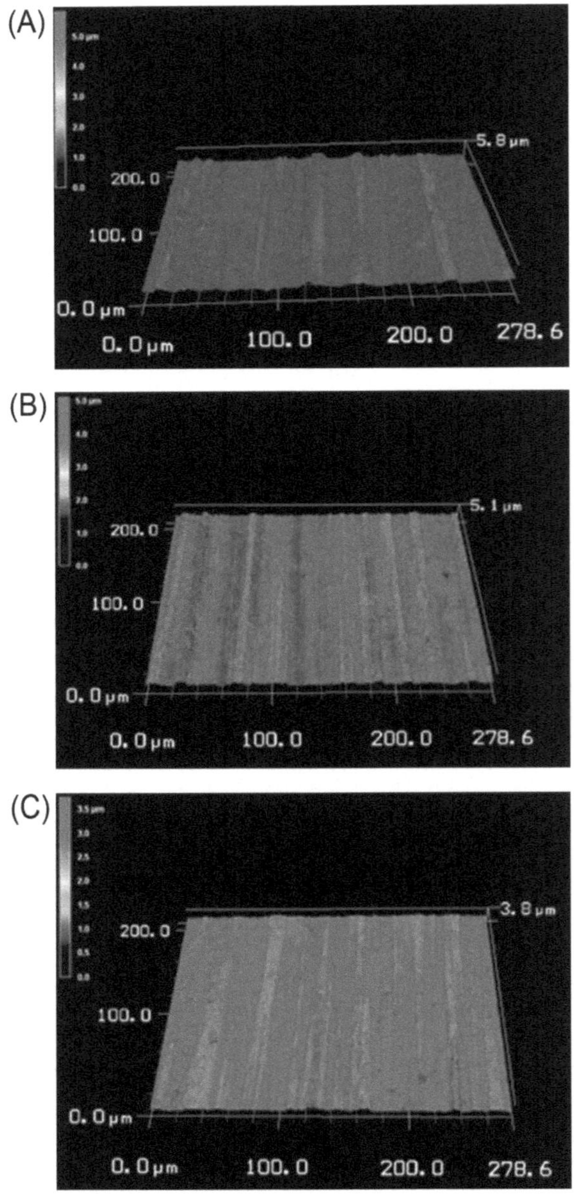

FIGURE 16.17 Rolled foil surface roughness under three different lubricating conditions: (A) dry friction, (B) oil lubrication, and (C) emulsion lubrication [16].

Friction and the surface quality of the rolled foil are closely related to the amount of oil drawn into the bite and the surface roughness on the rolls and the in-going foil.

Fig. 16.19A shows the surface roughness distribution along the rolled foil surface. The surface roughness of the initial foil and rolled foil samples taken

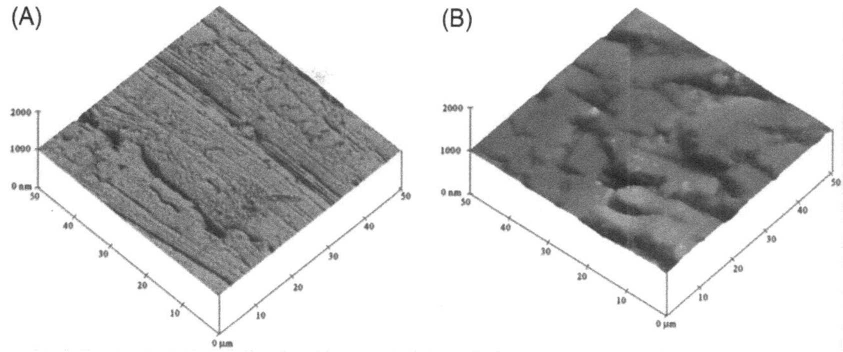

FIGURE 16.18 AFM micrograph of rolled foil with and without lubricating rolling.

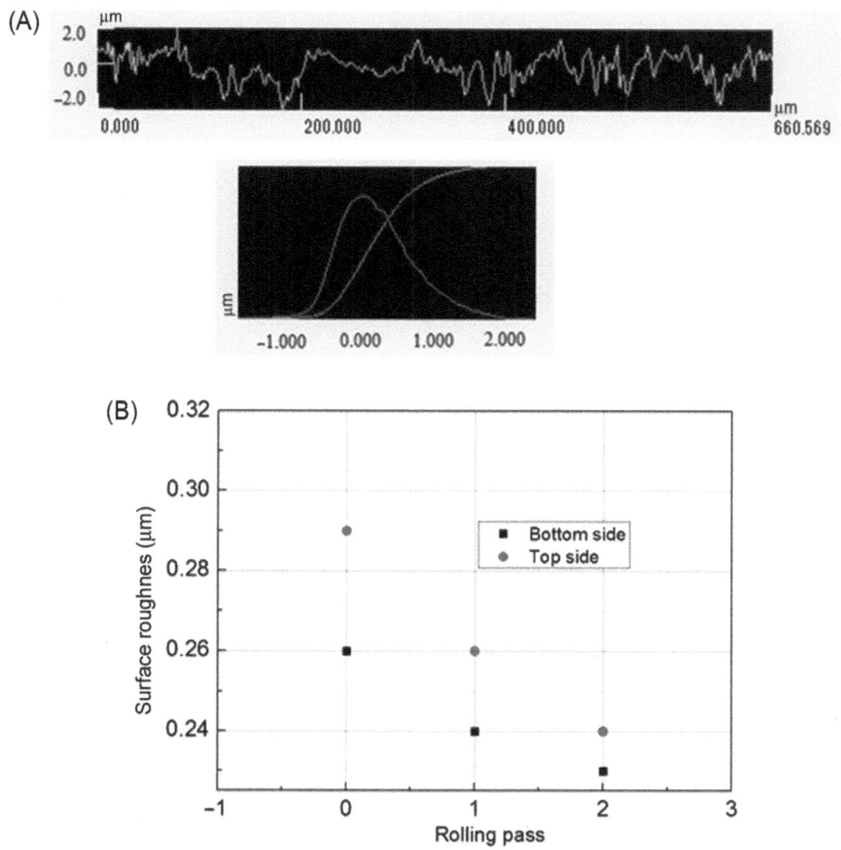

FIGURE 16.19 Surface roughness evolution during rolling process: (A) roughness distribution along the foil surface and (B) roughness change with rolling pass.

after the first and second passes is shown in Fig. 16.19B. Each point represents an average of 20 measurements. Results show that the roughness on the bottom surface of the initial foil is significantly greater than that on the top surface. Nevertheless, both sides of the surface have nearly conformed to the roll surface after the first pass, so that the difference between the two sides is then reduced. In the second pass, the surface roughness is further reduced slightly.

The surface quality continuously changes in rolling process, and the surface quality is of great difference with different oil-film thickness. Because there is a thin layer of oil-film thickness in rolling deformation zone, which not only prevents the adhesion between the roll and sample surfaces, but also makes sample surface fully crushed, so the rolled surface is bright and the surface quality is improved.

Micro pits play an important role in the surface finish of stainless steel, understanding of their evolution during forming processes is very limited. Fig. 16.20 illustrates the effects of rolling speed on oil pit distribution under oil lubricating rolling, it can be seen that more oil pits appeared on foil

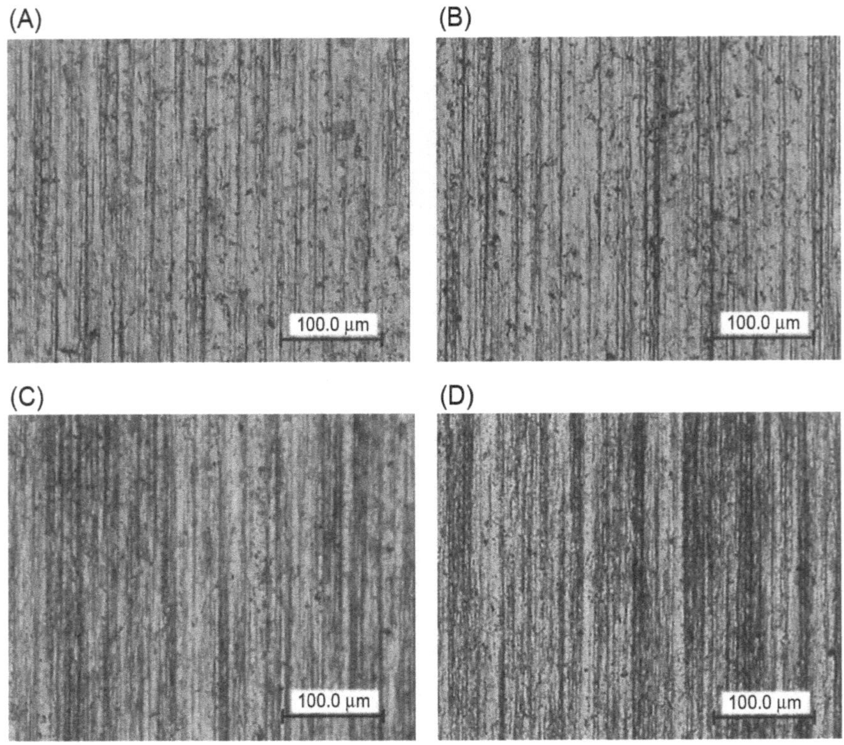

FIGURE 16.20 Effects of rolling speed on oil pit distribution under oil lubricating rolling (reduction 15%): (A) 15 rpm in rolling speed, (B) 30 rpm in rolling speed, (C) 40 rpm in rolling speed, and (D) 50 rpm in rolling speed [16].

surface with increasing rolling speeds. In other word, a lot of oil pits are formed in high speed rolling and make the dull surface. It seems that the pits observed on the foil surface are associated with surface roughening in regions where there is a significant oil film separating the roll and the foil. Under low rolling speed zone, the speed effect on oil pit is insignificant, but there is a great difference in high rolling speed zone.

Fig. 16.21 shows the oil pit area change with rolling speed. It can be seen that the oil pit area increases with increasing rolling speed. The rolling speed effect is more significant in oil lubricating rolling than that in emulsion lubricating rolling under high rolling speed zone.

In summary, the micro rolling deformation characterization has been investigated through foil rolling in micro rolling mill. The roll Young's modulus and speed rate have significant influence on rolling reduction limit. The rolled material profile has been compared between experimental values and theoretical analysis results. Tribological size effects are also observed for different foil thickness of 100 and 50 μm. The oil-film thickness is higher in the thicker foil than that in thinner foil. The oil-film thickness decreases with increasing pass reduction and decreasing rolling speed. The lubrication conditions influence the surface roughness evolution during rolling. In relation to increase in oil-film thickness, rolling pressure reduces; in contrast, the product surface becomes dull. The most preferable lubricant for foil rolling is emulsion, and the rolling force decreases dramatically in lubricating rolling. Lubricating condition is remarkably influenced by roll coating formation, especially in using the smooth surface of rolls. The amount of micro oil pits increases with rolling speed. These findings will provide useful information for micro rolling of ultrathin foil.

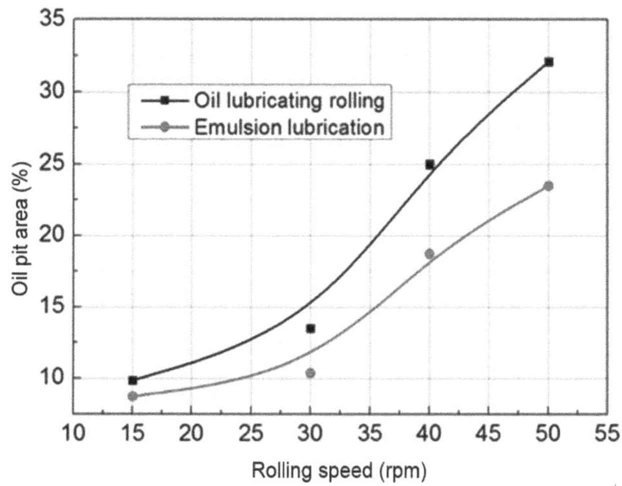

FIGURE 16.21 Percentage of oil pit area for different lubricants versus rolling speed [16].

REFERENCES

[1] J. Tateno, T. Hiruta, S. Katsura, A. Honda, T. Miyata, A. Kamimaru, Experimental analysis of thickness reduction limits in ultra thin stainless steel foil rolling, ISIJ Int. 51 (2011) 788–792.

[2] Z.Y. Jiang, S.W. Xiong, A.K. Tieu, Q.J. Wang, Modelling of the effect of friction on cold strip rolling, J. Mater. Process. Technol. 201 (2008) 85–90.

[3] S. Shida, Y. Komatsu, K. Yasuda, K. Narita, Minimum thickness in cold thin strip rolling, J. JSTP 37 (1999) 13–14.

[4] N. Fleek, K. Johnson, Towards a new theory of cold rolling thin foil, Int. J. Mech. Sci. 29 (7) (1987) 507–524.

[5] N. Fleek, K. Johnson, M. Mear, L. Zhang, Cold rolling of foil, J. Eng. Manufact. 206 (6) (1992) 119–131.

[6] M.P.F. Sutcliffe, P.J. Rayner, Experimental measurements of load and strip profile in thin strip rolling, Int. J. Mech. Sci. 40 (9) (1998) 887–899.

[7] H. Keife, J.Y. Shi, D.L.S. McElwain, T.A.M. Langlands, Foil rolling: a new interpretation of experimental results, J. Mater. Process. Technol. 121 (2012) 50–59.

[8] K. Dick, J. Lenard, The effect of roll roughness and lubricant viscosity on the loads on the mill during cold rolling of steel strips, J. Mater. Process. Technol. 168 (1) (2005) 16–24.

[9] W. Wilson, J. Walowit, An isothermal hydrodynamic lubrication theory for strip rolling with front and back tension, Proceedings of the Tribology Convention, 27, Institution of Mechanical Engineers, London, 1972, pp. 169–172.

[10] A.G. Atkins, Hydrodynamic lubrication in cold rolling, Int. J. Mech. Sci. 16 (1974) 1–19.

[11] N. Parit, H.S. Cheng, An average flow model for determining effects of three-dimensional roughness on parial hydrodynamic lubrication, ASME J. Lubric. Technol. 100 (1978) 12–17.

[12] H.R. Le, M.P.F. Sutcliffe, Analysis of surface roughness of cold-rolled aluninium foil, Wear (2000) 71–78.

[13] M.P.F. Sutcliffe, P. Montmitonnet, Numerical modelling of lubricated foil rolling, La Rev. Metall.-CIT/Sci. Gen. Mater. (2001) 435–442.

[14] W. Wilson, J. Walowit, An isothermal hydrodynamic lubrication theory for strip rolling with front and back tension, Proceedings of the Tribology Convention, 27, Institution of Mechanical Engineers, London, 1972, pp. 169–172.

[15] H.B. Xie, K. Manabe, T. Furushima, K. Tada, Z.Y. Jiang, An experimental and numerical investigation on micro rolling for ultra-thin strip, Int. J. Mater. Form. 9 (2016) 405–412.

[16] H.B. Xie, K. Manabe, T. Furushima, K. Tada, Z.Y. Jiang, Lubrication characterisation analysis of stainless steel foil during micro rolling, Int. J. Adv. Manufact. Technol. 82 (2016) 65–73.

Chapter 17

Practice of Micro Deep Drawing

17.1 EQUIPMENT AND TOOLS FOR MICRO DEEP DRAWING

17.1.1 Heat Treatment

Stainless steel 304 sheets with a thickness of 50 μm were heated at different temperatures under an argon gas protection atmosphere, and microstructures of these sheets were observed under a metallurgical microscope. With different heat treatment temperatures, different grain sizes can be obtained. In order to compare the grains sizes and shapes of these differently treated sheets, three heat treatments at temperatures of 975°C, 1050°C, and 1100°C for 2 min, respectively, were selected. Based on the heating temperatures, sheets were classified as material groups 1, 2, and 3 correspondingly. Table 17.1 shows the classification of each material group and their general grain sizes. The grains on these treated sheets were of equiaxed crystals and their average grain sizes were 10, 20, and 40 μm, respectively, as shown in Fig. 17.1 [1].

Table 17.2 lists parameters of three-group materials' properties, including strength coefficient, strain hardening index, and elastic modulus.

17.1.2 Micro Tensile Tests

Tensile tests were conducted to obtain mechanical properties on a designed micro tensile tester. Taking size effects into account, the tensile test samples were scaled down, and the width of each sample is only 1 mm, as shown in Fig. 17.2A. Due to the small size, traditional elongation measurements cannot be used. Therefore, an image processing program was developed in MATLAB to calculate the strain as a noncontact elongation measurement. A parallel area on the middle of the tensile samples was firstly marked with a contrasting color to the tensile samples. A video focusing on this area was recorded by the microscope and then separated to individual images by MATLAB. Subsequently, the colorful images were transferred to gray level images and then to binary images. During the transformation, noise reduction process was conducted to eliminate small areas where the marked color was lost during the tensile tests. Therefore, errors introduced by the color

Microforming Technology. DOI: http://dx.doi.org/10.1016/B978-0-12-811212-0.00017-0

TABLE 17.1 Heating Temperatures and Average Grain Sizes After Heat Treatments

Group	Heating Temperature (°C)	Heating Time (min)	Average Grain Size (μm)
1	975	2	10
2	1050	2	20
3	1100	2	40

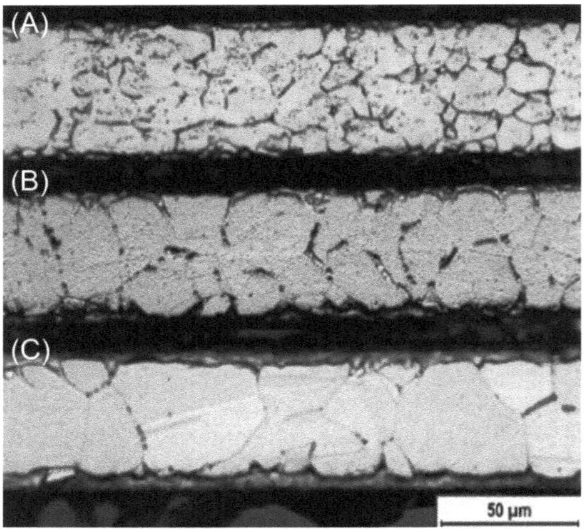

FIGURE 17.1 Microstructures of sheets heated under different temperatures: (A) 975°C, (B) 1050°C, and (C) 1100°C [1].

TABLE 17.2 Parameters of Materials' Properties

Group	Strength Coefficient K/MPa	Strain Hardening Index N	Elastic Modulus E/GPa
1	1771 ~ 1908	0.324 ~ 0.405	189.0 ~ 220.0
2	959.3 ~ 1073	0.212 ~ 0.249	185.7 ~ 221.7
3	957.1 ~ 1068	0.227 ~ 0.294	183.3 ~ 216.9

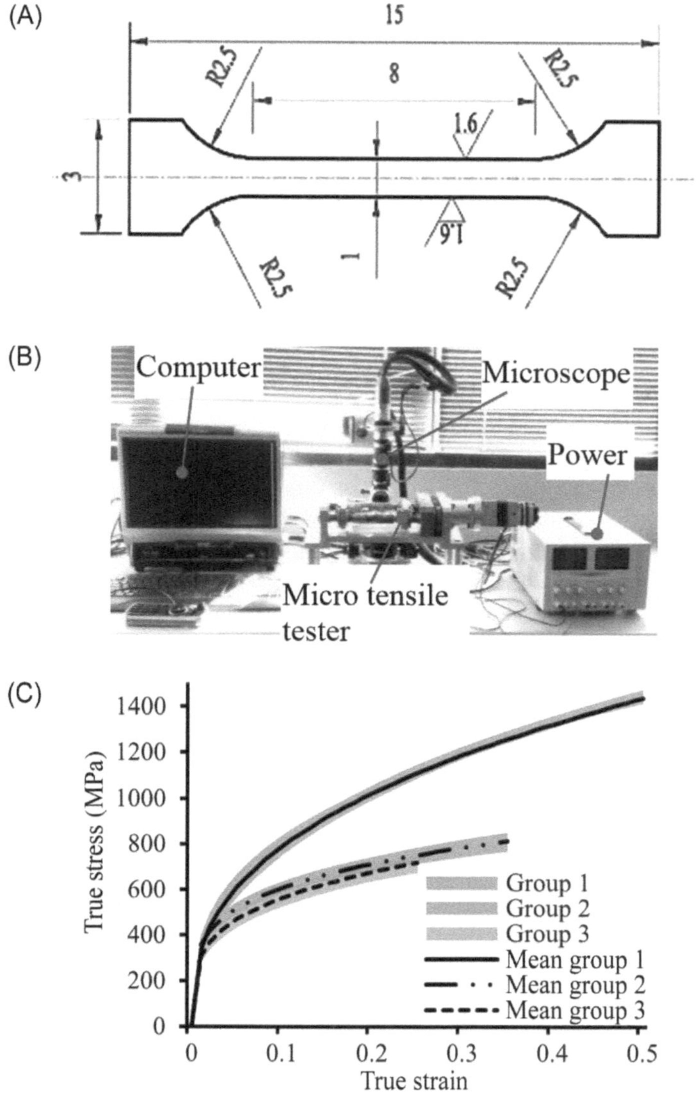

FIGURE 17.2 (A) Tensile sample, (B) micro tensile test system, and (C) strain–stress curves of each material group [1].

loss in the marked area during the tensile tests can be eliminated. Consequently, boundaries of the marked area on the micro tensile samples can be obtained through color value comparison between adjacent pixels. If the color values, being zero or one for a pixel in the binary image, of neighboring pixels are different, the pixel was selected as a boundary point. Once all boundary points were found, the points were sorted according to their x- and y-coordinates sequentially. Thus four groups of points can be divided as four boundaries of the marked area. The length, along the length direction of the tensile sample, of the marked area can then be calculated based on these boundaries. By comparing the length of the marked area in each image, the strain can be obtained. Furthermore, the real length of the marked area can be obtained via a scale factor from the scale bar on the image. The scale factor is the ratio of the number of pixels that a scale bar has to be displayed the length value on the scale bar. Simultaneously, a force sensor recorded and transferred the tensile force signal to the computer. By matching the video and the force recording frequencies, both engineering and true strain–stress curves can be obtained.

Fig. 17.2B shows micro tension test system, including micro tensile tester, computer, microscope, and power. Fig. 17.2C displays the true strain–stress curves of stainless steel 304 covering 10 repeated tests and their mean value of each material group. With a low heating temperature, grains were small and the flow stress was high. This is because that the annealing effect, which eliminates the work hardening of the raw sheets, was weak at low-temperature heating condition. Moreover, small grain size means large quantity of grain boundaries where dislocation movement is blocked and dislocation is piled up. The accumulation of dislocation thus increased the deformation resistance which is presented as the tensile strength in the tensile tests. Therefore, the tensile strength of the material group 1 was the highest among all material groups. Moreover, with a few grains in the thickness direction, randomness of each grain's properties significantly impacts the overall properties of the samples and leads to a great scatter of samples' properties. The total elongation also decreased with the increase of grain size as a few grains in thickness direction caused the fast development of micro cracks and thus the early fracture of tensile samples.

To obtain the surface roughness information, the sheets of all material groups were tested by a high-resolution laser-assisted digital microscope (VK-X1000/X2000 series from ©KEYENCE). A 3D digital image was generated by a series of images at sequential focusing heights, and then a virtual surface was fitted according to the observed 3D image. Finally, the surface roughness parameters were calculated from the projection of the 3D image onto the virtual surface. Thus the surface roughness information can be separated from the geometry of the samples. The Universal Tribometer from ©Rtec Instruments was used to test the coefficient of friction (COF). A piece of SUS304 sheet was firstly fixed on a testing disk, and then installed into the tribometer platform. For accurate COF tests, the same materials of the punch and the die were chosen as the friction pair in the Tribometer. Moreover, a similar relative speed of the blank during the drawing process is transferred to the rotation speed in the friction tests. The COF was therefore obtained from the mean values of the repeated friction tests.

17.1.3 Equipment of Micro Deep Drawing

The micro deep drawing (MDD) experiments have been conducted on a MDD system, as shown in Fig. 17.3. Fig. 17.4 shows (A) the overview/back and (B) the die set attached. Firstly, the blanking die and the blanking holder moved downwards at a speed of 0.1 mm/s and the die stayed still as a blanking punch. A raw blank for the following drawing process was cut at the first half stroke. Subsequently, the punch moved down continuously and contacted with the blank, whereas the die stayed still. Finally, a micro circular cup was drawn by the punch at the end of the second half stroke. Due to this design of the MDD system, the press machine performing one stroke can fulfill the blanking and the MDD processes subsequently. Table 17.3 lists parameters of MDD machine and process.

17.2 MICRO DEEP DRAWING PRACTICE

17.2.1 Microstructure Effect

The sheets were drawn into cups. During the deep drawing process, the punch force obtained from 10 experiments for each material group were recorded, and their average values are shown in Fig. 17.5. The punch force has a relatively slow increase initially, and a fast yet constant increase until it reaches a peak value. After remaining this peak value for a short period, punch force of each material group decreased to a nonzero value at the end of the process. Initially, the resistance of bending dominated the punch force, while other forces were small. As the process continued, large deformation caused high-flow stress, and simultaneously friction force increased due to increasing contact forces. Hence, the punch force increased significantly during the following period. Furthermore, with smaller elastic module, material group 1 had slower growth of punch force initially, whereas in the later period, the higher strain hardening rate increased flow stress significantly,

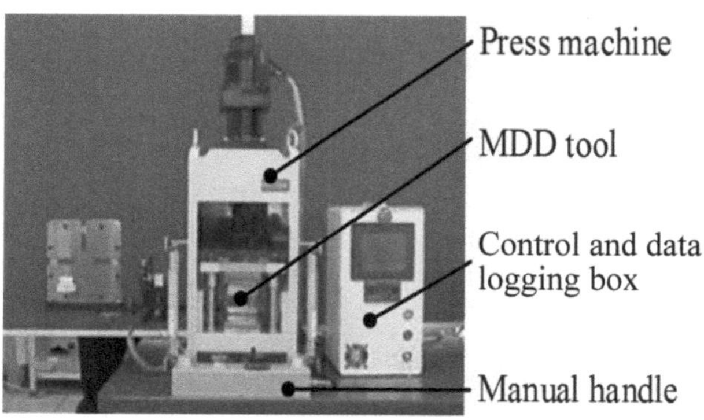

FIGURE 17.3 Micro deep drawing system [1].

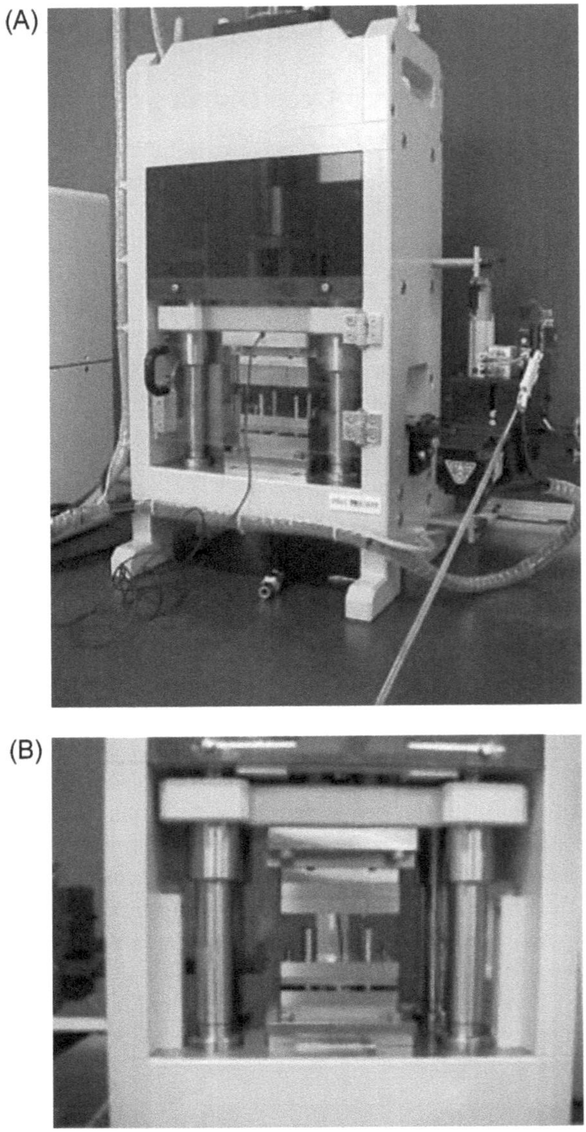

FIGURE 17.4 (A) The overview/back of micro deep drawing device and (B) the die set attached.

TABLE 17.3 Parameters of MDD Machine and Process

Punch Diameter	Die Diameter	Radius of Punch Fillet	Radius of Die Fillet	Drawing Speed	Blander Gap	Friction Condition	Initial Blank Diameter
0.8 mm	0.975 mm	0.3 mm	0.3 mm	0.1 mm/s	0.055 mm	Dry friction	1.6 mm

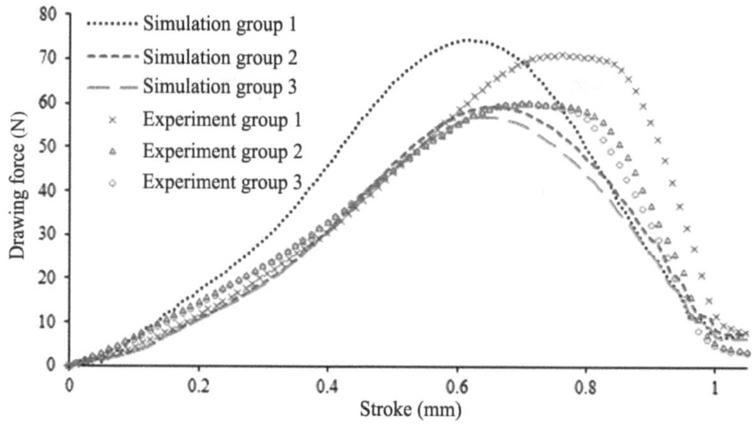

FIGURE 17.5 Punch forces in simulation and experiments [1].

TABLE 17.4 Key Parameters of MDD Process and Formed Cups (Unit: μm)

Group	Min. Inner Diameter	Max. Inner Diameter	Outer Diameter	Max. Distance	Min. Distance	Wrinkles (%)
1	787.083	816.645	941.092	77.004	62.223	24.546
2	780.558	814.277	938.137	78.789	61.930	27.334
3	789.532	831.042	950.033	80.251	59.496	35.091

resulting in the faster growth of punch force compared with that of the other two material groups. Finally, the peak punch force of the material group 1 was the highest among all material groups. Thereafter, the peak punch force remained for a period and decreased until the end of the drawing process. Moreover, due to the stored strain energy in the drawn cups, the punch force at the end of the process was not zero. Therefore, after the drawing process, the drawn cups underwent springback and stored stain energy released gradually. Moreover, the higher the residual punch force is, the more significant the springback is.

Table 17.4 lists the average geometrical values of drawn cups for each material group and their definitions are shown in Fig. 17.6. Due to the limitation of the die, the blank wrinkled toward the punch. Therefore, the outer surface of the drawn cups was relatively smooth and the wrinkles were mainly on the inner surface of the drawn cups. Correspondingly, the difference between the outer and minimum inner diameters was defined as the maximum distance, and the deviation between the outer and maximum

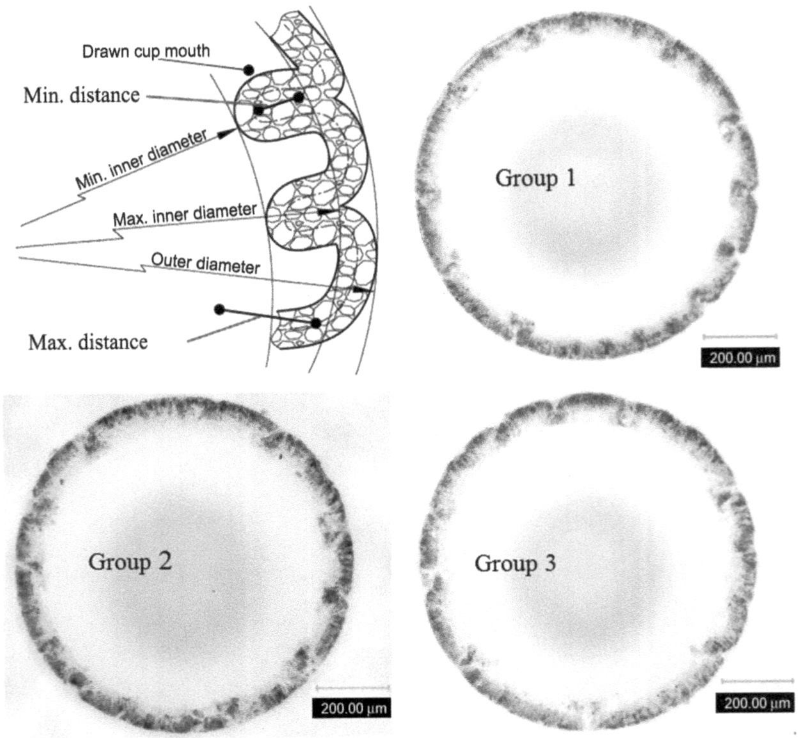

FIGURE 17.6 Definitions for wrinkling parameters and top views of drawn cups with different materials [1].

inner diameters was defined as the minimum distance. Then, their relative difference, calculated from Eq. (17.1), was employed as a judgment of wrinkles. With the increase of grain size, wrinkling phenomenon becomes significant. This is because that a few grains on the thickness direction make the deformability decrease, and consequently, compression stability on the flange of the blank is weak. Therefore, the blank is easy to wrinkle and needs great wrinkles to compensate this compression instability.

$$\text{Wrinkles} = \frac{\text{Max. distance} - \text{Min. distance}}{\text{Max. distance}} \times 100\% \qquad (17.1)$$

The thickness was measured based on the colorful images from a 3D profile microscope. A 3D structure of a drawn cup was combined from a series of images at different focusing heights. Moreover, the drawn cup was colored according to the height, and therefore the drawn cup's mouth can be distinguished based on the color, as shown in Fig. 17.7 [2]. The diameters of the cup mouth were displayed in Fig. 17.7D. In addition, the wrinkled area was avoided for the thickness measurement. Therefore, the thickness was between the maximum and minimum distances for each material group.

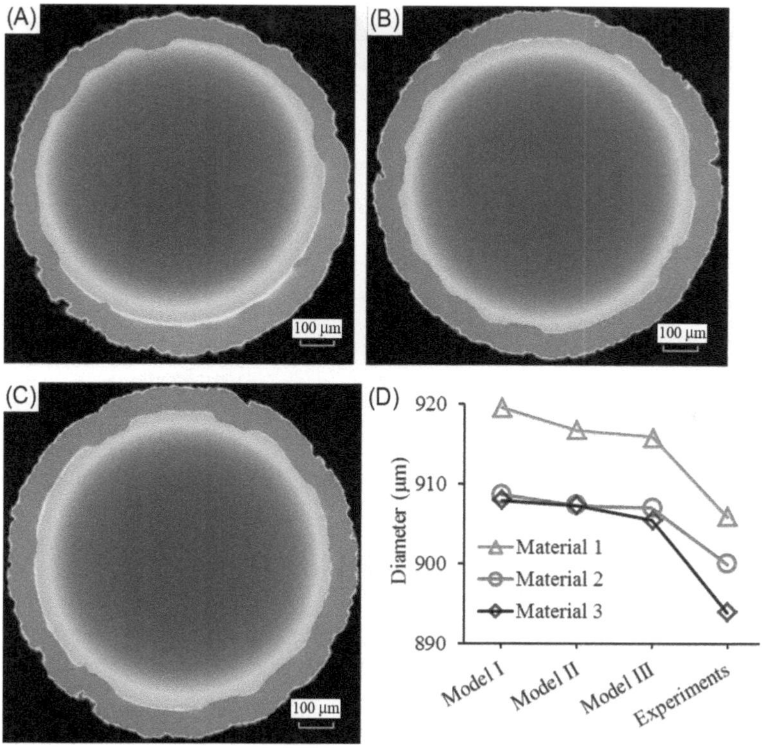

FIGURE 17.7 Drawn cups in experiments: (A) material 1, (B) material 2, (C) material 3, and (D) diameters of the cup mouth [2].

As there are limited grains on the drawn cups' mouth, thickness was determined by these few grains with different mechanical properties. As a result, there was no obvious trend regarding the average thickness at the cup mouth due to the randomness of grains properties and grain distributions.

Numerical and experimental results were then compared as listed in Table 17.5. The punch force in experiment increases slower than simulation results. That lies in that randomly distributed grains with different mechanical properties on the blank cannot identically represent actual blank deformation behavior. The earlier occurrence of the peak punch force in simulation than that of experimental one comes from an over estimated strain hardening effect in simulation model. According to Eq. (17.2), flow stress, $\overline{\sigma_i}$, increases faster in simulation due to the greater strain hardening effect than that in experiments, whereas the radial strain, $\ln\frac{R_t}{r_0}$, has almost the same decreasing speed to that in experiments. As the radial tension is a primary force of punch force, two factors of the radial stress have an overall result as the early occurrence of the peak radial tension and the peak punch force. This also explains the higher peak punch force in simulation than the experimental one. Due to the change of friction condition at the end of the drawing

TABLE 17.5 Comparison Between Experimental and Simulation Results

Items	Experimental	Simulation	Error (%)
Force slope I (N/mm)	27.27	25.49	6.53
Force slope II (N/mm)	124.80	148.50	−18.99
Force slope III (N/mm)	−212.00	−207.10	2.31
Peak force (N)	64.88	69.18	−6.63
Peak force stroke (mm)	0.71	0.61	13.99
Residual force (N)	16.35	6.07	62.86
Average wall thickness at cup mouth (μm)	62.223	66.199	−6.01

process, the residual punch force in experiment is much higher than that in simulation.

$$\sigma_{\mathrm{rmax}} = 1.1\overline{\sigma_i}\ln\frac{R_{\mathrm{t}}}{r_0} \tag{17.2}$$

where σ_{rmax} is the maximum radial stretch stress, $\overline{\sigma_i}$ is the average flow stress, R_{t} is the current radius of the blank, and r_0 is the inner radius of flange on the blank.

The average wall thickness at the cup mouth in simulation is quite close to that of experimental one. Moreover, the thickness distribution at cup mouth is irregular which is similar to the simulation results.

17.2.2 Effect of Lubrication

The lubricant viscosity is one parameter for optimizing drawing processes. The influence is due to the velocity depending friction behavior between the punch and the blank, as well as the effective process forces and resulting pressures in the lubricant. High viscosity lubricant reduces the occurring friction significantly. Fig. 17.8 lists the sketch of dry friction and lubrication in MDD. When lubricating, there exists closed lubricant pockets and open lubricant pockets in contact area where the friction status is affected during MDD [3].

Fig. 17.9 shows the effective COF, which is calculated from the maximum punch force [4]. The method delivers an average value for the COF and has the advantage that the effect of process variations (as punch speed variation) can be traced easily. It can be seen that the effective COF decreases with an increase of punch velocity. High effective COF occurs when the punch diameter is lower.

Fig. 17.10 illustrates the size-dependent COF in cup deep drawing [5]. They are in average 0.129 for punch diameter of 1 mm, 0.075 for punch

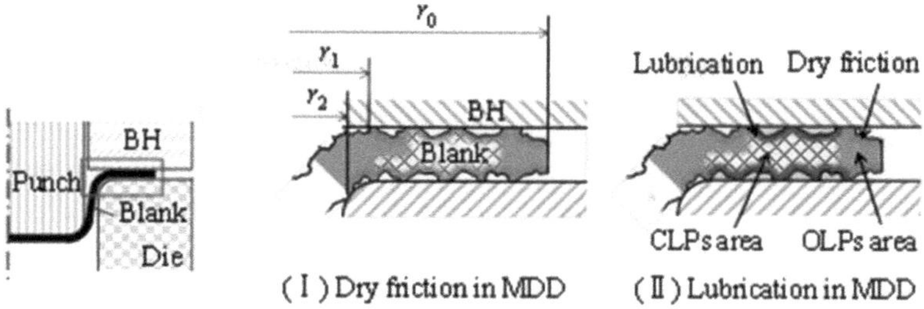

FIGURE 17.8 The sketch of dry friction and lubrication in MDD [3].

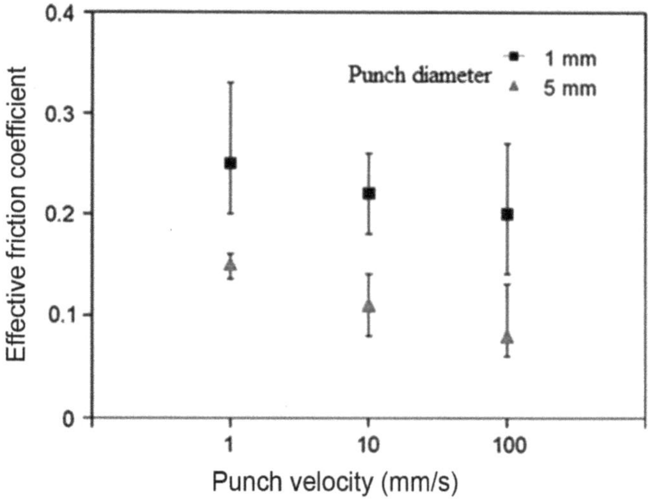

FIGURE 17.9 Change of the effective COF with different punch speeds [4].

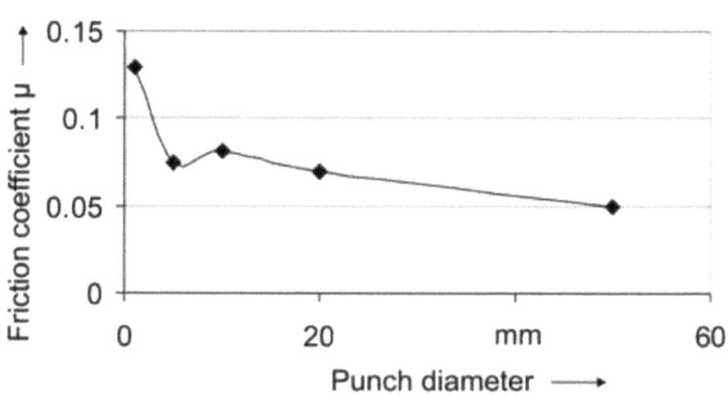

FIGURE 17.10 Size-dependent friction coefficients in deep drawing [5].

diameter of 5 mm, and 0.065 for punch diameter of 20 mm. It can be seen that the COF is in an increasing trend when the process dimension decreases.

The relationship of the drawing punch stroke and punch force under different lubricant conditions is shown in Fig. 17.11 [6]. The tendencies of the curves are almost the same. During deep drawing process, the punch force first increases and then decreases, and the first peak value occurs. After that, the flange wrinkled due to the lack of the sufficient blank holder force and the punch compelled the wrinkled blank drawing into the clearance between the punch and die, then the second peak value of the punch force occurred. With the increase of the deep drawing ratio, the wall of the cup cracked if an improper lubricant was used, and the deep drawing process was terminated.

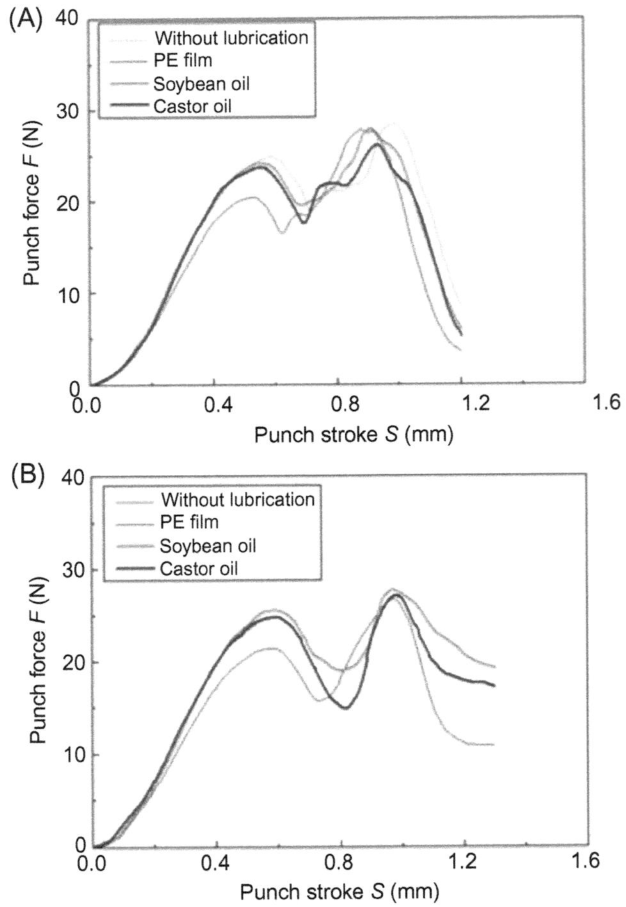

FIGURE 17.11 Punch stroke–punch force curves under different drawing ratio: (A) 1.8, (B) 1.9, (C) 2.0, and (D) 2.1 [6].

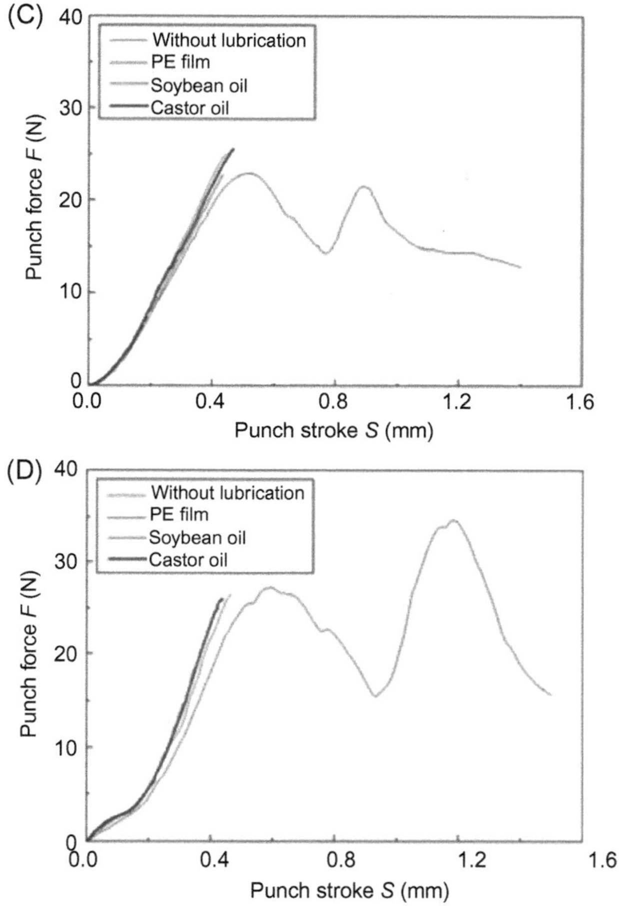

FIGURE 17.11 (Continued).

The drawing ratio can be calculated by the following equation [6]:

$$K = \frac{D}{d} \tag{17.3}$$

where D is the diameter of the drawing blank and d is the diameter of the drawing punch.

Fig. 17.12 shows the SEM images of micro cups with a drawing ratio of 1.8 under three different lubricating conditions: the dry, soybean oil, and castor oil lubrications. It is clear that there are few scratches in the micro cups, especially in the rim of the cups compared to that under the lubrication of polyethylene film (see Fig. 17.12A). This is because there is no lubricant left or the lubricant is overflowed. In this case, the blank and die contacted directly, scratches were inevitable under those serious large deformation and friction.

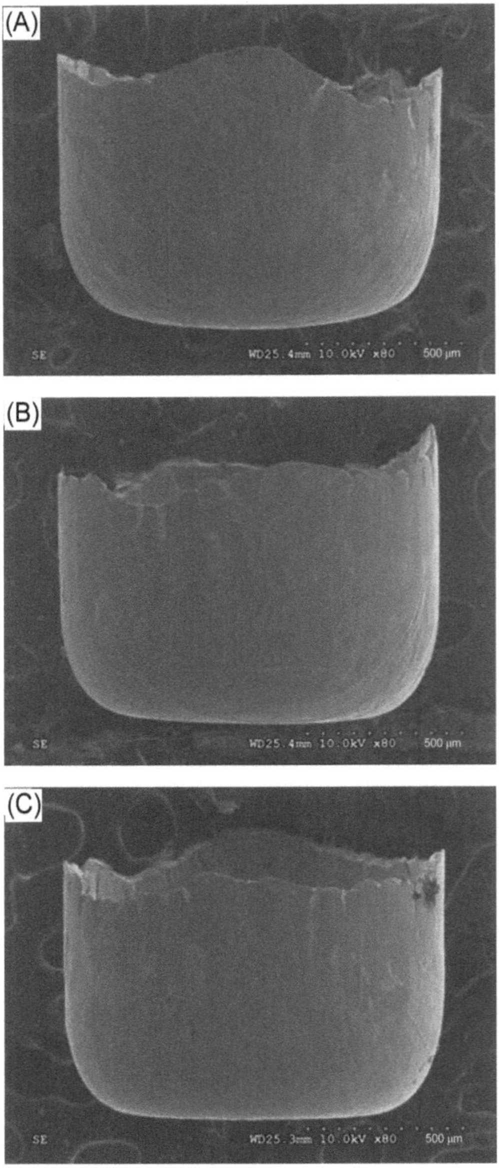

FIGURE 17.12 SEM images of micro cups formed under different lubrication conditions: (A) without lubrication, (B) soybean oil, and (C) castor oil [6].

17.2.3 Punch Diameter Influence and Size Effects

In microscale forming process, product accuracy and process parameters can be influenced by the ratio of surface to volume or the ratio of grain size to foil thickness. For the analysis of these effects, deep drawing experiments had been conducted for CuZn37 foils with punch diameter from 8.0 to 1.0 mm and a foil thickness ranging from 0.3 to 0.04 mm [7].

Fig. 17.13 shows the normalized load-stroke curves of the several forming cases with different material states (*t*, foil thickness) [7]. It can be seen that the deformation load decreases with the increase of grain size, while the as-received copper sheet shows the highest deformation load. These phenomena can be explained based on both the change of surface grain fraction and the grain boundary strengthening behavior. The surface layer model is developed based on the fact that the surface grains have less constraints and their flow stress is lower than that of the inner grains. The decrease of blank thickness and the increase of grain size in turn increase the fraction of surface grains. It can be seen that the deformation load decreases with the increase of punch radius, due to the fact that a larger bending force is needed in the case with the small punch radius. Such a size effects become less obvious in the smallest size-scaled case with coarse grain size. In addition, it can be seen that the maximum load occurs at the longer stroke with the increase of punch radius. The plastic deformation satisfies the incompressibility condition, in such a way that a longer stroke is needed to form the full cup without the flange with the increase of punch radius.

Fig. 17.14 shows the geometries of the drawn cups [7]. According to Table 17.1, the grains grow larger at a higher annealing temperature. The large grains in the thin sheets lead to the uneven deformation and the rough surface finish in the drawn cups. For the copper sheet annealed at 500°C, the grain size is in the range of 15–21 μm. The distribution of different grains is more even in the blank with the thickness of 0.4 mm than those in the cases with the thicknesses of 0.2 and 0.1 mm. Therefore, both the surface finish and the formed part geometry are better than those of the latter two cases. For the sheets annealed at 650°C and 800°C, the grain sizes approach to the corresponding sheet thicknesses and only a few grains or single grain exist over the thickness in each case. Grain distribution and orientation highly affect the deformation behavior, which further affects the profile and accuracy of the formed cups. The formed cup geometry and surface finish tend to be irregular and rough with the increase of grain size and the decrease of the formed cup size. Moreover, the anisotropic deformation appears based on the wave-shaped rim, i.e., earing, of the formed cups. It is found that the earing in the group with a large punch radius is much obvious compared to that in the group with a small punch radius.

17.2.4 Initial Gap Influence

Fig. 17.15 illustrates three physical cups produced with the initial gaps: −100, 0, and 100 μm [8]. It can be seen that the initial gaps affect the final cup profiles, in particular at the shoulder corner radius. In addition, the heights of the final cups are slightly different where the cup height increases with the same initial gap.

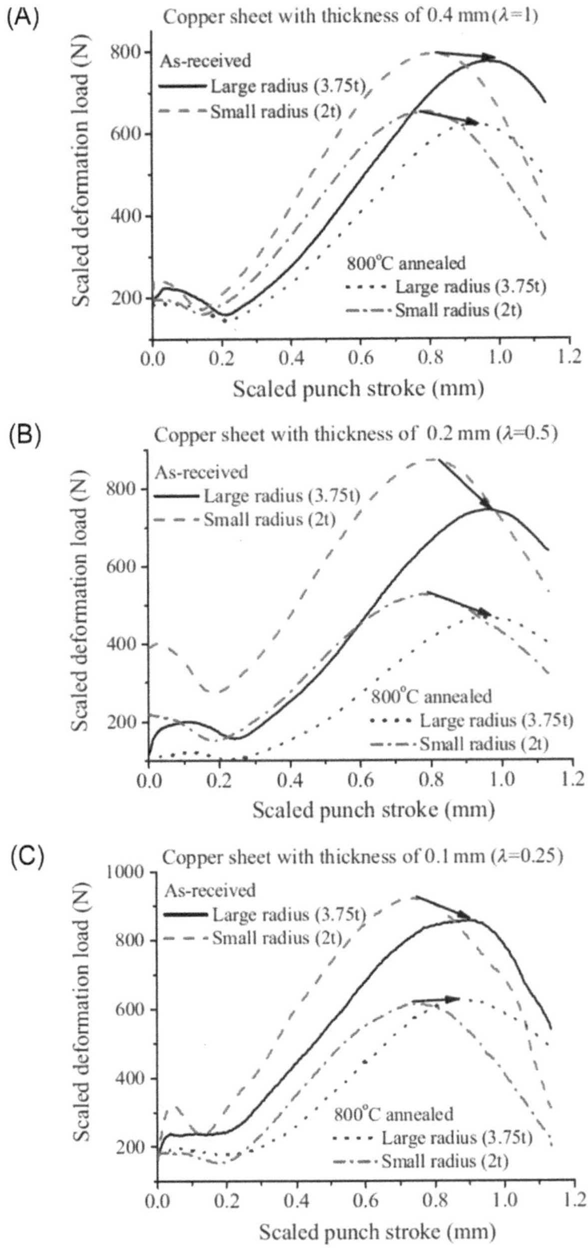

FIGURE 17.13 The influence of punch radius on the deformation load in different size-scaled cases: (A) $\lambda = 1$, (B) $\lambda = 0.5$, and (C) $\lambda = 0.25$ [7].

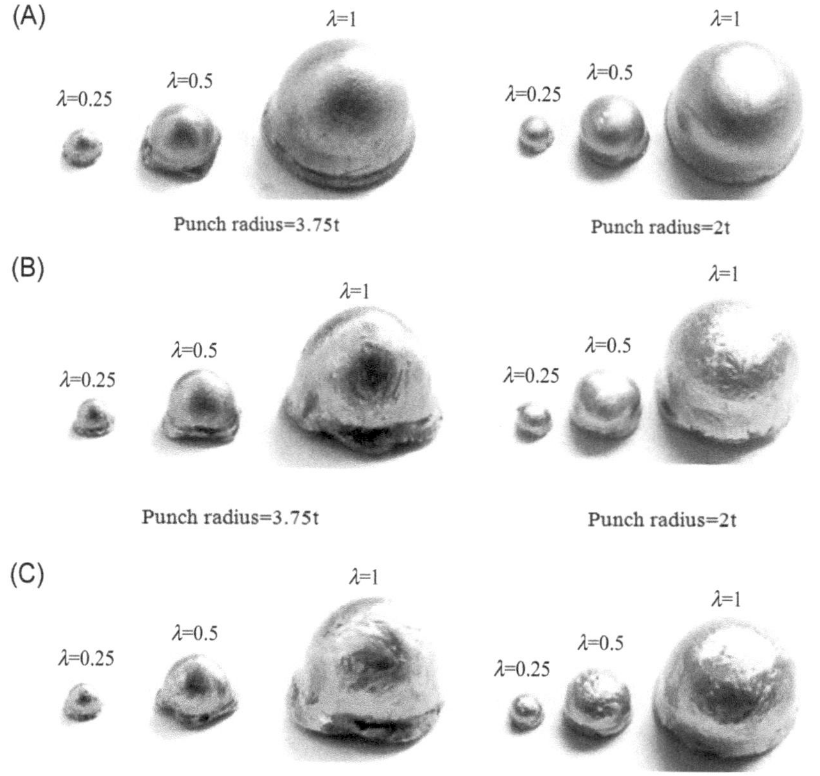

FIGURE 17.14 Comparison of the different size-scaled cups for blanks under different annealing temperatures: (A) 500°C, (B) 650°C, and (C) 800°C [7].

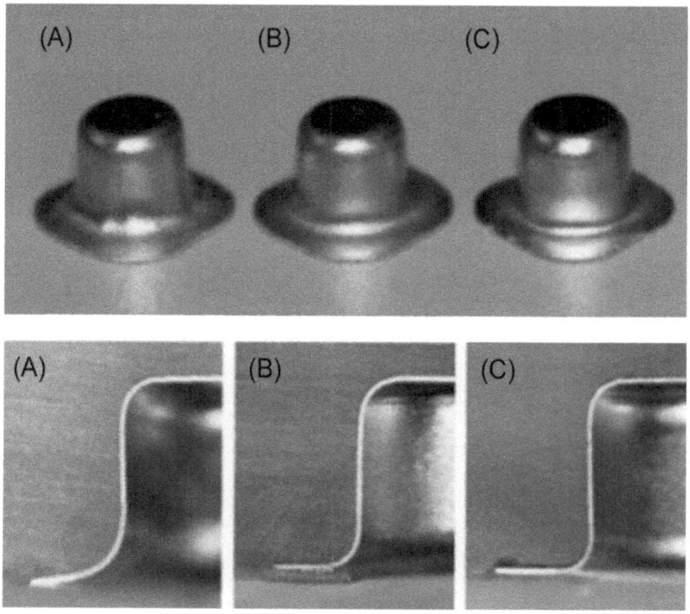

FIGURE 17.15 Cups formed with different initial gaps: (A) −100 μm, (B) 0 μm, and (C) 100 μm [8].

The thickness distributions along the transverse direction, which predicted by the simulations with the different gaps are presented in Fig. 17.16 [8]. Reducing the initial gaps causes an increase in thickness reduction especially at the side wall of the formed cup. The results detect that the maximum reduction in the side wall thickness decreases from 15.8% to 11.9% when the initial gap increases from 0 to 100 μm, as well as increases from 15.8% to 28% by reducing the gap from 0 to −100 μm. Additionally the location of the maximum thinning changes from the nose radius to the upper part of the side wall of the final cups when reducing the initial gaps. Higher tensile stresses generated at the side wall and shoulder radius of the cup result in an increase in thickness reduction. It is easy to lead to failure by fracture. Therefore, higher initial gap can improve the formability of the workpiece.

17.2.5 Thickness Size Effects

Fig. 17.17 shows a comparison of the thickness strain distribution from different measuring planes [9]. In general, the smallest reduction in the thickness strain occurs in the plane 90 degrees to the rolling direction (RD), and a large thickness strain reduction was observed in the planes 0 and 45 degrees to the RD. This is because that there exists a lower plastic anisotropy in the RD. During the deep drawing process, because of the low plastic anisotropy in the RD, the thickness in the RD first decreases owing to the stretch bending of the punch corner. Then, owing to the interaction between the normal anisotropy in the plane 45 degrees to the RD and that in the RD, the thickness reduction in the RD is compensated, and the thickness in the plane 45 degrees to the RD gradually decreases to a value to that in the RD.

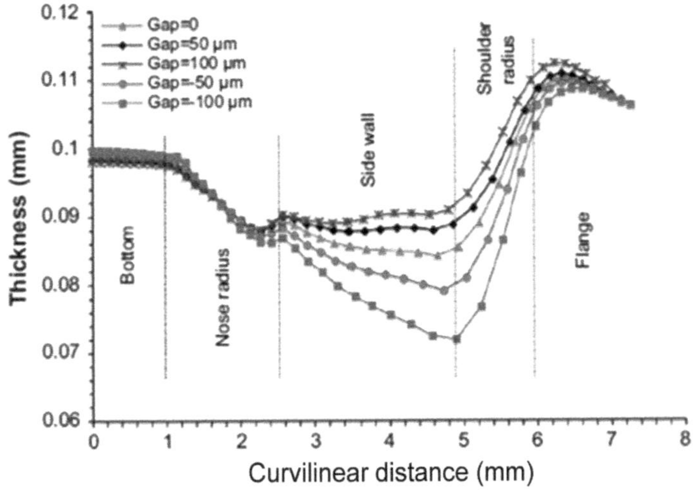

FIGURE 17.16 Thickness distribution at transverse direction for different gaps [8].

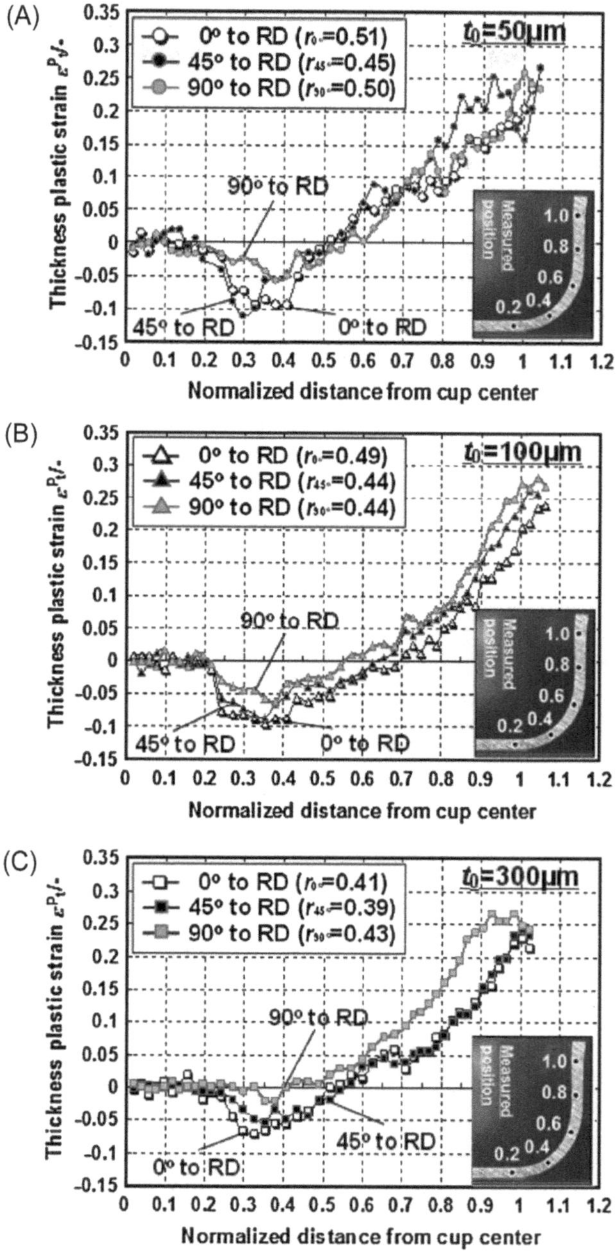

FIGURE 17.17 Comparisons of thickness strain distribution for planes with different angles to the RD: (A) $t_0 = 50$ μm, (B) $t_0 = 100$ μm, and (C) $t_0 = 300$ μm [9].

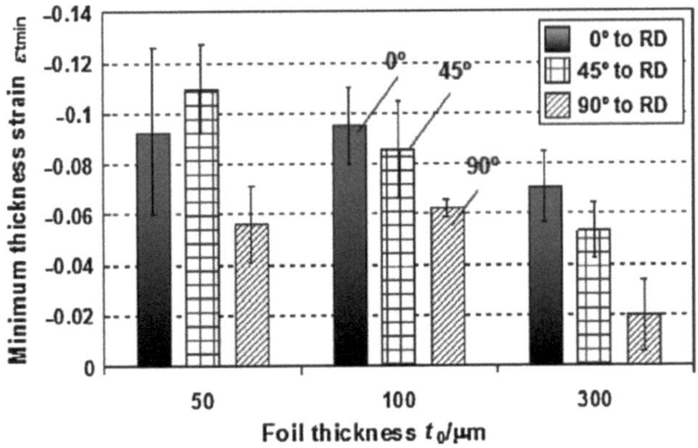

FIGURE 17.18 Comparison of the minimum thickness strain for three different measuring planes and each foil thickness [9].

TABLE 17.6 Surface Roughening Characteristics and Nonuniformity of the Thickness [9]

Initial thickness t_0 (μm)	50.38	99.21	303.99
Initial roughness R_{z0} (μm)	**0.91**	**1.16**	**1.17**
Minimum thickness at cup corner t_{min} (μm)	45.60	89.76	281.76
Roughness at cup corner R_z (μm)	**2.37**	**2.23**	**1.76**
Nonuniformity of thickness f_R	0.896	0.950	0.987
Roughening rate α	**16**	**11**	**9**
Relative roughening rate t_0/α	3	9	35

Comparing the minimum thickness strain for each foil thickness (see Fig. 17.18), there is no correlation between the minimum thickness strain and the mean normal plastic anisotropy of the foil. Although the highest plastic anisotropy was obtained for the 50 μm-thick foil, the thickness reduction at the cup corner was larger than that for the 300 μm-thick foil, which had the lowest plastic anisotropy.

The roughness data for each thickness are summarized in Table 17.6. The nonuniformity of the thickness f_R in the table is defined by the following equation:

$$f_R = \frac{t_B}{t_A} = 1 - \frac{2(\alpha \cdot \bar{\varepsilon} + R_0)}{t_A} \tag{17.4}$$

where t_A is the thickness of the peak profile, t_B is the thickness of the valley profile, α is the roughening rate, R_0 is the initial surface roughness, and ε is the equivalent strain. In addition, the roughening rate α is the rate of increase in surface roughness with increasing equivalent strain, obtained from the following equation:

$$R_z = C \cdot d \cdot \bar{\varepsilon} + R_0 = \alpha \cdot \bar{\varepsilon} + R_0 \tag{17.5}$$

where C is a constant and d is the grain size.

The lowest value of f_R at the cup corner was obtained for the 50 μm-thick foil. This value of f_R is equivalent to a large nonuniformity of about 10% between the peak and valley profile thicknesses. Corresponding to the lowest f_R value, the highest a value was obtained for the 50 μm-thick foil. This suggests that the largest nonuniformity of the thickness profile of the 50 μm-thick foil was due to the highest roughening rate during the process.

Considering the mechanism of the roughening behavior, α value is also affected by the material inhomogeneity of the grains, such as the grain size, crystal orientation, and crystal rotation during the process. However, resulting geometrical inhomogeneity such as roughened surface topography strongly affects the instability and limit strain of ultrathin metal foils as above. Therefore, in addition to the material anisotropy based on the crystal texture, the geometrical anisotropy such as surface topography and defects should be taken into account to discuss the unstable deformation of ultrathin rolled metal foils.

In summary, the heat treatments effectively change the material's microstructures in terms of grain sizes and mechanical properties. Correspondingly, the flow stress and the total elongation increase with the increase of heat treatment temperatures. The material properties considering size effects can be obtained through the developed micro tensile test system, and the MDD can be successfully conducted by the one-stroke blanking-drawing MDD system. The peak punch force increases according to the heat treatment temperatures, whereas the wrinkling increases with the growth of the grains caused by the high-heat temperature. The significant effects of grain size, lubrication, punch diameter, initial gap, and foil thickness have been discussed and validated.

REFERENCES

[1] L. Luo, Z.Y. Jiang, D.B. Wei, K. Manabe, X.B. He, P. Li, An experimental and numerical study of micro deep drawing of SUS304 circular cups, Manufact. Rev. 2 (2015) 27.

[2] L. Luo, Z.Y. Jiang, D.B. Wei, K. Manabe, Xi Zhao, D. Wu, et al., Effects of surface roughness on micro deep drawing of circular cups with consideration of size effects, Finite Elem. Anal. Des. 111 (2016) 46–55.

[3] H. Sato, K. Manabe, D.B. Wei, Z.Y. Jiang, S. Alexandrov, Tribological behavior in microsheet hydroforming, Tribol. Int. 97 (2016) 302–312.

[4] F. Vollertsen, Effects on the deep drawing diagram in micro forming, Prod. Eng. Res. Dev. 6 (2012) 11–18.

[5] Z. Hu, F. Vollertsen, Modelling of friction with respect to size effects, Int. J. Mater. Form. 1 (2008) 1231–1234.

[6] F. Gong, B. Guo, C.J. Wang, D.B. Shan, Effects of lubrication conditions on micro deep drawing, Microsyst. Technol. 16 (2010) 1741–1747.
[7] M.W. Fu, B. Yang, W.L. Chan, Experimental and simulation studies of micro blanking and deep drawing compound process using copper sheet, J. Mater. Process. Technol. 213 (2013) 101–110.
[8] I. Irthiea, G. Green, S. Hashim, A. Kriama, Experimental and numerical investigation on micro deep drawing process of stainless steel 304 foil using flexible tools, Int. J. Mach. Tools Manufact. 76 (2014) 21–33.
[9] T. Shimizu, M. Ogawa, M. Yang, K. Manabe, Plastic anisotropy of ultra-thin rolled phosphor bronze foils and its thickness strain evolution in micro-deep drawing, Mater. Des. 56 (2014) 604–612.

Chapter 18

Practice of Micro Hydromechanical Deep Drawing

18.1 EQUIPMENT AND TOOLS FOR MICRO HYDROMECHANICAL DEEP DRAWING

18.1.1 Design of MHDD Apparatus

The design of the new micro hydromechanical deep drawing (MHDD) apparatus includes simplification of the forming processes and apparatus structure through (1) one-stroke forming avoiding the need for material transfer and positioning control, (2) a constant-gap method to avoid the need for highly accurate force control for preventing wrinkling, (3) sealing and generation of stable fluid pressure by decreasing the clearance between tools to avoid the need for ultrafine flow-rate control, and (4) a multiaxis system inside the die to avoid the need for a double-action mechanical press. A simple apparatus can be realized by decreasing the number of control targets based on the measures mentioned above. This apparatus enables the fabrication of components with a high dimensional accuracy.

Fig. 18.1 shows the schematic of the MHDD die and hydraulic system [1]. In general, a multiaxial servo press is used for one-stroke forming, whereas the developed MHDD apparatus adopts an uniaxial servo press to perform one-stroke forming. Therefore, a die structure will be designed to reproduce the behavior of a multiaxial servo press inside the die and simplify the MHDD apparatus structure. Each side of the MHDD die developed is as small as ~100 μm. The drawing die (blanking punch), blanking die, drawing punch, or bush is independently replaceable, accommodating various dimensions of components simply by changing individual parts. In addition, tools with a high dimensional accuracy were used; the tolerance of the blanking and drawing tools are ±0.001 and ±0.010 mm at the flat and curved parts, respectively. To measure the punch force, an ultra small compression-type load cell with a rating capacity of 50 N was incorporated in the die so that the force directly applied to the drawing punch is transferred to the load cell. The hydraulic system consists of a relief valve to control the maximum

Microforming Technology. DOI: http://dx.doi.org/10.1016/B978-0-12-811212-0.00018-2

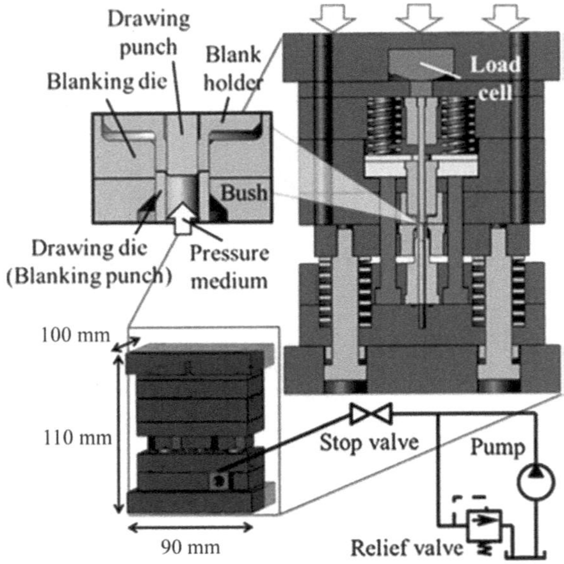

FIGURE 18.1 The sketch of developed micro hydromechanical deep drawing tool set and hydraulic system [1].

counter pressure, a stop valve to suppress the rate of the flow into the die, and a hydraulic pump with a maximum counter pressure of 20 MPa.

The desktop servo screw press machine shown in Fig. 18.2 was used with the MHDD apparatus. The specifications of the servo screw press machine were as follows: load capacity of 50 kN, die height of 130 mm, maximum stroke of 37 mm, and positioning accuracy of 0.4 μm. Table 18.1 lists the specifications of the MHDD apparatus.

18.1.2 Hydraulic Equipment

The hydraulic station was linked to the micro deep drawing (MDD) system to supply the hydraulic pressure. Its systematic view is shown in Fig. 18.3. Mechanical hydraulic oil with a viscosity of 46 mm^2/s was utilized as the working medium. The handle-operated directional valve (#10) was of a three-position four-path valve with the "M" function. At the start of the experiments, the valve was at its middle functional position, and the oil could flow back to the oil tank (#1) through the handle-operated valve without loading. When preparing for the working pressure, the handle-operated valve was moved to its left functional positon, and the oil supplied by the pump (#6) was blocked. Once the pressure reached the designed working pressure controlled by the relief valve (#9), the oil flowed back to the oil tank through the relief valve. When all settings were ready, the handle-operated valve was set to its correct functional position, and the pressed oil

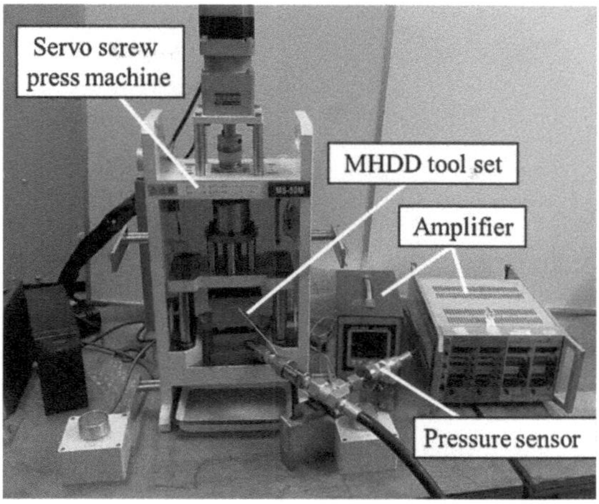

FIGURE 18.2 Newly developed MHDD system with servo screw press apparatus [1].

TABLE 18.1 Specification of Developed MHDD Apparatus

Main MHDD system	Servo screw press machine	Load capacity (kN)	50
		Motion resolution (nm)	400
	Hydraulic system	Pump pressure (MPa)	20
Tooling	Forming process	One-stroke forming	
	Blank holder	Constant-gap method	
	Pressure generation	1 μm clearance between tools	
	Tolerance	Flat part (mm)	±0.001
		Radius part (mm)	±0.01

was supplied for the MHDD experiments. Once a cup is drawn, the hydraulic station can be separated from the MDD system through quick coupling (#13) without stopping the pump. Thus, there is no need for resetting the working pressure for the next cup. One of the pressure gauges (#7.1) was used for pressure setting while the other one (#7.2) was utilized for monitoring the hydraulic pressure during the MHDD experiments. The throttle valve (#11) was employed for adjustment of the oil-flow rate. Because of the minute dimensions of the MHDD system, tiny gaps between the die/blank holder and the blanking die/blanking holder were used to seal the pressurized oil. As the gaps were about 2–4 μm, oil leakage was small and the amount of pressure lost was limited. Because of the increase in the hydraulic pressure caused by the fluid compression from the deformation of the blank, a relatively stable hydraulic pressure can be maintained during the MHDD experiments.

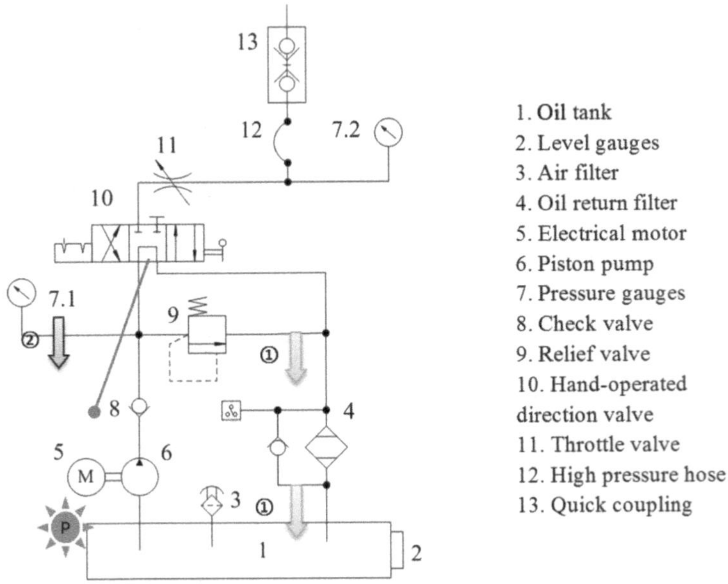

FIGURE 18.3 Sketches of the hydraulic system.

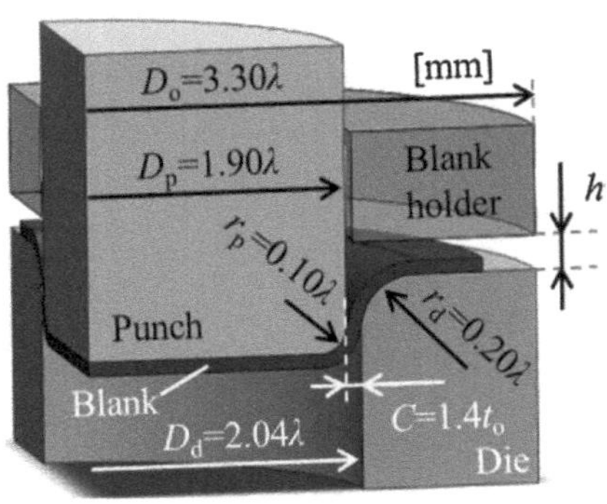

FIGURE 18.4 Tool dimensions for micro deep drawing ($\lambda = 1/5, 1$) [1].

18.1.3 Determination of Tool Dimensions

The performance test of the MHDD apparatus was carried out to examine its operation and determine appropriate tool dimensions to prevent wrinkling and generate a counter pressure [1].

Phosphor bronze foils (C5191-H) with thickness t_0 of 20 and 50 μm were used in the experiment. Fig. 18.4 shows the dimensions of the die and the scale factors $\lambda = t_0/t_{\mathrm{ref}} = 2/5$ and 1 were adopted on the basis of the law of geometric similarity. At $\lambda = 2/5$, a punch shoulder radius r_p of 0.04 or

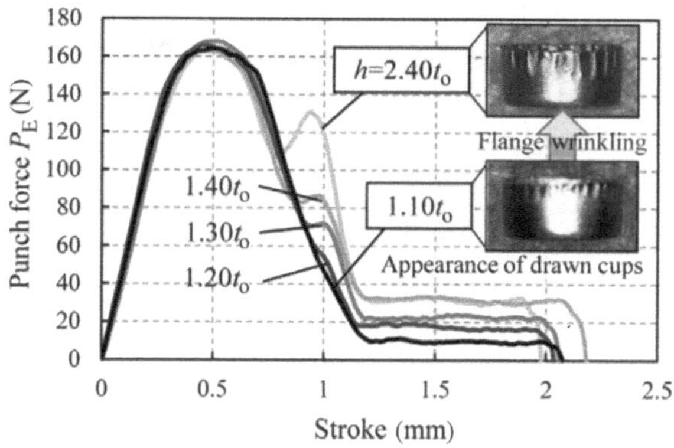

FIGURE 18.5 Effect of gap between blank holder and drawing die on punch force–stroke curve (C5191-H).

0.25 mm was used. Cups with a drawing ratio D_0/D_p of 1.74 were targeted. To examine the effect of the constant gap on the wrinkling, experiments were carried out under five conditions, the constant gap $h = 1.10t_0$, $1.20t_0$, $1.30t_0$, $1.40t_0$, and $2.40t_0$ by conventional MDD in dry friction. For MHDD, a forced pressurization was adopted using hydraulic oil (40°C, 44 mm^2/s) as a pressure medium. The temperature was controlled by cooling the hydraulic oil with water. The drawing rate was 0.4 mm/s.

Fig. 18.5 shows the effect of the constant gap between the blank holder and the drawing die on the punch force–stroke curves. At $h = 1.10t_0$, a normal punch force–stroke curve is obtained. The constant punch force observed in the latter half of the forming process is the sliding force generated as a result of the sliding of the drawn cup edge against the die side wall. At $h > 1.10t_0$, ironing force of wrinkles is observed after the force reaches the maximum value; the ironing and the sliding forces increase with increasing constant gap. When the cups formed at $h = 1.10t_0$ and $2.40t_0$ are compared, flange wrinkles are observed in the cups formed at $h = 2.40t_0$, whereas wrinkles are generated only at the cup edge at $h = 1.10t_0$, with reduced wrinkles height. This result indicates that for large gaps, the blank holder does not function properly, leading to the generation of flange wrinkles. When the flange wrinkles are crushed, the ironing force is generated and the sliding force increases. For small gaps, in contrast, the blank holder functions properly, which reduces the wrinkles height, leading to the suppression of the ironing force and reduction in the sliding force. Under the forming conditions adopted, the appropriate constant gap was determined to suppress the wrinkling, although it was impossible to completely prevent wrinkling by controlling the constant gap. For the developed MHDD apparatus, $h = 1.10t_0$ was adopted.

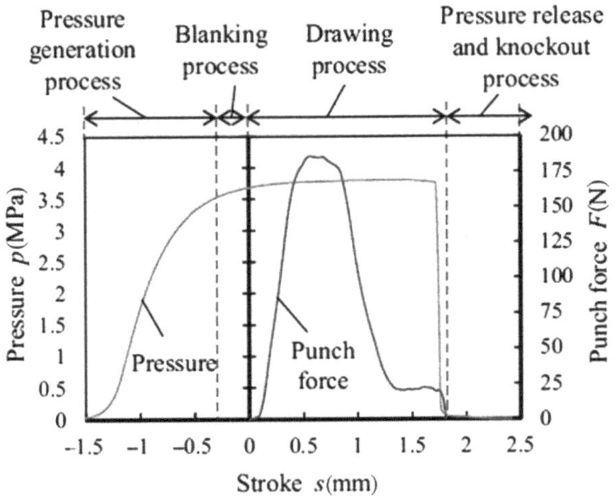

FIGURE 18.6 Counter pressure- and punch force–stroke curves during MHDD process.

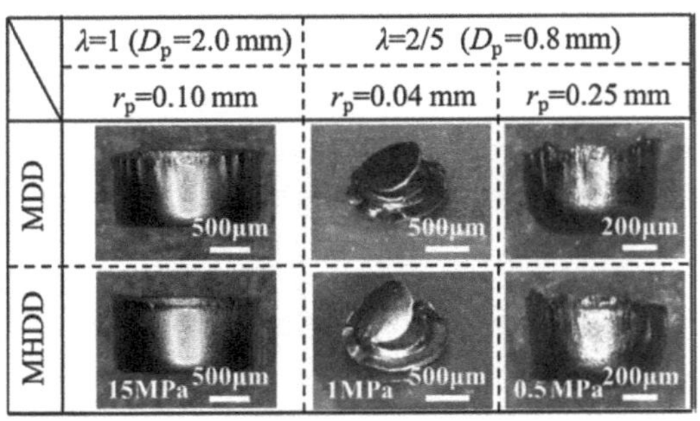

FIGURE 18.7 Drawn cups under different scale factors and punch shoulder radii in MDD and MHDD [1].

Fig. 18.6 illustrates the counter pressure–stroke curve and the punch force–stroke curve during the MHDD process. As shown in Fig. 18.2, the forming processes consist of (1) counter pressure generation, (2) blanking, (3) drawing, and (4) pressure release and knockout. Stable counter pressure was found to be generated during drawing when the clearance between the drawing die and the bush was set to 1 μm. The counter pressure did not decrease during the drawing process, which proves that the pressure medium did not leak from the blank rim. Namely, the process is a radial pressure-aided deep drawing by an indirect method in which a compressive force is applied from the blank rim radially inward. By this method, the friction force is reduced by the leakage of pressure medium between the blank and blank holder and the meridional stress is also reduced.

Fig. 18.7 shows drawn cups obtained under different scale factors (λ) and punch shoulder radii (r_p) for MDD and MHDD. From this figure,

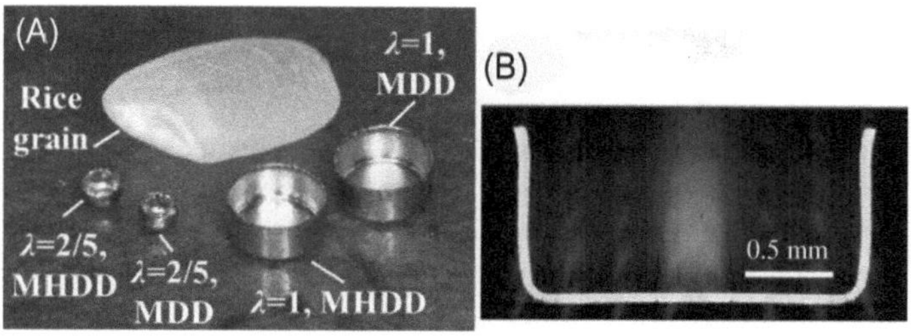

FIGURE 18.8 Appearance of drawn cups in MDD and MHDD (A) different scale cups around a rice grain and (B) axial cross section of drawn cup ($\lambda = 1$, $D_P = 2.0$ mm) [1].

it can be seen that the MHDD apparatus can form micro cups with diameters ≤ 1 mm under a certain condition. At $\lambda = 1$ and $r_p = 0.10$ mm, the micro cups were successfully fabricated. However, with scaling down to $\lambda = 2/5$ and $r_p = 0.04$ mm based on the law of geometrical similarity, the micro cups were fractured at punch shoulder. It seems that the difference comes from size effects, such as the decrease of ductility with decreasing thickness. Within creasing the punch shoulder radius $r_p = 0.04$ mm to $r_p = 0.25$ mm at $\lambda = 2/5$, the micro cup can be successfully fabricated. It implies that the fracture occurs at $\lambda = 2/5$ and $r_p = 0.04$ mm with severe bending deformation. Fig. 18.8A shows the appearance of micro cups obtained by MDD and MHDD. The drawn micro cups with diameters of 0.8 and 2 mm are smaller than a rice grain. Fig. 18.8B shows a cross section of a drawn cup. The micro cup was successfully drawn even at $\lambda = 1$ and a sharp punch shoulder of $r_p = 0.10$ mm. The above results confirmed that the MHDD apparatus with a simple tooling structure and forming processes realized the required performance without the need for large and expensive equipment.

18.2 MICRO HYDROMECHANICAL DEEP DRAWING PRACTICE

18.2.1 Effect of Counter Pressure in MHDD

Phosphor bronze (C5191-H), stainless-steel (SUS304-H), and pure titanium (TR270C-H) foils with a thickness $t_0 = 50$ μm were used. Table 18.2 lists the mechanical properties of the three materials obtained by the tensile test. The tool dimension shown in Fig. 18.4 was used, and scale factor $\lambda = 1$ and constant gap $h = 1.10t_0$ were adopted. For MHDD, forced pressurization was adopted using machine oil as the pressure medium. MDD was carried out without lubricant. The drawing rate was 0.4 mm/s. Effective punch force

TABLE 18.2 Mechanical Properties of Materials Used

Material	Young's Modulus, E (GPa)	Yield Stress, σ_y (MPa)	Tensile Strength, σ_B (MPa)	Elongation, δ (%)
Phosphor bronze (C5191-H)	110	610	682	0.6
Stainless-steel (SUS 304-H)	193	1217	1331	2.4
Pure titanium (TR270C-H)	117	601	807	1.4

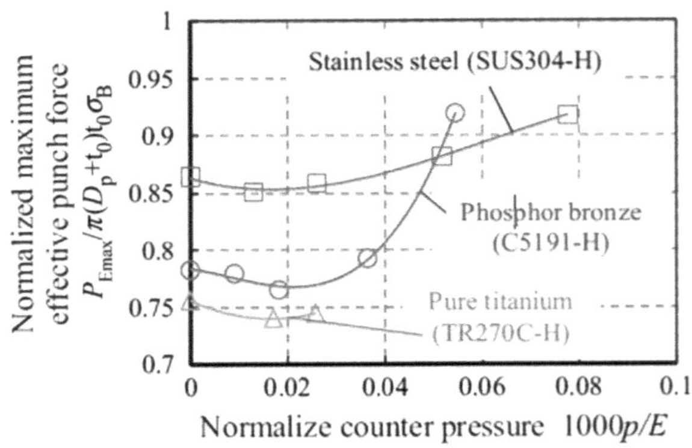

FIGURE 18.9 Effect of counter pressure on normalized effective punch force for various materials.

(P_E) was obtained by subtracting the counter pressure force to push back the punch (F_H) from the measured punch force (P_H) as

$$P_E = P_H - F_H = F_S + F_B + (F_{FDie} + F_{FBH}) \tag{18.1}$$

where F_S is the pure drawing force at the flange, F_B is the bending force at the die shoulder, F_{FDie} is the friction force between the blank and the die, and F_{FBH} is the friction force between the blank and the blank holder. Under the same tooling conditions, F_S and F_B can be considered to be almost constant even if the fluid pressure increases. Therefore, the change in effective punch force can be estimated as the change in friction force. Thus, even though F_S and F_B are unknown, the change in the friction force can be evaluated by the change in P_E.

Fig. 18.9 shows the effect of normalized counter pressure on the maximum effective punch force normalized by the tensile strength for the each material. The effective punch force is normalized to cup cross section at side wall ($\pi^*(D_p + t_0)^* t_0$) and tensile strength σ_B. The maximum possible punch force is expressed by ($\pi^*(D_p + t_0)^* t_0)\sigma_B$. Thus, the normalized maximum effective punch force can be written for different blank materials by the above maximum possible punch force. For the stainless-steel foil, the maximum effective punch force decreases with applying the counter pressure. The change in the effective punch force represents the change in the friction force as shown in Eq. (18.1). By applying the counter pressure, the friction force decreases, and the lubrication is improved. However, the maximum effective punch force increases when the counter pressure is further increased. This is because the blank is pushed against the blank holder by the counter pressure, which increases the friction force. From this result, the fracture at the punch shoulder under high pressures is considered to be caused by the excessive friction force at the flange. This behavior is observed in all of the three materials.

Fig. 18.10 shows the effect of counter pressure on the occurrence of wrinkling and fractures during MHDD. When the counter pressure normalized by Young's modulus (*p*/*E*) is smaller than a certain value, wrinkling occurs (wrinkling zone), whereas the cups fracture when *p*/*E* is larger than a certain value (fracture zone). When *p*/*E* falls between these two values, cups without wrinkles or fractures are obtained (success zone). For the pure-titanium foil, the size of wrinkles decreases with applying the counter pressure, however, the cups fracture before wrinkles are completely removed, not showing the success zone. This is because the friction force for the pure-titanium foil becomes too large and the fracture zone shifts to the low-pressure side, leading to the fracture zone merging with the wrinkling zone and the disappearance of the success zone. The success zone for pure-titanium foils may be obtained by improving the lubrication to decrease the friction force and shift the fracture zone to the high-pressure side. The limit blank holding pressure

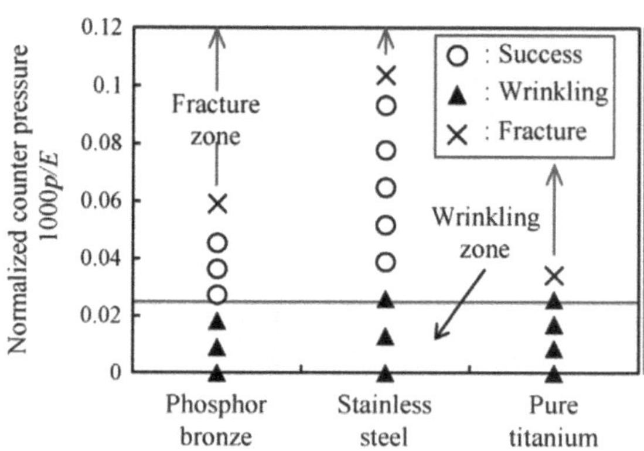

FIGURE 18.10 Effect of normalized counter pressure on occurrence of wrinkling and fracture in MHDD.

in cylindrical deep drawing normalized by Young's modulus is predominantly determined by the tool dimensions alone. The existence of a wrinkling zone at a certain *p*/*E* or less under the same tool conditions means that the counter pressure in MHDD functions equivalently to the blank holder. From the above, the applying appropriate counter pressure in MHDD is considered to prevent the wrinkling and reduce the friction force, leading to the improvement of the forming limit.

Fig. 18.11 shows the effect of applying counter pressure on the shape accuracy at the bottom of cups. The bottom obtained by conventional MDD is convex with respect to the punch, whereas the bottom obtained by MHDD with the application of counter pressure is concave; thus, the bottom shape obtained by MDD is different from that obtained by MHDD. In addition, the concave deformation toward the punch increases with increasing counter pressure during MHDD. In conventional MDD, the bottom of the cup is subjected to bulge deformation by the bending moment during deep drawing, resulting in a convex shape with respect to the punch, as shown in Fig. 18.12.

In contrast, when the fluid pressure that counters the bending moment is applied to the bottom during MHDD, the convex bulge deformation is suppressed and a flat cup bottom is formed. However, the degree of adhesion between the punch and the blank is increased by increasing the counter pressure and a negative pressure is generated during the knockout process. The counter pressure is continuously applied until the end of the knockout process, leading to a concave shape with respect to the punch. The negative pressure can be suppressed by making a hole at the center of the punch or a small groove on the side wall of the punch. The concave cup bottom obtained by MHDD can be made flat by controlling the tool and forming conditions. These results confirmed that the shape accuracy of the micro cups is improved by applying counter pressure during MHDD.

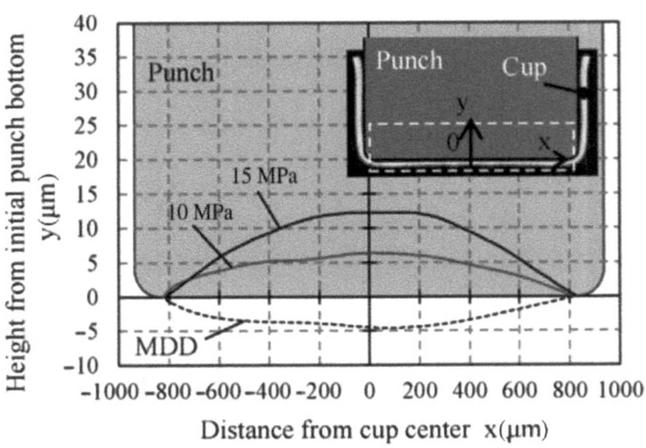

FIGURE 18.11 Effect of counter pressure on drawn cup profile at the bottom of the cup (stainless steel).

18.2.2 Tribological Behavior in MHDD

18.2.2.1 Friction Model in MHDD

In microforming, the material surface cannot be considered as smooth but instead, contains many peaks and valleys called "roughness." The roughness and the extent of the valleys become relatively large as compared to the size of the workpiece with scaling down. The valleys which connect to the edge of the blank cannot retain the lubricant. This area is called open lubricant pockets (OLPs), as shown in Fig. 18.13. With the decrease in specimen size, the fraction of OLPs increases. Thus, the lubricant cannot be kept during conventional microforming and it results in an increase of the friction force. This means that the OLPs must be considered when the tribological behavior of microforming is studied.

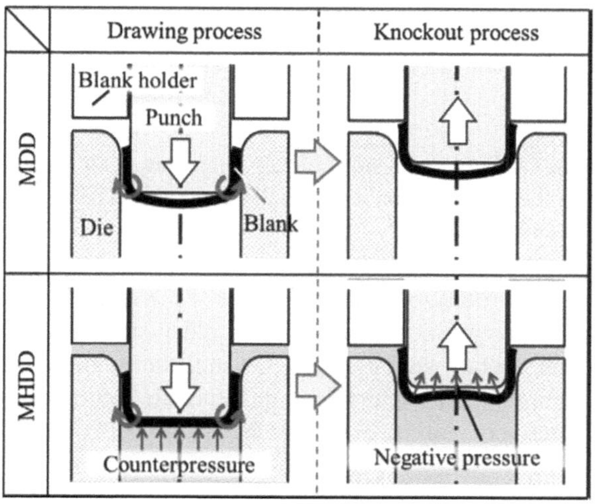

FIGURE 18.12 Comparison of blank deformation at punch bottom in MDD and MHDD.

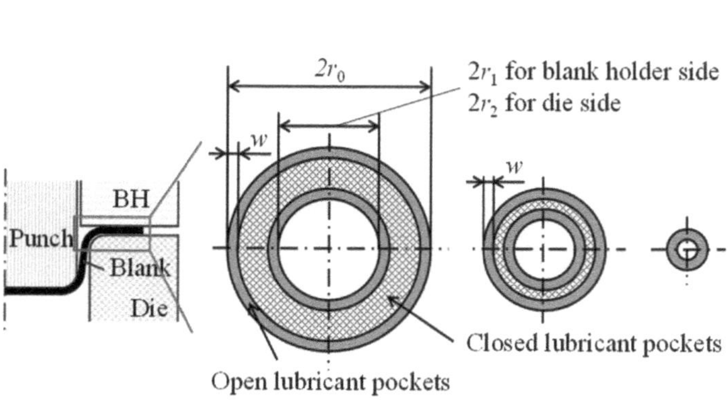

FIGURE 18.13 The change of fraction of OLPs in flange area with the decrease of blank size [2].

The relationship among OLPs, closed lubricant pockets (CLPs) and real contact area (RCA) can be expressed as

$$\alpha_o + \alpha_c + \alpha_{RC} = 1 \tag{18.2}$$

where α_o, α_c, and α_{RC} are the fractions of OLPs, CLPs, and RCA, respectively. In the real contact state, RCA exists in both OLPs and CLPs areas, hence, the nominal OLPs include OLPs itself and RCA in OLPs area. In the same manner, the nominal CLPs does CLPs itself and RCA. Using the nominal OLPs and CLPs, the relationship among OLPs, CLPs, and RCA in Eq. (18.2) can be simplified and rewritten by

$$\overline{\alpha}_o + \overline{\alpha}_c = 1 \tag{18.3}$$

where $\overline{\alpha}_o$ and $\overline{\alpha}_c$ are the fractions of nominal OLPs and CLPs, respectively.

The scale factor λ can be expressed by the ratio of width of OLPs w to blank radius R_0 as follows,

$$\lambda = R_0/w \tag{18.4}$$

The width of OLPs w does not change when the scale becomes small. In the macro case, the scale factor λ is usually larger than 10, and in the micro-case, λ is close to 0.

For MHDD, the double side friction acts because the contact state in flange area should be modeled as shown in Fig. 18.14. Furthermore, the nominal contact area at die and blank holder sides are different. Here, the total nominal contact area on both die and blank holder sides in flange part A_t, the nominal OLPs and CLPs on both die and blank holder sides A_c and A_0, respectively, can be expressed by

$$A_t = \pi\{2r_0{}^2 - r_1{}^2 - r_2{}^2\} \tag{18.5}$$

$$A_o = 2\pi w\{2r_0 + r_1 + r_2\} \tag{18.6}$$

$$A_c = A_t - A_o \tag{18.7}$$

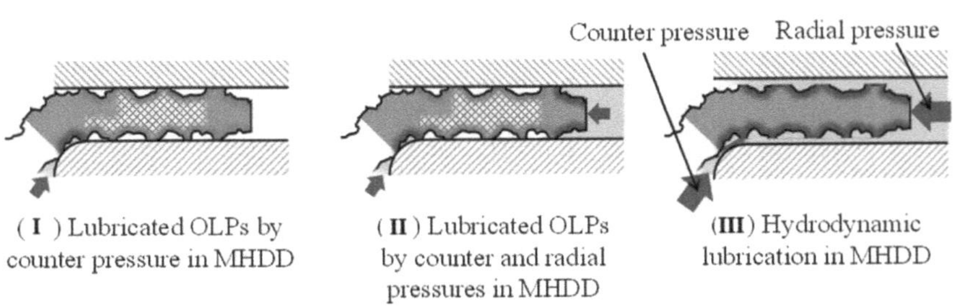

FIGURE 18.14 Examples of lubrication conditions considering OLPs and CLPs in MHDD [2].

The fraction of nominal OLPs $\overline{\alpha}_o$ can be obtained by

$$\overline{\alpha}_o = \frac{A_o}{A_t} = \frac{2w(2r_0 + r_1 + r_2)}{2r_0^2 - r_1^2 - r_2^2} = \frac{2(2\beta_0 + \beta_1 + \beta_2)}{\lambda\left(2\beta_0^2 - \beta_1^2 - \beta_2^2\right)} = \frac{1}{\lambda\beta^*} \tag{18.8}$$

where the relative current blank radius $\beta_0 = r_0/R_0$, the relative radius at entrance of die shoulder radius $\beta_1 = r_1/R_0$, the relative radius at exit of die shoulder $\beta_2 = r_2/R_0$, and $\beta^* = \left(2\beta_0^2 - \beta_1^2 - \beta_2^2\right)/\{2(2\beta_0 + \beta_1 + \beta_2)\}$. When $\lambda \leq 1/\beta^*$, only OLPs exists in the flange area and $\overline{\alpha}_o = 1$.

Considered the scale factor λ, the average coefficient of friction (COF) in flange area μ can be calculated as follows,

$$\mu = \frac{A_o\mu_{dry} + (A_t - A_o)\mu_c}{A_t} = \frac{\mu_{dry} + (\lambda\beta^* - 1)\mu_c}{\lambda\beta^*} \tag{18.9}$$

where μ_c is the COF in CLPs and μ_{dry} is the COF for dry friction in OLPs.

For MHDD, the fluid medium may insert into the OLPs by applying the fluid pressure as shown in Fig. 18.14. In this area, the friction behavior is different from the dry friction. As the fluid pressure increases, the COF in lubricated OLPs by fluid pressure shown in Fig. 18.14 decreases, and the hydrodynamic lubrication can be obtained when the fluid pressure exceeds the certain value. Thus, in MHDD, the lubricated OLPs by fluid pressure exist in addition to CLPs and OLPs with dry friction. To model this friction model, Eq. (18.9) is rewritten as,

$$\mu = \frac{A_{o\text{-dry}}\mu_{dry} + A_{o\text{-fluid}}\mu_{fluid} + (A_t - A_o)\mu_c}{A_t} = \frac{(1-\overline{\alpha}_f)\mu_{dry} + \overline{\alpha}_f\mu_{fluid} + (\lambda\beta^* - 1)\mu_c}{\lambda\beta^*} \tag{18.10}$$

where the dry friction area in nominal OLPs $A_{o-dry} = (1 - \overline{\alpha}_f)A_o$, the lubricated area by fluid pressure in nominal OLPs $A_{o-fluid} = \overline{\alpha}_f A_o$, μ_{fluid} is the COF in lubricated OLPs by fluid pressure, and $\overline{\alpha}_f$ is the ratio of the lubricated nominal OLPs by fluid pressure to whole nominal OLPs. $\overline{\alpha}_f$ is changed by pressurization methods as shown in Fig. 18.15. $\overline{\alpha}_f = 0$ for no fluid pressure, $\overline{\alpha}_f = 1/4$ for counter pressure, $\overline{\alpha}_f = 1/2$ for radial pressure, $\overline{\alpha}_f = 3/4$ for counter and radial pressures. Moreover, $\overline{\alpha}_f = 1$ when $\overline{\alpha}_o = 1$ at any pressurization methods. When the hydrodynamic lubrication occurs, $\mu = \mu_h$ which is the COF in hydraulic lubrication.

18.2.2.2 Required Fluid Pressure for Hydrodynamic Lubrication

The hydrodynamic lubrication occurs when the fluid medium leaks between the blank and tools, and the blank does not contact with the tools. Therefore, when the fluid pressure exceeds the contact pressure at die shoulder, the fluid pressure leaks between the blank and tools and the hydrodynamic lubrication can be obtained. The required fluid pressure for hydrodynamic lubrication p_h is given by the contact pressure at die shoulder q_d can be obtained by [2]

$$p_h = q_d = \frac{t\sigma_\varphi}{r_d + t/2} - \frac{t\sigma_\theta}{r}\sin\Phi \tag{18.11}$$

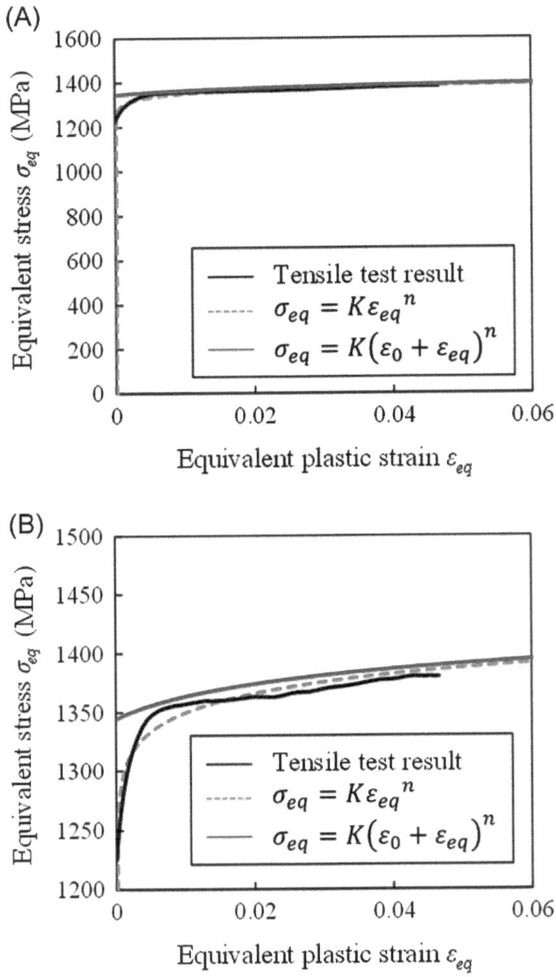

FIGURE 18.15 Stress–strain curves obtained by tensile test and used for calculation (A) overall view and (B) enlarged view [2].

Assuming that the friction resistance under the hydrodynamic lubrication is quite small and can be ignored, Eq. (18.11) can be rewritten as [2]

$$
\begin{aligned}
p_\mathrm{h} &= 1.1\sigma_\mathrm{eq}\left\{\frac{t}{r_\mathrm{d}+t/2}\ln\frac{r_0}{r_2}+\frac{t}{r_2}\left(1-\ln\frac{r_\mathrm{o}}{r_2}\right)\sin\Phi\right\}\\
&= 1.1\sigma_\mathrm{eq}\left\{\frac{1}{r_\mathrm{d}/t+1/2}\ln\frac{r_0}{r_2}+\frac{1}{D_\mathrm{p}/2t+c/t+(1-\sin\Phi)r_\mathrm{d}/t}\left(1-\ln\frac{r_\mathrm{o}}{r_2}\right)\sin\Phi\right\}
\end{aligned}
\tag{18.12}
$$

18.2.2.3 Comparison of Tribological Behavior in MHDD and Conventional HDD

Generally, the fluid pressure can induce hydrodynamic lubrication and improve the tribological behavior in conventional HDD. It is important in MHDD to clarify that the effect of fluid behavior on tribological behavior and its differences between MHDD and conventional HDD. The experimental results of MHDD are compared with those of HDD and the effect of D_p/t on fluid behavior and hydrodynamic lubrication are theoretically investigated here.

18.2.2.4 Analytical Conditions

Stainless-steel foil (SUS304-H) with a thickness of 50 μm was used. The stress–strain curves obtained by a tensile test and calculated by the Swift equation are shown in Fig. 18.15. The stress–strain curve was also calculated using power law as a comparison. The tool dimensions are presented in Fig. 18.4. The angle of the no contact area at the entrance of the die shoulder is experimentally measured and is set to $\Phi_e = 12$ degrees. Two types of lubrication conditions were used in the calculation as shown in Fig. 18.16. Dry friction conditions were adopted in MDD and lubricated OLPs by counter pressure was assumed in MHDD. The COF for different lubrication used in calculation are listed in Table 18.3. The COF for each lubrication case were determined with reference to the COF shown in the several papers related with tribology [3–5]. Based on the experimental results as mentioned below, the width of OLPs w used in calculation is set up at 175 μm.

18.2.2.5 Experiment of MHDD

The experiments of MDD and MHDD were carried out with the load cell capacity of 20 kN to measure the punch force. A pump with a maximum pressure of 20 MPa was employed. The tool dimensions for MDD and MHDD are presented in Fig. 18.7. The constant-gap method in which the gap between the blank and the blank holder is fixed was used. The experiments were performed under dry friction in MDD and machine oil was used in MHDD. The drawing speed was set up to 0.4 mm/s.

The punch force P was used to evaluate the change of friction force by applying the counter pressure in experiment and theory. From the meridional stress at the exit of die shoulder, P can be presented as

$$P = 2\pi r_2 t \sin\Phi \left[e^{\mu\Phi} \left\{ 1.1\sigma_{eq} \ln\frac{r_0}{r_2} + \frac{\mu(q + p_c)(r_0 - r_1)}{t} - \frac{r_0}{r_2} p_r \right\} + \left(1 + e^{\mu\Phi}\right) \frac{1.1\sigma_{eq} t}{4(r_d + t/2)} \right] \tag{18.13}$$

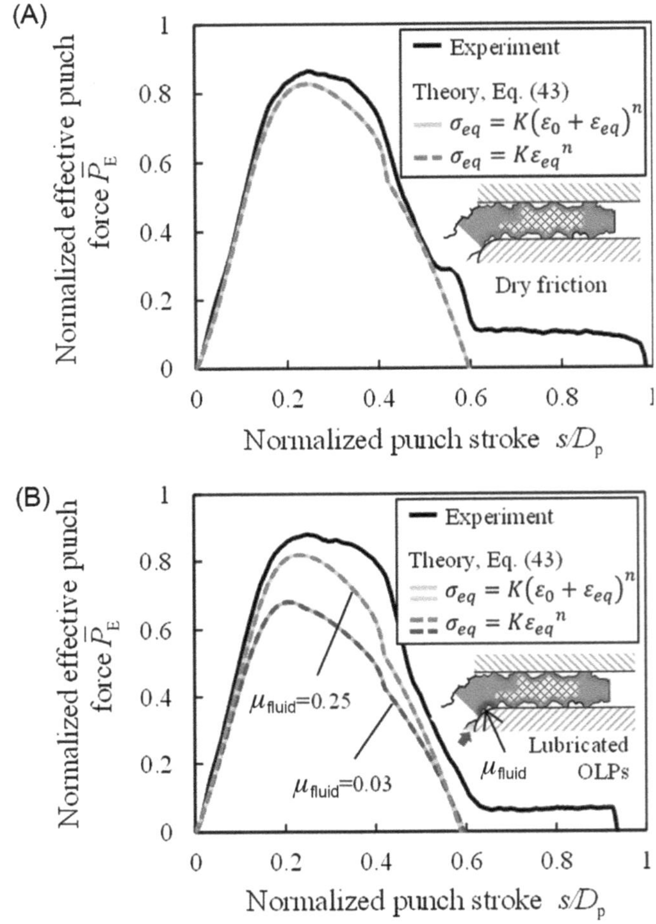

FIGURE 18.16 Comparison of experimental and theoretical results of normalized punch force–stroke curves (A) in MDD ($\mu_c = \mu_{dry} = 0.3$, $\overline{\alpha}_f = 0$) and (B) in MHDD ($p_c = 10$ MPa, $p_r = 0$MPa, $\mu_c = \mu_{dry} = 0.3$, $\overline{\alpha}_f = 1/4$) [2].

TABLE 18.3 The COF for Each Lubrication and Area Used in Calculation

Dry Friction in OLPs μ_{dry}	Friction Coefficient in OLPs μ_c	Lubricate OLPs by Fluid Pressure μ_h	Hydrodynamic Lubrication μ_h
0.3	0.30, 0.03	0.03, 0.25	0.005

This punch force consists of the drawing, friction, bending, and unbending forces P_S, P_F, P_B, and P_{UB}. In the same tooling dimension, P_S, P_B, and P_{UB} are approximately constant even if the counter pressure is changed. The change of counter pressure only influences the friction force as shown in Eq. (18.13). Therefore, the different punch forces for each counter pressure shows the change of friction force. In the experiment, the force acting on the punch oppositely by counter pressure P_p is included in the measured punch force P_M. For this reason, the effective punch force P_E in which P_M is subtracted from P_p is used to evaluate the change of friction force by applying counter pressure. It can be expressed as

$$P_E = P_M - P_p = P_S + P_F + P_B + P_{UB} \quad (18.14)$$

This P_E similarly consists of P_S, P_F, P_B, and P_{UB}. Therefore, the difference of P_E at each counter pressure can evaluate the change of friction force. The punch force normalized by the tensile strength $\overline{P}_E$ can be expressed by

$$\overline{P}_E = P_E/\pi(D_p + t)t\sigma_B = P/\pi(D_p + t)t\sigma_B \quad (18.15)$$

18.2.2.6 Results and Analysis

18.2.2.6.1 Model Validation

Fig. 18.18 shows the comparison of experimental and theoretical results of normalized effective punch force–stroke curves. In the experimental results of MDD and MHDD, the punch force–stroke curves increase at the early stage ($s/D_p < 0.2$) and reach the maximum value at $s/D_p = 0.2$. At the middle stage ($0.2 < s/D_p < 0.6$), the punch force decreases and the sliding force occurs during passing thought the straight die part in Fig. 18.16 at the late stage ($s/D_p > 0.6$). The theoretical results can represent the increase and decrease of punch force although it cannot represent the sliding force. The maximum punch forces of the theoretical results in Fig. 18.16B are almost as large as the experimental results in MDD and MHDD of $\mu_{fluid} = 0.25$. This means that the theoretical results agree with the experimental results. For MHDD of $\mu_{fluid} = 0.03$, the effective punch force is much smaller than that in experimental result and theoretical result of $\mu_{fluid} = 0.25$. It means that the COF in OLPs cannot be sufficiently reduced by applying counter pressure of 10 MPa. Moreover, it shows that the COF in OLPs significantly influences the effective punch force in MHDD. The punch force–stroke curves calculated by Swift equation are completely the same as the power law. This is because the strain-stress curves are almost the same for material with a low work hardening exponent. The punch force–stroke is not significantly changed by the approximation of stress–strain curve [6,7].

18.2.2.6.2 Effect of Fluid Pressure on Drawability in HDD and MHDD

Fig. 18.17 shows the effect of counter pressure on normalized maximum effective punch force. In the experimental result, the maximum punch force of MDD decreases once when the counter pressure is applied. As the counter pressure is increased, the maximum punch force increases when the counter pressure exceeds $p_c = 3$ MPa. The change of normalized maximum effective punch force shows the change of friction force. This means that the friction force increases as the counter pressure increases in MHDD. As a result, the fracture occurs at the punch shoulder when the high counter pressure is applied, as shown in Fig. 18.18.

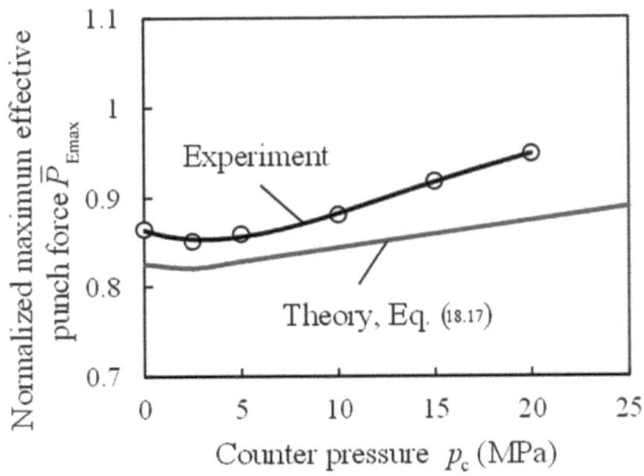

FIGURE 18.17 Effect of fluid pressure on normalized maximum effective punch force in experiment and theory ($\mu_c = \mu_{dry} = 0.3$, $\overline{\alpha}_f = 0$ for MDD, $\mu_c = \mu_{dry} = 0.3$, $\mu_{fluid} = 0.25$, $\overline{\alpha}_f = 1/4$, $p_r = 0$ for MHDD).

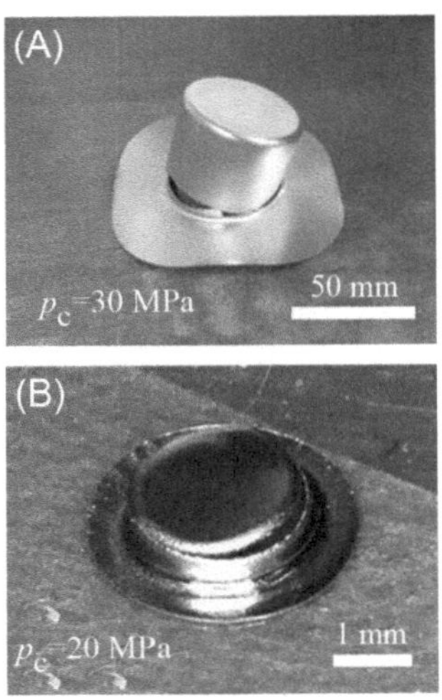

FIGURE 18.18 Appearance of drawn cups fractured (A) at die shoulder in conventional HDD and (B) at punch shoulder in MHDD [2].

The theoretical result shows the same tendency with the experimental result. This is because the application of counter pressure lubricates the blank and at the same time increases the contact pressure between the blank and blank holder as shown in Eq. (18.13). The counter pressure plays a role of blank holder pressure in the constant-gap method. This means that the application of counter pressure almost linearly increases the friction force and the fracture occurs at punch shoulder due to the excessive punch force in MHDD using the constant-gap method.

On the other hand, in conventional HDD, the fracture typically occurs at punch shoulder at high counter pressure as shown in Fig. 18.18. High-fluid pressure induces the hydrodynamic lubrication which improves the tribological behavior and the friction holding effect, which can reduce the applied force to blank at the punch shoulder. These mean that the fracture does not occur at the punch shoulder, but rather at the die shoulder when the blank is subjected to excessive reverse bulging deformation at the die shoulder. It appears that the characteristics of fluid pressure are different in HDD and MHDD and these differences cause the difference in the type of fracture which occurs [3,4] (Fig. 18.18).

18.2.2.6.3 Effect of Relative Punch Diameter to Thickness on Fluid Behavior

The differences in the tribological behavior and fracture types are resulted from fluid behavior. In conventional HDD, the fluid pressure temporarily decreases at the early and middle stages which show the leakages between the blank-die and the blank-blank holder, respectively, as shown in Fig. 18.19. These leakages cause the hydrodynamic lubrication in the flange area and reduce the friction force but there is no decrease in fluid pressure in MHDD. This shows that there is no hydrodynamic lubrication in MHDD. This is the reason for the high friction force in MHDD.

According to Eq. (18.12), the required fluid pressure for hydrodynamic lubrication p_h can be obtained as shown in Fig. 18.20. The fluid pressure required for hydrodynamic lubrication increases with the decrease of D_p/t. In particular, quite high-fluid pressure is required to leak the fluid medium in microscale in which D_p/t approximately ranges from 10 to 100. This is because the contact pressure at the die shoulder and sealablity become high at small D_p/t, in which the die shoulder radius is relatively small compared to the blank thickness.

18.2.2.7 Verification and Size Effects Prediction on Lubricated OLPs by Fluid Pressure in MHDD

In MHDD, not only the hydrodynamic lubrication, but also boundary and mixed lubrications exist during the MHDD process. In general, the lubricant cannot be kept in OLPs which connect to the edge of the blank in boundary

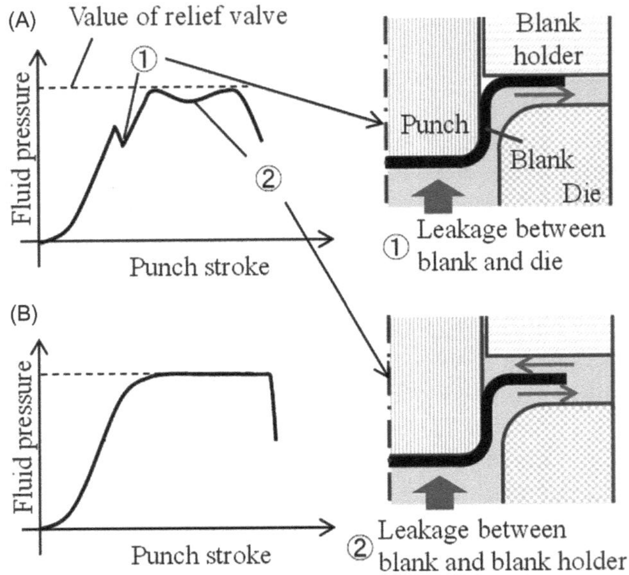

FIGURE 18.19 Comparison of fluid behavior between (A) MDD and (B) MHDD [1].

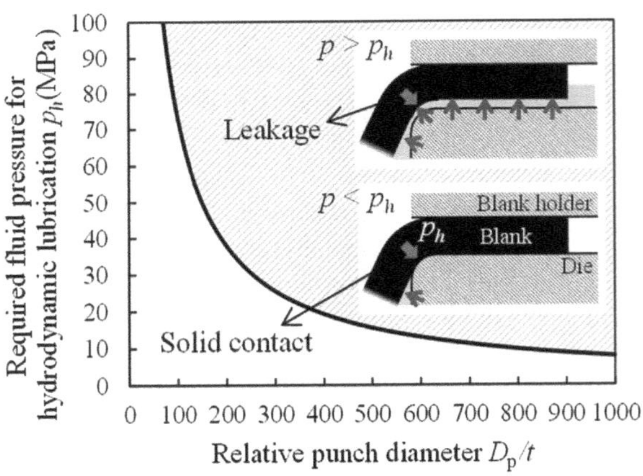

FIGURE 18.20 Effect of D_p/t on required fluid pressure for hydrodynamic lubrication introduced in Eq. (18.12) [2].

and mixed lubrications in conventional microforming. On the other hand, if the fluid medium can be filled in the OLPs in MHDD as shown in Fig. 18.16, the lubricant can be kept in the OLPs and the COF can be reduced in microscale. To confirm this phenomenon, an evaluation test for OLPs utilizing liquid was carried out and the size effects of lubricated OLPs by fluid pressure is theoretically investigated.

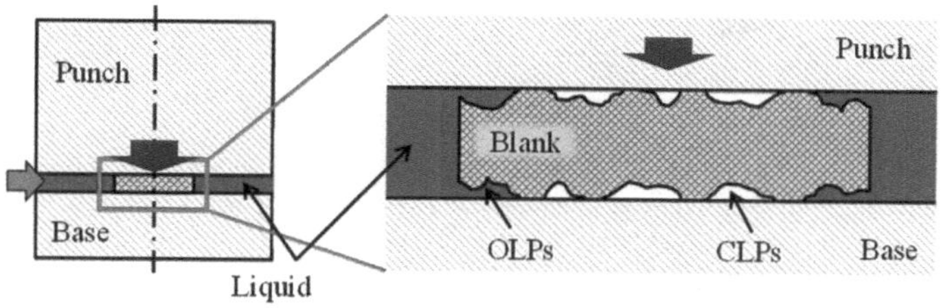

FIGURE 18.21 Schematic of evaluation test for OLPs utilizing liquid.

18.2.2.8 Evaluation Test for OLPs Utilizing Liquid

The same material used in the MHDD experiment was used. The blank diameters are $D_0 = 1.7$ and 15 mm. The blank is compressed by the tools under the contact pressure of approximately 20 MPa and the liquid is filled in the tool as shown in Fig. 18.21. The area into which the liquid intruded is colored. After the intruded liquid dries out, the blank is taken out. Thus, the area into which the liquid intruded can be visualized. The appearance of the blank was observed using a digital microscope and the pictures were digitized.

18.2.2.9 Lubrication Model by Fluid Pressure in MHDD

To confirm the tribological size effects in MHDD, the lubricated OLPs by counter and radial pressure was adopted. As a comparison, dry friction and lubrication in MDD, and hydrodynamic lubrication in MHDD were compared. For MHDD with lubricated OLPs by counter and radial pressures, the counter and radial pressures $p_c = p_r = 100$ MPa and the COF in lubricated OLPs by fluid pressure $\mu_{fluid} = 0.03$. For hydrodynamic lubrication, they are set up at 180 MPa based on Eq. (18.12). The COF for each condition is listed in Table 18.3.

To evaluate the tribological size effects, the friction force P_F was calculated. From Eq. (18.13), the friction force can be obtained by

$$P_F = 2\pi r_2 t \sin\Phi(1 - e^{\mu\Phi})\left\{1.1\sigma_{eq}\ln\frac{r_0}{r_2} + \frac{\mu(q+p_c)(r_0 - r_1)}{t} - \frac{r_0}{r_2}p_r + \frac{1.1\sigma_{eq}t}{4(r_d + t/2)}\right\} \tag{18.16}$$

The punch force normalized by tensile strength $\overline{P}_F$ is expressed by

$$\overline{P}_F = P_F/\pi(D_p + t)t\sigma_B \tag{18.17}$$

18.2.2.10 Discussion on Lubricated OLPs by Fluid Pressure in MHDD

18.2.2.10.1 Appearance of OLPs in Fluid Medium

Fig. 18.22 shows the appearance of OLPs with the intruded liquid for different blank diameters. For both blank diameters, the outer edge of the blank was colored. This colored area contacts with the blank edge and does not

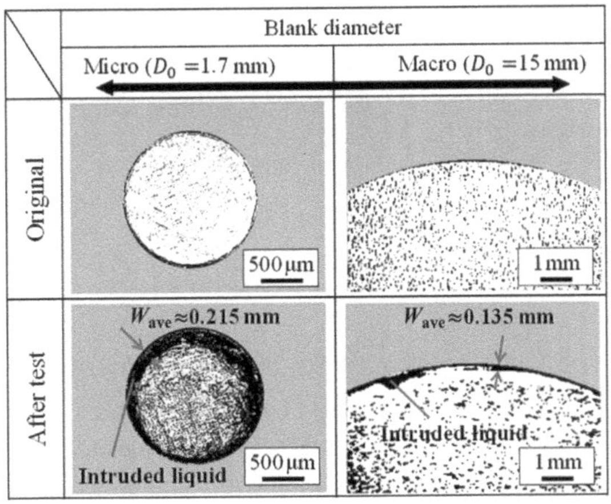

FIGURE 18.22 Appearance of OLPs with intruded liquid for different blank diameters.

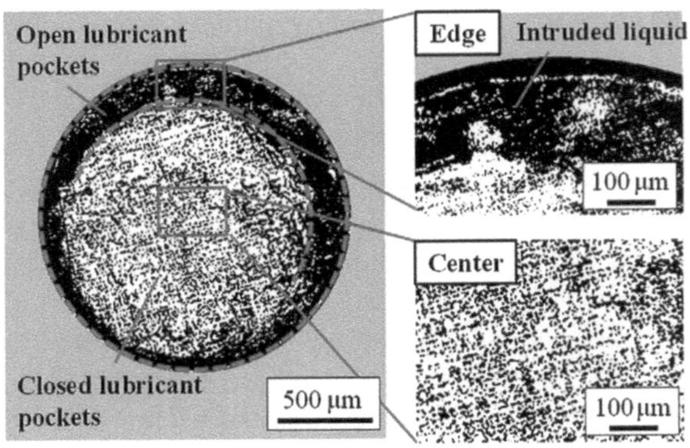

FIGURE 18.23 Appearance of OLPs and CLPs on blank (D_0 = 1.7 mm).

appear in the inner area of the blank as shown in Fig. 18.23. This means that the fluid medium was poured into the OLPs. The average widths of the OLPs for $D_0 = 1.7$ and 15 mm are approximately $w_{ave} = 215$ and 135 μm, respectively. The scale factors for both blank diameters are approximately $\lambda = 4.0$ and 55.6, respectively. Thus, the fraction of OLPs increases as the blank diameter decreases. Similar results were reported in a conventional compression test for a small specimen. For a blank diameter of 1.7 mm, the width of the OLPs is not uniform. This is because the tool surface is not exactly parallel to the blank surface. This means that the contact area between the blank and the tool is displaced from the center position and alters the shape of the circle. At the blank edge, the shear drop exists because the blank is fabricated by the blanking. In the area of the shear drop, the blank does not have contact with the tool and this area is included in the OLPs. From these results, it could be seen that the fluid medium can be kept at the OLPs by applying the fluid pressure and it induces lubrication in OLPs.

18.2.2.10.2 Friction Force of Lubricated OLPs by Fluid Pressure

As mentioned above, the OLPs can be filled with the fluid medium by applying fluid pressure. This means that the lubricant can be kept and the COF can be reduced on OLPs in MHDD. Based on this result, the scale dependences of maximum friction force for different lubrication conditions in MDD and MHDD were calculated as shown in Fig. 18.24A. In MDD with dry friction, the friction force is constant because the COF in the OLPs and CLPs is the same. The COF does not change in macroscale and microscale. For MDD with lubrication, the friction force increases with scaling down to microscale because the fraction of OLPs increases because it cannot keep the lubricant. This causes the increase of the COF and the friction force in MDD. The same behavior has been reported in previous research into conventional microforming. On the other hand, in MHDD with lubricated OLPs by counter and radial pressures, the friction force decreases with scaling down. This behavior is the opposite of the tribological size effects in conventional microforming. The OLPs can be filled with the fluid medium by applying fluid pressure. In addition, the fraction of the OLPs increases with scaling down, and the COF can be reduced. Therefore, the average COF in the flange area decreases with the decrease of specimen size in MHDD with lubricated OLPs by counter and radial pressures as shown in Fig. 18.24B. In comparison with hydrodynamic lubrication conditions, the friction force in MHDD with lubricated OLPs by counter and radial pressures is large in the macro scale, but is not very different in the microscale in which λ is under 5. It was found that the friction force can be reduced in the microscale without hydraulic lubrication by applying fluid pressure to the OLPs.

18.2.3 Effect of Microstructure in MHDD

The drawn cups from the experiments were measured under a 3D high-resolution digital microscope. The drawn cups mouths and side views are exhibited in Fig. 18.25. From the experimental results, under low hydraulic pressures, the wrinkling is not obvious, however, when the hydraulic pressure increases, obvious wrinkling occurs. Furthermore, the 1100°C heated foils have more wrinkles than that of the other two kinds of foils [8]. The earing has a similar development trend to the wrinkling phenomenon. These trends are in accordance with the simulation results.

In summary, in this chapter, an MHDD apparatus has been developed with the concept of realizing high dimensional accuracy of drawn components by minimizing the number of control targets, using an apparatus with

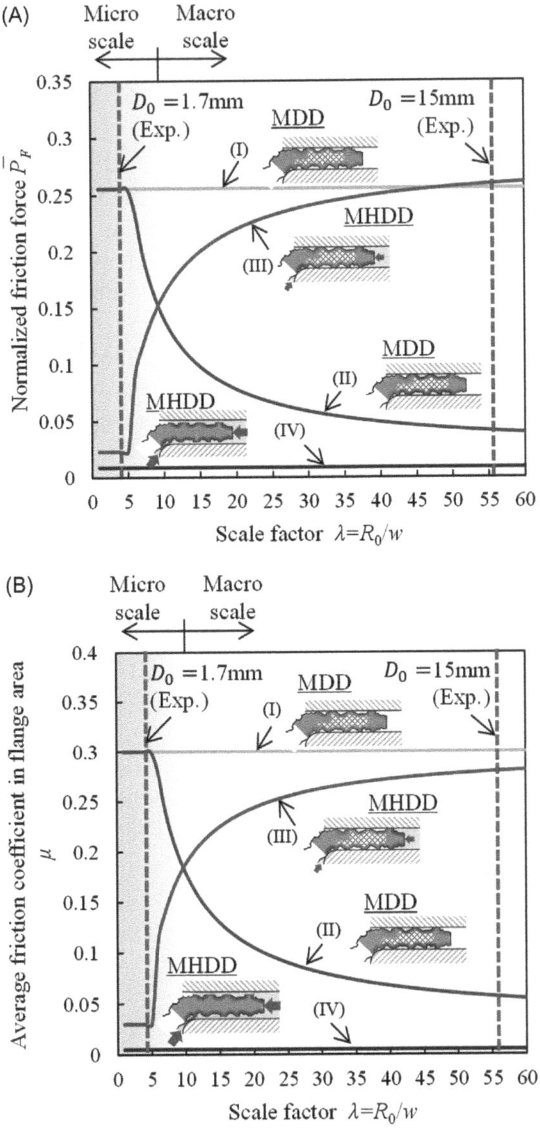

FIGURE 18.24 Scale dependences of (A) normalized friction forces and (B) average coefficient of friction in flange area at the maximum punch force for different lubrication conditions in MDD and MHDD introduced in Eqs. (18.10) and (18.17) ((I) dry friction, (II) lubrication, (III) lubricated OLPs by counter and radial pressures, and (IV) hydrodynamic lubrication).

simple forming processes and tooling structure. A multiaxial pressure can be reproduced by one-stroke forming and the constant-gap method. The micro cups with a diameter of 0.8 mm and a sharp cup with a diameter of 2.0 mm and a punch shoulder radius of 0.1 mm were successfully formed.

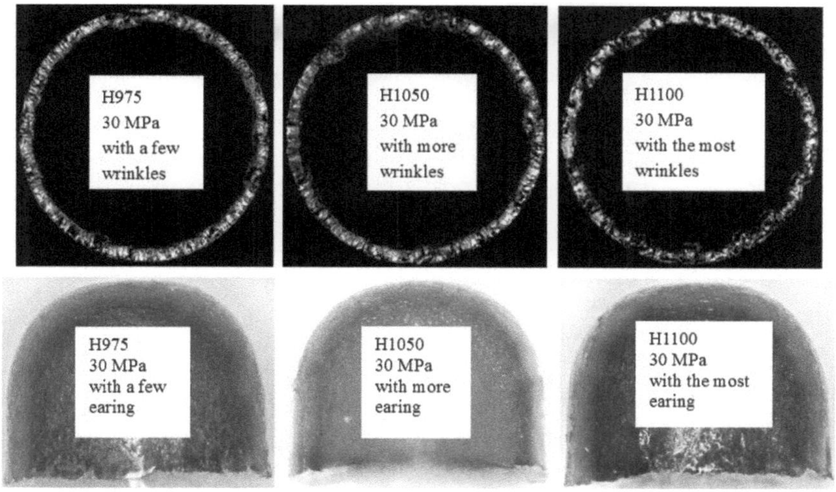

FIGURE 18.25 Drawn cups under 30 MPa hydraulic pressures [8].

The appropriate counter pressure can eliminate the occurrence of wrinkles, reduce the friction force, and improved forming limit. The theoretical simple friction model considered the fluid pressure and both OLPs and CLPs is proposed. At high-fluid pressure, the fracture occurred at the punch shoulder in MHDD. The tribological behavior in MHDD is characterized by the high sealablity to occur hydrodynamic lubrication and lubricated OLPs by fluid pressure which is caused by the relative punch diameter to thickness and surface topography. MHDD can improve the tribological behavior by inducing these lubrications through the application of appropriate fluid pressure. The grain size has significant influence on the formation of drawn cups.

REFERENCES

[1] H. Sato, K. Manabe, K. Ito, D.B. Wei, Z.Y. Jiang, Development of servo-type micro-hydromechanical deep-drawing apparatus and micro deep-drawing experiments of circular cups, J. Mater. Process. Technol. 224 (2015) 233–239.

[2] H. Sato, K. Manabe, D.B. Wei, Z.Y. Jiang, S. Alexandrov, Tribological behaviour in micro-sheet hydroforming, Tribol. Int. 97 (2016) 302–312.

[3] F. Vollertsen, Z. Hu, Tribological size effects in sheet metal forming measured by a strip drawing test, CIRP Ann. 55 (2006) 291–294.

[4] T. Shimuzu, M. Yang, K. Manabe, Classification of mesoscopic tribological properties under dry sliding friction for microforming operation, Wear 330-331 (2015) 49–58.

[5] K. Manabe, T. Shimizu, H. Koyama, M. Yang, K. Ito, Validation of FE simulation based on surface roughness model in micro-deep drawing, Int. J. Mater. Process. Technol. 204 (2008) 89–93.

[6] K. Manabe, H. Sato, T. Furushima, D. Wei, N. Mathew, Z. Jiang, Deformation Behavior in Micro Sheet Hydroforming Process, Special edition, Steel Research International, 2012, pp. 651–654.

[7] H. Sato, K. Manabe, D.B. Wei, Z.Y. Jiang, Analysis of axisymmetric cup forming of metal foil and micro hydroforming process, Proceedings of the ASME: International Mechanical Engineering Congress and Exposition, Micro- and Nano-Systems Engineering and Packaging, 2013. p. V010T11A040.

[8] L. Luo, Z.Y. Jiang, D.B. Wei, K. Manabe, Influences of Hydraulic Pressure on Wrinkling and Earing in Micro Hydro Deep Drawing with Consideration of Size Effects, International Conference on Materials Processing Technology, March 28–29, 2016 Tokyo, Japan.

Chapter 19

Practice of Micro Bending

19.1 INTRODUCTION

Micro parts made from nonferrous alloys, such as copper, aluminum, and magnesium alloys, are widely applied in electronic components, and medical apparatus and devices due to their high electrical and thermal conductivity, good ductility and excellent wear resistance, fatigue strength, and bearing properties. The increasing demand for microscaled products with high accuracy and high quality has inspired the development of microforming technology. In the micro world, size effects play a significant role in affecting materials' mechanical properties, texture, microstructure, and formability. In practice, the most versatile technique to present grain size effects is to obtain a given recrystallized grain size via heat treatment after which a scale factor (specimen dimension/grain size) can be achieved.

Several investigations have been conducted to explore the role that grain size effects played in texture evolution in conventional and microforming processes with various nonferrous alloys. Park et al. [1] found that the tensile properties of friction stir welded magnesium alloy AZ61 were strongly affected by crystallographic orientation distribution and grain size. Valle et al. [2,3] carried out equal channel angular pressing (ECAP) and large-strain hot rolling experiments on magnesium alloy and concluded that both texture and gran size had a great influence on work hardening as well as ductility. Llanes et al. [4] found that much more pronounced cyclic hardening and high saturation stresses were linked to some certain fiber textures in coarse-grained copper rather than fine-grained copper. Furthermore, Hoseini et al. [5] proposed that after ECAP processing, the grain size of pure titanium had no impact on the corrosion resistance when the (0002) planes parallel to the surface. Janssen et al. [6] found that the ratio of thickness to grain size could be viewed as a clue for flow stress when texture in samples was similar. Yin et al. [7] concluded that in Mg AZ31 alloy with grain size of 0.8 μm or below could eliminate large yield asymmetry caused by $\{10\bar{1}2\}$ twinning by tensile, compressive and ECAP testing, and Guo et al. [8] also found that the work-hardening rate of AZ31 alloy decreased with the decrease of grain size, but was irrelevant to (0002) pole intensity. Mabuchi et al. [9] proposed that the grain size dependence was related to the distribution of grain boundary misorientations, and Leffers and Juul Jensen [10]

Microforming Technology. DOI: http://dx.doi.org/10.1016/B978-0-12-811212-0.00019-4

described that both texture and microstructure changed dramatically when a brass plate was severely bent during rolling. Zhang et al. [11] studied the improvability of texture of AZ31B Mg alloy and found that texture control had more effect on bendability at low temperature, and Kaneko and Eguchi [12] also verified the decisive effect of texture on bendability. Song et al. [13] explained these results in-depth by investigating texture evolution during bending process. Sutou et al. [14,15] investigated the damping properties and pseudoelasticity of Cu-Al-Mn-based shape memory alloys by studying the effect of grain size and texture. Barnett et al. [16] found that grain size was critical when identifying twinning-slip transition by compressing extruded samples. Vaidya and Sinha [17] expressed the electromigration response of thin film aluminum by a function of grain size and the degree of {111} fiber texture.

In this chapter, the practice of micro bending is introduced, with investigations of grain size effects on texture evolution of annealed and micro V-bended phosphor bronze foils. Heat treatment experiment is first carried out on phosphor bronze foils to obtain different scale factors which are expressed as thickness (T)/average grain size (D) >1, <1 and ≈ 1. Micro V-bending is then conducted three times under the same experimental condition. Microtexture analyses of the annealed and micro V-bended specimens are carried out by employing high-resolution electron backscatter diffraction (EBSD) system. Texture characterizations during micro V-bending process are investigated systematically by comparing the texture data from the annealed specimens with that from the micro V-bended ones with consideration of the ratio T/D.

19.2 MICRO BENDING PRACTICE

19.2.1 Material Preparation

Cold-rolled C5191 phosphor bronze foil is chosen as the experimental material in this chapter. The chemical compositions of C5191 are displayed in Table 19.1. The dimensions of the as-received foil are $1000 \times 100 \times 0.07\ \text{mm}^3$. For micro V-bending tests, the phosphor bronze foil is sectioned via a wire-cut electrical discharging machine to be the dimensions of $2 \times 1 \times 0.07\ \text{mm}^3$.

TABLE 19.1 Chemical Compositions of Phosphor Bronze C5191 (wt%)

Sn	P	Pb	Fe	Zn
5.87	0.22	0.004	0.001	0.004

Specimens measuring $2 \times 1 \times 0.07\ mm^3$ are subjected to heat treatments to acquire different grain sizes. To accomplish the aim of different grain size effect ratios, the materials are annealed in a KTL1400 vacuum tube furnace under argon protection with the same temperature but different holding times. The annealing conditions are presented in Table 19.2. Although the range of grain size effects is small (from 0.68 to 1.23), what these three ratios reflect three different special microstructural features: a few grains exist in thickness direction, one grain exists in thickness direction, and one partial grain in thickness direction. So from this point, the focus of this chapter is to investigate the effect of three microstructural features and their deformation mechanisms during micro bending. This is distinctive from other research which has a large range of ratio values, while only one microstructure is expressed: several grains in specimen's thickness direction.

The microstructures of three different grain sizes in thickness direction are shown in Fig. 19.1, which also illustrates the crystallographic orientation information in terms of inverse pole figure (IPF). It can be seen that the

TABLE 19.2 Heat Treatment and Average Grain Size

Temperature (°C)	550	550	550
Time (h)	4	6	8
Thickness, *T* (μm)	70	70	70
Average grain size, *D* (μm)	57.1	65.7	103.1
T/D	1.23	1.07	0.68

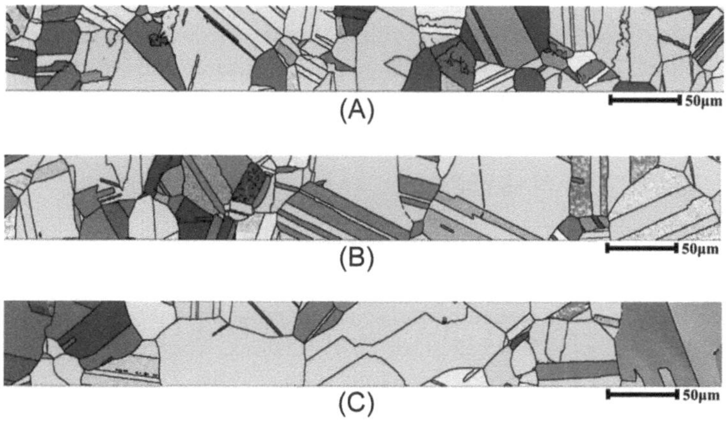

FIGURE 19.1 IPF maps of annealed phosphor bronze foils with different holding times: (A) 4 h, $T/D > 1$; (B) 6 h, $T/D \approx 1$; and (C) 8 h, $T/D < 1$.

grain size increases with holding time under the same annealing temperature. After annealing, the specimens are divided into two groups: one group is for recrystallization texture analysis and the other is for micro V-bending experiments and then to be scanned for obtaining grain orientation maps of bended specimens.

19.2.2 Micro V-bending Setup

To investigate the relationship between the ratio *T/D* and the change of distribution of crystallographic orientations during bending process, micro V-bending of the annealed specimens is conducted on the desk-top servo press machine DT-3AW, as shown in Fig. 12.11. Micro V-bending tests are performed at room temperature with punch speed of 0.1 mm/s and punch stroke of 0.43 mm, and other testing parameters are the same as that described in Chapter 12, Simulation of Micro Bending.

19.2.3 Texture Acquisition

The cross sections of the specimens along rolling direction (RD)—normal direction (ND) with three different *T/D* values are mechanically polished. Due to the small geometry and the softness of specimens, an ion beam milling system (Leica EM RES102) is adopted to finalize polishing, qualifying the surface of specimens for EBSD data acquisition. EBSD mapping is carried out on a JEOL-JSM7001F field emission gun scanning electron microscope (SEM), fitted with a Nordly-II (S) camera at 15 kV, ~5 nA, and 10 mm working distance. A constant step size of 0.25 μm is maintained for all the maps. In all maps, a minimum of 3 pixels are used to identify grain structures, and misorientations (θ) <2 degrees are disregarded. The orientation distribution functions (ODFs) are calculated via HKL Channel-5 and the α-fiber intensities are determined with a 10 degrees deviation from the ideal skeleton line. The grain size in IPF maps is measured independently of the twin grains according to the standard of ASTM.

19.3 RESULTS ANALYSIS

19.3.1 Effect of *T/D* on Recrystallization Texture

The ideal texture components of face-centered cubic (FCC) materials which are involved in this research are schematically shown in Fig. 19.2 and Table 19.3 with $\phi_2 = 0$, 45 degrees ODF sections. The coloring orientation maps of annealed phosphor bronze foils with various *T/D* values have been illustrated in Fig. 19.1.

Moreover, $\phi_2 = 0$, 45 degrees ODF sections of the selected areas in Fig. 19.1 with increasing *T/D* factors are given in Fig. 19.3. The variations in

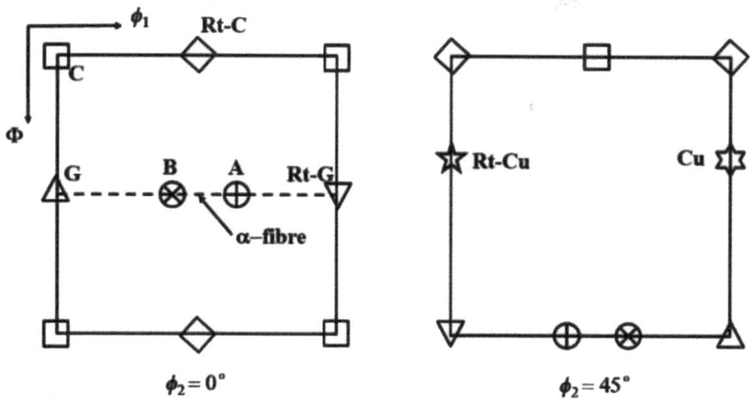

FIGURE 19.2 A schematic representation of the important texture components in FCC material.

TABLE 19.3 Euler Angles and Miller Indices for Common Texture Components in FCC Materials

Texture Component	Symbol	Euler Angles			Miller Indices	Fiber
		ϕ_1	Φ	ϕ_2		
Cube (C)	□	45°	0°	45°	{001}⟨110⟩	–
Goss (G)	△	90°	90°	45°	{110}⟨001⟩	α
Brass (B)	⊗	55°	90°	45°	{110}⟨112⟩	α/β
A	⊕	35°	90°	45°	{110}⟨111⟩	α
Rotated Goss (Rt-G)	▽	0°	90°	45°	{011}⟨011⟩	–
Rotated Cube (Rt-C)	◇	0/90°	0°	45°	{001}⟨110⟩	–
Copper (Cu)	✡	90°	34°	45°	{112}⟨111⟩	β
Rotated Copper (Rt-Cu)	☆	0°	35°	45°		–

intensity ($f(g)$) of the individual texture components along the various fibers are illustrated in Fig. 19.4. It is well established that grain orientation is related to strain-induced recrystallization, but that is not the point in this chapter since the material used is a final product and no more plastic strain will occur before heat treatment. With respect to phosphor bronze (FCC structure) which is classified as low stacking fault energy (SFE) material [18], the annealing process results in the development of the characteristic α-texture and {110} texture which is parallel to the transverse direction (TD).

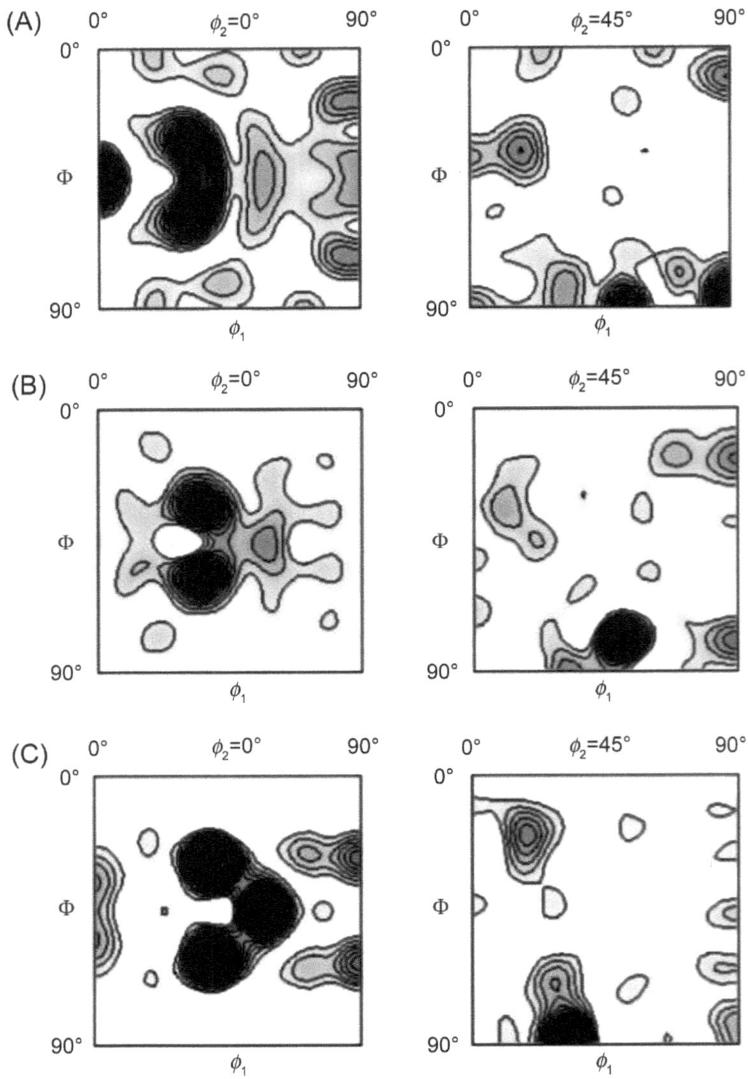

FIGURE 19.3 $\phi_2 = 0$, 45 degrees ODF sections: (A) $T/D > 1$, (B) $T/D \approx 1$, and (C) $T/D < 1$. Contour levels = 1 ×.

As seen in Fig. 19.3A, the texture in the material with $T/D > 1$ comprises an α-fiber from the strong Goss (G) orientation to the strong Brass (B) orientation and the slight A orientation, and then the Rotated Goss (Rt-G) orientation along $\phi_1 = 0–90$ degrees at $\Phi = 45$ degrees, $\phi_2 = 0$ degrees. This can be understood according to the study by Hirsch and Lücke [19] that with reducing the SFE, the intensity of the Cu orientation decreases and the B orientation enhances with a spread toward the G orientation. Among all the

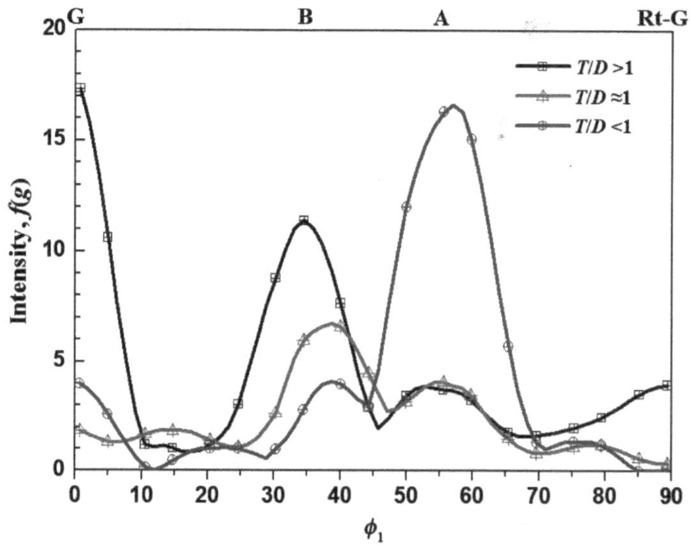

FIGURE 19.4 The variation in texture components intensity ($f(g)$) along α-fiber.

components along α-fiber, the most perceptible feature in the material with this scale factor is the threshold of the twinning of Brass as it can be seen that both intensity peaks of B orientation are twinning and moving to the opposite direction. With longer annealing time which means when the ratio of thickness to average grain size is approximately equal to one, a noticeable decrease in the intensity of the G and Rt-G orientations can be observed in Fig. 19.4. The B orientation slightly shifts toward the A orientation during this stage along with its intensity weakening. Moreover, two completely separate B components shown in Fig. 19.3B indicate that the twinning of brass texture is developing and the intensity of the A orientation remains almost unchanged. The most distinctive revolution in the material with $T/D < 1$ is the significant surge of the A orientation in terms of the intensity which arrives at $f(g) = 16.6$. Meanwhile, the twinning of B components is close to finish so that the B orientation totally deviates from its ideal coordinate in Euler space. This results in the intensity of the B orientation keeping deceasing along α-fiber. The above intensity variation can be explained in terms of twinning such that the Rt-G and A components have a first-order twin relationship with the G/B orientation while the first-order twinning of the B orientation leads to crystallographically identical variants and explains its retention on recrystallization [20]. Furthermore, it is noticeable for the disappearance of Rt-Cu component from $T/D > 1$ to $T/D < 1$. The Rt-Cu is the only recrystallization component that does not exhibit the first-order twin relationship among these texture components, and it has a $\sim\Sigma9$ (37.3 degrees $\langle 101 \rangle$) relationship with the A orientation and a 30 degrees $\langle 111 \rangle$

relationship with the B orientation. This indicates that when the second order of twinning of the B orientation occurs (when $T/D < 1$), it may lead to the disappearance of the Rt-Cu component. Similarly, it also explains the surge in A intensity and the plunge in B intensity as the ratio $T/D < 1$. It can be seen that the parameter T/D is closely linked to the relationship between the grain size effects and recrystallization texture components and their intensities.

19.3.2 Effect of *T/D* on Deformation Texture in Micro V-bending

The IPF maps of the deformation zone of bended specimens are illustrated in Fig. 19.5. By comparing Figs. 19.1 with 19.5, it can be seen that after micro V-bending the crystallographic orientations change in all the three specimens. More specifically, the material with $T/D > 1$ has the most obvious color change, which indicates that this material has experienced the most grain rotation during micro V-bending. The material with $T/D \approx 1$ and that with $T/D < 1$ undergo less and the least grain rotation, respectively, since the former displays an average degree color change and the latter has the least color change. When $T/D > 1$, there are several grains consisted in the specimen's thickness direction. During micro bending process, it is understood that some grains which reach the critical resolved shear stress, as a threshold value, will firstly initiate dislocation slip and plastic deformation. Since these grains are surrounded by other grains in the specimen due to $T/D > 1$, their deformation cannot be isolated and arbitrary. This means that the first-initiated grains have to coordinate their rotation and deformation with the adjacent grains to ensure the consistency of the specimen's deformation. On the basis of grain coordination, driven by the applied stress on the specimen and the stress concentration which is caused by dislocation pile-up groups from glided grains, increasing grains in the material will undergo rotations and plastic deformation. When $T/D \approx 1$, only one complete grain ideally exists along the thickness direction. The above grain coordination will still exist in the bending process, but because of the increase of grain size, the influence of grain coordination deformation will be weakened. Conversely, as the grain size is close to the thickness geometry, grains' deformation is likely to be affected by the upper and lower boundaries of the specimen. According to crystallography, the grains' movement can be divided into two patterns according to the direction of the applied force. One is the grain slip movement due to the resolved shear stress along the slip direction, and the other is rotating toward the stress direction because the stress component is normal to slip plane. The most obvious difference between these two microstructures ($T/D > 1$ and $T/D \approx 1$) when plastic deformation is initiated is that the second type of grain movement will be comparatively limited by the upper and lower specimen boundaries. Fig. 19.6 illustrates how the grain

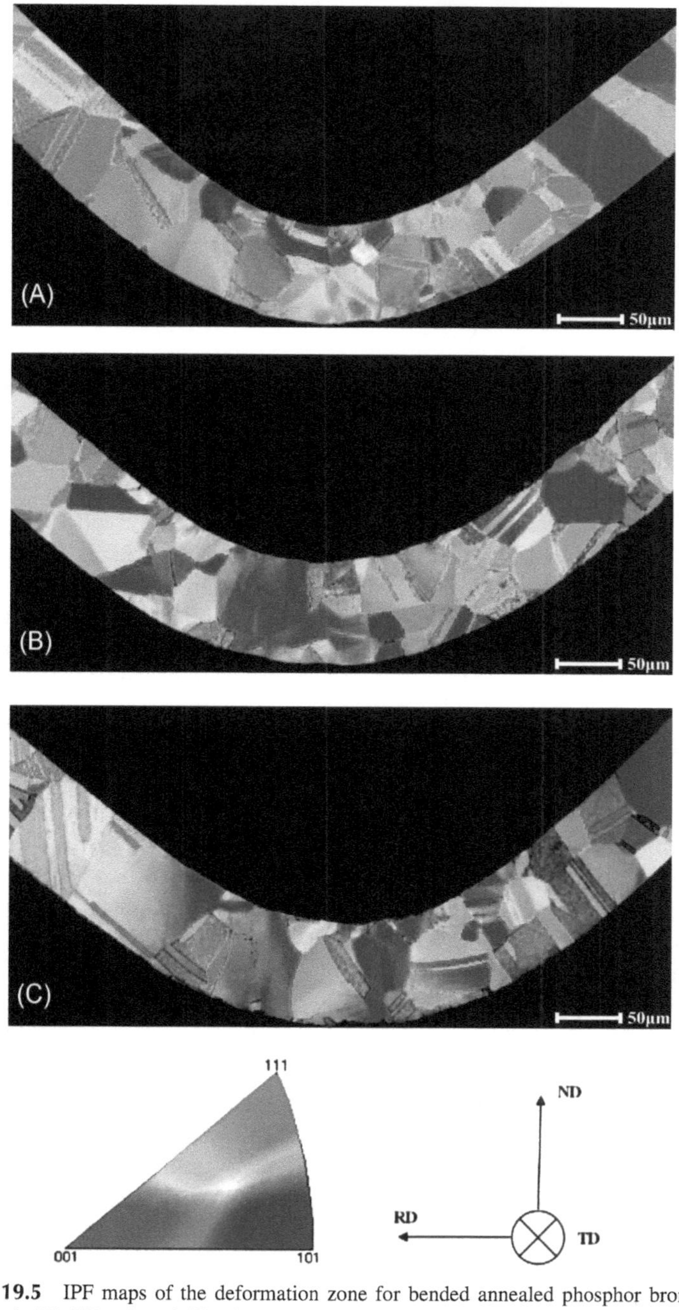

FIGURE 19.5 IPF maps of the deformation zone for bended annealed phosphor bronze foils: (A) $T/D > 1$, (B) $T/D \approx 1$, and (C) $T/D < 1$.

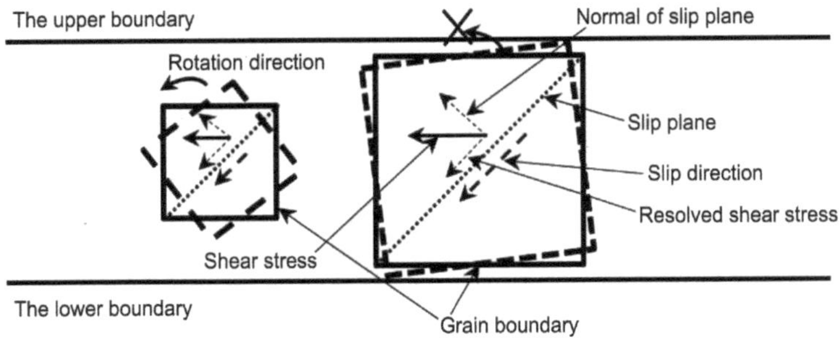

FIGURE 19.6 Schematic illustration of how grain rotation is restricted by grain boundary.

which is not located in the vicinity of material boundaries (when $T/D > 1$) can rotate easily while the grain rotation in the material with $T/D \approx 1$ will be restricted by the upper and lower boundaries.

When $T/D < 1$, the grain rotation will be affected and restricted further by the upper and lower boundaries along the thickness direction. Moreover, during micro V-bending process, the movement of slip bands in each grain will be terminated not only in the vicinity of grain boundaries, but also in the vicinity of the upper and lower boundaries, since the grain size is so coarse when $T/D < 1$ that grain boundaries in vertical direction are replaced by the material's upper and lower ones. This can explain why grain's motion is severely affected by the material's upper and lower boundaries when grain size is larger than the thickness value. This impact is more profound than that in the material with $T/D \approx 1$, because the grains have relatively free space to slip and rotate in the latter condition. In the case of V-bending, the experimental conditions for each T/D sample are completely the same, including punch stroke, punch speed, bending angle, and bending radius. The free edges in the bending specimens are identical. The upper boundary contacts with punch and the lower boundary contacts with V-die. The noncontact areas belong to free boundary, while the contact areas, no matter with punch or V-die, belong to constraint boundary. Therefore, the only difference in this design is how the size of grain takes its effect on free and constraint boundaries. From Fig. 19.7, it can be seen that the punch contacts with a limited number of grains due to the tiny size of punch radius.

According to the bending theory, the inside layer will experience compression deformation and tensile strain will occur in the outer layer. So grain coordination deformation will become more important at the free and punch constraint boundaries with the decrease of grain size. The other issue needs to be considered is the constraint boundary contacting V-die, and the difference at the constraint boundary is the number of grain boundaries which are in touch with V-die. Based on the theory of open lubricant pockets (OLPs) and closed lubricant pockets (CLPs) [21], when there are more grain boundaries (from $T/D < 1$ to $T/D > 1$), there will be more OLPs, resulting high-

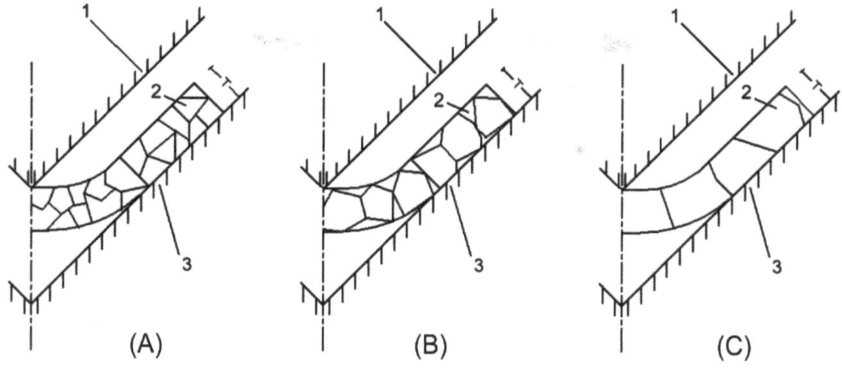

FIGURE 19.7 Schematic illustration of micro V-bending: (A) $T/D < 1$, (B) $T/D \approx 1$, and (C) $T/D > 1$. "1"-punch, "2"-grain, and "3"-V-die.

friction force between the specimens and V-die. This difference at constraint boundary will increase the corresponding bending force of specimens with different *T/D* ratios.

19.3.3 Effect of *T/D* on {110} Texture in Micro V-bending

The abovementioned statements are based on the relationship between the grain size effects and crystallographic grain rotations during micro V-bending process. With respect to grain orientation in bended specimens, it can be seen from Fig. 19.8A–C that the areas indicated by the points labeled ①–③ in the deformation zone increase with the decrease of *T/D*, which can be interpreted that the grain has a tendency to stay its original crystallographic plane under plastic deformation when the size of grain is close enough to the thickness value or even scale over it. In order to further confirm this judgment, several points in the areas in main deformation zone have been randomly chosen, and their orientations have been illustrated in the corresponding (110) pole figures, as shown in Fig. 19.8D.

It is shown that no matter the value of *T/D* is, the grain orientations almost stay at the initial orientations of their annealing textures: the Brass component and the A component. However, the quantity of these components changes. So, it is concluded that for phosphor bronze during micro V-bending, the stability of {110} texture increases comparing to other texture components when the ratio of *T/D* decreases from >1 to <1. This judgment can also be explained in terms of crystallography. The slip system of FCC materials is {111}⟨110⟩, while the Miller indices for the B component and the A component are {110}⟨112⟩ and {110}⟨111⟩, respectively. Table 19.4 delivers how the family of crystallographic planes {111} and {110} intersects, in which "–" means nonperpendicular while "P" represents two planes are perpendicular.

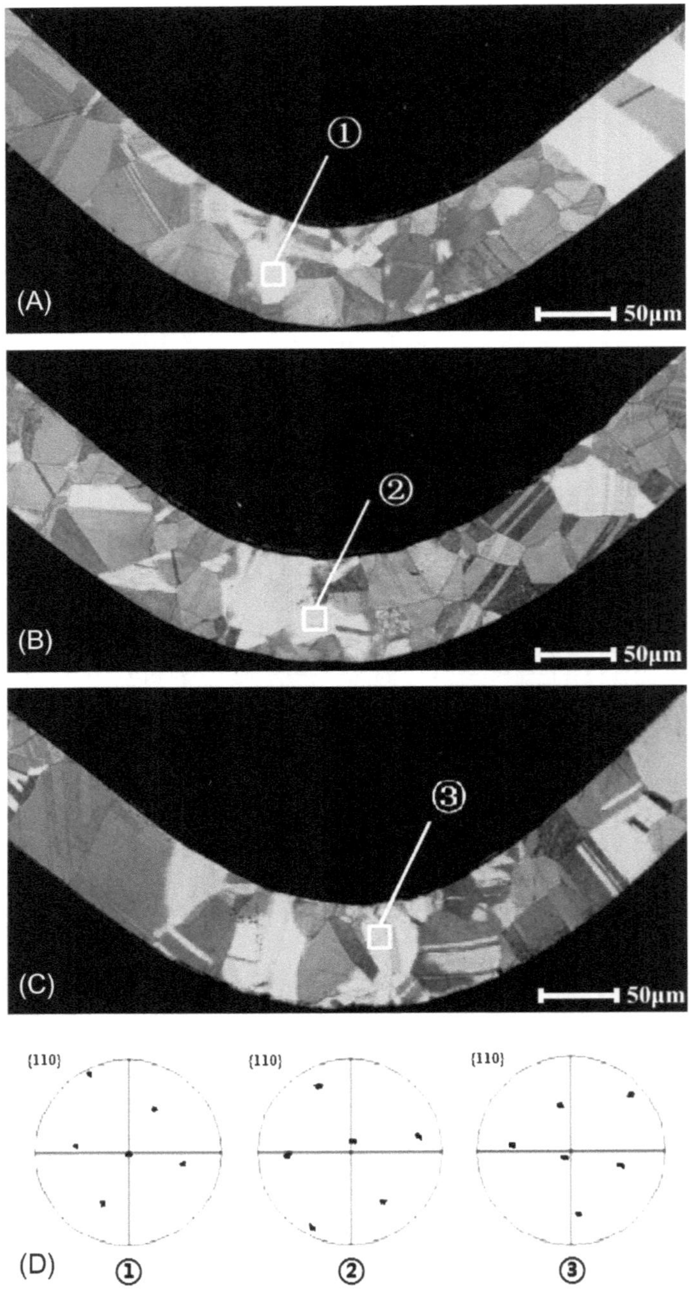

FIGURE 19.8 {110} portion in bended specimens: (A) $T/D > 1$, (B) $T/D \approx 1$, (C) $T/D < 1$, and (D) (110) pole figures for the points labeled ①–③.

TABLE 19.4 The Intersection of the Family of Crystal Planes {111} and {110}

	(110)	(101)	(011)	$(\bar{1}10)$	$(\bar{1}01)$	$(0\bar{1}1)$
(111)	–	–	–	P	P	P
$(\bar{1}11)$	P	P	–	–	–	P
$(1\bar{1}1)$	P	–	P	–	P	–
$(11\bar{1})$	–	P	P	P	–	–

It is shown that the half of two crystallographic planes {111} and {110} are mutually perpendicular. Therefore, when the resolved shear stress reaches to its critical value and initiates the slip on slip plane {111}, the slip on {110} plane is hard to be initiated since it is vertical to {111} plane and no resolved shear stress can work on {110} plane.

REFERENCES

[1] S.H.C. Park, Y.S. Sato, H. Kokawa, Effect of micro-texture on fracture location in friction stir weld of Mg alloy AZ61 during tensile test, Scr. Mater. 49 (2003) 161–166.

[2] J. Del Valle, F. Carreño, O.A. Ruano, Influence of texture and grain size on work hardening and ductility in magnesium-based alloys processed by ECAP and rolling, Acta Mater. 54 (2006) 4247–4259.

[3] J. Del Valle, O. Ruano, Separate contributions of texture and grain size on the creep mechanisms in a fine-grained magnesium alloy, Acta Mater. 55 (2007) 455–466.

[4] L. Llanes, A. Rollett, C. Laird, J. Bassani, Effect of grain size and annealing texture on the cyclic response and the substructure evolution of polycrystalline copper, Acta Metall. Mater. 41 (1993) 2667–2679.

[5] M. Hoseini, A. Shahryari, S. Omanovic, J.A. Szpunar, Comparative effect of grain size and texture on the corrosion behaviour of commercially pure titanium processed by equal channel angular pressing, Corros. Sci. 51 (2009) 3064–3067.

[6] P. Janssen, T.H. De Keijser, M. Geers, An experimental assessment of grain size effects in the uniaxial straining of thin Al sheet with a few grains across the thickness, Mater. Sci. Eng. A 419 (2006) 238–248.

[7] S. Yin, C. Wang, Y. Diao, S. Wu, S. Li, Influence of grain size and texture on the yield asymmetry of Mg-3Al-1Zn alloy, J. Mater. Sci. Technol. 27 (1) (2011) 29–34.

[8] L. Guo, Z. Chen, L. Gao, Effects of grain size, texture and twinning on mechanical properties and work-hardening behavior of AZ31 magnesium alloys, Mater. Sci. Eng. A 528 (2011) 8537–8545.

[9] M. Mabuchi, Y. Chino, H. Iwasaki, T. Aizawa, K. Higashi, The grain size and texture dependence of tensile properties in extruded Mg-9Al-1Zn, Mater. Trans. JIM 42 (2001) 1182–1189.

[10] T. Leffers, D. Juul Jensen, The effect of bending and straightening on rolling texture and microstructure in brass, Scr. Metall. Mater. 24 (1990) 2431–2435.

[11] H. Zhang, G. Huang, D. Kong, G. Sang, B. Song, Influence of initial texture on formability of AZ31B magnesium alloy sheets at different temperatures, J. Mater. Process. Technol. 211 (2011) 1575–1580.

[12] H. Kaneko, T. Eguchi, Influence of texture on bendability of Cu-Ni-Si alloys, Mater. Trans. 53 (2012) 1847–1851.

[13] B. Song, G. Huang, H. Li, L. Zhang, G. Huang, F. Pan, Texture evolution and mechanical properties of AZ31B magnesium alloy sheets processed by repeated unidirectional bending, J. Alloys Compds. 489 (2010) 475–481.

[14] Y. Sutou, T. Omori, N. Koeda, R. Kainuma, K. Ishida, Effects of grain size and texture on damping properties of Cu–Al–Mn-based shape memory alloys, Mater. Sci. Eng. A 438 (2006) 743–746.

[15] Y. Sutou, T. Omori, K. Yamauchi, N. Ono, R. Kainuma, K. Ishida, Effect of grain size and texture on pseudoelasticity in Cu–Al–Mn-based shape memory wire, Acta Mater. 53 (2005) 4121–4133.

[16] M. Barnett, Z. Keshavarz, A. Beer, D. Atwell, Influence of grain size on the compressive deformation of wrought Mg–3Al–1Zn, Acta Mater. 52 (2004) 5093–5103.

[17] S. Vaidya, A. Sinha, Effect of texture and grain structure on electromigration in Al-0.5% Cu thin films, Thin Solid Films 75 (1981) 253–259.

[18] B. Peters, The evolution of annealing textures in a heavily-deformed bronze, Metall. Trans. 4 (1973) 757–763.

[19] J. Hirsch, K. Lücke, Overview no. 76: Mechanism of deformation and development of rolling textures in polycrystalline fcc metals—I. Description of rolling texture development in homogeneous CuZn alloys, Acta Metall. 36 (1988) 2863–2882.

[20] A.A. Saleh, E.V. Pereloma, A.A. Gazder, Texture evolution of cold rolled and annealed Fe–24Mn–3Al–2Si–1Ni–0.06 C TWIP steel, Mater. Sci. Eng. A 528 (2011) 4537–4549.

[21] T. Sobis, U. Engel, M. Geiger, A theoretical study on wear simulation in metal forming processes, J. Mater. Process. Technol. 34 (1992) 233–240.

Chapter 20

Practice of Micro Compression

20.1 INTRODUCTION

Compression test is usually used to investigate the flow stress and other deformation behavior in large deformation. Micro compression test can be used to study the effects of both the specimen size and grain size. The loading condition within a compression test is relevant to the actual state of stress in most metal forming processes. Additionally, the preparation and handling of cylindrical metal specimens is less complicated. Finally, material data could be obtained up to high levels of strain; thus, avoiding an excessive extrapolation of material flow curve in the eventual numerical analysis.

Currently, two approaches to compensate or reduce the influence of size effects are to increase the temperature or refine the microstructure of the deformed parts during the forming process. Heat-assisted microforming can be accomplished by heating the workpiece using, e.g., laser heating, and microstructural refinement can be achieved by, e.g., severe plastic deformation, as discussed in Chapter 2, Size Effects in Microforming. In this chapter, methods for micro compression tests at both room and elevated temperatures using the desk-top servo-press machine are introduced, which will supply the theoretical basis for the more complicated microforming techniques. A series of micro compression tests with assistance of laser heating are conducted using cylindrical specimens made of pure copper with different grain sizes, and the influences of temperature in micro compression are investigated. Specimens with various grain sizes and different cylindrical diameters are used to investigate the effects of feature size on the material's response.

20.2 MICRO COMPRESSION PRACTICE

20.2.1 Material Preparation

Experiments are performed with cylinders made from pure copper. The original wire samples are cold drawn and carefully subdivided into pieces of cylindrical specimens by wire cutting, as shown in Fig. 20.1. The ratio of height to diameter of the specimen is 1.5. There are three kinds of geometrical dimensions including $\Phi 0.8 \times 1.2\ \mathrm{mm}^2$, $\Phi 1.0 \times 1.5\ \mathrm{mm}^2$, and $\Phi 1.5 \times 1.8\ \mathrm{mm}^2$. These specimens are measured with digital spiral micrometer in order to have an accurate measurement of the geometries. Both of

Microforming Technology. DOI: http://dx.doi.org/10.1016/B978-0-12-811212-0.00020-0

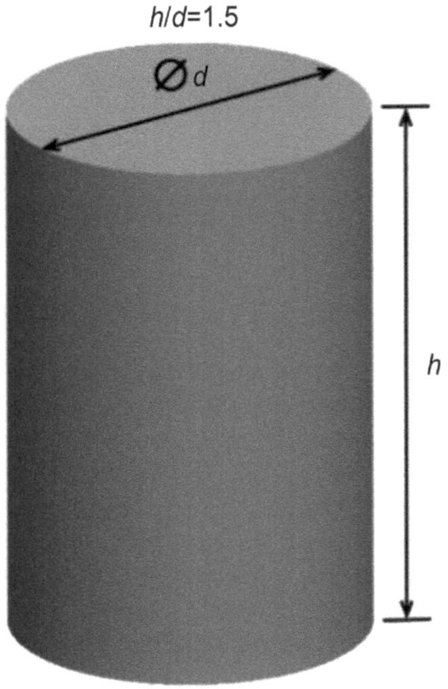

FIGURE 20.1 Schematic illustration of cylindrical specimen for micro compression test.

the cross sections of the specimen are rubbed by #3000 sandpaper to minimize the influence of friction at interface of tools and workpiece. When investigating the variation of specimen side surface during deformation, it is necessary to strip off the disturbance of initial surface asperity of specimen. Thus, the side surfaces of all the specimens are polished using chamois before tests.

In order to investigate size effects, the variation of the ratio between the geometrical dimensions of the specimen and the average grain size is proposed. There are two approaches to study size effects. One is to vary the grain size via heat treatment while the specimen dimensions are kept constant. The other is to prepare the specimens with different scales of dimensions; meanwhile the grain size is not changed. In this chapter, the former one is applied to investigate the size effects on flow stress and side surface asperity. The latter one is utilized to estimate the influence of specimen effect on friction. In order to vary grain size, the specimens are annealed at 973 K in a tube furnace for different times. The commercial software of iTools (Fig. 20.2) is used to control the whole process. Argon is used as a protecting gas to prevent the specimens from oxidation.

Specimens undergone heat treatments are etched in the solution listed in Table 20.1. Average grain sizes of the heat-treated specimens are measured according to the ASTM Standard E112-88.

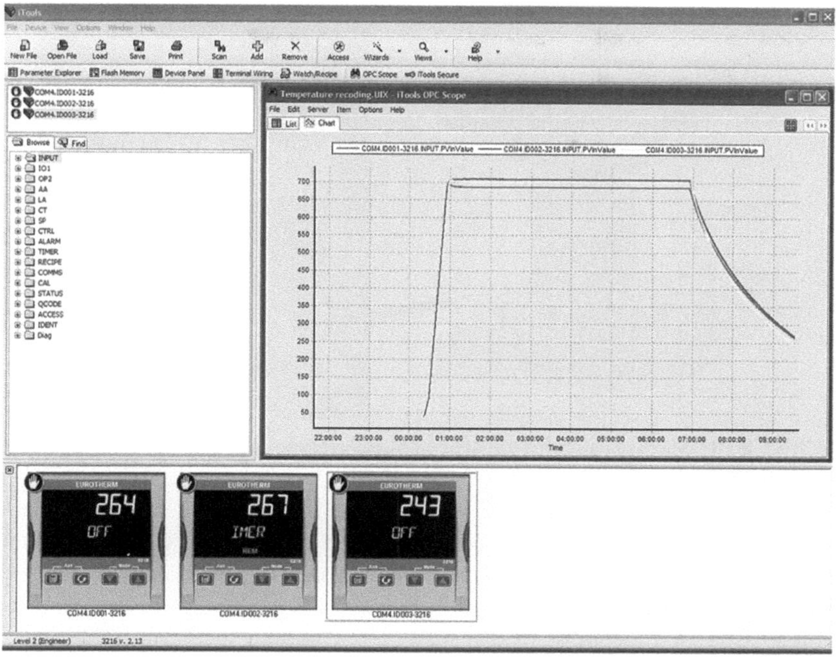

FIGURE 20.2 Controlling process of heat treatment with iTools.

TABLE 20.1 Etching Strategy of Copper

Solution	Soaking Time
$FeCl_3$: 5 g, HCl: 50 ml, H_2O: 100 ml	10–15 s

20.2.2 Micro Compression Setup

Micro compression tests are conducted on the desk-top servo-press machine. A scaled experimental setup for the investigation of size effects occurred in micro compression processes established in the die set is illustrated in Fig. 20.3. The server motor provides force to drive the punch downwards and compress the metal specimen as designed procedures. The punch force applied on the specimen is measured and recorded during each test using a load cell (Kyowa LMR-S-20KNSA2). The force–displacement curves are recorded and analyzed by a computer. Once the displacement of punch achieves the target value, the specimens are removed for measurement of surface roughness and observation of microstructure.

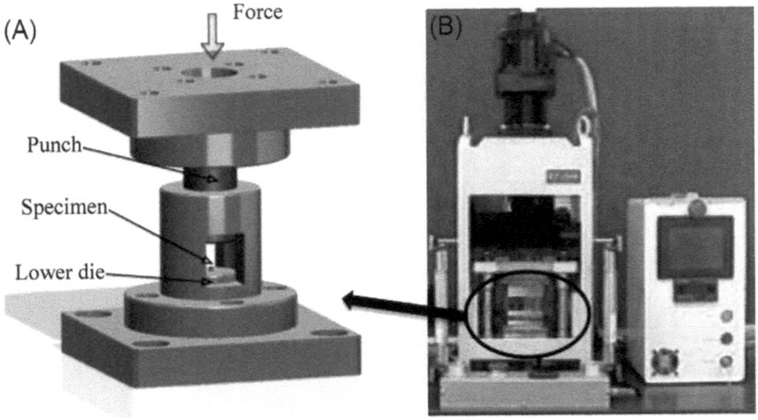

FIGURE 20.3 Micro compression test setup: (A) compression die set and (B) desk-top servo-press machine.

During micro compression tests, the speed of punch is set at a constant value of 0.1 mm/s. The contact surface between the tool and specimen is lubricated in order to reduce the influence of friction. The specimens are compressed by reductions of 20–80%. Even though the strain rate control during forming operations is known to be an important factor that may affect the overall deformability of the specimens, it is also commonly known that the material properties of most metals and metal alloys are not quite sensitive to the strain rate in the cold-forming processes as compared to a warm/hot-forming operation. Therefore, for the compression tests conducted at room temperature, the effect of strain rate on the material properties is not considered. However, when compression is performed at elevated temperatures, the strain rate will be taken into account as a critical factor.

20.2.3 Laser Heating Strategy

With assistance of laser heating, micro compression tests with pure copper at elevated temperatures are carried out to investigate the role of warm forming on the size effects of material flow during micro compression.

20.2.3.1 Laser Heating Equipment

During micro compression process, laser is used to supply energy to warm the parts to be deformed, as shown in Fig. 1.15, which can alter the deformability of metals and homogenize the grain heterogeneity that becomes significant in microforming. A micro laser soldering system, which is a Class 4 product based on IEC standard, is employed to elevate the temperature of specimen. The entire system is mainly composed of monitor, laser generator, controller, laser eye protector, laser head, and manual state, as shown in Fig. 20.4. The specifications of this system are listed in Tables 20.2 and 20.3. One of the most two common lasers used in metal processing is the Nd: YAG laser (the other one is the CO_2), operating in the near infrared just outside the visible wavelength region as demonstrated by Bergstrom et al. [1]. The Nd: YAG laser is used in the current work to heat

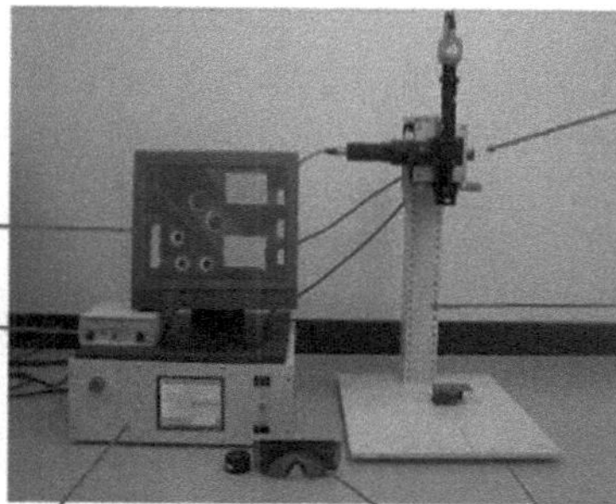

FIGURE 20.4 Laser heating system.

TABLE 20.2 Specification of Micro Laser Solders System

Classification	Class 4 Laser (IEC60825–1)
Wavelength	1064 nm
Oscillation	CW (Continuation Wave)
Maximum output	20 W (Light Source)
Laser beam diameter	See Table 20.3
Focal length	f 200 mm/f 400 mm
Work distance	170 mm (f 200)/370 mm (f 400)

TABLE 20.3 Laser Beam Diameter and Focal Length

Focal Length (mm)	Working Distance (mm)	Zoom Mechanism	
		Maximum Beam Diameter (μm)	Minimum Beam Diameter (μm)
200	180	100	50
400	380	200	100

workpieces. The spot size of the laser is controlled by the distance from the lens to the specimen surface. In the current work, the focus length applied is 400 mm, and the beam diameter is between 100 and 200 μm. The distance in each experiment is kept a constant value.

As demonstrated in Fig. 20.5, the incidence of laser beam on copper workpiece is finely focused based on the approximate data in Table 20.3. Moreover, in order to achieve the desirable product quality, the intrinsic parameters of laser system including laser power, heating time, and heating velocity should be adjusted and optimized as well. A major processing control parameter is the target laser power, which can be represented in terms of fluence or radiance. In laser material processing, the power can be controlled either internally to the laser using the inherent excitation and light amplification characteristics to advantage, or externally by utilization of a light value and modulator. These parameters of the laser system are controlled by the external controller.

20.2.3.2 Laser Heating Practice

In laser heating tests, the specimen's geometrical dimensions are kept constant while grain size is changed. Consequently, the influence of specimen effect on friction is eliminated and the influence of the ratio of specimen surface to volume on heat loss is neglected as well. The diameter and height of the cylindrical specimen are 0.8 and 1.2 mm, respectively. At beginning, the upper punch is moved to the initial position where there is just no gap between the workpiece and punch. Then, the specimen is compressed by 20–80% in height at a punching speed of 0.1 mm/s, which is performed on the desktop servo-press machine with the assistance of laser heating.

The nature of laser heating is to change the internal microstructure of metal by the energy absorption. The metal absorptivity of laser light is a critical factor that influences the deformation process, and the copper absorptivity of laser light is almost independent on temperature [2]. When the micro specimen is heated up by laser beam, it is assumed to achieve an even distribution of temperature due to the excellent thermal conductivity of pure copper.

The copper specimen is heated up to the expected temperature prior to compression process but the heat loss is quite significant due to the relatively

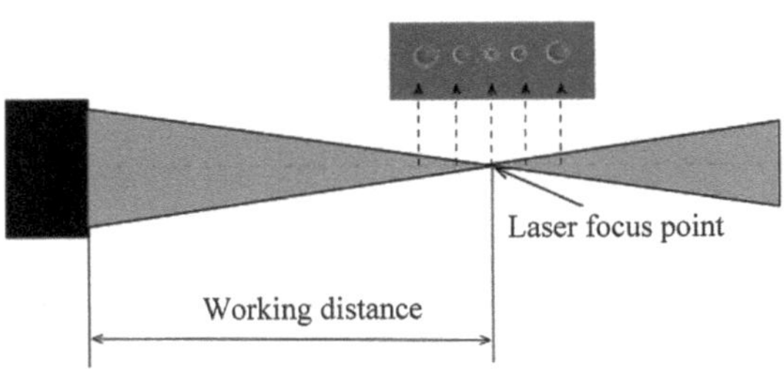

FIGURE 20.5 Adjustment of the beam diameter with focus.

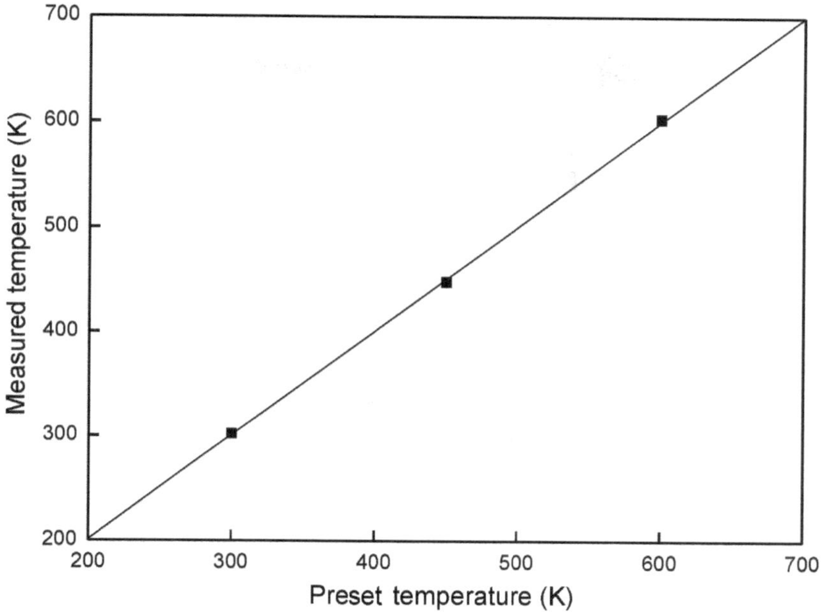

FIGURE 20.6 Comparison between the preset and measured temperatures of t_2.

large ratio of specimen surface to volume. During deformation, the laser power is adjusted to an appropriate value to compensate the energy loss due to heat convection and radiation from specimen to atmosphere which is especially significant in microforming. The strategy of laser heating illustrated in Fig. 2.18 is used in this chapter. Fig. 20.6 shows the comparison between the preset and measured temperatures of t_2. It can be seen that for the preset temperatures of 300, 450, and 600 K, the measured values of t_2 are 302, 448, and 603 K, respectively, indicating that the temperature can be well controlled by using the laser heating method.

As mentioned, warm forming has been used as a practical method to improve material properties in microscale/mesoscale. However, this method is confronted with some inevitable problems such as oxidation due to the elevated temperatures. In order to prevent specimens from oxidation, the upper heating temperature of pure copper sample is set to be 650 K, because the surface quality would be inadequate for micro compression due to the generation of massive oxide layer when the forming temperature is above 650 K, as indicated in Fig. 20.7.

20.3 RESULTS ANALYSIS

The micro compression experimental results at room temperature, including relationship between punch load and displacement, stress–strain curves with

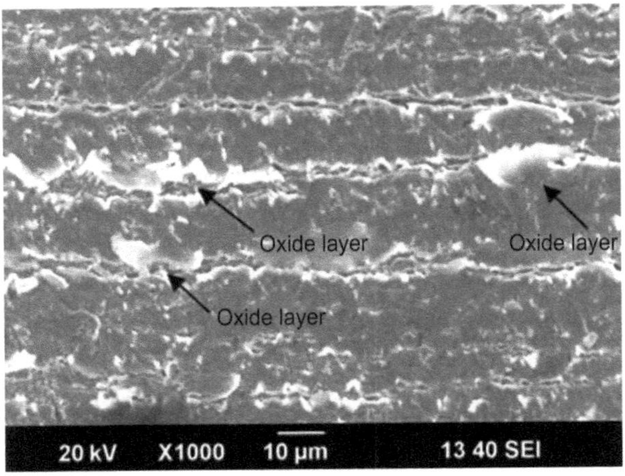

FIGURE 20.7 Oxidized layer of copper sample after laser heating above 650 K.

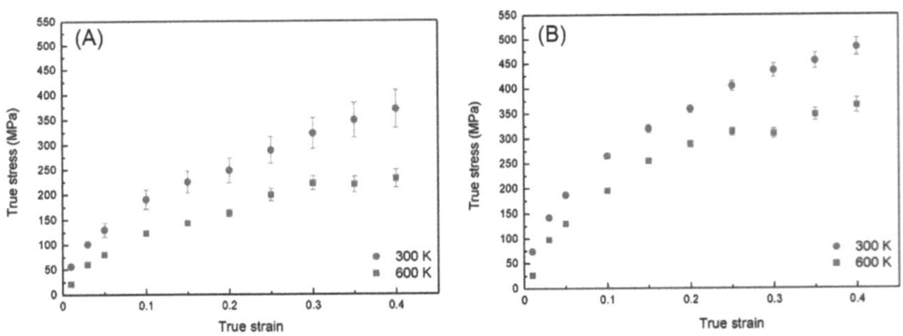

FIGURE 20.8 Dependence of stress on strain at elevated temperatures: (A) grain size of 200 μm and (B) grain size of 20 μm.

different λ values, profiles of specimens after compression tests, and influence of λ on H_{as}, have been presented in Figs. 13.7, 13.8, 13.11, and 13.13, respectively. Fig. 20.8 shows the dependence of stress on strain at elevated temperatures, which indicates a decreasing trend of flow stress with an increase of forming temperature. This tendency is similar to that in conventional metal forming and can be explained by the dislocation theory. During plastic deformation of metals, dislocations move and accumulate, and have to overcome the obstacles involving among the crystal structure. For the typical face-centered cubic (FCC) metal, copper, the primary short-range barriers are other dislocations which intersect the slip plane and impede the motion of gliding dislocations [3,4]. The increased forming temperature of deformed part can activate the constrained slip system resisted by short-range obstacles, which accounts for the decrease of flow stress with an increase of forming temperature as demonstrated by both the simulated and experimental results. The number of grains involved in the micro deformed part decreases in consequence with the miniaturization of specimen dimension. Therefore, the mechanical properties of individual grain play a more significant role in the material behavior and the inhomogeneity of grain characteristics such as grain size, shape, and orientation, and may result in the scatter of the measured mechanical properties. As shown in Fig. 20.8, the scatter of flow stress reduces as the temperature of specimen increases from 300 to 600 K as a result of thermal activation of more dormant slip systems, which reveals the same trend as the results announced by Farrokh and Khan [5]. The homogenizing effect of grain refinement can also be seen by comparing Fig. 20.8A and B. It is evident that there is a significant increase in the scatter of flow stress when the grain size increases from 20 to 200 μm.

There are mainly two effective approaches to minimize the influence of material heterogeneity in microforming process, i.e., refinement of the crystal structure of polycrystalline metal and elevation of the temperature of the deformed micro part. The interactions of grain size, temperature, and scatter of material mechanical properties are complicated [5]. A parameter sc_h is defined to assess the variation of flow stress scatter quantitatively:

$$sc_h = \frac{sc - sc_{mean}}{sc_{mean}} \tag{20.1}$$

where sc and sc_{mean} are the scatters and mean scatters, respectively. As shown in Fig. 20.9, the value of sc_h varies with grain size and forming temperature. The effect of grain size plays a dominative role in minimizing the scatter of flow stress, and the elevated temperature homogenizes the grain heterogeneity remarkably as well. The scattered value becomes more

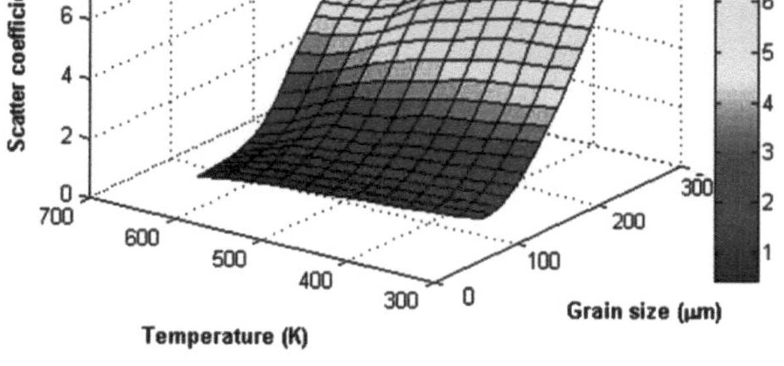

FIGURE 20.9 Variation of experimental stress scatter with grain size and forming temperature.

sensitive to heating temperature when the grain size is gradually increased. When the grain size is 200 μm, sc_h reduces by 4.5% at least with an increase of forming temperature ranging from 300 to 650 K. However, if grain size is around 20 μm, there is no significant change in stress scatter with the temperature variations. The results clearly indicate how the scatter of flow stress of coarse-grained copper exhibits a greater degree of sensitivity to forming temperature than that of fine-grained copper. With an increase of grain size, there is a decline in the density of geometrically necessary dislocations (GNDs) which are considered to participate in both short-range and long-range effects. Therefore, the influence of thermal activation on short-range barriers reduces, which accounts for a decreased homogenizing effect of forming temperature on material's mechanical behavior for the coarse-grained copper. With the refinement of grain size and elevation of forming temperature, it is possible to reduce inhomogeneous flow, and consequently minimize size effects in micro compression. Moreover, a plateau exists in Fig. 20.9 with grain size of around 200 μm and temperature of about 575 K, indicating that the scatter value may not be sensitive to grain size and temperature under such circumstances as the flow stress scatter in microforming is affected by a number of characteristics of grains, including size, shape, orientation, and stochastic distribution [6,7].

GNDs accumulate and cause strengthening dependence of grain size (the grain boundary strengthening). With an increase of grain size, the volume fraction of grain boundary in the global volume of specimen declines, which weakens GNDs' impact on the grain boundary strengthening in a certain extent. In contrast, statistically stored dislocations (SSDs) spreading evenly in the grain interior region contribute to the short-range interaction effect which can be influenced by the deformation temperature. Therefore, there is a positive correlation between the grain size and the thermal component of flow stress reflecting mainly short-range effect. Of the total flow stress, the σ_{th} fraction increases dramatically by approximately 80% when the grain size increases from 0.1 to 250 μm for all the three thermal conditions, as shown in Fig. 20.10. The increased heating temperature promotes the increase of thermal stress as well, although its impact on thermal stress is less than that of grain size.

The scatter of individual grain's plastic deformability affects the specimen profile. Consequently, it supplies another approach to penetrate the effect of forming temperature on the deformation behavior and the quality of produced micro parts. Side surface asperity of specimen after compression has been measured. As shown in Fig. 20.11, the surface asperities of the samples processed at different temperatures reveal significant distinction.

Fig. 20.12 demonstrates the effects of forming temperature on the side surface asperity of copper specimens with different grain sizes. It is noticed that there is a decrease in H_{sa} (see its definition in Section 13.3) with an increase of forming temperature for both of the materials with grain sizes of

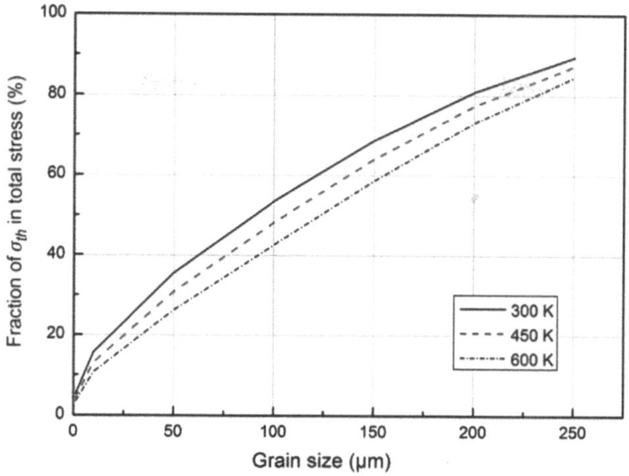

FIGURE 20.10 Dependence of fraction of thermal stress on grain size and forming temperature.

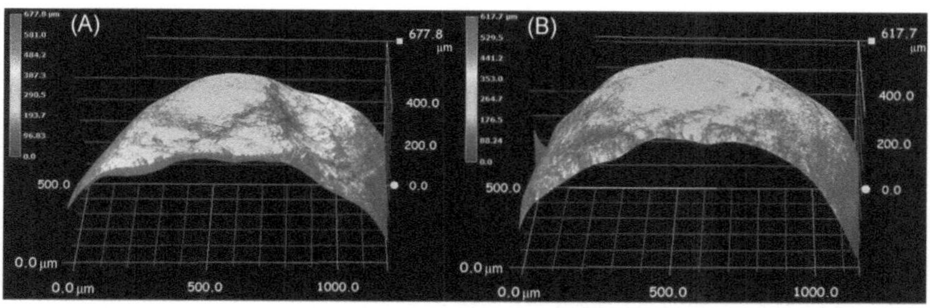

FIGURE 20.11 Profiles of specimens underwent compression tests at different temperatures: (A) 300 K and (B) 600 K.

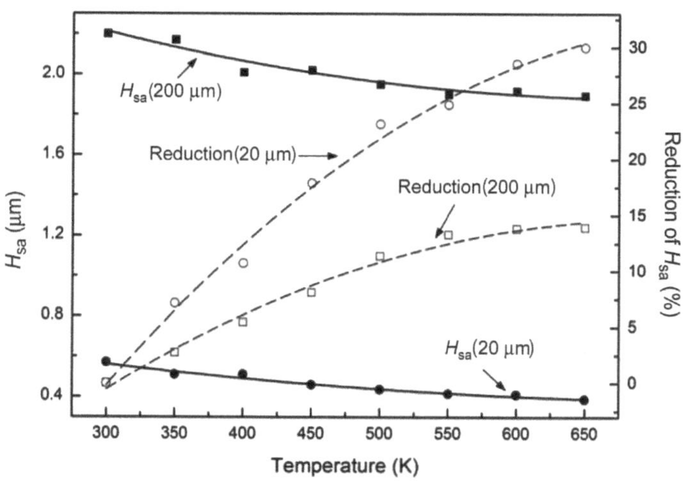

FIGURE 20.12 Influence of temperature on side surface of cylindrical sample.

20 and 200 μm. A remarkable difference between the two types of materials in terms of H_{sa} reduction with forming temperature can be observed. H_{sa} of the material with the grain size of 20 μm reduces by 30% which is nearly two times higher than that of the coarse-grained specimen. The results indicate that an increase in forming temperature benefits the enhancement of material-flow behavior, and the deformability of fine-grained metal is more sensitive to forming temperature from another point of view.

In summary, this chapter has demonstrated that the flow behavior of copper in micro compression can be improved through optimization of processing conditions, e.g., rising of forming temperature and refinement of material's microstructure. Such a practice is helpful for the study on size effects in microforming aimed at improving the quality of the formed micro products.

REFERENCES

[1] D. Bergstrom, J. Powell, A.F. Kaplan, Absorptance of nonferrous alloys to Nd:YLF and Nd:YAG laser light at room temperature, Appl. Opt. 46 (2007) 1290–1301.

[2] D. Bergstrom, The Absorption of Laser Light by Rough Metal Surfaces, Mid Sweden University, Sweden, 2008.

[3] L.P. Evers, W.A.M. Brekelmans, M.G.D. Geers, Non-local crystal plasticity model with intrinsic SSD and GND effects, J. Mech. Phys. Solids 52 (2004) 2379–2401.

[4] L.P. Evers, W.A.M. Brekelmans, M.G.D. Geers, Scale dependent crystal plasticity framework with dislocation density and grain boundary effects, Int. J. Solids Struct. 41 (2004) 5209–5230.

[5] B. Farrokh, A.S. Khan, Grain size, strain rate, and temperature dependence of flow stress in ultra-fine grained and nanocrystalline Cu and Al: synthesis, experiment, and constitutive modelling, Int. J. Plast. 25 (2009) 715–732.

[6] W.L. Chan, M.W. Fu, J. Lu, J.G. Liu, Modeling of grain size effect on micro deformation behaviour in microforming of pure copper, Mater. Sci. Eng. A 527 (2010) 6638–6648.

[7] B. Eichenhueller, E. Egerer, U. Engle, Microforming at elevated temperature – forming and material behaviour, Int. J. Adv. Manufact. Technol. 33 (2007) 119–124.

Index

Note: Page numbers followed by "*f*" and "*t*" refer to figures and tables, respectively.

M

S

V

W

Y

MIX

Papier | Fördert gute Waldnutzung

FSC® C083411

Kolibri 360 GmbH
Ferdinand-Jühlke-Straße 7
99095 Erfurt, Deutschland
produktsicherheit@kolibri360.de